T0351100

Flow Control

This book provides a comprehensive treatment of passive and active flow control in fluid dynamics, with an emphasis on utilizing fluid instabilities for enhancing control performance. Examples are given from a wide range of technologically important flow fields occurring in aerospace applications, from low-subsonic to hypersonic Mach numbers. This essential book can be used for both research and teaching on the topics of fluid instabilities, fluid measurement, and flow actuator techniques, and problem sets are provided at the end of each chapter to reinforce key concepts and further extend readers' understanding of the field. The solutions manual is available as an online resource for instructors. The text is well suited for both graduate students in fluid dynamics and practicing engineers in the aerodynamics design field.

Thomas C. Corke is the Clark Chair Professor of Engineering, founding Director of the Center for Flow Physics and Control (FlowPAC), and Director of the Hessert Laboratory for Aerospace Research at the University of Notre Dame. He is a fellow of the American Institute of Aeronautics and Astronautics, the American Society of Mechanical Engineers, and The American Physical Society. He is the author of the books *Design of Aircraft* (2003) and *Wind Energy Design* (2018).

Flow Control

A Fluid Instability Approach

THOMAS C. CORKE
University of Notre Dame

Shaftesbury Road, Cambridge CB2 8EA, United Kingdom

One Liberty Plaza, 20th Floor, New York, NY 10006, USA

477 Williamstown Road, Port Melbourne, VIC 3207, Australia

314–321, 3rd Floor, Plot 3, Splendor Forum, Jasola District Centre,
New Delhi – 110025, India

103 Penang Road, #05–06/07, Visioncrest Commercial, Singapore 238467

Cambridge University Press is part of Cambridge University Press & Assessment,
a department of the University of Cambridge.

We share the University's mission to contribute to society through the pursuit of
education, learning and research at the highest international levels of excellence.

www.cambridge.org
Information on this title: www.cambridge.org/9781108832366

DOI: 10.1017/9781108955935

First published 2024

A catalogue record for this publication is available from the British Library

A Cataloging-in-Publication data record for this book is available from the Library of Congress

ISBN 978-1-108-83236-6 Hardback

Additional resources for this publication at https://www.cambridge.org/9781108832366.

In a large way, this text book represents my 40-year journey as an academic, research scientist and engineer. Throughout, it would have been impossible to devote the time needed to pursue my research interests without the support of my wife, Bobbie, and daughters, Catherine, Laura, and Sarah.

I would also be remiss in not mentioning the support I have received from my sons-in-law, Anthony, Michael, and Clinton, and the joy that has come from my grandchildren, Patrick, Allison, Madeline, Claire, and Gabriel.

This book is dedicated to all of them.
—TCC

Contents

Preface

This book is intended to be a text for a graduate fluid dynamics course on passive and active flow control. A special emphasis is given to utilizing fluid instabilities as a means of enhancing control performance. It is based on my 40 years of experience applying flow control techniques to a large variety of applications. The topics in the book cover a wide range of fundamental flow fields, including wakes, jets, free shear layers, and boundary layers over a range of Mach numbers from low-subsonic to hypersonic. Applications include lift control, dynamic stall control, enhanced mixing, separation control, drag reduction, and shock–boundary-layer control. These topics form the basic organization of the book. This is presented in ten chapters.

Chapter 1 provides a historical introduction and motivation to flow control. This leads to background on flow sensors and actuators that is presented in Chapter 2. The chapters that follow are broken down into general topics based on applications. Chapter 3 deals with bluff body wakes. A bluff body can be generally categorized as one in which its length in the flow direction is approximately the same as its height, perpendicular to the flow in a 2-D representation. Such shapes exhibit a wide wake on the scale of the body (2-D) height, and aerodynamic drag is dominated by the low pressure region that forms in the near wake of the body. Bluff body wakes are complex, involving boundary layer separation and multiple shear layer interactions. Chapter 4 involves flow control to prevent or control flow separation. Flow separation occurs in a large number of external and internal flow fields and is generally detrimental to aerodynamic performance. As a result, its prediction and control has been a major topic area. The chapter considers both steady and dynamic flow separation, with the latter resulting from pitching or plunging motion.

A component of separated flows is a free shear layer, which is an elevated region in a flow field with an inflexional mean velocity profile. Such a mean profile is inviscidly unstable to disturbances, making it highly sensitive to unsteady actuator forcing. The control of free shear layers is the topic of Chapter 4. This relates to co-flowing streams, and jets and wakes, with applications that include enhanced mixing and noise control.

Chapters 6 and 7 involve 2-D and 3-D laminar boundary layers. In contrast to free shear layers, boundary layers form over surfaces where viscosity plays a major role. An important part of boundary layers is the prediction and control of laminar to turbulent transition. At subsonic and supersonic Mach numbers, transition is primarily important to viscous drag. Promoting turbulent transition at these Mach numbers can be used to prevent flow separation. At hypersonic Mach numbers however, turbulent transition governs the degree of surface heating, which can be as much as five times that in the laminar boundary layer. Therefore, there is a substantial advantage if it can be prevented. In both of these chapters, the full range of Mach numbers from incompressible to hypersonic is presented.

In almost all applications with respectable Reynolds numbers, turbulence is inevitable. This is the case with boundary layers. Thus, Chapter 8 focuses on fully turbulent boundary layers. As with separated flows, the study and control of turbulent boundary layers is one of the most frequent topics in the fluid dynamics literature. The primary emphasis of turbulent boundary flow control is toward reducing viscous drag. Of late, there have been significant breakthroughs toward this objective that are thoroughly covered in the chapter.

Supersonic and hypersonic propulsion systems involve the use of generated shock waves to lower the local Mach number, and increase the static pressure to enhance combustion. In many cases, this involves leading edges, joints, or ramps on opposite side walls. Such mechanical shock generators are heavy and inflexible to changing conditions. This has prompted the development of active shock generation methods that are presented in Chapter 9. In addition to shock control, the chapter considers shock–boundary-layer interactions that result in local flow separation and unwanted unsteadiness. Adaptations of methods for boundary layer separation control that are discussed in Chapter 4 are included as part of Chapter 9.

In a majority of applications in the literature, flow control is introduced to *correct a problem*. Chapter 10 presents a different concept where active flow control is included in the design of the application geometry *from the beginning*. As an example, an airfoil trailing edge geometry that introduces a flow separation that is receptive to separation control can provide lift control without a moving surface. Although such geometries and actuator placement might be designed intuitively, Chapter 10 presents a method based on an adjoint form of the Navier–Stokes that couples geometric changes and flow actuator location to maximize flow control authority.

The learning objectives for the book are (1) to understand the general background and objectives of flow control with successful and unsuccessful applications, (2) to understand the general methods and tools, and where they can best be applied, (3) how to make the best use of fluid instabilities to enhance flow control performance, (4) assessing the flow control performance merits and limitations, and (5) designing for flow control from the beginning.

The book can be used in either of two ways. First, it can be used as a textbook in a course to develop a broad understanding of the concepts of flow control across many applications and Mach numbers. This is the way that it was envisioned while developing the book. This approach makes the best use of the problem sets at the end of each chapter.

The second use of the book is as a reference text where a specific application or flow regime is considered. An outcome of this approach can be the design of an active flow control system, for example, to replace an existing passive approach. The effect of different active input parameters can be investigated with the object to find an optimum based on a defined merit function.

As a complete course, the chapters are intended to be followed in chronological order. A conscious attempt has been made to make each chapter self-contained. This is particularly useful if the book is used as a reference text.

different [illegible] methods can be investigated with the [illegible] based on a defined merit function.

As a [illegible], the chapters are [illegible] chronological order [illegible] has been made to make each chapter [illegible]. It is hoped [illegible] useful as [illegible] reference text.

1 Introduction

Many of the adverse outcomes of technologically important applications that involve fluid flows find their root in a fluid instability. Examples include shear layer instabilities that result in large acoustic levels (e.g., jet noise); boundary layer instabilities that lead to turbulence onset causing increased drag and surface heating, with the latter being extreme at hypersonic Mach numbers; unsteady flow separation behind bluff bodies (e.g., suspended cables, bridges, buildings, ground vehicles, etc.) that leads to large unsteady loads and extreme amplitude vibration; and even in turbulent boundary layers where instabilities in the laminar sublayer increase viscous drag and enhance turbulence production. However, flow control that *exploits* fluid instabilities can, with minimum energy input, produce beneficial effects. In the case of a free shear layer instability associated with jet noise, passive flow control involving trailing-edge geometries that modify the instability in basic flow is effective. If jet mixing is the objective, it can be substantially enhanced by introducing periodic disturbances that key on shear layer and jet core instabilities. An extreme example is "blooming jets" presented in Chapter 5 that was aimed at highly enhanced fluid mixing. Since free shear layers are associated with separated flows, introducing unsteady disturbances that cause the shear layer to accelerate turbulence onset is an effective method for flow reattachment. Bluff body flows discussed in Chapter 3 are in a special class as they can exhibit a global instability that is highly sensitive to even small modifications in the basic flow. As a result, passive modifications to near-wake bluff body flows have been especially effective.

1.1 Flow Control: An Instability Approach

Given these examples, the flow control approach embraced in this book is to recognize relevant fluid instabilities of a given flow field and then utilize them to seek to modify the flow to achieve a positive outcome. Figure 1.1 illustrates the basic physics of a fluid instability. It is important to understand that a fluid instability is that of the basic (time-independent) flow. An example of an unstable basic flow field that is illustrated in the figure is a free shear layer that exhibits an inflectional mean profile. The instability is promoted by unsteady disturbances from the environment. These disturbances might contain a broad spectrum of frequencies; however, the instability mechanism acts like a band-pass-filter amplifier that amplifies a selected band of frequencies based on the basic flow. If the amplitudes of the environment disturbance are within levels assumed by linear stability analysis (generally below 1 percent of the maximum mean flow velocity), the amplified bands of frequencies are those whose integrated growth, or N-factor, is the largest based on linear stability analysis. Linear stability predicts that unstable disturbances grow exponentially in time or space. The amplification rate depends on the frequency or wave number.

In general, there exists a controlling parameter for a fluid instability that needs to be exceeded for disturbances to be amplified. In buoyancy-driven flows, this parameter is the Rayleigh number. In centrifugally driven flows, the parameter is the Taylor number. In open flows driven by either a pressure gradient or moving surfaces, the parameter is the Reynolds number. For any of these, there is a critical value below which linear disturbance will decay exponentially. The critical value depends on the disturbance frequency or wave number. The particular frequency or wave number with the lowest critical parameter value is called the minimum critical value. The right-most box in Figure 1.1 illustrates the amplitude response of a convective instability that is growing in space. R in this case is the Reynolds number based on a spatial development length scale, x, where then $R = Ux/\nu$. As illustrated, the amplitude of disturbances decays until the critical value, $R_{\mathrm{I}} = Ux_c/\nu$, is exceeded. Past the critical value, the amplitude increases, reaching a maximum and then decaying. In some flows, such as a Blasius boundary layer, there exists a second instability Reynolds number, R_{II}, beyond which linear disturbances decay. This does not exist in the example free shear flow. However, as the instability amplitude grows, it brings on nonlinear effects that generally cause the growth to asymptote. In some cases, the nonlinear effects include a mean flow distortion that triggers rapid turbulence onset.

Flow control approaches that utilize a fluid instability to amplify *controlled* disturbances perform best when they are introduced near the minimum critical value of the controlling parameter, for example, near R_I in a boundary layer. This approach can then fully exploit the region of exponential growth, as well as to not have to compete with instability modes resulting from uncontrolled background disturbances that may be more amplified than the controlled instability.

There are two basic approaches to flow control that are based on a fluid instability approach. These are illustrated in Figures 1.2–1.4. Since fluid instabilities are an instability of the basic (time-independent) flow, the first approach is to *modify the basic flow*. Figure 1.2 provides an example of this approach as applied to a free shear layer. In this example, a sinusoidal arrangement of peaks and valleys is added to the otherwise straight trailing edge of the flow splitter plate. The amplitude and spanwise wavelength of the peaks and valleys are parameters that could be selected based on linear stability analysis of the distorted mean flow. The general purpose of the sinusoidal trailing edge is to introduce streamwise vorticity that modifies the mean flow and therefore suppresses the growth of the initially most amplified two-dimensional (2-D) instability wave. As shown in Figure 1.3, this approach is widely used on the jet engines of commercial aircrafts to lower the sound levels that are intrinsically tied to the free shear layer instability. Other similar approaches described in Chapter 5 include spanwise-periodic chevron-shaped deformations of the trailing edge and

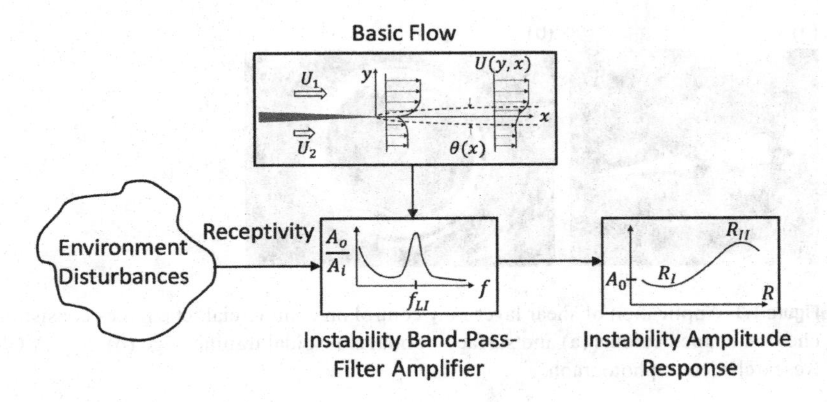

Figure 1.1 Illustration of a natural fluid instability process.

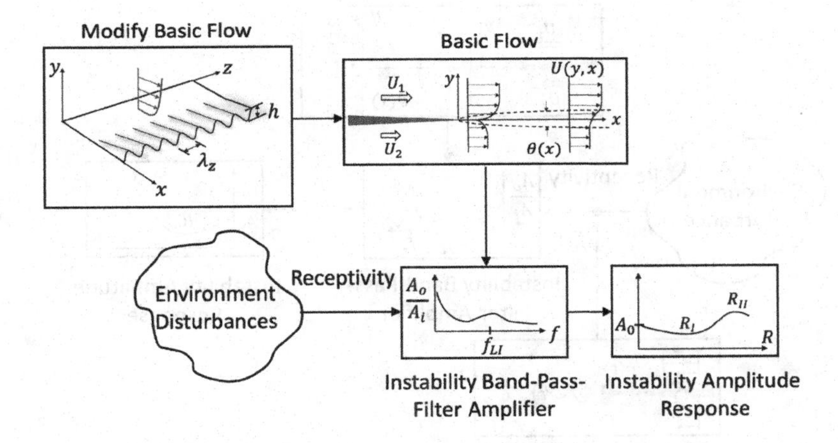

Figure 1.2 Example of an instability approach to flow control that involves a method that modifies the basic (time-independent) flow field.

surface streamwise vortex generators (VGs). In the free shear layer that exits from round jets, changes in the basic state have included introducing an azimuthal variation in the length of the jet nozzle to produce an azimuthal variation in the initial shear layer thickness. Since the shear layer instability frequency scales with the initial shear layer thickness, this modification of the mean flow can result in azimuthal bifurcations in the otherwise azimuthally contiguous (ring) vortices.

The second approach to flow control is based on *actively* interfering with the natural instability frequency, wavelength, or initial amplitude; for example, to introduce unsteady disturbances that either enhance or delay the growth of the natural instability. This approach is illustrated in Figure 1.4 which again considers the free shear layer as an example. In this case, disturbances with a prescribed amplitude and frequency are introduced through an oscillation of the trailing edge of the splitter plate. In the simplest case, the trailing-edge oscillation would not vary in the spanwise direction and

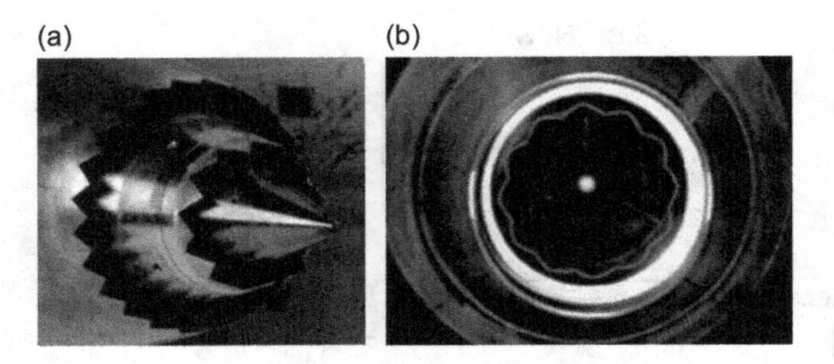

Figure 1.3 Application of shear layer flow control on commercial jet engines consisting of chevron-shaped cutouts (a) and surface-normal sinusoidal trailing edge (b). NASA Glenn Research Center photograph.

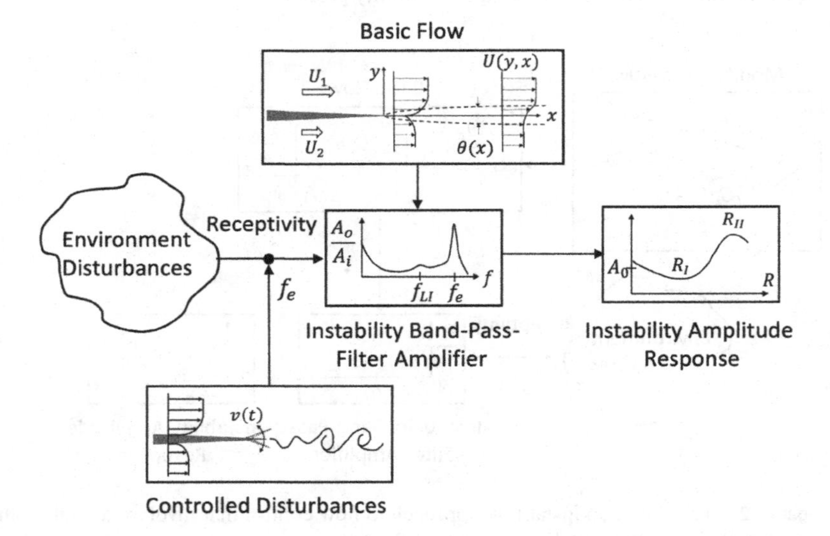

Figure 1.4 Example of an instability approach to flow control that involves a method that introduces an unsteady disturbance that can be amplified by flow field.

thereby would only introduce 2-D disturbances. At incompressible Mach numbers, the 2-D instability wave is most amplified. At compressible Mach numbers, three-dimensional (3-D) oblique waves are most amplified and would require controlled 3-D disturbances.

If the objective is to enhance the growth of the instability waves, the trailing-edge frequency should be close to the most amplified shear layer frequency based on linear stability theory. This would correspond to f_{LI} on the frequency axis of the amplitude spectrum shown in the middle of Figure 1.4. One quite spectacular example of enhanced growth of a 2-D free shear layer that dates back more than 150 years from Tyndall (1864) is shown in Figure 1.5. In this example, the introduction of a monochromatic sound source caused the gas jet to dramatically spread. As discussed in Chapter 5, acoustic pressure is transformed into equivalent motion of the trailing edge through an "acoustic receptivity" mechanism.

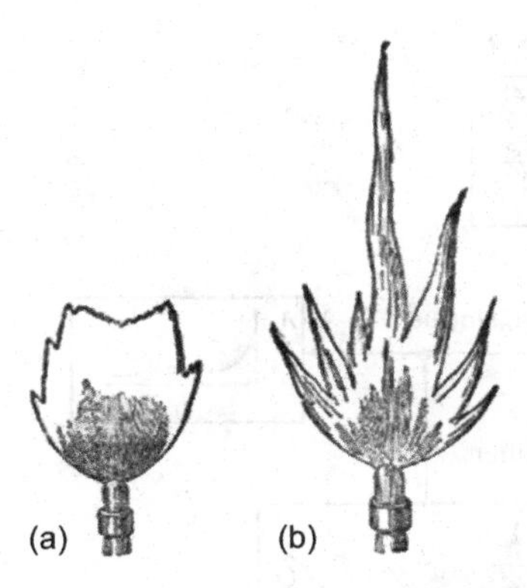

Figure 1.5 Images of a gas flame in a quiescent state (a) and when excited by monochromatic sound "whistle." (b) Taken from Tyndall (1864).

If the objective is to suppress the natural development of the shear layer instability, one approach involves introducing a disturbance at a frequency that is *less amplified*. In that case, by virtue of the controlled excitation, its higher initial amplitude can cause the less amplified excited mode to dominate over more amplified frequencies having lower initial amplitudes. This is illustrated in Figure 1.4, where f_e is the excitation frequency. The slower growing mode will eventually reach an amplitude where non-linear effects will emerge. This will include a modification of the mean flow, which will further suppress the possibility of the growth of more amplified frequencies.

The previous approach to instability control is an example of *open-loop* control. If, however, the amplitude of an instability wave is low enough to satisfy linear theory assumptions, it is possible to suppress its spatial growth by the addition of a second instability wave whose amplitude matches that of the oncoming wave, and whose periodic motion is phase-shifted by 180°. The approach is "linear phase cancelation." An example is illustrated in Figure 1.6 where an oscillating airfoil is placed within the shear layer at a downstream location. The airfoil oscillating motion would be selected to produce linear-amplitude vorticity waves that could linearly cancel the approaching shear layer instability wave. The effectiveness of linear phase canceling depends on the ability to match the amplitude of the incoming wave as well as to produce an opposing phase shift. This generally requires some form of closed-loop control.

Linear phase cancelation has mainly been attempted to control the growth of Tollmien–Schlichting (T-S) instability waves in low-speed boundary layers. Illustration examples are shown in Figure 1.7. In the setup shown in Figure 1.7(a), two vibrating ribbons are used to individually excite T-S waves (Schubauer and Skramstad, 1947). One ribbon is located at an upstream location near the T-S Branch I neutral growth curve. This ribbon is used to excite the primary T-S wave. The other ribbon is located some distance downstream where the T-S wave still exhibits linear growth. Knowing the amplitude and phase of the upstream vibrating ribbon motion, it is relatively easy to adjust the amplitude and phase of the downstream ribbon to cancel the incoming traveling wave. However, when the primary T-S wave results from uncontrolled background disturbances, the method is more challenging.

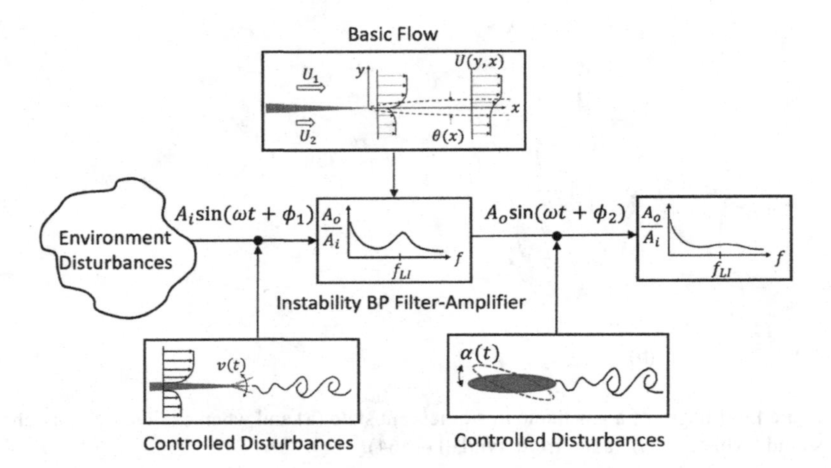

Figure 1.6 Example of an instability approach to flow control that involves linear phase canceling of a downstream traveling instability wave.

Phase cancelation attempts have also utilized sensors that were located upstream of the phase-canceling wave device. Figure 1.7(b) illustrates that approach. The upstream sensors are intended to measure the amplitude and phase of the incoming wave. A trained control system then constructs a conjugate waveform to be produced by the flow actuators to suppress the oncoming instability wave. In some cases, downstream sensors are used to measure the amplitude of any residual wave(s) to adjust the actuator motion while seeking to minimize any residual wave motion. Overall, with uncontrolled initial conditions, these approaches do not provide much benefit in delaying turbulence onset. A further complication in the attempts to phase cancel 2-D T-S wave is that even asymptotically small amplitude residual 2-D waves can interact with 3-D waves having the same phase speed that results in the resonant growth of the 3-D waves and turbulence onset (Craik, 1971).

The previous examples have focused on 2-D mean flows; however, in many applications, the mean flow is 3-D. The flow over a swept wing that is typical of most high-speed aircraft is an example of a 3-D mean flow that results in a cross-flow instability that dominates over other potential instabilities leading to turbulence onset (Saric and Reed, 2002). The cross-flow arises from a combination of pressure gradient and wing sweep that causes the inviscid-flow streamlines at the boundary layer edge to deflect inboard. The cross-flow boundary layer exhibits an inflectional mean velocity profile that is a characteristic of free shear layers and is inviscidly unstable. The cross-flow instability results in both traveling and stationary cross-flow modes. Although the traveling cross-flow modes are more amplified, the stationary modes are extremely sensitive to surface roughness, and generally dominate turbulence onset. The stationary modes appear as a regular pattern of corotating vortices that are easily visible in surface flow visualization or surface heat flux images. As an example, Figure 1.8 shows an image of the surface heat flux over a 70° swept fin at Mach 6 in which discrete roughness elements have been applied to excite stationary cross-flow

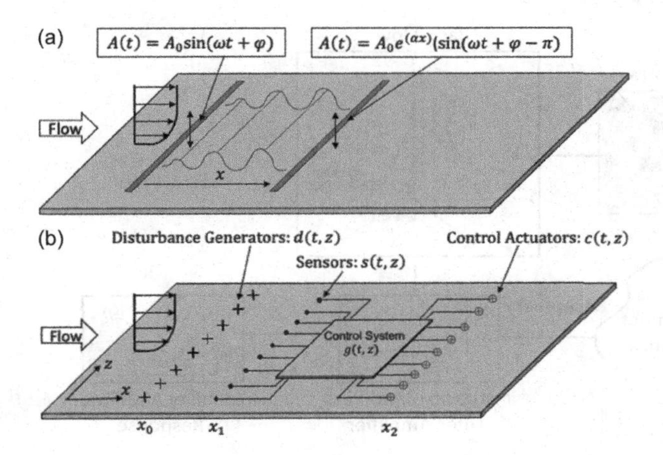

Figure 1.7 Examples of a laminar boundary layer 2-D (a) and (b) 3-D (b) T-S wave phase cancelation setup.

modes. The presence of the stationary cross-flow vortices is evident in the image by the periodic light and dark streaks that respectively correspond to regions of high and low surface heat flux.

Having this knowledge of the cross-flow instability, the approach to control the instability and subsequently delay turbulence onset involves introducing a controlled disturbance that will excite a less amplified mode that can dominate over other more amplified modes. In this case, it is done by applying discrete roughness having a wavelength that will excite a less amplified stationary cross-flow mode. This process, which was pioneered by Saric et al. (1998a), is illustrated in Figure 1.8. In that, the growth of the most amplified stationary cross-flow mode with "critical" wavelength λ_C is circumvented by the introduction of an evenly spaced discrete roughness with a wavelength, λ_{SC}. The growth of the less amplified stationary cross-flow modes ultimately inhibits the growth of the more amplified stationary modes by modifying the basic flow. Saric et al. (1998c) found that patterned surface roughness in the form of hemispherical "dots" with a height of 50 μm provided an effective transition delay in low-speed swept wing experiments. The method has also proved to be effective in experiments at Mach 3.5 (Schuele et al., 2013) and Mach 6 (Corke et al., 2018).

1.1.1 Free Shear Layers and Jets

In all of these examples, the flow control approach is tied to an instability of a specific basic flow that depends on such factors as (1) the flow condition such as the Mach and Reynolds numbers, (2) the geometry over which the basic flow develops, and (3) the receptivity of the basic flow to the disturbance environment. The examples in Figures 1.1–1.6 utilized a free shear layer to demonstrate basic flow control approaches that are linked to the control of a fluid instability. Free shear layers of the kind that are formed by the merging of two streams or a single stream into a quiescent fluid are

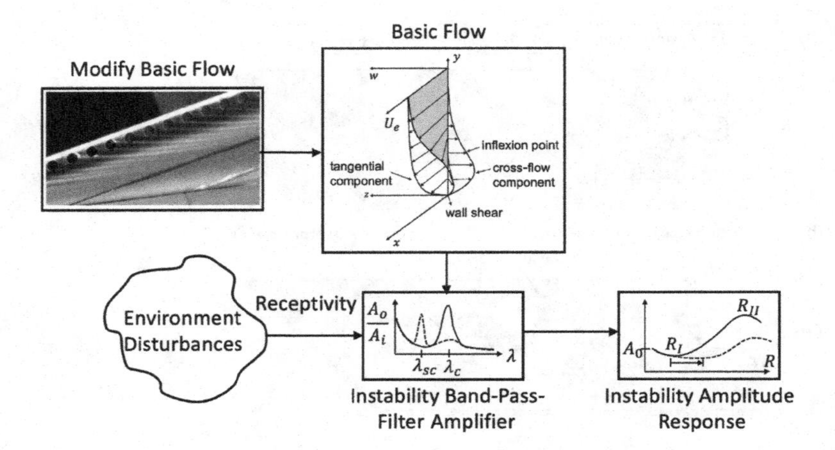

Figure 1.8 Example of an instability approach to flow control that involves a method that introduces an unsteady disturbance that can be amplified by flow field. Taken from Middlebrooks (2022).

simple flow configurations that arise in numerous natural phenomena as well as many engineered devices. In most of the engineered applications, the objective is to enhance fluid mixing.

One of the most striking and distinct features of free shear layers is the emergence of subharmonic components of the initial shear layer instability frequency. This results in a sequence of successive mergings (pairings) of the discrete vortices formed by the initial role-up of the instability wave. A visual example of this sequence was captured by Winant and Browand (1974). In this process, the mean vorticity of the basic flow is sequentially redistributed and spread across the shear layer. Controlled conditions that encourage the successive vortex pairings can greatly enhance shear layer spreading and mixing.

Ho and Huang (1982) observed that exciting a higher initial level fundamental shear layer frequency can temporarily suppress the growth of the subharmonic, frequency, and therefore delay vortex pairing. Similarly, when exciting a frequency, f_e, such that $f_n/3 < f_e < f_n/2$, the frequency that emerged in the shear layer jumped to the first harmonic, $2f_e$, that came closest to the natural fundamental shear layer frequency, f_n. In this case, the excitation became the subharmonic of the initial vortex passage frequency, and in contrast with the previous fundamental excitation, vortex pairing was promoted. Ho and Huang (1982) also documented that further reductions in the excitation frequency could lead to successive frequency-locking stages in which the resulting shear layer frequency became the second and third subharmonics of the excitation frequency, resulting in the coalescence of as many as three or four vortices. These results provide a prime example of efficient instability-based flow control. By this approach, Ho and Huang (1982) indicate that only very low excitation levels, on the order of 0.01–0.1 percent of $\overline{U}$, were required.

Acoustic pressure disturbances that occur in the formation and pairing of shear layer vortices can feed upstream and be felt at the splitter plate trailing edge, the acoustic receptivity site. As a result, this pressure feedback can reenforce a narrow band of instability frequencies and lead to a feedback resonance. The sensitivity to acoustic disturbances increases with increasing trailing-edge sharpness (i.e., decreasing trailing-edge radius).

The most cited evidence of the feedback mechanism comes from Kibens (1980) in which the shear layer of a low Mach number axisymmetric jet was acoustically excited. The exit shear layers of round jets have all of the characteristics of 2-D shear layers when the ratio of the initial momentum thickness to the jet exit radius of curvature is small, $\theta/r < 0.01$. While smoothly scanning through a range of excitation frequencies, f_e, Kibens (1980) documented stair-step-like frequency "lock-in" regions corresponding to excitation frequencies $f_e, f_e/2, f_e/4$, and $f_e/8$ that were indicative of a feedback resonance. They indicated that the frequency stair-step jumps corresponded to vortex pairing events, which was consistent with the factor of two frequency changes at each jump.

In a somewhat surprising revelation, the evolution of vortical structures in *turbulent* shear layers is governed by essentially the same dynamical processes as their laminar counterpart. Brown and Roshko (1974) subsequently confirmed that large-scale coherent structures are indeed intrinsic features of turbulent mixing layers at high Reynolds numbers. Furthermore, sequential mergings of vortices provide the primary mechanism for the spreading of the layer in the downstream direction, as underscored by the experiments of Winant and Browand (1974).

A majority of the literature on shear layer instabilities, like those cited, have involved incompressible Mach number flows in which the most amplified instability modes are 2-D. At compressible Mach numbers, the most amplified instability modes in free shear layer are 3-D, or more specifically oblique waves. Instability control in this regime then focuses on introducing controlled 3-D disturbances. Some of the most successful approaches have involved passive modifications to the basic flow, again using 3-D ramps and tabs. In supersonic jets, the so-called intermediate-origin nozzles (Wlezien and Kibens, 1988) have been very successful in enhancing spreading and vectoring of the jet core flow. Active approaches are more limited, which has generally been due to the limits in flow actuator frequency response and/or amplitudes. However, some success has been reported by Samimi et al. (2007) using arc-filament plasma actuators.

1.1.2 Laminar Boundary Layers

In contrast to free shear layers, wall-bounded flows such as boundary layers and channel flows undergo a viscous instability. Although Taylor (1915) had previously indicated that viscosity can destabilize a flow that is otherwise stable, it remained for Prandtl (1921) to independently make the same discovery and set in motion investigations that led to a viscous theory of boundary layer instability. Any doubt that instability and transition to turbulence were synonymous in boundary layers was erased by the low value of its instability-critical Reynolds number. The seminal experiment of Schubauer and Skramstad (1947) fully demonstrated the existence of instability waves in the boundary layer, their connection with turbulent transition, and the quantitative description of their behavior to the theory of Tollmien (1935) and Schlichting (1940). Although the instability waves are referred to as Tollmien–Schlichting waves, or T-S waves, based on their contribution in validating the theory, it might be fitting to refer to them as Tollmien–Schlichting–Schubauer–Skramstad waves.

The T-S modes are traveling vorticity waves. The waves can be 2-D or oblique. At subsonic Mach numbers, the 2-D waves have the lowest critical Reynolds number and are most amplified. At supersonic Mach numbers, this reverses, with 3-D waves being most amplified. In either case, the wave phase velocity is always less than the free-stream velocity so that at some height in the boundary layer, the mean velocity is equal to the phase velocity. The height at which this occurs defines the "critical layer," which plays a central role in the mathematical theory, as well as being *important in T-S wave control*.

The critical Reynolds number varies with frequency or wave number. Being a dispersive medium, different frequencies propagate with different phase velocities so that individual harmonic components in a group of waves will displace from each other over time. The group velocity can be considered to be a property of the individual waves. If an observer is moving at the group velocity, the wave in the moving frame of reference will appear to grow in time (temporal growth). In contrast, in the stationary frame of reference, the wave appears to grow in space. This insight yields a relation between temporal and spatial growth that is attributed to Gaster (1962).

A focus of boundary layer studies has been on predicting turbulence onset. The conjecture is that it occurs when the instability mode amplitude reaches a *critical* level. Linear stability theory can only predict the growth in amplitude, whereas the absolute instability *amplitude* can only be predicted when the initial amplitude is known. This disjoint led to a turbulence prediction method based on the spatial integration of the linear growth of a disturbance of arbitrary initial amplitude. The method, which is a hallmark for boundary layer turbulence onset prediction, is referred to as the e^N method. Values of N have come from experiments and range from 7 to 10.

The e^N method is extensively used in industry because its low computational overhead makes it suitable for iterative design approaches. The reason the method works is that in boundary layers, the development length of the linear instability process is long compared to that of the nonlinear development prior to turbulence onset. This reduces the sensitivity to the choice of N in these predictions based on this method. From a boundary layer transition control point of view, maintaining N values below those predicted from experiments can provide a metric of merit for preventing turbulence onset.

Boundary layer instability growth rates and N-factors strongly depend on the streamwise pressure gradient. Favorable pressure gradients stabilize the boundary layer, whereas adverse pressure gradients have a destabilizing effect. The former is the basis of laminar control airfoils whose section shape moves the maximum thickness point aft to maintain a favorable pressure gradient over a larger extent of the airfoil chord length.

In an effect similar to a favorable pressure gradient, wall-normal suction stabilizes boundary layers. Wall suction is even effective in stabilizing boundary layers that are close to separating and exhibiting highly unstable (inviscidly) inflectional mean velocity profiles. Conversely, similar to an adverse pressure gradient, wall-normal blowing destabilizes boundary layers. Braslow (1999) provides a history on the use of suction-type laminar flow control on flight research.

There are a wide range of *passive* boundary layer "tripping" techniques used to accelerate the streamwise development to a fully turbulent regime. These include distributed roughness particles, dimpled surfaces, and 2-D protrusions. A critical parameter for any form of roughness is the roughness Reynolds number based on a length scale that corresponds to the maximum roughness height that *does not affect* the turbulent onset location. The velocity used in the roughness Reynolds number is the friction velocity based on the wall shear stress at the location of the roughness.

The criteria for turbulent trips made up of distributed "sand paper" roughness are not as clear as that for 2-D roughness. For example, the roughness height can refer to the mean height of the particles, the peak height of the particles, or the root mean square of the roughness height distribution. From a boundary layer instability perspective, one might consider each roughness particle as an unsteady disturbance source with broad frequency content. These disturbances will be amplified according to the stability of the basic flow. By this perspective, the distributed roughness does not modify the mean flow. Rather, its effect is to increase the initial amplitude of disturbances that feed the T-S waves, causing them to reach nonlinear amplitudes in a shorter streamwise distance.

Active transition delaying approaches that have focused on linear phase cancelation have at best produced marginal results. The practical limitations of 2-D T-S wave cancelation for transition control were brought out in experiments by Thomas (1983). Although a reduction in the amplitude of the T-S wave was observed, smoke-wire flow visualization revealed that with the reduction in the plane T-S wave, 3-D waves emerged likely through a triad resonance mechanism predicted by Craik (1971).

In an attempt to account for the growing 3-D instability waves, Li and Gaster (2006) constructed a multi-input multi-output control system based on a 3-D array of sensors and actuators. The arrangement was shown in Figure 1.7(b). The control system was found to reduce but not completely cancel out the disturbance waves.

In an approach to minimize the computational overhead of a feedback boundary layer transition control system, Fan et al. (1995) utilized a neural network controller. Backpropagation (feedback control) from a sensor located downstream of the control actuator was used to train the neural network. Fan (1995) observed that with an initial training period of about 100 T-S cycles, almost complete T-S wave cancelation occurred. In addition, a low residual level could be maintained if the flow conditions change on a timescale that was larger than the initial training period.

In contrast to boundary layers at subsonic Mach numbers, in supersonic boundary layers, the most amplified T-S modes are *oblique* waves. The wave angles range from 45° to 75°, where the wave angle increases with increasing Mach number. In 2-D boundary layers up to slightly below Mach 5, the oblique T-S modes are most responsible for turbulence onset. However, at approximately Mach 5, an inviscid instability first identified by Mack (1984) becomes dominant over the viscous instability that governs the oblique T-S waves. The inviscid instability results in a family of 2-D "Mack modes" whose mode numbers increase with increasing Mach number. Physically, the Mack modes correspond to sound waves that reflect inviscidly between the solid wall and the relative sonic line in the boundary layer.

Both the T-S and second Mack mode amplification rates are sensitive to the wall temperature, which provides a degree of instability control. Wall heating will increase the amplification of T-S modes but decrease the amplification of Mack modes. The reverse occurs with wall cooling.

Most of the instability control in 2-D hypersonic boundary layers has focused on passive approaches. This is undoubtedly the result of the limitations of active disturbance actuators to produce frequencies (order of 500 kHz) needed to interact with dominant second Mack mode waves. In addition to wall heating, examples of passive approaches include leading-edge bluntness, acoustic absorbing wall liners, and strategically placed surface roughness or wall waviness.

Fedorov and Malmuth (2001) were the first to propose that the growth of the second Mack mode could be suppressed through a suitably designed ultrasonically absorptive porous coating. Based on an analytical model, they predicted that a 25 percent porous coating having 10–20 blind pores per second mode wavelength would result in a factor-of-two reduction in the second mode growth rate. Experimental validation of the concept was first performed by Wartemann et al. (2011). This utilized the natural porosity of a fiber-reinforced ceramic (carbon–carbon) material that was applied to the surface of a 7° half-angle right-circular cone at Mach 7.5. The experiments documented a 70 percent reduction in the second mode amplitude compared to an uncoated smooth surface.

In a related approach, it is known that CO_2 gas has acoustic damping properties, which suggests that injection of the gas into the boundary layer could suppress the growth of the second mode in a manner similar to acoustic absorbing liners. Evidence of this was reported by Jewell et al. (2013) in which a gas mixture consisting of 50 percent air and CO_2 by mass was found to delay turbulence by 30 percent.

Another passive approach aimed at suppressing the growth of the second Mack mode has involved the placement of 2-D roughness elements at the T-S/second mode "synchronization point" where the phase speeds of the viscous and inviscid modes are very close to each other (Federov, 1997). This was initially investigated through numerical flow simulations (Duan et al., 2013) and validated in experiments at Mach 5 and 8 by Casper et al. (2016). The experimental test article consisted of a 7° half-angle cone that was installed with a series of roughness strips. The effects of both single and multiple roughness strips were investigated. The height of the roughness strips was half of the boundary layer thickness, which was within the recommended 50–60 percent of the boundary layer thickness (Fong, 2017). Also as recommended (Fong, 2017), the streamwise width of the strips was twice the boundary layer thickness. When multiple strips were used, they were spaced 10 strip widths apart. The results were mixed. At Mach 8, the experiment documented a reduction in the second

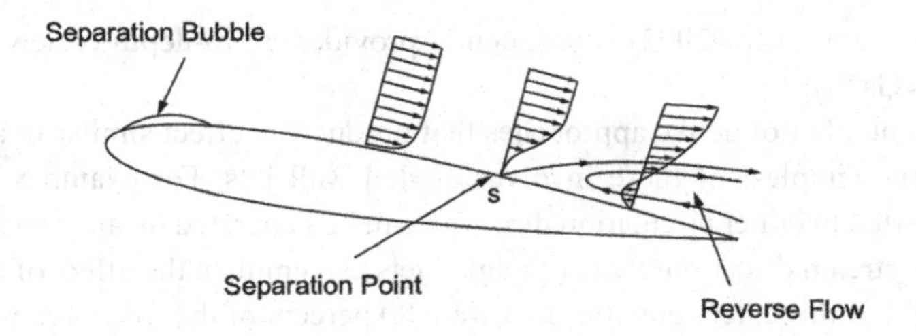

Figure 1.9 Example of flow separation that can occur on airfoils at moderate or large angles of attack.

mode growth. However, at Mach 5, the roughness strips effectively "tripped" the flow causing immediate turbulence onset.

1.1.3 Separated Flows

Boundary layer separation results when there is insufficient momentum in the flow to overcome an adverse pressure gradient or when viscous dissipation occurs along the flow path. Boundary layer separation is almost always associated with some form of aerodynamic penalty, including a loss of lift, an increase in drag, a loss of pressure recovery, and an increase in entropy. Although boundary layer separation is often viewed as 2-D and *steady*, experiments have shown the process to be strongly 3-D and highly unsteady. As with their free shear layer counterpart, the separated shear layer instabilities are highly receptive to external unsteady excitation, which is a characteristic that can be exploited in flow separation control.

In steady flows, boundary layer separation only occurs in regions of an adverse pressure gradient. As illustrated in Figure 1.9, the separation location corresponds to the stagnation point on the wall that separates the downstream moving and upstream moving (reverse) flows. The boundary layer velocity profile at that point exhibits an inflection that is therefore inviscidly unstable to external disturbances. These disturbances can be a combination of vortical and acoustic (pressure) fluctuations whose internal source could be convective instability waves if the upstream boundary layer is laminar or turbulent fluctuations if the upstream boundary layer is turbulent.

There are a number of boundary layer separation control methods that are aimed at increasing the momentum in wall-bounded flows. One of the most commonly used passive devices is a VG. VGs are designed to introduce coherent streamwise vortices that transport high momentum fluid from the free-stream toward the wall. Passive VGs generally consist of a row of small plates or airfoils that are mounted on the wall surface and set at an angle of incidence to the local mean flow direction. In early applications of VGs used for separation control, their height was on the order of the boundary layer thickness. Kuethe (1972) developed and examined nonconventional "vane-type" VGs with heights that were from 42 percent to only 27 percent of the boundary layer thickness. Rao and Kariya (1988) suggested that passive VGs with $h/\delta \leq 0.625$ could be just as effective in flow separation control but with much lower parasitic device drag. Lin (2002) subsequently provided an in-depth review of these "low-profile VGs."

There are a number of active approaches that produce an effect similar to a passive VG. One of the simplest of these involves angled wall jets. For example, a single streamwise vortex in either circulation direction can be generated by angling a wall jet in either cross-stream direction. Pairs of angled jets can emulate the effect of a passive vane-type VG. For these, jet velocities that were 80 percent of the free-stream velocity have been found to be effective in controlling boundary layer separation (Johnston and Nishi, 1990).

The previous approaches can be used to both *prevent* flow separation as well as to *cause* a separated flow to reattach. In these cases, the flow control is aimed at modifying the mean (basic) flow.

In the case of an already separated flow, the instability characteristics of the separated shear layer offer a low-energy active approach to reattach the flow. For example, it has been demonstrated (Seifert et al., 1996) that oscillatory blowing introduced just upstream of the separation point is an effective means of reattaching the flow. In this, an important observation was that the most effective excitation frequency was the one having a reduced frequency, F^+, that scaled with the length of the separated flow region and local free-stream velocity of which the optimum was $F^+ = 1$. In this case, the optimum excitation frequency clearly acted on a separated shear layer instability mechanism. Assuming that the separated shear layer mean profile has a representative hyperbolic-tangent velocity distribution, then linear stability analysis predicts a wave phase speed of $c_r = 0.45U_\infty$. For the optimum reduced frequency, $\omega L/U_\infty = 1$, in which L is the length of the separated region, and noting that the frequency of the traveling waves is $\omega = c_r/\lambda$, where λ is the wavelength, the optimum frequency to attach a separated flow corresponds to that in which there are two instability wavelengths spanning the separated flow region. With regard to methods for introducing the periodic disturbances for flow separation control, they have included acoustic excitation (Nishioka et al., 1990; Nishizawa et al., 2003) and a periodic body force produced by a pulsed plasma actuator (Kelley et al., 2014).

The receptivity of the separated shear layer to external or internal disturbances and controlled excitation can be exploited to provide a method for detecting separation onset. For example, in the analysis of acoustic receptivity of a boundary layer over a parabolic leading edge, Haddad et al. (2005) found that there was a 100-times growth in the receptivity coefficient just prior to flow separation. As a result, just prior to flow separation, the boundary layer is significantly more responsive to the disturbances (background or controlled). Based on this, He (2008) and Lombardi et al. (2013) devised a flow separation detection approach that became an integral part of a closed-loop separation control system. The approach utilized a pressure sensor located a short distance downstream of the leading edge on the suction side of an airfoil. A periodic disturbance (perturbation) was introduced upstream of the pressure sensor location. At low airfoil angles of attack where the flow remained attached, the amplitude of the perturbations sensed by the pressure transducer was low. However, when the angle of attack was increased to just before the flow would separate, as a result of the enhanced receptivity of the boundary layer, the perturbations sensed by the pressure transducer significantly increased. The method thus provided an indication of the incipient flow separation. In a closed-loop control system, the indication of imminent flow separation could then initiate any of the separation control approaches previously discussed.

1.1.4 Shock–Boundary-Layer Interaction

Shock–boundary-layer interactions (SBLI) appear in numerous high-speed flows including those of supersonic engine intakes, transonic gas turbine blade tip gaps, transonic turbine blade passages, scramjet isolator ducts, transonic and supersonic flight vehicle surfaces, and surfaces of rockets, missiles, and reentry vehicles. It is of particular interest because it can greatly affect the boundary layer development including causing large temporal and spatial pressure fluctuations and *flow separation* that greatly affects aerodynamic performance.

A schematic that illustrates the features of an incident oblique shock wave interacting with a wall boundary layer is shown in Figure 1.10. The incident shock can originate either from an external surface above the wall or from a shock wave that reflected from an upper wall as part of a duct. If the incident shock is sufficiently strong, the pressure gradient across the shock can cause the boundary layer to separate, forming a recirculating separation bubble. The separation bubble causes the mean flow to deflect away from the wall. This results in an adverse pressure gradient that communicates upstream through the subsonic portion of the boundary layer to a point where a "reflected" shock is formed. The location of the reflected shock is well upstream of where it would have been located if the flow were inviscid. Finally, an expansion fan forms over the top of the separation bubble. This is quickly followed by compression waves that form near where the separated flow reattaches. Further downstream, the boundary layer can gradually return to a fully developed equilibrium condition. However, depending on the size of the separation bubble, the downstream distance needed to reach equilibrium can take an order of 10–20 boundary layer thicknesses.

Wall cooling can reduce the effect of the adverse pressure gradient that forms downstream of the incident shock and thereby delay or reduce the extent of boundary layer flow separation (Kepler and O'Brien, 1962). In a turbulent boundary layer, if the ratio of the wall temperature to recovery temperature is less than 1, it will lower the viscosity compared to an adiabatic wall condition. This has the effect of increasing the local Reynolds number and increasing the wall shear stress, similar to adding momentum near the wall. By the same arguments, heating the wall has the opposite effect and can therefore hasten flow separation.

Strategic wall suction has shown some benefits toward alleviating the adverse effects of a SBLI. If applied downstream of an impinging shock, the increase in the wall shear stress in the thinned boundary layer can offset the adverse pressure gradient produced by the incident shock, and thereby limit the formation of the separation bubble. Experiments at Mach 2.8 and 3.78 (Mathews, 1969; Seebaugh and Childs, 1970) indicated that suction levels from 2-5 percent of the mass flow of the approaching boundary layer were sufficient to prevent flow separation.

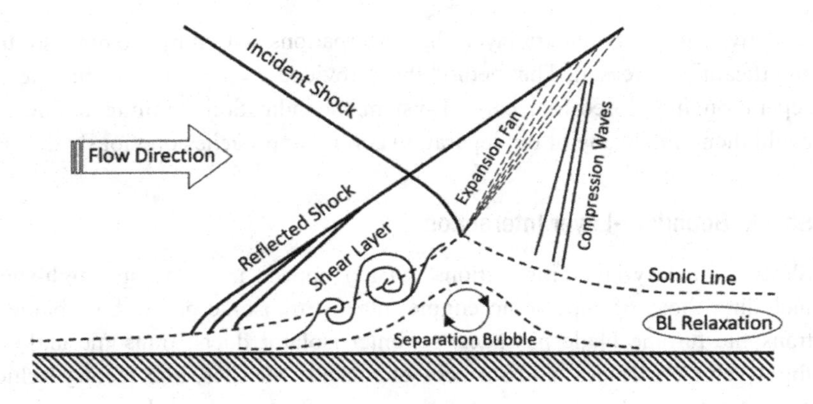

Figure 1.10 Schematic illustration of features of an incident oblique shock wave interacting with a wall boundary layer.

Wall suction applied upstream of the impinging shock could have a detrimental *or* a beneficial effect. Wall suction that thins the approaching boundary layer will move the sonic line closer to the wall and thereby make the shock near the wall stronger. This could more strongly promote flow separation downstream of the shock and commensurate flow unsteadiness. However, based on the "sonic point criterion" (Li, 2007), the stronger oblique shock would be more stable and thereby be less receptive to unsteady disturbances. Assuming that the flow unsteadiness that occurs from SBLI is due to an instability of the incident shock, then the thinning of the approaching boundary layer would reduce the unsteady shock motion.

One passive SBLI control approach exploits the pressure gradient across the shock (Barn et al., 1983). This approach locates a wall cavity covered by a porous screen under the impinging shock. The concept is that the pressure difference across the incident shock will circulate a flow through the cavity and exit upstream of the incident shock where it will energize the incoming boundary layer. Another passive approach involves the placement of wall-mounted streamwise VGs upstream of the incident shock. The intent is to enhance mixing throughout the boundary layer and thereby reduce the potential for incident shock-induced flow separation. McCormick (1993) performed a comparison of the effectiveness of the two techniques. The conclusion was that the covered wall cavity performed better at reducing the pressure loss across the incident shock.

With the same objective for the use of passive VGs, Souverein and Debieve (2010) investigated the use of a spanwise array of angled wall jets as an active approach to introduce streamwise vortices into the boundary layer for SBLI control in a Mach 2.5 flow. The wall-jet array was located five boundary layer thicknesses upstream of the incident shock location. The mass flow of the jet array was about 3 percent of the mass flow of the boundary layer. As expected, the angled jet array acted to thicken the approaching boundary layer. This was found to reduce the size of the shock-induced separation bubble but not to eliminate it.

Valdivia et al. (2009) sought to improve the effect of vortex generating wall jets by combining them with passive VGs with the specific motivation was to control "unstart" in an scramjet inlet isolator at Mach 5. The passive VGs were located upstream of the wall jets. Although each type of VG was effective by themselves, the combination of the two was found to be most effective with the reason being that the passive VGs mitigated the blockage produced by the bow shock that otherwise formed upstream of the wall-jet orifices.

Narayanaswamy et al. (2012) investigated the use of pulsed plasma-jet actuators as a replacement for air-driven wall jets. Two wall-jet orientations were examined: one in which the jet injection was normal to the wall surface, and another in which the injection angle was pitched and skewed to match a previously used orientation of air-driven wall jets (Souverein and Debieve, 2010). A spanwise array of three plasma jets was located six boundary layer thicknesses upstream of the reflected shock. The pulsing frequency corresponded to a Strouhal number that matched that of the characteristic unsteadiness associated with an SBLI. This was found to reduce the magnitude of pressure fluctuations by 30 percent.

In an approach that more directly utilizes the electro-magneto-hydrodynamics properties of the ionized air (plasma), experiments have been performed in which the electrodes are exposed to the primary flow and operated to generate long plasma filaments that extend downstream. In this arrangement, the primary function of the plasma discharge is to generate heat that locally lowers the Mach number and thereby weakens the incident shock strength (Leonov et al., 2010). Wang et al. (2009) combined the plasma filament generation with a magnetic field in order to seek to enhance the control of SBLI. The addition of a magnetic field coupled with the plasma discharge current results in a Lorentz force that can act on the flow field. In SBLI experiments at Mach 2.2 (Wang et al., 2009), the incident shock strength was reduced by as much as 11 percent.

All of the previous SBLI control approaches have focused on the boundary layer approaching the incident shock with the general intention of weakening the incident shock in order to reduce the strength and size of the flow recirculation bubble that forms downstream of the shock. However, a number of techniques used to generally control boundary layer separation can be applied. One example is steady wall-tangential blowing. Applying this approach to a shock-induced separation bubble, Viswanath et al. (1983) found that the distance to reattachment was significantly reduced.

1.1.5 Bluff Body Wakes

A bluff body can generally be categorized as one in which its length in the flow direction is approximately the same as its height perpendicular to the flow in a 2-D representation. Such shapes exhibit a wide wake on the scale of the body (2-D) height, and aerodynamic drag that is dominated by the low-pressure region that forms in the near wake of the body. Bluff body wakes are complex, involving boundary layer separation and multiple shear layer interactions.

The most widely studied bluff body shape is a large aspect ratio (2-D) circular cylinder. The flow around a circular cylinder can be considered as a prototype of bluff body wakes because of the simplicity of the boundary conditions and the complexity of the physical processes involved. A boundary layer over the cylinder surface originates at the upstream (windward) stagnation point. The boundary layer eventually separates as it approaches the adverse pressure gradient that exists near the downstream (leeward) side of the cylinder. The separation location will depend on the Reynolds number and surface roughness that determine whether the boundary layer state is laminar or turbulent. A low-pressure region is formed between the pair of separated shear layers. This low-pressure region includes a symmetric pair of flow recirculating cells. The flow recirculation in these cells has implications on the global stability of the wake flow.

A hallmark of bluff body wakes is the unsteady vortex shedding. von Kármán (1912) analyzed the stability of vortex street configurations and established a link between the vortex street structure and the aerodynamic drag. The relation between the vortex shedding frequency to the cylinder diameter and external flow conditions is attributed to the classic early experiments of Strouhal (1878). Investigations on the relation between the Strouhal number and the Reynolds number (Kovasznay, 1949; Roshko, 1954) led to characterizing three Reynolds number ranges: "Stable" $40 < Re_D < 150$, "Transitional" $150 < Re_D < 300$, and "Irregular" $300 < Re_D < 10,000+$. Over a range of cylinder diameter Reynolds numbers where the separated shear layers are laminar, the Strouhal number is relatively constant with a value of 0.21. This is referred to as the "subcritical" Reynolds number range.

A general characteristic of bluff bodies is that the pressure or wake drag is much larger than the viscous drag of boundary layers over its surface. Morkovin (1964) found that the overall drag coefficient was inversely proportional to the Strouhal number. The drag coefficient abruptly changes for $Re_D > 200,000$, which is where the previously laminar separated shear layers become turbulent. This results in an abrupt contraction of the near-wake width that results in a substantial decrease in the drag coefficient referred to as the "drag crisis." In general, the separated boundary layer and resulting free shear layers that develop off of the cylinder are convectively unstable. Therefore, any disturbances that would excite instabilities of the shear layer will amplify with downstream distance.

Landmark transient experiments (Mathis et al., 1984; Strykowski, 1986) have indicated that at low Reynolds numbers, the vortex shedding results from a global instability. As a result, disturbances grow in space and time rather than in space alone as with a convective instability. This has profound implications with respect to the control of low Reynolds number bluff body flows.

For example, Strykowski (1986) demonstrated that the addition of a small *control* cylinder into the near-wake region of a circular cylinder could completely suppress the vortex shedding. The addition of the control cylinder is thought to introduce asymmetry in the basic flow that changes the local stability properties of the wake, particularly those governing the absolute instability. Resistively heating the control cylinder was found to dramatically widen the region of vortex shedding suppression (Strykowski and Sreenivasan, 1990).

One of the most common methods to suppress the wake shedding is with the addition of a splitter plate. This has included single and segmented rigid and flexible plates. In most cases, these have been attached to the trailing portion of the cylinder. Investigations by Roshko (1954) on the streamwise placement of splitter plates to suppress vortex shedding of cylinders in the subcritical Reynolds number range found them to be effective when placed as far as four cylinder diameters downstream. Within that range, the effective length of the bluff body extended to the trailing edge of the splitter plate (Roshko, 1954). As a result, the process can be considered as "virtual streamlining."

Cardell (1993) studied the effect of a *permeable* splitter plate on cylinder wake. He noted that a permeable splitter plate that was carefully chosen to minimize geometrical effects would interfere with communication across the wake center plane, and by varying the permeability, it was possible to vary the magnitude of the interference created by the splitter plate. Hinged-rigid splitter plates have been recently studied (Assi et al., 2009; Shukla et al., 2009). For these, the control parameter is the damping in the hinge motion. With a relatively large hinge damping (Assi et al., 2009), the splitter plates did not oscillate but assumed a stable position at an angle to the flow direction, somewhat like a rigid plate. This was found to effectively suppress the wake shedding and provide a commensurate amount of drag reduction.

Shukla et al. (2013) considered the effects of a *flexible* splitter plate. In that case, the control parameter was the flexural rigidity (EI) of the splitter plate. Two Reynolds number regimes of periodic oscillations of the flexible splitter plate were observed. Within the two regimes, the normalized tip oscillation frequency was close to the natural (subcritical Reynolds number regime) Strouhal number (0.21). The bounds of these regimes depended on the amount of flexural rigidity.

Another of the early methods of bluff body wake control is "base bleed," in which some fraction of fluid is injected at the aft part of the body. An early demonstration (Wood, 1964) of the effect of base bleed on an airfoil blunt trailing edge produced a suppression of vortex shedding along with a substantial drag reduction. The critical bleed coefficient was found to be in good agreement with stability calculations (Monkewitz and Nguyen, 1987) that showed the wake to be everywhere convectively unstable below the critical value.

A number of active bluff body flow control approaches have involved the introduction of unsteady disturbances on the body. This has included surface motion using a piezoelectrically active actuator (Wehrmann, 1967), an unsteady 2-D slotted surface jet (Huang, 1996), and single dielectric barrier discharge plasma actuators (Thomas et al., 2008). In each approach, the choice of excitation frequencies could either suppress or reenforce the natural wake vortex shedding. These experiments have generally focused on the subcritical Reynolds number regime where the shedding Strouhal number is nearly constant, and the separating shear layers are laminar.

On circular cylinders, the most sensitive location to introduce controlled unsteady disturbances is just upstream of where the shear layers separate from the cylinder surface. In the subcritical Reynolds number regime, shear layer separation occurs at an approximate angle of 85° (measured from the stagnation line) or just forward of

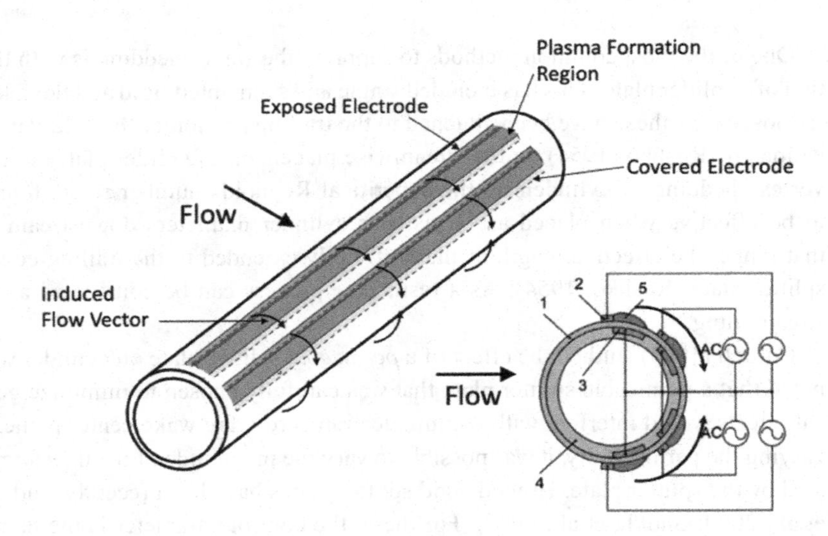

Figure 1.11 Illustration of four single dielectric barrier discharge plasma actuators used for active cylinder wake control. Taken from Thomas et al. (2008).

the cylinder apex. If periodic excitation is applied near that location, it can cause the shear layer separation location to oscillate, providing significant amplification of the disturbance amplitude.

Figure 1.11 shows an example of an arrangement of plasma actuators used to control vortex shedding of a 2-D circular cylinder. The cylinder had four plasma actuators located at ±90° and ±135° from the stagnation line. The plasma actuators were designed to induce a velocity component that was tangent to the cylinder surface and directed toward the downstream side of the cylinder.

Exploring a range of unsteady frequencies, Thomas et al. (2008) found that an optimum excitation frequency to suppress vortex shedding and subsequently to maximize the drag reduction occurred for an actuation frequency corresponding to St $=$ 1, or five times the natural shedding Strouhal number (0.21) for a cylinder in the subcritical Reynolds number regime. Examples of the wake structure under this excitation condition are shown in Figure 1.12, where (a) corresponds to the symmetric shedding condition in which the top-half and bottom-half plasma actuators are operating in phase and (b) corresponds to alternate shedding condition where the top-half and bottom-half plasma actuators are operating with a 180° phase shift. In both cases, the flow visualization indicates that the vortex street has been completely suppressed. Thomas et al. (2008) indicated that the effect persisted well beyond eight cylinder diameters downstream.

1.1.6 Turbulent Boundary Layers

Flow control strategies for turbulent boundary layers have generally focused on either removing or altering in some way mechanisms underlying coherent motions that have

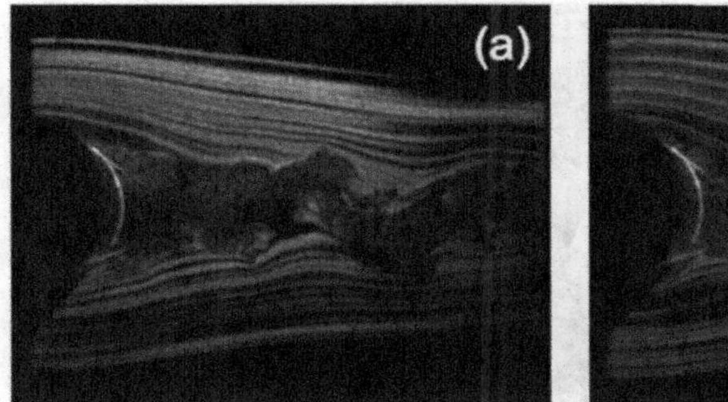

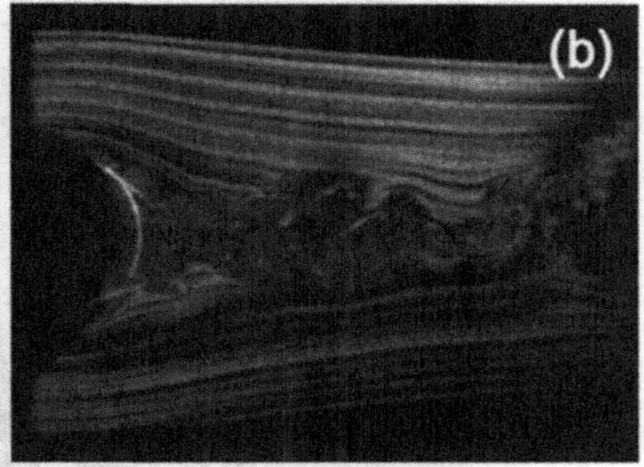

Figure 1.12 Flow visualization of the cylinder wake at $\mathrm{Re}_D = 3.3 \times 10^4$ with unsteady plasma actuations to excite symmetric shedding (a) and alternate shedding (b) at five times the natural Strouhal number, St = 1. Taken from Thomas et al. (2008).

been shown to contribute to sustaining turbulent energy production. This has generally involved either modifying the large vortical motions in the outer half of the boundary layer or modifying the small-scale vortical motions that occur in the lower third of the boundary layer, near the wall. Both passive and active flow control approaches have been examined and led to varying degrees of success.

In a landmark study using flow visualization, Kline et al. (1967a) observed surprisingly well-organized vortical motions in the near-wall region of a turbulent boundary layer. This appeared as a coalescence of the visualization markers into long streamwise oriented "streaks." Figure 1.13 provides an example image, where the flow direction is from the top. The spacing between the "streaks" was found to scale with the wall shear velocity, u_τ. Kline et al. (1967a) observed that the coherent streaks underwent a process of "gradual lift-up, sudden oscillation, bursting and ejection." This sequence became simply known as "bursting" and was felt to play an important role in the production of turbulence in the wall and buffer layers. Of singular importance, Kline et al. (1967a) noted a linear dependence of the viscous drag on the frequency of the bursting events.

The connection between the organized streamwise vorticity within the buffer layer and skin friction drag of turbulent boundary layers has been noted by Kim (2011). In this scenario within the buffer layer, coherent streamwise vortices transport near-wall fluid toward the wall on one flank and eject fluid away from the wall on the other flank. The wall-directed motions give rise to the so-called splatting events that steepen gradients of mean velocity, thereby resulting in higher time-mean friction drag. As noted by Schoppa and Hussain (1998a), the velocity gradient reduction effect of the outward directed motion is small in comparison to the enhancement of the velocity gradient produced by the wall-directed motion.

A general consensus has emerged (Jimenez and Moin, 1991; Waleffe et al., 1993; Hamilton et al., 1995; Jimenez and Pinelli, 1999), which is of an autonomous cycle for turbulence production involving the generation, growth, and instability of coherent streamwise vortices associated with the wall-layer "streak" structure. As part of

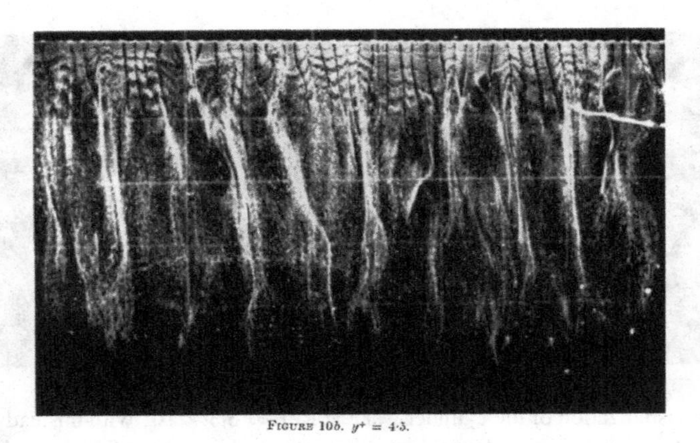

Figure 1.13 Flow visualization of wall-layer "streak" structure first discovered by Kline et al. (1967a).

this, Schoppa and Hussain (2002) demonstrated that near-wall turbulence production may originate from a sinuous instability of the coherent streamwise vortices through the process of "streak transient growth." A critical parameter to the instability is the wall-normal vorticity, ω_y, that flanks the coherent streamwise vortices.

An idealized diagram of an autonomous cycle based on a "streak transient growth" is shown in Figure 1.14. This begins with the quasi-steady coherent streamwise vortices associated with the wall "streak" structure. The pumping action of the coherent vortices results in a spanwise mean flow distortion (thickening and thinning) of the buffer layer that results in elevated levels of the wall-normal vorticity, ω_y. The levels of ω_y eventually reach a critical value that triggers a sinuous instability of the streamwise vortices causing them to distort, lift up, and break up in the process described by Kline et al. (1967a). Following their breakup, the cycle begins again with the next generation of coherent wall-layer streamwise vortices.

Active flow control that focuses on disrupting this autonomous turbulence production cycle has focused on reducing the spanwise distortion produced by the coherent wall-layer streamwise vortices. This directly focuses on reducing the levels of ω_y to below the critical value of the sinuous instability. The approach is therefore one that acts to modify the basic state. Since this process of vortex lift-up and break-up was also linked to the viscous drag (Kline et al., 1967a), the stabilization would also result in drag reduction.

Schoppa and Hussain (1998) demonstrated drag reduction in a Direct Navier Stokes simulation of a turbulent channel flow in which they imposed a spanwise velocity component along the channel wall through either a pair of counter-rotating streamwise vortices or opposed wall jets. Both approaches resulted in significant drag reduction. Corke et al. (2017) and Thomas et al. (2016) demonstrated unprecedented levels of drag reduction of up to 70 percent by applying a mean spanwise velocity component at the wall using an array of plasma actuators. Measurements by Duong (2019) confirmed that this also resulted in a significant reduction in the frequency of "burst" event, the turbulent Reynolds stresses, and turbulence production.

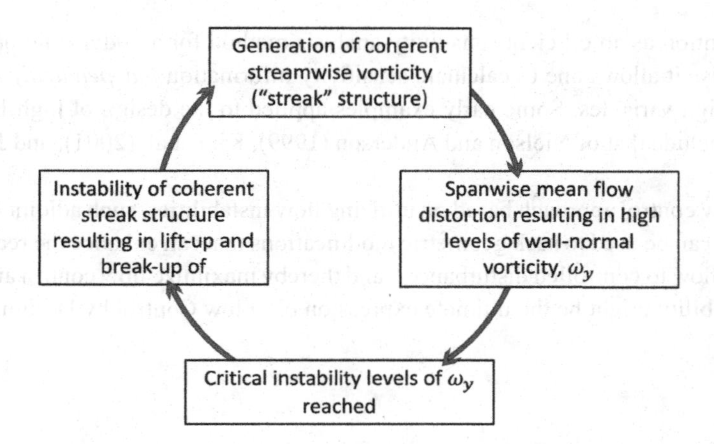

Figure 1.14 Autonomous wall-layer turbulence production cycle based on the concept of streak instability.

There are a large number of active and passive flow control approaches that focus on other mechanisms associated with coherent vortical structures in turbulent boundary layers. These are extensively covered in Chapter 8.

1.1.7 Flow Control by Design

In most cases, flow control is applied to an existing geometry that determines the basic flow. This can constrain flow control methods that are based on utilizing flow instabilities. As previously discussed, it is possible and sometimes advantageous to modify the basic flow through changes in the geometry. Examples included trailing-edge waviness, and the addition of passive elements such as splitter plates, VGs, ramps, cavities, and steps. In many of the previous examples, modification of the mean flow was used to prevent the growth of a fluid instability.

"Flow Control by Design" takes a broader approach in which a geometry is modified to make it more *receptive* to flow control that utilizes fluid instabilities. As an example, Patel et al. (2006) modified the trailing edge of an airfoil section by adding a 20° convex ramp that would cause the boundary layer to separate at that location. As presented earlier, a separated flow is highly receptive to disturbances and easily made to reattach through controlled periodic excitation. When the ramp was located on the pressure side of the airfoil, the separated flow at the trailing edge acted like a deflected trailing edge flap, increasing the aerodynamic lift. The added lift was removed when the separation bubble was forced to reattach. The result was lift control without a moving surface.

With the growing ability of computational fluid dynamics, it is now possible to formulate simulations that can systematically investigate multiple parameters that lead to a design that meets specific performance metrics. One approach utilizes an adjoint formulation of the Navier–Stokes (N-S) equations. The adjoint method has gained much attention as an efficient sensitivity analysis method for aerodynamic optimization because it allows one to calculate sensitivity information *independently* for each of the design variables. Some early examples applied to the design of high-lift wing sections include that of Nielson and Anderson (1999), Kim et al. (2001), and Jameson (2003).

In a flow control approach based on utilizing flow instabilities, such adjoint N-S formulations can be used to seek geometric modifications that can enhance the receptivity of a basic flow to controlled disturbances, and thereby maximize flow control authority. Such capability might be the ultimate expression of "Flow Control by Design."

2 Sensors and Actuators

The following sections provide a background on the types of flow sensors and actuators that are frequently used in establishing the instability of various flow fields and in the documenting of conditions before and after flow control. The emphasis here is to provide a basic overview of these sensors and actuators. Of particular interest is highlighting the conditions where they are best applied.

2.1 Velocity Sensors

This section presented a variety of fluid velocity measurement approaches that include Pitot probes, hot-wires, laser Doppler velocimetry (LDA), and particle image velocimetry (PIV). Three methods of hot-wire operation are presented with comparisons of their static and dynamic responses. With hot-wires, special emphasis is placed on the effects of sensor angularity, as well as ambient effects such as temperature and compressibility. A method of frequency compensating is also presented. The operating principles of LDA and PIV are presented along with best practices.

2.1.1 Pitot Probes

Pitot probes can offer a simple and generally accurate means of documenting the mean basic state that is necessary in the analysis of fluid instabilities. Assuming a one-dimensional (1-D) flow of an incompressible frictionless fluid, the velocity derived from a Pitot probe is simply

$$q = \sqrt{\frac{2(P_0 - P_s)}{\rho}}. \tag{2.1}$$

Deviations from this can occur due to misalignment of the probe axis with the mean flow. These generally follow a cosine dependence of the alignment angle. The measurement area relative to a velocity shear can also introduce errors by providing a spatial average. Since many of the fluid instability cases involve shear layers or boundary layers, this presents a veritable issue in documenting the mean basic state. As illustrated in Figure 2.1, Matlis (2004) flattened nominally round Pitot probes to reduce spatial averaging in boundary layer measurements over a 7° half-angle cone at Mach 3.5.

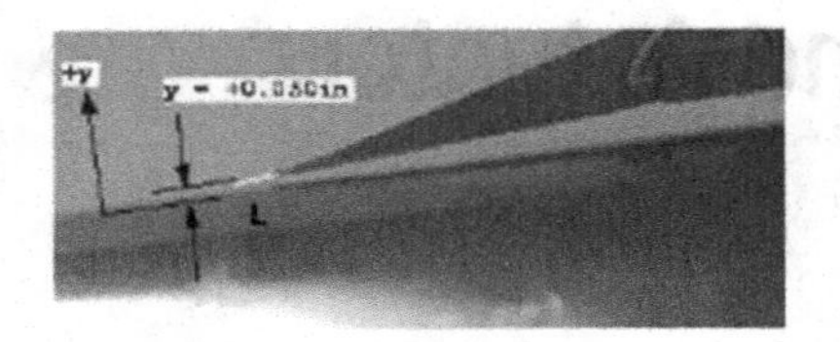

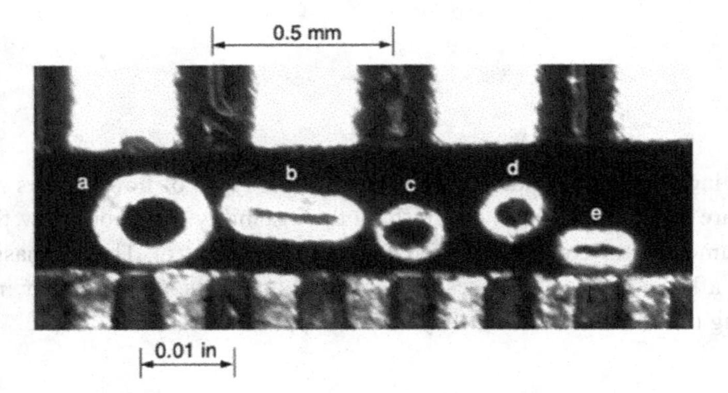

Figure 2.1 Photographs of the ends of Pitot probes used by Matlis (2004) to improve spatial resolution in a boundary layer over a 7° half-angle cone at Mach 3.5.

Taking compressibility into account, Binder (1951), determined the following relation for low Mach number flows:

$$q = \sqrt{\frac{2\gamma}{\gamma - 1}\frac{P_s}{\rho_s}\left[\left(\frac{P_0}{P_s}\right)^{(\gamma-1)/\gamma} - 1\right]}, \tag{2.2}$$

where γ is the ratio of specific heats, c_p/c_v. The relation between P_0 and P_s comes from the isentropic flow equation

$$\frac{P_0}{P_s} = \left(1 + \frac{\gamma - 1}{2}M^2\right)^{\gamma/(\gamma-1)}. \tag{2.3}$$

With supersonic Mach numbers, a bow shock forms ahead of the Pitot probe. Assuming this to be a normal shock, then the ratio P_0/P_s is given as (Binder, 1951)

$$\frac{P_0}{P_s} = M^2\left(\frac{\gamma + 1}{2}\right)^{\gamma/(\gamma-1)}\left[\frac{2\gamma M - \gamma + 1}{M^2(\gamma + 1)}\right]^{1-1/(\gamma-1)}. \tag{2.4}$$

The frequency response of a conventional Pitot probe is low, owing to long tubing lengths leading to a pressure transducer. Typical frequency response is in the range of 10–100 Hz. However, with the advent of more miniature pressure transducers, Corke et al. (2018) and Arndt et al. (2020) incorporated pressure transducers into a Pitot tube

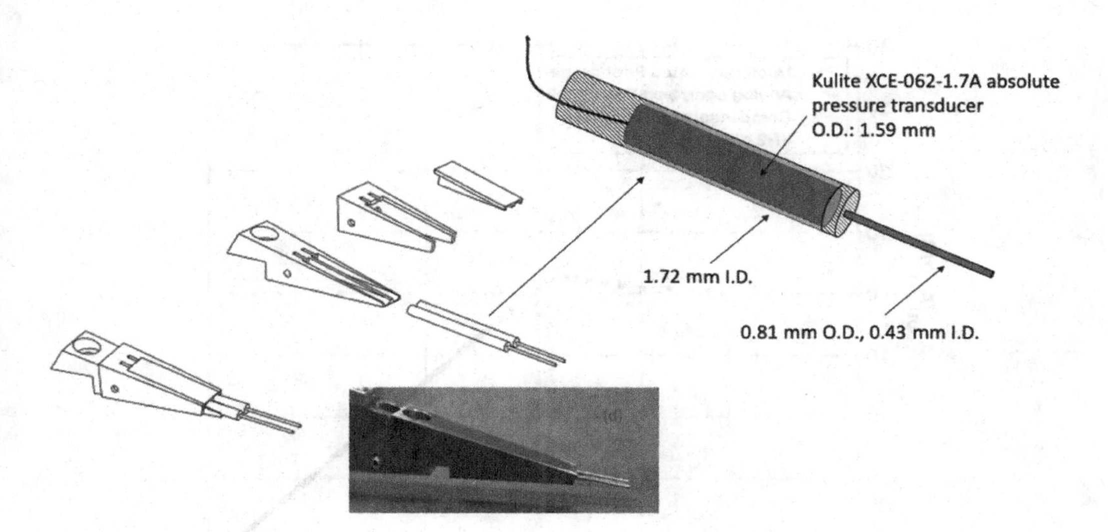

Figure 2.2 Illustration of "fast-response" Pitot probe design used by Corke et al. (2018) and Arndt et al. (2020).

in order to improve their dynamic response. An illustration of their "fast-response" Pitot probe design is shown in Figure 2.2. It utilized the Kulite model XCE-062-1.7A (1.7 bar absolute) pressure transducer. The outside diameter (OD) of the Kulite transducer was 1.59 mm. The transducer was inserted into a circular sleeve whose inside diameter (ID) closely matched the OD of the transducer body. Any gap was sealed. The design used a series of three telescoping hypodermic tubes that stepped down to a final ID of 0.43 mm, which was then the sensing diameter. As illustrated in Figure 2.2, a pair of the "fast-response" Pitot probes were mounted in a wedge assembly that was part of a traversing mechanism and used to measure pressure fluctuations associated with traveling cross-flow instability modes in the boundary layer over a 7° half-angle cone at an angle of attack at Mach 6.

An important consideration of the Pitot probe design in Figure 2.2 was its frequency response. The measured amplitude response of the uncompensated Pitot probe is shown by the dotted curve in Figure 2.3. This indicates a maximum frequency response, within ± 2 dB, of approximately 3.5 kHz. This is considerably less than that of the pressure transducer alone and is attributable to the small diameter tube that defines the measurement area. A custom (Corke et al., 2018) analog frequency compensation circuit was used (Corke et al., 2018) to increase the frequency response of the Pitot probe system. The resulting amplitude response is shown by the solid curve in Figure 2.3. This extended the uncompensated response to approximately 40 kHz, which was above the frequencies of the traveling cross-flow instability modes to be measured. The design of an analog frequency compensation circuit is presented in Section 2.1.2.

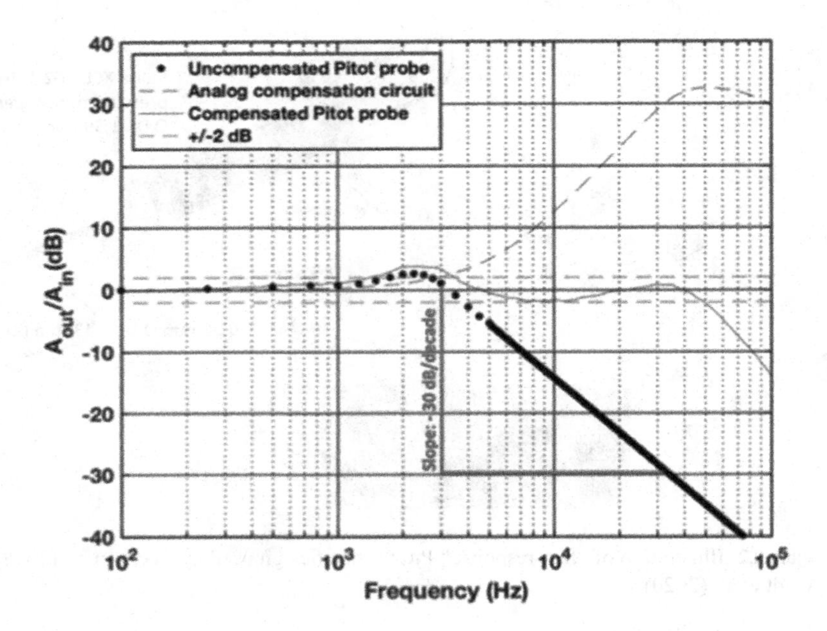

Figure 2.3 Frequency response of the uncompensated and frequency compensated Pitot probe illustrated in Figure 2.2. Taken from Corke et al. (2018).

2.1.2 Hot-Wires

Hot-wires are fast response sensors used to measure mean and fluctuating velocity and temperature components in fluid flows, mostly in gases. The sensors consist of thin metallic elements (with a generally cylindrical cross section) that are heated by an electric current (Joule effect) and cooled by the incident flow, which acts by virtue of its mass flux and its temperature through various effects but predominantly through forced convection. The sensors do not directly measure fluid velocity, but rather indirectly deduce velocity by the change in the sensor resistance, which is a function of the sensor temperature. The relation between the fluid velocity that controls heat transfer from the sensor, or to the sensor when used to measure temperature, generally involves a calibration relation.

Hot-wire sensors are generally metallic wires with diameters ranging from 0.5 to 5 μm. The wire sensor length generally ranges from 0.1 to 1 mm. The choice of the sensor length affects the measurement area or volume in the case of multiple sensor arrangements. An important parameter is the sensor length-to-diameter ratio which needs to be sufficiently large to minimize the heat conduction from the sensor to the wire supports. This would otherwise negate the assumption that changes in the sensor temperature result solely from forced convective heat transfer. Figure 2.4 illustrates a single-sensor hot-wire supported by a pair of tapered metal supports. The ends of the sensor wire are plated with a larger diameter metal (e.g., copper) to define a sensor length, which is the unplated wire length in the middle.

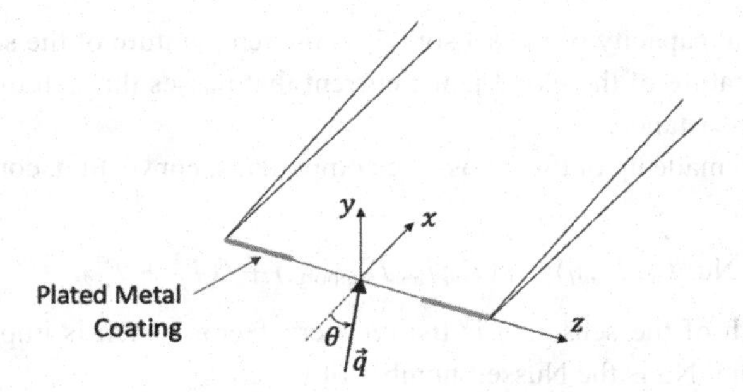

Figure 2.4 Illustration of a single-sensor hot-wire supported by a pair of tapered metal supports.

The materials used for hot-wires are ones with a high resistivity (resistance change with temperature) and strength, and minimum oxidation at elevated temperatures. The metals of choice are platinum, tungsten, platinum–rhodium, and iridium. Of these, platinum and tungsten are the most commonly used. Platinum can be soldered to the metal supports. Tungsten cannot and is either welded to the supports or plated with copper on the ends, which can be soldered.

The sensing portion of the hot-wire is heated by an electric current that passes through the wire. In order to maximize the sensitivity to air velocity (or mass-flux), the wire needs to be heated above the ambient temperature by as much as possible. For the types of materials and wire diameter used, the maximum sensor temperature is about 300 °C. Higher temperatures will lead to rapid oxidation and change (increase) in the wire resistance, which further increases heating in a destructive cycle.

For temperature measurements, a minimum electric current is used that will not cause the wire to resistively heat. In this case, the sensor is called a "cold-wire." The wire resistance is then a function of the mean and fluctuating ambient gas temperature. The low current reduces the static sensitivity. To counter that, smaller wire diameters are used, which reduces the thermal inertia.

Analysis and optimization of the performance of a hot-wire starts with the temporal heat balance based on the first law of thermodynamics, namely

$$\frac{dE}{dt} = W - H, \tag{2.5}$$

where E is the heat stored, where

$$E = C_w T_w, \tag{2.6}$$

W is the heat generated, where

$$W = I^2 R_w, \tag{2.7}$$

and H is the heat lost, where

$$H = hA(T_w - T_a). \tag{2.8}$$

In these, C_w is the heat capacity of the sensor, T_w is the temperature of the sensor, T_a is the ambient temperature of the gas, I is the current that passes through the sensor, and R_w is the sensor resistance.

The heat loss, H, is made up of three possible components, convection, conduction, and radiation, such that

$$H \equiv f(\mathrm{Nu}, T_w, T_a, \eta) + f(T_w, l_w, T_{\mathrm{supports}}) + f(T_w^4 + T_a^4), \tag{2.9}$$

where l_w is the length of the sensor, η is the recovery factor which is important in compressible flows, and Nu is the Nusselt number of which

$$\mathrm{Nu} = \frac{hd}{k} = f(\mathrm{Re}, \mathrm{Pr}, M, \mathrm{Gr}, \theta) \tag{2.10}$$

and Re is the Reynolds number based on the diameter of the sensor, Pr is the Prandtl number which is a function of the gas, M is the Mach number of the gas flow, Gr is the Grashof number which is based on the wire diameter, and θ is the off-perpendicular mean flow angle.

Assuming that the heat transfer to the supports is small, and that radiation can be neglected (a good assumption except when making hot-wire measurements near a surface), a formulation for H reduces to obtaining a general relation for the Nusselt number, where $\mathrm{Nu} = f(\mathrm{Re}, \mathrm{Pr}, M, \mathrm{Gr}, \theta)$. For the time being, the flow is considered to be incompressible ($M \ll 1$), with a mean vector perpendicular to the wire sensor ($\theta \simeq 0$).

In seeking a relation between the Nusselt number and the Reynolds and Prandtl numbers for a heated cylinder, under these assumptions, the most widely used form was proposed by King (1914) and is referred to as "King's law," where

$$\mathrm{Nu} = C_1 + C_2\sqrt{\mathrm{Re}_d}. \tag{2.11}$$

Kramer (1946) determined for $1 \leq \mathrm{Re}_d \leq 1000$ with "normal" fluids that

$$C_1 = 0.32\mathrm{Pr}^{0.2} \text{ and } C_2 = 0.57\mathrm{Pr}^{0.33}. \tag{2.12}$$

The sensor resistance as a function of temperature is

$$R_w = R_w(T_w) = R_0\left[1 + \alpha(T_w - T_0) + \alpha_2(T_w - T_a)^2 + \cdots\right], \tag{2.13}$$

where $R_0 = R_w$ at T_0 and α is the sensor resistivity (ohms/$A_{\text{cross-section}}$).

Neglecting the higher-order terms in $R_w(T_w)$ and incorporating King's law,

$$H = k\frac{\pi d}{4}\left[C_1 + C_2\sqrt{\frac{\rho q d}{\mu}}\right](T_w - T_a) \tag{2.14}$$

and substituting

$$T_w - T_a = \frac{R_w - R_a}{\alpha R_0}, \tag{2.15}$$

one obtains

$$H = k\frac{\pi d}{4}\left[C_1 + C_2\sqrt{\frac{\rho q d}{\mu}}\right]\left(\frac{R_w - R_a}{\alpha R_0}\right). \tag{2.16}$$

Considering the sensor *static response* in which $dE/dt = 0$,

$$I^2 R_w = k\frac{\pi d}{4}\left[C_1 + C_2\sqrt{\frac{\rho q d}{\mu}}\right]\left(\frac{R_w - R_a}{\alpha R_0}\right) \tag{2.17}$$

or more generally,

$$\frac{I^2 R_w}{R_w - R_a} = A + B\sqrt{q}. \tag{2.18}$$

Other forms are

$$\frac{I^2}{I_0^2} = \frac{\frac{R_w}{R_a} - 1}{\frac{R_w}{R_a}}\sqrt{\frac{q}{q_0}} + 1, \tag{2.19}$$

where

$$\frac{R_w}{R_a} \equiv \text{overheat ratio}, \tag{2.20}$$

and

$$\frac{I^2}{I_0^2} = \frac{a_w}{1 + a_w}\sqrt{\frac{q}{q_0}} + 1, \tag{2.21}$$

where

$$a_w = \frac{R_w - R_a}{R_a} \equiv \text{overheat parameter}. \tag{2.22}$$

There are three common approaches to hot-wire sensor operating:

1. Constant I referred to as a constant current anemometer (CCA)
2. Constant R_w referred to as a constant temperature anemometer (CTA)
3. Constant V_w referred to as a constant voltage anemometer (CVA)

These are reviewed in terms of their static sensitivity and frequency response in the following sections.

Constant Current Anemometer

A simple CCA circuit is shown in Figure 2.5. The circuit is powered by a DC voltage source, E_s. The circuit uses a ballast resistance, R_s, whose resistance is large compared to the hot-wire resistance, R_w, namely $R_s \gg R_w$. The output from the circuit is passed through a unity-gain-follower operational amplifier circuit which provides an infinite

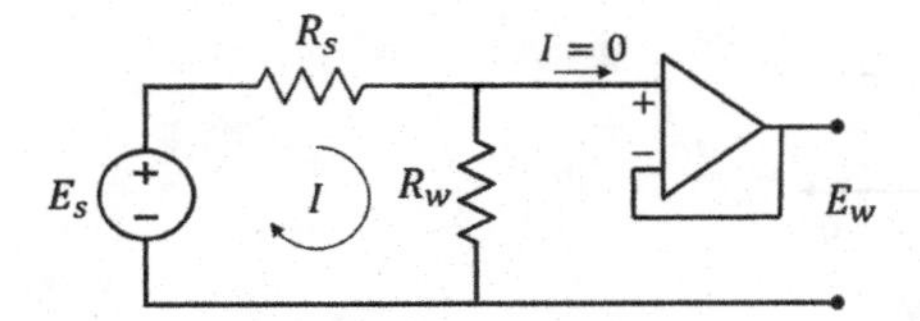

Figure 2.5 Simple constant current anemometer (CCA) circuit.

input impedance buffer with zero input current. As a result, the current that passes through R_w is

$$I = E_s/R_s, \tag{2.23}$$

which for $R_s \gg R_w$ is constant. Other approaches can substitute the voltage and ballast resistance with a constant current source.

Based on Eq. (2.18) for a constant current, I,

$$\frac{I^2 R_w}{R_w - R_a}\frac{I}{I} = A + B\sqrt{q} \tag{2.24}$$

or

$$\frac{E_w}{E_w - E_a} = \frac{A + B\sqrt{q}}{I^2}, \tag{2.25}$$

where solving for the voltage across the sensor, E_w,

$$E_w = \frac{E_a}{1 - \left(\frac{I^2}{A+B\sqrt{q}}\right)} \tag{2.26}$$

or where solving for the velocity, q,

$$q = \frac{1}{B^2}\left(\frac{I^2 E_w}{E_w - E_a} - A\right)^2. \tag{2.27}$$

The relation for the CCA voltage output, E_w, as a function of the velocity, q, is illustrated in Figure 2.6.

The static sensitivity for the CCA is

$$S_{CC} \equiv \left(\frac{\partial E_w}{\partial q}\right)_I \tag{2.28}$$

or

$$S_{CC} = -\frac{B I^2 E_a}{2\sqrt{q}\left(A + B\sqrt{q} - I^2\right)^2}, \tag{2.29}$$

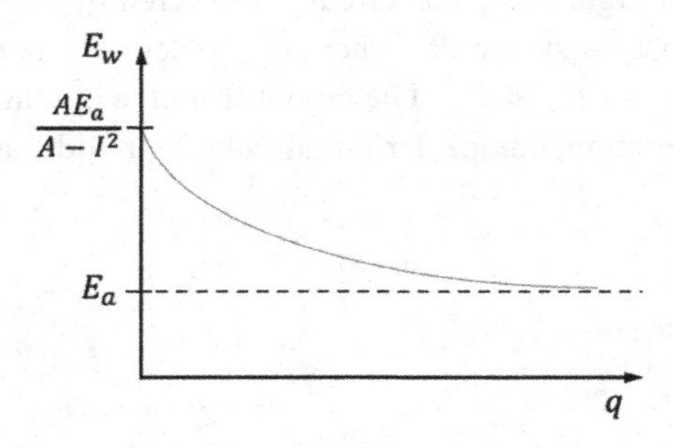

Figure 2.6 Example of constant current anemometer (CCA) voltage output, E_w, as a function of the velocity, q.

or incorporating the overheat parameter,

$$S_{CC} = -\frac{B\sqrt{a_w + 1}a_w^{3/2} R_a}{2\sqrt{q}\sqrt{A + B\sqrt{q}}}. \tag{2.30}$$

Here, we note that

$$\mathrm{L}_{q\to 0} S_{CC} = -\infty \text{ and } \mathrm{L}_{q\to\infty} S_{CC} = 0, \tag{2.31}$$

which is illustrated in Figure 2.6. In addition, we note that, as expected, the static sensitivity increases as the overheat parameter increases, namely as the sensor temperature increases above the ambient temperature.

The analysis of the frequency response starts with the energy balance given in Eq. (2.5), where in contrast to the steady analysis, now $dE/dt \neq 0$. The assumptions in the analysis are that fluctuations are small and occur about a mean operating point. This allows dropping higher-order terms in Taylor expansions of different quantities. A general form for the time response of the wire-based velocity sensors that results is

$$(\tau D + 1)\Delta T_w = F\Delta(\rho q) + G\Delta I + J\Delta T_a, \tag{2.32}$$

where

$$\tau = c_w/L \tag{2.33}$$

$$G = \frac{\partial W}{\partial I}/L \tag{2.34}$$

$$F = -\frac{\partial H}{\partial \rho q}/L \tag{2.35}$$

$$J = -\frac{\partial H}{\partial T_a}/L \tag{2.36}$$

$$L = \frac{\partial H}{\partial T_w} - \frac{\partial W}{\partial T_w}. \tag{2.37}$$

(2.38)

For the CCA, the heat energy is

$$E_w = IR_w \tag{2.39}$$

so that the change with changing resistance is

$$\Delta E_w = I\Delta R_w. \tag{2.40}$$

The sensor resistance is a function of temperature and resistivity, α, namely

$$R_w = R_0\left[1 + \alpha(T_w - T_0)\right] \tag{2.41}$$

or

$$\frac{R_w - R_0}{\alpha R_0} = T_w - T_0 \equiv \Delta T_w = \frac{\Delta R_w}{\alpha R_0} \tag{2.42}$$

so that

$$\Delta T_w = \frac{\Delta E_w}{I\alpha R_0}. \tag{2.43}$$

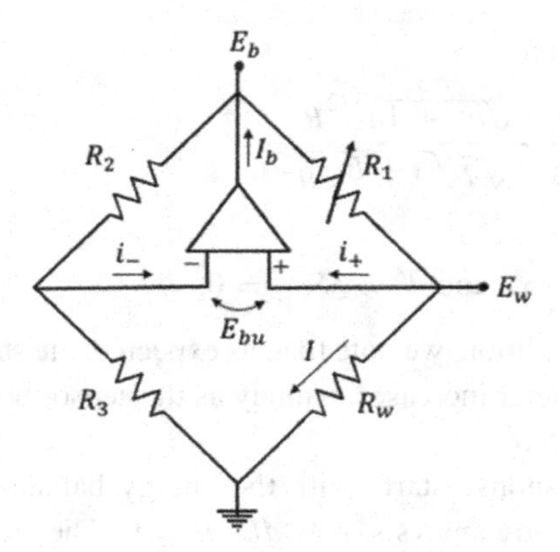

Figure 2.7 Simple constant temperature anemometer (CTA) circuit.

Substituting for the terms in Eq. (2.32) and putting the resulting equation into a general form for a first-order response, the CCA time constant is

$$\tau = \frac{c_w a_w}{I^2 \alpha R_0}, \tag{2.44}$$

where c_w is a material property of the wire sensor (for Tungsten, $c_w = 0.033$ Cal/gm°C), α again is the wire sensor resistivity (for Tungsten, $\alpha = 0.00481/°\mathrm{C}$), and a_w is the overheat parameter as before. As a first-order system, the response amplitude is constant to a frequency, $f = 1/\tau$, and then decreases at a rate of −20 dB/decade. Based on Eq. (2.44), the frequency response of a CCA decreases with increasing overheat. In contrast, the static sensitivity of a CCA increases with increasing overheat. Putting in some typical values for a Tungsten hot-wire sensor, $f \simeq 500$ Hz. Such a low-frequency response generally poses a problem, although analog frequency compensation can be used to extend the response into hundreds of kilohertz.

Constant Temperature Anemometer

A simple CTA circuit is shown in Figure 2.7. This consists of a full Wheatstone bridge that includes an operational amplifier. The standard characteristics of the operational amplifier are that the input currents, i_- and i_+, are both zero. In addition, the output of the operational amplifier, E_{bt}, is proportional to the difference in the voltage across its input, E_{bu}, which corresponds to the voltage imbalance between the left and right legs of the bridge. The current through the sensor is I, and the voltage across the sensor is E_w.

In order to balance the bridge,

$$\frac{E_w}{R_w} = \frac{E_{bt}}{R_w + R_1}, \tag{2.45}$$

but based on the bridge circuit,

$$E_w = IR_w = R_w \frac{E_{bt}}{R_w + R_1} = \frac{R_w}{R_w + R_1} E_{bt}. \tag{2.46}$$

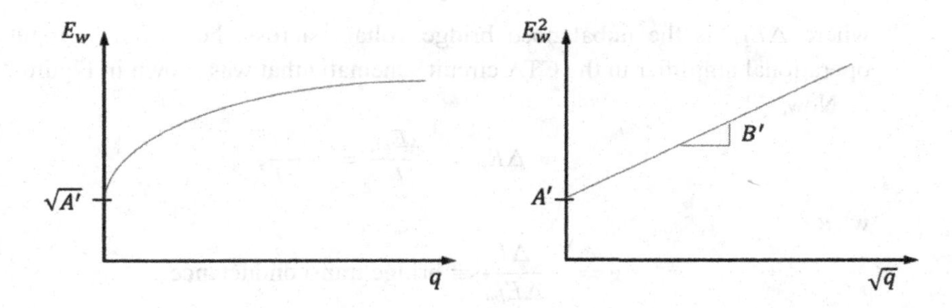

Figure 2.8 Examples of constant temperature anemometer (CTA) output voltage, E_w, as a function of velocity, q.

For the CTA, the current through the sensor is controlled by the adjustable resistance, R_1. The overheat ratio in this case is

$$\frac{R_w}{R_w + R_1}. \tag{2.47}$$

Now, given that

$$E_w^2 = I^2 R_w R_w \tag{2.48}$$

and comparing that to Eq. (2.18), one obtains

$$E_w^2 = R_w(R_w - R_a)\left(A + B\sqrt{q}\right). \tag{2.49}$$

Again with constant temperature operation, R_w is constant. If, in addition, the ambient temperature is constant, then

$$R_w(R_w - R_a) = \text{constant}, \tag{2.50}$$

and these constants can be absorbed into A and B in Eq. (2.49) to obtain the general calibration equation for a CTA, namely

$$E_w^2 = A' + B'\sqrt{q}, \tag{2.51}$$

where the prime notation signifies that A' and B' have values that are different from the original A and B coefficients. The relation for the CTA voltage output, E_w, as a function of the velocity, q, is illustrated in Figure 2.8.

The static sensitivity for the CTA is

$$S_{\text{CTA}} \equiv \left(\frac{\partial E_w}{\partial q}\right)_{R_w} = \frac{B'}{4}\left(A' + B'\sqrt{q}\right)^{-1/2} q^{-1/2}. \tag{2.52}$$

In general,

$$|S_{\text{CTA}}| < |S_{\text{CC}}|. \tag{2.53}$$

Considering the frequency response, to first order

$$\Delta E_{bu} = I \Delta R_w, \tag{2.54}$$

where ΔE_{bu} is the unbalanced bridge voltage across the + and − inputs to the operational amplifier in the CTA circuit schematic that was shown in Figure 2.7.

Now,

$$\Delta R_w = \frac{\Delta E_{bu}}{I} = -\frac{\Delta I}{gI}, \tag{2.55}$$

where

$$g = -\frac{\Delta I}{\Delta E_{bu}} \equiv \text{bridge transconductance.} \tag{2.56}$$

From before with the CCA,

$$\Delta R_w = \alpha R_w \Delta T_w \tag{2.57}$$

so that

$$\Delta T_w = -\frac{\Delta I}{\alpha R_0 g I}. \tag{2.58}$$

In the analysis of the CCA, because I was constant in Eq. (2.32), the term G was zero. In the case of the CTA,

$$G = \frac{\partial W / \partial I}{L} \tag{2.59}$$

where noting that R_0 is a constant with constant temperature operation, then

$$\frac{\partial W}{\partial I} = 2IR_0 \tag{2.60}$$

and, therefore,

$$G = \frac{2IR_0}{L}. \tag{2.61}$$

As with the CCA,

$$L = \left(A + B\sqrt{\rho q} - I^2 \alpha R_0\right). \tag{2.62}$$

Based on the steady heat energy equation,

$$A + B\sqrt{\rho q} = \frac{I^2 R_w}{T_w - T_a}, \tag{2.63}$$

which leads to the following form for G that is absent from the CCA, namely

$$G = \frac{2R_w a_w}{I \alpha R_0}. \tag{2.64}$$

Substituting for the terms in Eq. (2.32) and putting the resulting equation in a general form for a first-order response, the CTA time constant is

$$\tau_{\text{CTA}} = \frac{\tau_{\text{CCA}}}{1 + 2R_w a_w g}, \tag{2.65}$$

where the time constant in the numerator is that of the CCA. Now the term in the denominator,

$$2R_w a_w g > 1 \tag{2.66}$$

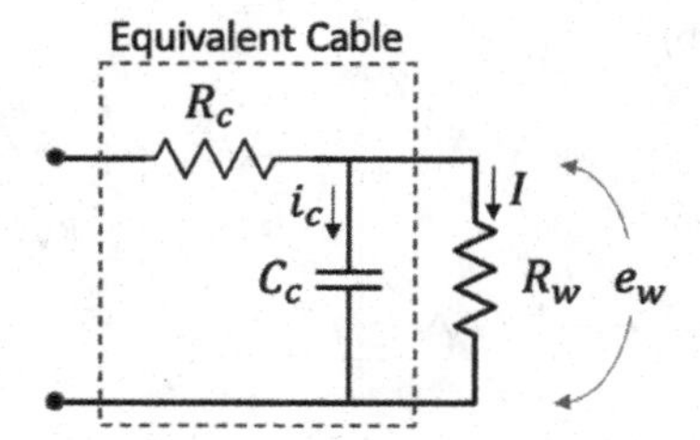

Figure 2.9 Circuit equivalent of a cable attached to a sensor.

so that

$$\tau_{\text{CTA}} < \tau_{\text{CCA}} \tag{2.67}$$

or in terms of frequency,

$$f_{\text{CTA}} > f_{\text{CCA}}. \tag{2.68}$$

The higher frequency response of the CTA compared to the CCA is due to the bridge transconductance that results from the operational amplifier in the bridge circuit. The typical frequency response of a CTA is on the order of 20 kHz.

Constant Voltage Anemometer

Why a CVA? The answer is that with long cables connecting the sensor to the anemometer electronics, the cable adds resistance and capacitance to the circuit. The circuit equivalence of a cable attached to a sensor with resistance, R_w, is shown in Figure 2.9. In the equivalent circuit, the cable capacitance is connected in parallel to the sensor resistance. The effect of the capacitance is to divide the current that otherwise goes to the sensor. The current through the capacitor is

$$i_c = C_c \frac{de_w}{dt}. \tag{2.69}$$

Therefore, if the voltage delivered to the sensor, de_w/dt, is constant, the current through the capacitor is zero, and, therefore, the capacitance effect of the cable is zero.

A simple CVA circuit is shown in Figure 2.10(a). Applying the properties for an operational amplifier, namely that the input current is zero, and the + and − input voltages are equal, one obtains the equivalent circuit that is shown in Figure 2.10(b). In the equivalent circuit, the sensor is located in a voltage divider network of resistances. The current through the sensor is $I = E_w/R_w$, where E_w is the voltage at the center point of the voltage divider. The currents, i_1 and i_2 are equal so that $E_w = E_{\text{ref}}(R_2/R_1)$. The voltage E_{ref} is a constant voltage supplied to the circuit that will ultimately control the sensor overheat parameter. As a result, the current through the sensor is

$$I = \frac{E_{\text{ref}}}{R_1} \frac{R_2}{R_w}, \tag{2.70}$$

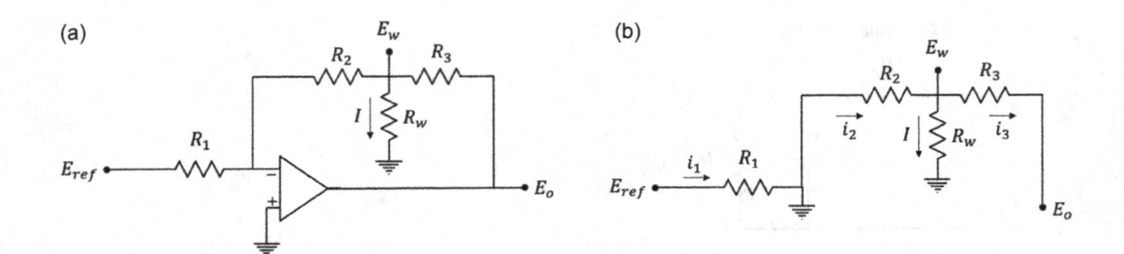

Figure 2.10 Simple constant voltage anemometer (CVA) circuit (a) and equivalent CVA circuit (b).

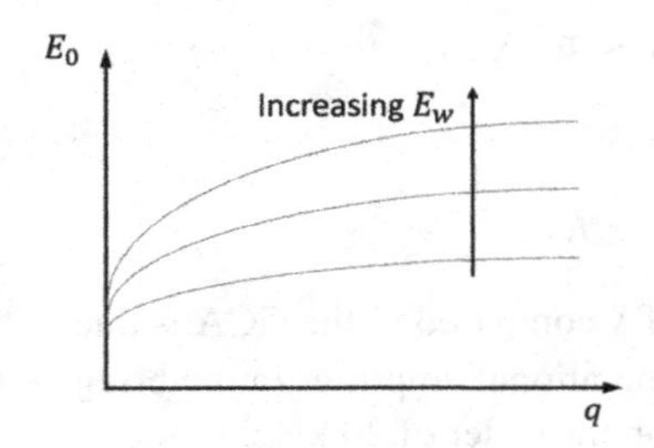

Figure 2.11 Illustration of a CVA output, E_o, as a function of velocity for different wire voltage values, E_w, that is equivalent to the overheat parameter, a_W.

and, therefore,

$$E_w = E_{\text{ref}}\frac{R_2}{R_1}. \tag{2.71}$$

It is noted that the sensor voltage, E_w, is not a function of the sensor resistance, R_w, as in the cases of the CCA and CTA. This is the outcome of the constant voltage operation.

Further analyzing the equivalent circuit, the circuit output voltage, E_o, is

$$E_o = \left(1 + \frac{R_3}{R_2} + \frac{R_3}{R_w}\right) E_w. \tag{2.72}$$

Note again that E_w is constant. Therefore, any change in the sensor resistance due to velocity-driven cooling in hot-wire operation, or due to temperature in cold-wire operation, will result in a change in the CVA output voltage, E_o. Figure 2.11 illustrates the CVA output, E_o, as a function of velocity for different wire voltage values, E_w, that is equivalent to the overheat parameter, a_w.

As before, the overheat parameter is

$$a_w \equiv \frac{R_w - R_a}{R_a} \tag{2.73}$$

which, in the case of the CVA, is

$$a_w = \frac{\frac{E_w}{I} - R_a}{R_a}. \tag{2.74}$$

Now,

$$\frac{E_w}{I} = \frac{E_w}{\frac{E_{ref}}{R_1} - \frac{E_w - E_o}{R_3}}. \tag{2.75}$$

Unfortunately, E_w cannot be eliminated from the expression for a_w. Therefore, it needs to be measured along with E_o for different values of E_{ref} which is used to set the sensor overheat. As an example, for the CVA circuit shown in Figure 2.10 with $R_1 = 5000\ \Omega$, $R_2 = R_3 = 500\ \Omega$, and $E_{ref} = 3.75$ V, $E_w = 0.375$ V, and $E_o = 17.8$ V. Therefore,

$$I = i_2 - i_3 = \frac{E_{ref}}{R_1} - \frac{E_w - E_o}{R_3} = 35.6 \text{ mA}. \tag{2.76}$$

A typical value for R_a for a 2-mm long, 3.81-μm diameter tungsten wire sensor is approximately 7 Ω. Therefore, substituting the values gives $a_w \simeq 0.51$, and an overheat ratio, $r = a_w + 1 = 1.51$. For hot-wire sensors, $1.3 \leq r \leq 1.8$. For cold-wire sensors, $r \leq 1.1$.

Comte-Bellot (1998) has derived the time response, τ, of the CVA relative to that of the CCA, and the result is

$$\tau_{CVA} = \frac{\tau_{CCA}}{1 + 2a_w}. \tag{2.77}$$

Since the overheat parameter, a_w, is less than 1 and on the order of 0.5,

$$\tau_{CVA} \simeq 0.5\tau_{CCA} \tag{2.78}$$

or

$$f_{CVA} \simeq 2f_{CCA}. \tag{2.79}$$

As presented, the frequency response of a CCA is relatively low, on the order of 500–600 Hz. The factor-of-two higher frequency response of the CVA still makes it low compared to the CTA. Therefore, in practical applications, as with the CCA, analog frequency compensation circuits are used.

Effect of Ambient Temperature

Since hot-wire sensors directly measure heat loss and not velocity, their accuracy is subject to changes in the ambient temperature that can be falsely interpreted as changes in velocity. To understand the dependence on temperature, we consider a basic hot-wire anemometer output voltage change with velocity,

$$E_{a_1}^2 = A_1 + B_1 q^{1/2}, \tag{2.80}$$

where the coefficients A and B depend on the calibration temperature, T_{a_1}. For a second temperature, T_{a_2}, the anemometer output voltage relation would be

$$E_{a_2}^2 = A_2 + B_2 q^{1/2}. \tag{2.81}$$

The effect of a change in the ambient temperature is a change in the temperature overheat ratio, defined here as

$$\theta = \frac{T_w}{T_{a_1}}. \tag{2.82}$$

Since the wire resistance is only a function of temperature, the change in temperature is also equivalent to a change in the resistance overheat ratio

$$r = \frac{R_w}{R_{a_1}}. \tag{2.83}$$

Since the calibration coefficients A and B are only functions of the difference between the wire and ambient temperatures ($T_w - T_a$), for a fixed velocity,

$$\frac{A_1}{A_2} = \frac{B_1}{B_2} = \frac{T_w - T_{a_1}}{T_w - T_{a_2}} = \frac{\theta - 1}{\theta - 1 + \epsilon}, \tag{2.84}$$

where

$$\epsilon = \frac{T_{a_1} - T_{a_2}}{T_{a_1}}. \tag{2.85}$$

Therefore, assuming that a velocity calibration was performed at T_{a_1}, the "correct" anemometer voltage at temperature T_{a_2} is

$$E_{a_1} = \left[\frac{\theta - 1}{\theta - 1 + \epsilon}\right]^{-1/2} E_{a_2}. \tag{2.86}$$

It is convenient to put Eq. (2.86) in terms of the sensor overheat ratio, r. This starts with the wire resistance dependence on temperature

$$R_w = R_{a_1}\left[1 + \alpha(T_w - T_{a_1})\right] \tag{2.87}$$

so that

$$\frac{R_w}{R_{a_1}} - 1 = \alpha(\frac{T_w}{T_{a_1}} - 1)T_{a_1}. \tag{2.88}$$

The equivalence between the wire temperature and resistance gives

$$\frac{T_w}{T_{a_1}} - 1 = \frac{a_w}{\alpha T_{a_1}} = \frac{r - 1}{\alpha T_{a_1}}. \tag{2.89}$$

Substituting into Eq. (2.86) gives

$$E_{a_1} = \left[1 - \frac{\alpha}{r - 1}\left(T_{a_1} - T_{a_2}\right)\right]^{-1/2} E_{a_2}. \tag{2.90}$$

Effect of Sensor Angularity

The velocity calibration of large length-to-diameter wire sensors is sensitive to the vector angle of the mean flow as well as fluctuation levels. A general velocity vector representation to a single wire sensor is illustrated in Figure 3.37. The effective heat transfer is the result of a velocity vector, $\vec{V}$

$$\vec{V} = \left[V_N^2 + k_T^2 VT^2 + k_N^2 V_{BN}^2\right]^{1/2}, \tag{2.91}$$

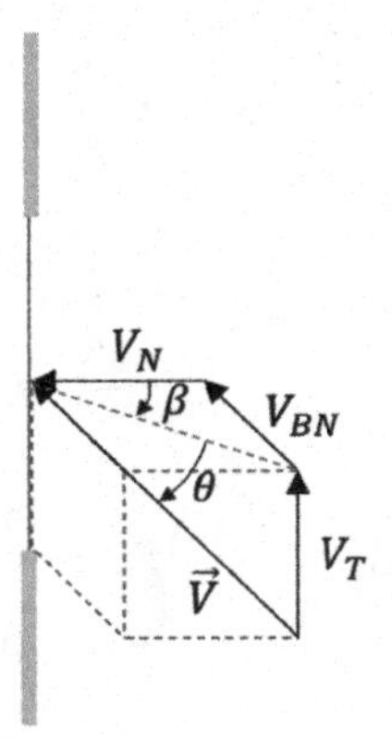

Figure 2.12 Illustration of a general velocity vector representation to a single wire sensor.

where V_N is the component normal to the wire sensor, V_T is the component along the long axis of the wire, and V_{BN} is the component normal to the wire into the page. The coefficients k_T and k_N are related to heat conduction along the axis of the wire sensor. They have values of $1.0 \leq k_N \leq 1.2$ and $0.0 \leq k_T \leq 0{:}2$ (Sandborn, 1972). The lower limits for k_N and k_T are when the ends of the wire are plated to a larger diameter as illustrated in Figure 3.37.

In order to estimate the effect of velocity fluctuations on hot-wire sensor effective heat transfer, the velocity components are decomposed into a mean and fluctuating part, namely,

$$V_i = \bar{V}_i + v_i \ \ i = 1, 2, 3, \tag{2.92}$$

where, for simplicity, $V_1 = V_N$, $V_2 = V_T$, and $V_3 = V_{BN}$. Therefore,

$$\vec{V}_{\text{effective}} = \left[(\bar{V}_1 + v_1)^2 + k_T^2(\bar{V}_2 + v_2)^2 + k_N^2(\bar{V}_3 + v_3)^2\right]^{1/2}. \tag{2.93}$$

If the sensor is nearly aligned with the mean flow vector, then $\theta \simeq 0$ in Figure 3.37. In that case, $\bar{V}_2 = \bar{V}_3 \simeq 0$ and $\bar{V}_1 = \bar{V}$. Then

$$\vec{V}_{\text{effective}} = \left[(\bar{V}_1 + v_1)^2 + k_T^2 v_2^2 + k_N^2 v_3^2\right]^{1/2}. \tag{2.94}$$

As indicated, k_T is small, and assuming small velocity fluctuations, the product $(k_T^2 v_2^2) \ll 1$ can therefore be neglected. In contrast, $k_N \simeq 1$, and so the product with v_3 fluctuations cannot be neglected. Therefore,

$$\vec{V}_{\text{effective}} = \left[(\bar{V}_1 + v_1)^2 + (v_3)^2\right]^{1/2}. \tag{2.95}$$

Expanding the first term on the right-hand side, and expressing $\vec{V}_{\text{effective}}$ as $(\bar{V} + v)$, then

$$\bar{V} + v = \left[(\bar{V}_1^2 + 2\bar{V}_1 v_1 + v_1^2 + v_3^2\right]^{1/2}. \tag{2.96}$$

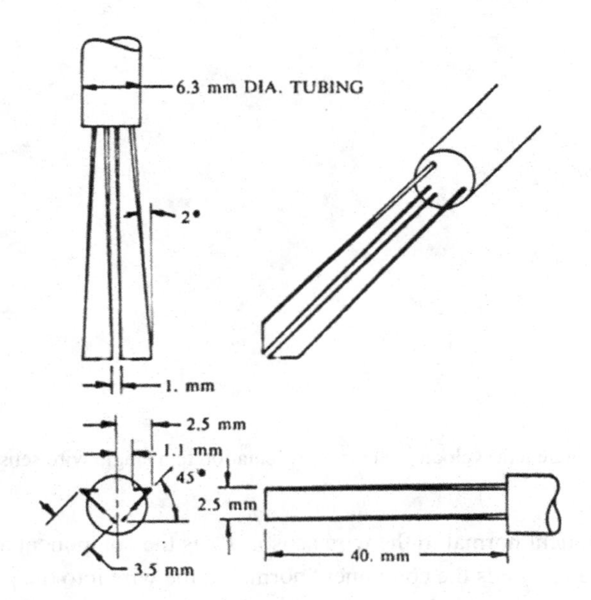

Figure 2.13 Illustration of a dual hot-wire sensor designed to measure the streamwise, U, and cross-stream, W, velocity components near the surface in a boundary layer.

Based on this,

$$\bar{V} = \bar{V}_1 + \text{Order}(2\bar{V}v_1) \tag{2.97}$$

and

$$v = v_1 + \text{Order}(v_3). \tag{2.98}$$

It is clear from Eqs. (2.97) and (2.98) that the accuracy of measurements of the mean and fluctuating velocity components depends on the level of the velocity fluctuations. These levels can be relatively large in flows undergoing turbulence onset, particularly in boundary layers, wakes, and shear layers. Therefore, this needs to be kept in mind with regard to these flow fields.

Multiple angled hot-wire sensors are often used to resolve multiple velocity components. An example of a dual hot-wire sensor probe is shown in Figure 2.13. The probe was designed to measure the streamwise, U, and cross-stream, W, velocity components near the surface in a boundary layer. The sensors are located on the same plane so that they will sense the same streamwise velocity, which, in the boundary layer, varies with distance from the surface. In addition, the sensor portions of the two hot-wires, which correspond to the smaller diameter region of the wires, are located as close as possible to the centerline of the probe in order to minimize the measurement area. The importance of this is evident in the manner in which the outputs from the two sensors are processed.

In the dual hot-wire sensor configuration shown in Figure 2.13, each sensor will measure an effective velocity component, q_e, based on its angle with respect to the

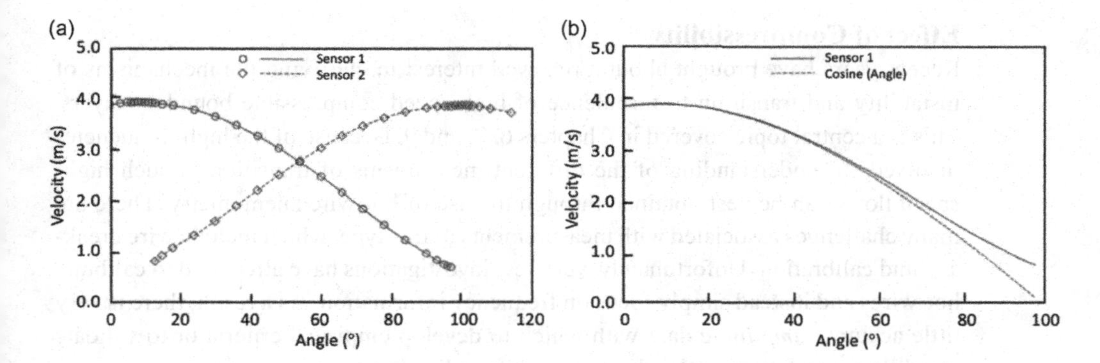

Figure 2.14 An example of an angular calibration performed on the dual hot-wire sensor illustrated in Figure 2.13.

mean velocity vector angle, θ. Assuming that β in the schematic in Figure 3.37 is small enough to be neglected, the effective velocity for a yawed wire sensor is

$$q_e = |\vec{q}| \cos\theta. \tag{2.99}$$

The simple cosine yawed response assumes that the hot-wire is not cooled at $\theta = \pi/2$. An example of an angular calibration performed on the dual hot-wire sensor illustrated in Figure 2.13 is shown in Figure 2.14(a). This was performed by rotating the probe body so that the hot-wires were swept through a range of angles in a steady, uniform flow field. This is a necessary step that determines the orientation angle of each of the sensors. The response of the hot-wire velocity reading versus the yaw angle indicates that the two sensors are at equal but opposite angles to each other and that with respect to the angle corresponding to the maximum velocity, the wire angles are at $\theta = \pm 53.77°$.

Figure 2.14(b) compares the angular dependence of one of the sensors to the simple cosine yaw response. It is evident that the cosine response does represent the yawed hot-wire response up to $\theta = 50°$ and within small error up to $\theta \simeq 60°$.

There have been a number of other "modified cosine laws" suggested (Friehe and Schwartz, 1968). One is

$$q_e = |\vec{q}|(\cos^2\theta + k^2 \sin^2\theta)^{1/2}, \tag{2.100}$$

where the second term is intended to account for cooling caused by the flow that runs parallel to the sensor. In this case, $0 \le k \le 0.2$, which applies to hot-wire length-to-diameter ratios of $600 \le l/d \le 200$.

Another modified cosine angularity relation is

$$q_e = |\vec{q}| \left[1 - k(1 - \cos^{1/2}\theta)\right]^2, \tag{2.101}$$

where

$$k = 1 = 2600 \left(\frac{d}{l}\right)^2. \tag{2.102}$$

Effect of Compressibility

Recent years have brought about a renewed interest in clarifying the mechanisms of instability and transition to turbulence of high-speed compressible boundary layers. This is a central topic covered in Chapters 6, 7, and 9. Because of the high frequencies involved, an understanding of the different mechanisms of transition in such high-speed flows can be best obtained through the use of hot-wire anemometry. There are many challenges associated with measurements of this type, which include wire breakage and calibration. Unfortunately, very few investigations have attempted to calibrate hot-wires and instead simply focus on frequency information. As a result, there is very little accurate *amplitude* data with which to develop empirical criteria or to validate instability simulations and turbulence onset predictions.

Calibration of a hot-wire in supersonic flows requires that the *simultaneous* sensitivity of the sensor to variations in velocity, temperature, and density be resolved. In practice, only mass-flux and temperature variation information is obtained. One approach to reduce the temperature dependence is to maintain a high enough overheat ratio so that the sensor is primarily sensitive to mass-flux fluctuations. In that case, it is possible to deduce the spatial amplification rate of linear modes by comparing the magnitudes of the sensor voltage fluctuations at different spatial locations. The difficulty in this approach is that it requires the sensor to be placed at locations where the mean mass flux is the same so that the dynamic sensitivity remains constant.

The standard approach to calibration involves placing the hot-wire in the free-stream at a fixed Mach number while changing stagnation pressure (P_0) and temperature (T_0) (hence mass flux) at multiple overheats. However, this may not be feasible in all facilities with limited control over P_0 and T_0. Matlis (2004) utilized an approach that made use of the variation of P_0 and T_0 within a laminar boundary layer at a compressible Mach number. This involved performing total-pressure surveys within the boundary layer to obtain Mach number profiles. This is illustrated in Figure 2.15.

The Mach number profile is found from a recursive solution to the identity

$$P_{0_2} = \underbrace{\frac{P_{0_2}}{P_{0_1}}}_{\mathcal{F}(M)} \underbrace{\frac{P_{0_1}}{P_1}}_{\mathcal{F}(M)} P_1. \tag{2.103}$$

The hot-wire was subsequently traversed through the boundary layer at the same distances above the wall as the total pressure measurements. These locations were all at heights where the local Mach number was supersonic. At each of these heights, the

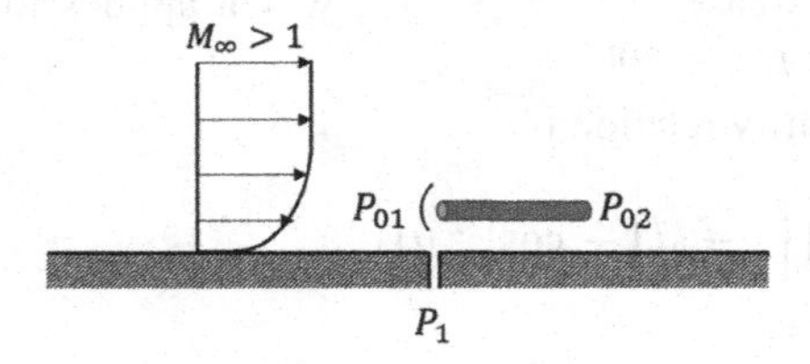

Figure 2.15 Illustration of the use of a total pressure probe used to determine the Mach number profile through a boundary layer where $M_\infty > 1$.

hot-wire anemometer output voltage is recorded for multiple overheat ratios. Matlis (2004) utilized 10 different overheat ratios.

The lowest overheat ratio was selected so that the sensor was primarily sensitive to temperature (i.e., a cold-wire). This cold-wire state was used to measure the recovery stagnation temperature T_w, where based on the temperature calibration of the sensor,

$$T_w = T_{\text{ref}} + \left(1 + \alpha(\frac{R_w}{R_{\text{ref}}})\right). \tag{2.104}$$

The stagnation temperature is then

$$T_0 = T_w/\eta, \tag{2.105}$$

where η is the recovery factor.

Knowing T_0, the static temperature, T, is found from the isentropic relation

$$T = \frac{T_0}{1 + \frac{\gamma - 1}{\gamma} M^2}. \tag{2.106}$$

The average mass at each measured height within the boundary layer is then found from

$$\overline{(\rho U)} = \frac{P_{0_2} M}{\sqrt{T}} \sqrt{\frac{\gamma}{R}}. \tag{2.107}$$

As indicated, a cold-wire sensor in a supersonic flow does not fully recover the stagnation condition T_0, but rather some temperature T_w that depends on the recovery factor, $\eta(M, \text{Re})$. Matlis (2004) investigated the variation of recovery factor with distance from the wall in a laminar boundary at Mach 3.5. It was found to vary from 1.020 near the wall to 0.965 at the edge of the boundary layer, which agreed with Lowell (1950).

Implicit in such an in-situ calibration approach is the need to *accurately measure* the mean quantities through the boundary layer. This raises the issue of *probe interference* in such measurements. In compressible boundary layer measurements, probe interference has been observed to lead to measured deviations from the well-established theoretical boundary layer Mach number profile. These deviations include an overshoot at the edge of the boundary layer and an undershoot in the lower part of the boundary layer. These two types of experimental deviations had been observed by Kendall (1957) and Morkovin and Bradfield (1954). At the outer edge of the boundary layer, Kendall (1957) found that the overshoot could be eliminated when the Pitot probe diameter was no larger than 12 percent of the boundary layer thickness. Morkovin and Bradfield (1954) associated the deviation in the lower part of the boundary layer with an upstream rise in the static pressure produced by the sensor and communicated upstream in the subsonic portion of the boundary layer. They determined an upstream influence distance, s, where the pressure disturbance diminishes by e^{-1} given as

$$s = \frac{1.3}{\sqrt{M^2 - 1}} \frac{T_{\text{wall}}}{T_1} \text{Re}^{1/6} s^{-1/3} - \frac{M^2 + 2}{2\sqrt{M^2 - 1}} \left(\frac{T_{\text{wall}}}{T_1}\right)^{0.7}. \tag{2.108}$$

For the experimental conditions of Matlis (2004), $s = 10\delta$. Following these recommendations (Morkovin and Bradfield, 1954; Kendall, 1957), the total pressure probe used by Matlis (2004) had a probe height of approximately 0.08δ ($< 0.12\delta$ guideline) and a probe length that extended approximately 10δ upstream of its wedge support. The resulting measured boundary layer Mach number profile was in excellent agreement with theory (Matlis, 2004).

Matlis (2004) used a constant-current anemometer like that shown in Figure 2.5 because the overheat ratio could be easily controlled by changing the voltage, E_s. The anemometer was fully computer controlled so that for any E_s, the wire voltage E_w was monitored, allowing determination of the cold and hot sensor resistance values that are needed to determine the overheat ratio.

The calibration approach begins in a similar manner to that for an incompressible Mach number, namely King's law given by Eq. (2.11), that relates the Nusselt number to the Reynolds number for steady heat transfer to/from the sensor wire, namely

$$\text{Nu} = hd/k = A + B\text{Re}^n. \tag{2.109}$$

Thermal equilibrium occurs when there is a steady balance between the power supplied to heat the sensor and the convective (power) heat loss, which can be expressed as

$$A_s h \left(T_w - T_e\right) = I^2 R_w, \tag{2.110}$$

where A_s is the sensor surface area, h is the convection heat transfer coefficient, T_w is the elevated sensor temperature, T_e is the sensor recovery or equilibrium temperature, I is the sensor current, and R_w is the sensor resistance at the elevated temperature.

When operated in constant current mode, the current is proportional to the supply voltage, E_s, and independent of the sensor resistance, R_w. Furthermore, $R_w \propto E_w/E_s$. Thus, these dependencies for the CCA shown in Figure 2.5 can be incorporated into Eq. (2.109) giving

$$\left(T_w - T_e\right) \frac{A_s k}{d} \left(A + B\text{Re}^n\right) = \frac{E_s E_w}{R_w}, \tag{2.111}$$

where $\text{Re} = \frac{\mu U d}{\rho}$.

At compressible Mach numbers, two terms in Eq. (2.111) depend on temperature. These are the thermal conductivity coefficient, k, and the absolute viscosity, μ, where

$$k = k_{\text{ref}} \left(\frac{T_o}{T_{\text{ref}}}\right)^a \tag{2.112}$$

and

$$\mu = \mu_{\text{ref}} \left(\frac{T_o}{T_{\text{ref}}}\right)^a. \tag{2.113}$$

In these expressions, a, k_{ref}, μ_{ref}, and T_{ref} are each constants.

Introducing into Eq. (2.111) the dependence of R_w on T_w and grouping constants produces the following:

$$C_1\overline{T_0}^a\overline{E_s}^{-2} + C_2\overline{T_0}^a\overline{(E_sE_w)}^{-1} - C_3\overline{T_0}^{(1+a)}\overline{(E_sE_w)}^{-1} + C_4\overline{(\rho U)}^n\overline{T_0}^{(a-an)}\overline{E_s}^{-2} + C_5\overline{(\rho U)}^n\overline{T_0}^{(a-an)}\overline{(E_sE_w)}^{-1} - C_6\overline{(\rho U)}^n\overline{T_0}^{(1+a-an)}\overline{(E_sE_w)}^{-1} = 1, \quad (2.114)$$

where n and a are known constants, $C_1 \cdots C_6$ are coefficients to be determined from calibration at different overheats, $\overline{E_s}$ and $\overline{E_w}$ are the mean anemometer input and output voltages, and $\overline{\rho U}$ and $\overline{T_o}$ are known mean flow conditions.

Solving this equation requires applying a minimum of six overheat voltages to the anemometer at each $\overline{\rho U}$, $\overline{T_o}$ calibration point. Matlis (2004) utilized 10 sensor overheat ratios to maximize the fidelity of the calibration. The approach yields a unique set of coefficients for each overheat voltage used during the calibration. It is possible to simplify Eq. (2.114) by absorbing the V_s terms into the calibration constants, by using a repeated set of overheat voltages, E_s, for every $[(\overline{\rho U}),\overline{T_o}]$ calibration data pair.

The unknown coefficients are determined through a least-square error minimization. This leads to the following general matrix, where $a_{1i}^j \cdots a_{6i}^j$ represent the terms multiplied by the respective coefficients for the ith $[(\overline{\rho U}),\overline{T_o}]$ data pair and the jth overheat voltage.

$$\begin{bmatrix} \sum_i (a_{1i}^2)^j & \cdots & \sum_i (a_{1i}a_{6i})^j \\ \sum_i (a_{2i}a_{1i})^j & \cdots & \sum_i (a_{2i}a_{6i})^j \\ \sum_i (a_{3i}a_{1i})^j & \cdots & \sum_i (a_{3i}a_{6i})^j \\ \vdots & & \vdots \\ \sum_i (a_{6i}a_{1i})^j & \cdots & \sum_i (a_{6i}^2)^j \end{bmatrix} \begin{bmatrix} (C_1)^j \\ (C_2)^j \\ (C_3)^j \\ \vdots \\ (C_6)^j \end{bmatrix} = \begin{bmatrix} \sum_i (a_{1i})^j \\ \sum_i (a_{2i})^j \\ \sum_i (a_{3i})^j \\ \vdots \\ \sum_i (a_{6i})^j \end{bmatrix}. \quad (2.115)$$

With the coefficients determined from the calibration, the time averaged sensor voltage, $\bar{E}_w$, can be decomposed into $(\overline{\rho U})$ and $\overline{T_0}$ by solving the following matrix:

$$\begin{bmatrix} \sum_j m_{1j}^2 & \sum_j m_{1j}m_{2j} & \cdots & \sum_j m_{1j}m_{4j} \\ \sum_j m_{2j}m_{1j} & \sum_j m_{2j}^2 & \cdots & \sum_j m_{2j}m_{4j} \\ \vdots & \vdots & & \vdots \\ \sum_j m_{4j}m_{1j} & \sum_j m_{4j}m_{2j} & \cdots & \sum_j m_{4j}^2 \end{bmatrix} \begin{bmatrix} \overline{T_0}^a \\ \overline{T_0}^{1+an} \\ \overline{(\rho U)}^n\overline{T_0}^{a-an} \\ \overline{(\rho U)}^n\overline{T_0}^{1+a-an} \end{bmatrix}$$

$$= \begin{bmatrix} -\sum_j m_{1j}m_{5j} \\ -\sum_j m_{2j}m_{5j} \\ \vdots \\ -\sum_j m_{4j}m_{5j} \end{bmatrix}, \quad (2.116)$$

where $m_{1j} = C_{1j}\overline{V_o}^j + C_{2j}$, $m_{2j} = C_{3j}$, $m_{3j} = C_{4j}\overline{V_o}^j + C_{5j}$, $m_{4j} = C_{6j}$, and $m_{5j} = C_{7j}\overline{V_o}^j$. Here, the j index represents the overheat voltages, E_s, used in calibration. As a best practice, the same overheat voltages are repeated for each $[(\overline{\rho U}),\overline{T_o}]$ calibration data pair so that in Eq. (2.116), the overheat voltages are absorbed into the calibration constants.

In the formulation, there are two constants that need to be determined. These are the exponent, a, in the temperature dependence of μ and k and the exponent on the Reynolds number in a generalized King's law, n. Matlis (2004) determined these coefficients

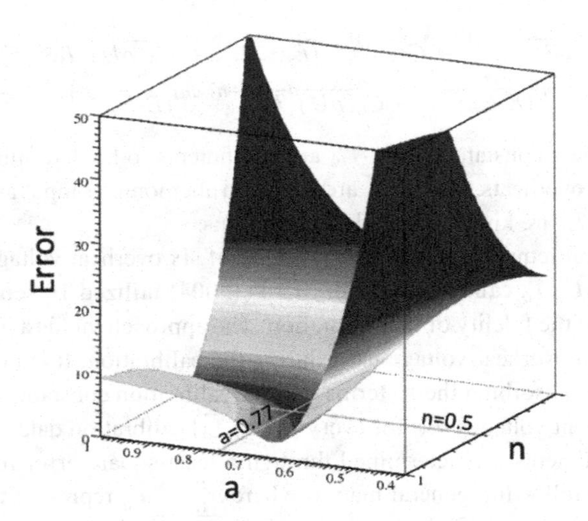

Figure 2.16 Result of error analysis to determine the "best" values of the power law coefficient, a, in the modeling of the temperature dependence of μ and k, and of the exponent on the Reynolds number, n, in a generalized King's law. Taken from Matlis (2004).

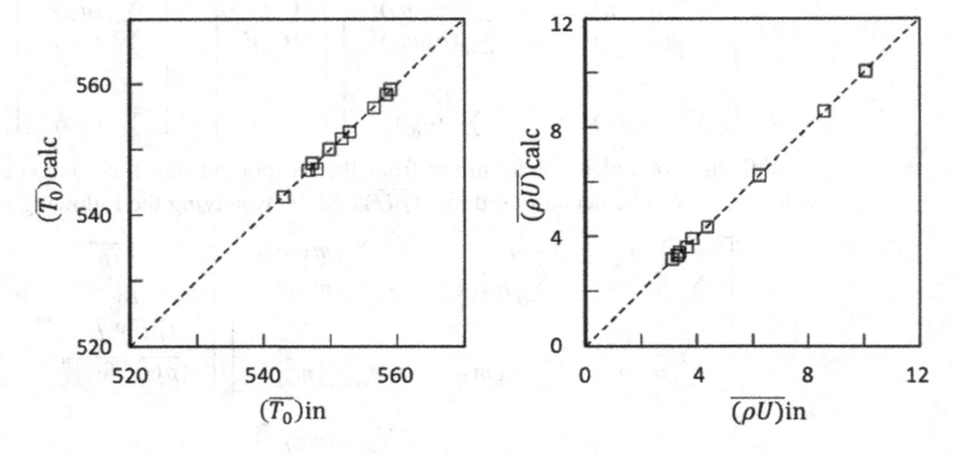

Figure 2.17 Comparison of mass-flux and temperature values used in the compressible Mach number sensor calibration to those values produced from Eq. (2.114). Taken from Matlis (2004).

through an iterative approach aimed at minimizing the total error in the calibration. The result is shown in Figure 2.16, where the values that minimized the calibration error are $a = 0.77$ and $n = 0.50$. At the onset, this step could be skipped by simply using $a = 2/3$ that commonly appears in the literature and $n = 1/2$ that is the standard for King's law.

Figure 2.17 illustrates the excellent ability of the compressible Mach number sensor calibration method to replicate the mass-flux and temperature data used for calibration. A further indication of the efficacy of the calibration method comes by comparing a measured boundary layer mass-flux profile to the theoretical boundary layer profile.

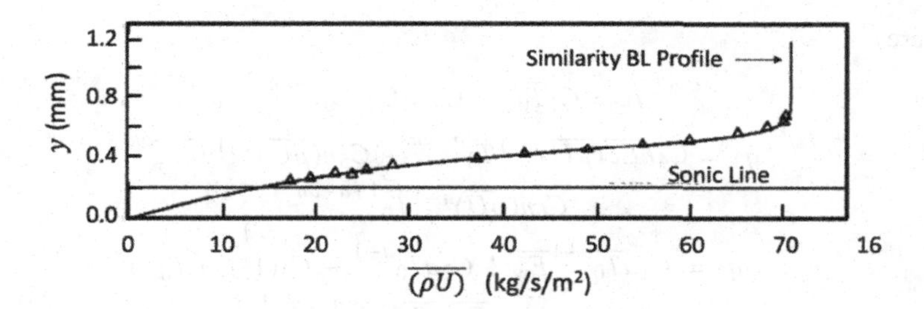

Figure 2.18 Comparison of theoretical and measured mass-flux profiles in the boundary layer over a 7° half-angle right-circular cone at zero angle of attack in a Mach 3.5 free-stream. Taken from Matlis (2004).

This is shown in Figure 2.18. The profiles correspond to a flow over a 7° half-angle right-circular cone at zero angle of attack in a Mach 3.5 free-stream. The solid curve corresponds to a similarity profile for the boundary layer at a streamwise location along a ray of the cone. The symbols correspond to the measured values from a hot-wire that was calibrated following the method described in this section. It is obvious that the agreement is excellent.

Obviously for fluid instability measurements, there is a need to also have calibrated values of fluctuation amplitudes. For this, the same calibration coefficients determined with the time-averaged quantities, $C_1 \cdots C_6$, are used for the determination of the fluctuation amplitudes. Here, the variance of the fluctuating flow quantities, $\overline{(\rho U)'^2}$ and $\overline{T_0'^2}$, are posed in terms of the variance of the anemometer voltage, $\overline{E_w'^2}$.

In the formulation, the quantities T_0, (ρU), and E_w are decomposed into their mean and fluctuating parts, namely

$$T_0 = \overline{T_0} + T_0' \tag{2.117}$$

$$(\rho U) = \overline{(\rho U)} + (\rho U)' \tag{2.118}$$

$$E_w = \overline{E_w} + E_w', \tag{2.119}$$

where the primed terms represent the mean-removed fluctuations of each of the quantities.

In order to relate these statistics to the anemometer voltage fluctuations, Eq. (2.114) is rewritten in a more general form as

$$C_1 T_0^a V_0 + C_2 T_0^a - C_3 T_0^{1+a} + C_4 E_w (\rho U)^n T_0^{a-an} + C_5 (\rho U)^n T_0^{a-an} - C_6 (\rho U)^n T_0^{1+a-an} - 1 = 0, \tag{2.120}$$

where the overheat controlling voltage, E_s, has been absorbed into the coefficients.

After substitution of the mean and fluctuation terms, the resulting equation is expanded and linearized. Within this equation, the terms that make up the mean equation Eq. (2.114) cancel out, leaving the following:

$$q_1 E_w' + q_2 (\rho U)' + q_3 T_0' = 0, \tag{2.121}$$

where

$$q_1 = C_1\overline{T_0}^a + C_4(\overline{\rho U})^n\overline{T_0}^{a-an} - 1 \tag{2.122}$$

$$q_2 = C_4 n\overline{E_w}(\overline{\rho U})^{n-1}\overline{T_0}^{a-an} + C_5 n(\overline{\rho U})^{n-1}\overline{T_0}^{a-an} - C_6 n(\overline{\rho U})^{n-1}\overline{T_0}^{1+a-an} \tag{2.123}$$

$$q_3 = C_1 a\overline{T_0}^{a-1}\overline{E_w} + C_2 a\overline{T_0}^{a-1} - C_3(1+a)\overline{T_0}^a + C_4(a-an)\overline{E_w}(\overline{\rho U})^n\overline{T_0}^{a-an-1} + C_5(a-an)(\overline{\rho U})^n\overline{T_0}^{a-an-1} - C_6(1+a-an)(\overline{\rho U})^n\overline{T_0}^{a-an}. \tag{2.124}$$

Taking the variance of the terms in Eq. (2.121) and rearranging gives the following:

$$\overline{E_w'^2} = \left(\frac{q_2}{q_1}\right)^2\overline{(\rho U)'^2} + \left(\frac{q_3}{q_1}\right)^2\overline{T_o'^2} + 2\frac{q_2}{q_1}\frac{q_3}{q_1}\overline{(\rho U)'T_0'}. \tag{2.125}$$

As a general form, Eq. (2.125) can be written as

$$k_1 D_1 + k_2 D_2 + k_3 D_3 - k_0 = 0, \tag{2.126}$$

where

$$k_1 = \left(\tfrac{q_2}{q_1}\right)^2 \quad k_2 = \left(\tfrac{q_3}{q_1}\right)^2 \quad k_3 = 2\tfrac{q_2}{q_1}\tfrac{q_3}{q_1},$$

which are functions of the sensor overheat, and

$$D_1 = \overline{(\rho U)'^2} \quad D_2 = \overline{(\rho U)'T_0'} \quad D_3 = \left(\tfrac{q_3}{q_1}\right)^2\overline{T_0'^2},$$

which are the unknown variances of the flow quantity fluctuations.

The unknown quantities, D_s, are found by minimizing the least-square error defined as

$$\sigma^2 = \sum_{j=1}^{N}\left[k_{1j}D_1 + k_{2j}D_2 + k_{3j}D_3 - k_{0j}\right]^2, \tag{2.127}$$

where, as before, j represents the sensor overheat ratios.

The unknowns are then found by solving the following matrix:

$$\begin{bmatrix} \sum_j(k_{1j}^2) & \sum_j(k_{1j}k_{2j}) & \sum_j(k_{1j}k_{3j}) \\ \sum_j(k_{1j}k_{2j}) & \sum_j(k_{2j}^2) & \sum_j(k_{2j}k_{3j}) \\ \sum_j(k_{1j}k_{3j}) & \sum_j(k_{2j}k_{3j}) & \sum_j(k_{3j}^2) \end{bmatrix}\begin{bmatrix} \overline{(\rho U)'^2} \\ \overline{T_0'^2} \\ \overline{(\rho U)'T_0'} \end{bmatrix} = \begin{bmatrix} -\sum_j m_{0j}m_{1j} \\ -\sum_j m_{0j}m_{2j} \\ -\sum_j m_{0j}m_{3j} \end{bmatrix}. \tag{2.128}$$

An indication of the accuracy of the method is shown in Figure 2.19. The data points are the wall-normal distribution of the root mean square (RMS) of the mass-flux fluctuations for the boundary layer over the 7° half-angle right-circular cone at a zero angle of attack. The curve is the theoretical Tollmien–Schlichting mode eigenfunction

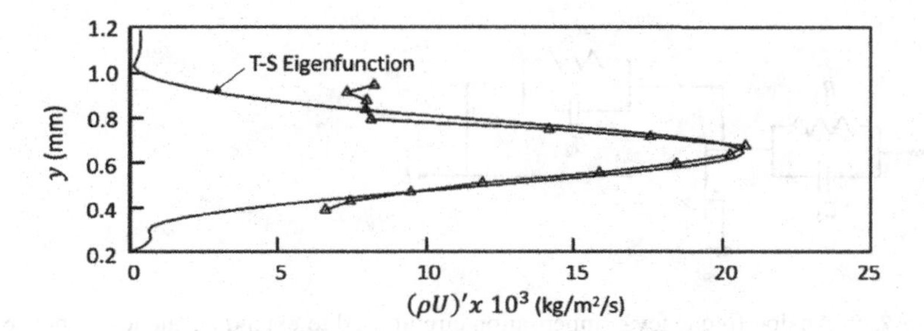

Figure 2.19 Comparison of theoretical and measured mass-flux wall-normal eigenfunction for Tollmien–Schlichting mode in the boundary layer over a 7° half-angle right-circular cone at zero angle of attack in a Mach 3.5 free-stream. Taken from Matlis (2004).

for the most-amplified frequency based on linear stability analysis. For comparison, the peak amplitude value of the theoretical eigenfunction was matched to the measured peak value. Therefore, the shapes of the distributions were compared. This is necessary since linear stability provides no information of the initial amplitude. It is evident that the agreement in the shape of the eigenfunction is very good, which gives credibility to the sensor calibration method.

Analog Frequency Compensation

The necessity for varying the overheat ratio with hot-wires used in measurements in supersonic and hypersonic flows motivates the use of constant-current and constant-voltage anemometers. In these anemometers, the sensor overheat is controlled by an input voltage, E_s, for the CCA and E_{ref} for the CVA. This is easily automated for computer control. Constant temperature anemometers by contrast control the sensor overheat through a resistance in the bridge. This makes automated control more difficult. In addition, the stability of the bridge varies with the overheat, which further complicates changing overheat values.

The advantage of a CTA is frequency response, which is especially important in studying fluid instabilities in high Mach number flows. As previously shown, the CCA and CVA have much lower frequency response compared to the CTA. Therefore, compensation is needed to extend their frequency response. One example of an analog frequency compensation circuit is shown in Figure 2.20. Analysis of the circuit provides a relation for the amplitude response in terms of the resistance and capacitance components, namely

$$\frac{E_o}{E_i}(D) = -\frac{R_2}{R_1}\frac{(R_1C_1D+1)}{(R_2C_2D+1)}, \tag{2.129}$$

where $D = d/dt$.

Figure 2.21 illustrates the amplitude response, E_o/E_i, given by Eq. (2.129). The design for frequency compensation starts with knowing the response of the sensor. An idealized example is shown in Figure 2.21(a). The abscissa is the frequency, $\omega(\text{s}^{-1})$. The ordinate is the natural log of the amplitude response, $\ln(E_o/E_i)$, so that exponential amplification (+ or −) can be represented in decibels (dB). In this example, the

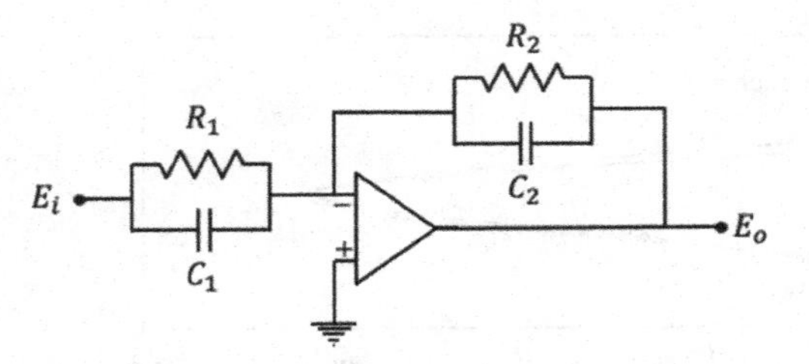

Figure 2.20 Analog frequency compensation circuit used to extend frequency response of constant-current or constant-voltage hot-wire anemometers.

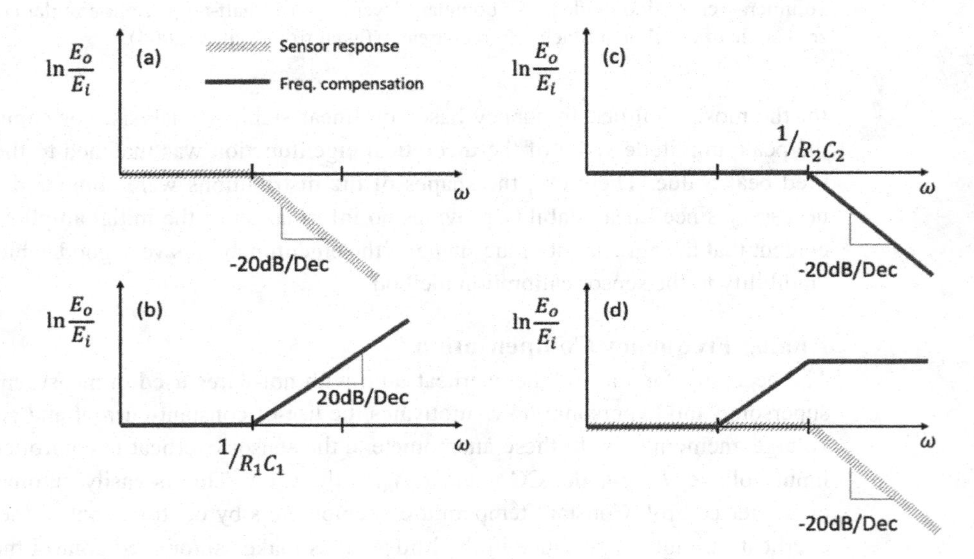

Figure 2.21 Example of frequency compensation of a constant-current or constant-voltage hot-wire anemometer. (a) The initial anemometer response. (b) The frequency response of the numerator term in Eq. (2.129). (c) The frequency response of the denominator term in Eq. (2.129). (d) The total response of frequency compensation and the result to the response of the sensor.

sensor response is constant out to its "cut-out" frequency, at which point the response is decreasing. For a first-order ($i\omega$) system, the decay rate is 20 dB per decade in frequency. The product R_1C_1 in the compensation circuit sets a frequency break-point. Because it appears in the numerator in Eq. (2.129), it results in a 20 dB/dec gain beyond the break-point frequency. This is illustrated in Figure 2.21(b). This break-point should be set to match the cut-out frequency of the sensor. The product R_2C_2 in the compensation circuit sets a second frequency break-point. Because it appears in the denominator in Eq. (2.129), it results in a −20 dB/dec gain beyond the break-point frequency. This is illustrated in Figure 2.21(c). This break-point should be set to be the desired cutoff frequency of the sensor. The result is a linear summation of the frequency response (a) through (c) that is illustrated in Figure 2.21(d). Finally, the resistance ratio, R_2/R_1, sets an overall gain across all frequencies of the output. Note that the negative sign indicates that the output is inverted.

If frequency compensation of a higher-order sensor (second, third, etc.) is desired, multiple compensation circuits like that in Figure 2.20 can be connected in series. The Resistor-Capacitor products can be adjusted to compensate for less ideal sensor response characteristics as well.

2.1.3 Laser Doppler Velocimetry

LDV is a direct-measure fluid velocity approach. This is in contrast to hot-wires which are an indirect (heat transfer not velocity) measurement approach. The LDV function is to measure the time-of-flight of particles that traverse a structured and fully characterized intensity distribution in space.

The principle behind LDV begins by considering a linearly polarized light wave, $u(x, z, t)$, with a single wavelength, λ, traveling in a uniform medium in the $\vec{k}$ direction that runs parallel to the x–z plane, namely

$$u(x, z, t) = u_0(\vec{r}) \exp[i(n\vec{k} \cdot \vec{r} - \omega t + \phi_0)]. \tag{2.130}$$

Here, u_0 is the wave amplitude, $k = |\vec{k}| = \omega/c = 2\pi/\lambda$, ω is the angular frequency, c is the light speed, ϕ_0 is the phase, and n is the refractive index of the medium through which the wave passes. Based on this complex waveform, the "intensity" is the complex magnitude

$$I(x, z, t) = uu^*, \tag{2.131}$$

where $*$ indicates complex conjugate.

An LDV system uses two coherent plan waves:

$$u_1(x, z, t) = u_{0_1}(z, z, t) \exp[i(nk_1(z \cos\alpha_1 + x \sin\alpha_1) - \omega_1 t + \phi_{0_1})] \tag{2.132}$$

$$u_2(x, z, t) = u_{0_2}(z, z, t) \exp[i(nk_2(z \cos\alpha_2 + x \sin\alpha_2) - \omega_2 t + \phi_{0_2})]. \tag{2.133}$$

If the two waves intersect at a point in space, it produces an intensity field, $I(x, z, t)$, given as

$$I(x, z, t) = uu^* = (u_1 + u_2)(u_1^* + u_2^*) \tag{2.134}$$

$$I(x, z, t) = u_{0_1}^2 + u_{0_1}^2 + 2u_{0_1}u_{0_2} \cos[nz(k_1 \cos\alpha_1 - k_2 \cos\alpha_2) + nx(k_1 \sin\alpha_1 - k_2 \sin\alpha_2) - \Delta\omega t + \Delta\phi]. \tag{2.135}$$

The resulting $I(x, z, t)$ given by Eq. (2.136) describes a set of alternate bright and dark fringes that are not stationary in time as a result of $\omega_1 \neq \omega_2$.

If frequencies of the two waves are the same so that $\omega_1 = \omega_2$, then assuming the two waves have the same wave speed, $k_1 = k_2$, and if $\Delta\phi = 0$, Eq. (2.136) reduces to

$$I(x, z, t) = u_{0_1}^2 + u_{0_1}^2 + 2u_{0_1}u_{0_2} \cos k[nz(\cos\alpha_1 - cos\alpha_2) + nx(\sin\alpha_1 - \sin\alpha_2)], \tag{2.136}$$

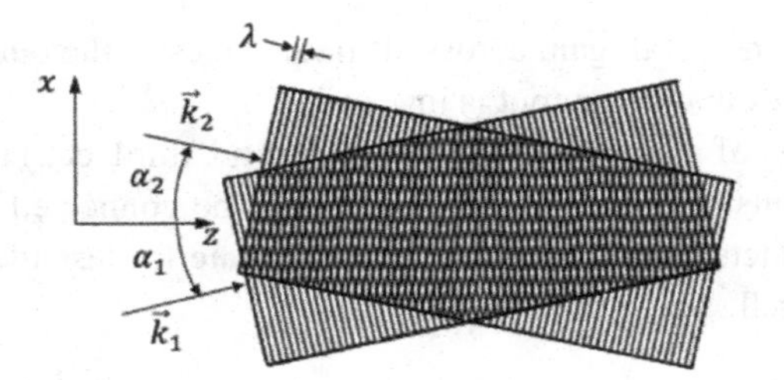

Figure 2.22 Stationary light–dark fringe pattern produced by the intersection of two waves, $\vec{k}_1$ and $\vec{k}_1$, having the same wave length, λ.

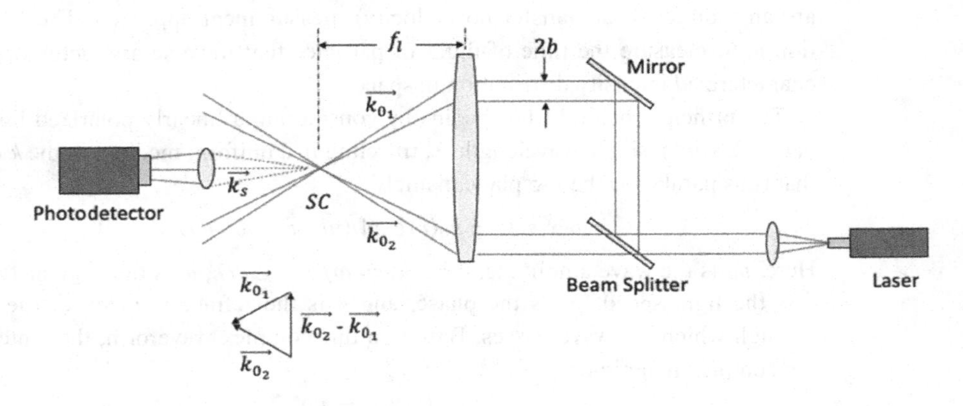

Figure 2.23 Example of an LDV setup utilizing a laser light source and photomultiplier.

with the result being a set of stationary light–dark fringes as illustrated in Figure 2.22. The fringe planes are parallel to the z–y plane and given by

$$x = \left[\frac{\cos\alpha_1 - \cos\alpha_2}{\sin\alpha_1 \sin\alpha_2} \right] z + \text{constant}, \tag{2.137}$$

and the plane of observation is normal to the bisector of the angle between the two beams.

Figure 2.23 shows an example of an LDV setup. Figure 2.24 provides fringe model equations that are used in its design. In this example, the laser source provides a coherent light beam with a given wavelength, λ. The beam is split into two beams that ultimately intersect at a shallow angle. As discussed, the intersection of the two monochromatic waves results in phase interference that appears as light and dark bands like those illustrated in Figure 2.22. In practice, the laser beam intensity is a maximum at the center of the beam and decreases toward the edges of the beam. This is modeled as a Gaussian distribution circular cross section. As a result, as illustrated in Figure 2.24, the fringe pattern viewed in the x–y plane appears as a circle. The diameter is a function of the laser wavelength, λ, and the angle between the waves at the cross section, namely

$$D = \frac{4}{\pi}\frac{\lambda}{\theta}. \tag{2.138}$$

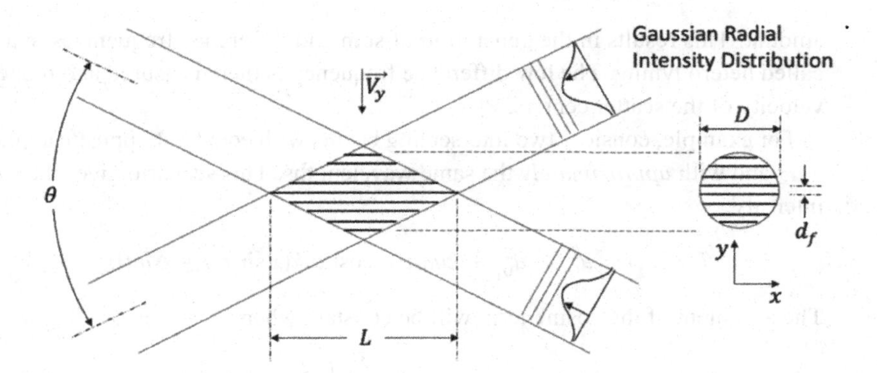

Figure 2.24 Details of LDV design based on two waves with equal wavelengths crossing at equal but opposite angles to form an interference pattern where the velocity of particles ("scatter centers") can be measured.

The fringe spacing, d_f, is

$$d_f = \frac{\lambda}{2\sin(\theta/2)}. \tag{2.139}$$

Finally, the length, L, of the fringe pattern is

$$L = \frac{4\lambda}{\pi\theta\sin(\theta/2)}. \tag{2.140}$$

The combination of D and L determines the sample volume of the LDV measurement. Generally, this should be kept as small as possible to increase spatial resolution. However, reducing the sample volume will reduce the rate of detection of particles ("scatter centers") passing through the measurement volume and scattering light to the photodetector.

A scatter center that passes through the sample volume will scatter light with a frequency, f_D, that is based on the time-of-flight, namely

$$t = \frac{d_f}{\bar{V}_y} \tag{2.141}$$

so that

$$f_d = \frac{1}{t} = \frac{\bar{V}_y}{d_f}. \tag{2.142}$$

Putting this in terms of the LDV design,

$$f_d = \frac{2\bar{V}_y}{\lambda}\sin(\theta/2). \tag{2.143}$$

As an example, if $\theta = 6°$, $\lambda = 0.5$ μm (Argon laser), and from Figure 2.23, $f_l/2b = 200$, then $D = 0.013$ cm, $L = 25$ cm, and $f_d = 2$ kHz/cm/s.

The particles scatter light uniformly in all directions and in proportion to the intensity to which they are exposed. The frequency of the scattered light is high, which can make it hard to measure. A solution is to separate the wave frequencies by a small amount. This results in the generation of sum and difference frequencies in a process called heterodyning. The low difference frequency is then measured and related to the velocity of the scatter centers.

For example, consider two intersecting beams with equal but opposite angles, $\alpha_1 = -\alpha_2$ and with *approximately* the same wavelengths. This scenario gives the following intensity:

$$I(x,z,t) = u_{0_1}^2 + u_{0_1}^2 + 2u_{0_1}u_{0_2}\cos[2nk(x\sin\alpha) - \Delta\omega t]. \tag{2.144}$$

The argument of the cosine term will be constant when

$$x = \frac{\Delta\omega t}{2nk\sin\alpha}, \tag{2.145}$$

which corresponds to a motion of the fringe planes perpendicular to themselves (x direction) with a velocity

$$v = \frac{\Delta\omega}{2nk\sin\alpha}. \tag{2.146}$$

The effect is the same as a strobe light with a frequency that is slightly off the frequency of motion. This approach is also the basis for determining the *direction* of a scatter center.

Figure 2.25 shows examples of intensity time series, $I(t)$, from an LDV. Figure 2.25(a) shows an ideal $I(t)$ in which a single scatter center passes through the measurement volume. The time series is referred to as a "burst." The corresponding frequency spectrum shows a well-resolved amplitude peak at f_d that is separated from broadband low-frequency energy that is associated with random light scattering.

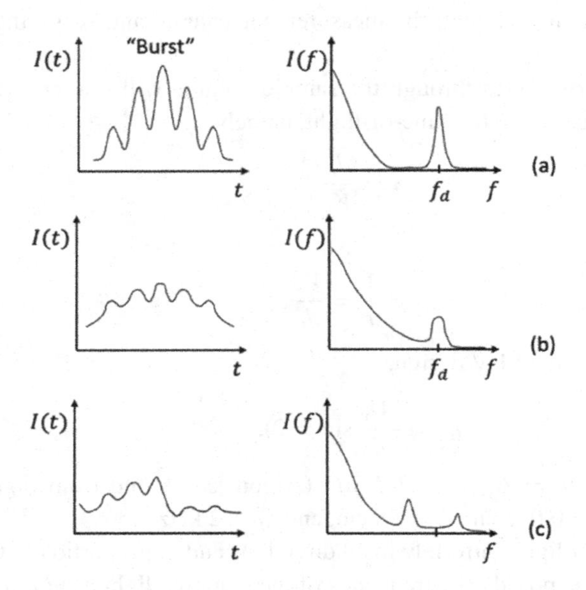

Figure 2.25 Examples of intensity time series, $I(t)$, from an LDV.

Figure 2.25(b) shows a less than ideal time series in which the waveform amplitude is less, and there is a larger background intensity. The spectrum still indicates a peak at f_d, although it is not as well-resolved above the background intensity fluctuations. There are three possible reasons for this less than ideal result:

1. The scatter center (particle) size is comparable to the fringe spacing.
2. There are unequal beam intensities.
3. The beam angle is too wide, making the fringe spacing too large.

Figure 2.25(c) shows an unacceptable time series, where more than one frequency appears in the intensity time series. This is the result of having more than one scatter center in the measurement volume at one time. This time series would be rejected.

The average intensity of light that strikes the scatter center is

$$I_0 = \frac{8P}{\pi D^2}, \tag{2.147}$$

where P is the laser power, and D is given by Eq. (2.138).

The amount of scattered radiation is

$$I_s = \frac{\sigma(\theta) I_0 N_{SC}}{R^2}, \tag{2.148}$$

where $\sigma(\theta)$ is the scattering cross section, SC is the number of scatter centers (ideally one), and R is the distance to the photomultiplier.

Based on these, the signal-to-noise ratio (Brayton and Goether, 1970) is

$$\frac{S}{N} = \frac{\pi^2 n \sigma(\theta) P N_{SC} \psi^4}{4hcB\lambda}, \tag{2.149}$$

where ψ is the scattered light collection angle, n is the photomultiplier quantum efficiency, B is the system bandwidth, h is Planck's constant, c is the speed of light, and, as before, λ is the wavelength. It is noted that S/N decreases as the particle velocity, V, increases since in this instance B appearing in the denominator increases as V increases.

The LDV signal is "digital-like" with *random arrivals*. This can lead to a high-speed *bias error* as a result of a higher probability that higher-speed sample will arrive at the measurement volume more frequently. In order to correct for this, the expected velocity, $< V >$, is

$$< V > = \sum_{i=1}^{N} \frac{V_i t_i}{T}, \tag{2.150}$$

where V_i is the velocity of the ith particle, t_i is the transit time of the ith particle, and T is the average transit time. The S/N of the higher-speed particles is lower due to the lower residence time in the measurement volume. This could bias measurements toward lower velocities.

As a final point, modern LDV systems utilize two and three lasers to simultaneously measure multiple velocity components. Different wavelength (color) lasers are used to separate the different velocity components.

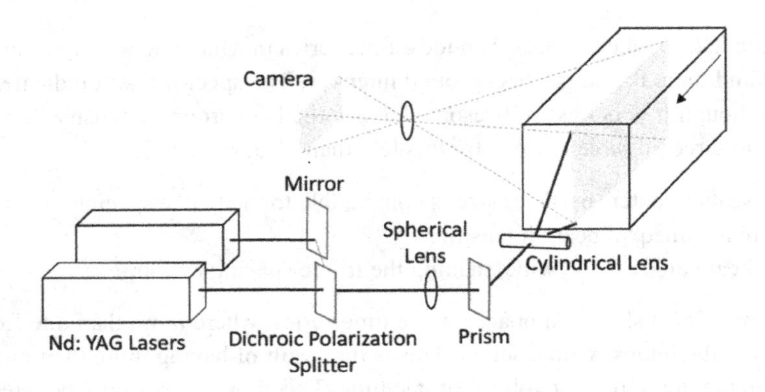

Figure 2.26 Example of a planar particle image velocimeter setup.

2.1.4 Particle Image Velocimetry

Like LDV, particle imaging velocimetry is a member of a broader class of velocity measuring techniques, which determines the motion of small, marked regions of a fluid by observing the changing locations of a field of markers (scatter centers) at two or more times. The determination of motion velocity is based on the fundamental definition

$$\vec{u}(\vec{x}, t) \doteq \frac{\Delta \vec{x}(\vec{x}, t)}{\Delta t}, \tag{2.151}$$

where $\Delta \vec{x}$ is the displacement of a marker located at $\vec{x}$ at time t over a short interval, Δt, that separates the observations of the marker images. The particles (scatter centers) can be the same as those used in LDV, except they are applied to fill a larger, full-field area. The instantaneous locations of the markers at various instants in time are recorded optically by short-duration pulses of light. The images are recorded on an optical recording medium. Early PIVs use photographic film to record the images. Modern PIVs uses digital cameras. A number of excellent early contributions to PIV are due to Adrian (1991), Willert and Gharib (1991), Westerweel (1997), Forliti et al. (2000), and Westerweel et al. (2013).

A basic two-dimensional (2-D) PIV setup is shown in Figure 2.26. Particles introduced into the fluid are illuminated by a sheet of laser light that is pulsed. The setup shows two lasers, although the laser beams are superimposed so that at each laser pulse, the exact same area is illuminated. The lasers in the sample setup are a commonly used Neodyme-YAG (Nd:YAG) laser having a 5–10 ns pulse duration, with up to 400 mJ/pulse, and a repetition rate of up to 10 Hz. The particles scatter light into a photographic lens located at 90° to the particle sheet so that its in-focus object plane coincides with the illuminated slice of the flow field. Images are formed on a photographic film or CCD array and subsequently transferred to a computer for automatic analysis.

The mean number of particles per unit volume, C, determines the spatial resolution of the measured velocities. If particles are modeled as randomly distributed, the probability of finding k particles in volume, V, is (Adrian, 1991)

$$\mathcal{P}(k) = \frac{(CV)^k}{k!} e^{-CV}. \tag{2.152}$$

Adrian (1984) defined a source density of particles as a mean number in a cylindrical volume formed by the intersection of an illuminating light sheet with a circle whose diameter, d_r/M_0, is that of the projected particle image. In that, d_r is the particle diameter, and M_0 is the image magnification. The volume is called a "resolution cell" where the value of the *source density*, N_s, is given by

$$N_s = \frac{C\Delta z_0}{M_0^2} \frac{\pi}{4} d_r^2 \tag{2.153}$$

and in which Δz_0 is the thickness of the light sheet (in m), and C is the tracer particle concentration (in m^{-3}). If two particles were to lay within a resolution cell, their images might overlap in the image plane. The probability of that occurrence becomes significant as $N_s \gg 1$. In contrast, a small N_s implies solitary images, indicating a particle-image limit (Adrian, 1984).

Adrian and Yao (1984) applied similar reasoning to the number of particle images within an "interrogation cell." The importance of an interrogation cell to PIV will become apparent.

The interrogation cell was defined as the intersection of a light sheet with a circle whose diameter, d_I/M_0, is equal to the diameter of an interrogation spot of the projected image. The ratio d_I/M_0 can also refer to the maximum 2-D displacement of particles, $|\Delta x|_{\max}$. Adrian and Yao (1984) defined the *image density* parameter, N_I, as

$$N_I = \frac{C\Delta z_0}{M_0^2} \frac{\pi}{4} d_I^2. \tag{2.154}$$

If $N_I \ll 1$, the probability of finding *more than one* particle in an interrogation cell is small. Conversely, if $N_I \gg 1$, there is a high probability of having many particles in an interrogation cell.

To ensure particle displacement accuracy, Keane and Adrian (1992) recommend for high-intensity images that

$$N_I > 7 \tag{2.155}$$

$$\frac{M|\Delta u|\Delta t}{d_I} < 0.03 \tag{2.156}$$

and

$$\frac{M|\Delta u|\Delta t}{d_r} < 0.03, \tag{2.157}$$

where Δu is the velocity difference within an interrogation spot, and Δt is the time between image exposures. Equations (2.156) and (2.157) provide constraints on the velocity gradient that can be tolerated within an interrogation spot. One further constraint is that out-of-plane particle displacements be less than one-quarter of the thickness of the light sheet, Δz_0 (Westerweel, 1997). While satisfying these constraints, valid measurements in over 90–95 percent of the interrogation spots are possible with only small bias errors (Adrian, 1991).

The PIV technique effectively samples the velocity of the fluid at the random location of each particle. If the particles were regularly located on a three-dimensional (3-D) periodic grid, the entire velocity field could be perfectly reconstructed from the velocity samples. This is provided that the Nyquist sampling criterion was satisfied, which requires that the sample spacing be less than one-half of the smallest wavelength in the field. However, since particle spacing is not uniform, the Nyquist sampling criterion does not apply in the usual sense. Results concerning random sampling of 1-D processes indicate that reconstruction will be inaccurate for scales smaller than several times the mean data spacing (Adrian and Yao, 1984).

The time separation of light pulses, Δt, is the single most important adjustable variable in a PIV system, as it determines the maximum and minimum velocities that can be measured. The duration of the light pulses, δt, determines the degree to which an image is frozen during the pulse exposure.

Pulsed lasers emit energy in collimated beams that can be efficiently formed into 2-D light sheets. For example, pulsed metal-vapor lasers produce green 10-ns pulses repetitively at free-running frequencies in the 5–20 kHz range and energies on the order of 10 mJ. Frequency doubled-pulsed Nd:Yag lasers fire 10-ns pulses at 532-nm light repetitively at rates up to 50 Hz. Two of these lasers can be combined (see Figure 2.26) to produce periodic trains of double pulses suitable for multiframing at 10–50 Hz with virtually any separation between the double pulses. The lower pulse energy of the Nd:Yag laser relative to other lasers is largely compensated by its shorter wavelength, which leads to improved scattering, thinner light sheets, and smaller images.

Although it is possible to use film to record particle images, modern PIV systems use Complimentary-metal-oxide-semiconductor-based digital high-speed cameras with frame rates at the highest resolution on the order of several kilohertz. Picture element (Pixel) resolutions as high as 13.5×13.5 μm in a 2048×1952 array are available with a 12-bit digital output. These are generally suitable for most fluid instability experiments.

As a result of recording the images in a CCD camera array, each pixel is a discretely addressable location in an array. The intensity values of each pixel are discretized based on the number of bits representing the full range of light intensity. For example a 12-bit conversion yields 2^{12} or 4096 quantized levels of particle illumination levels.

Starting with two digital images of a 2-D array of particles taken at times t and $t + \Delta t$, the first step in the processing is to divide the image pairs into small interrogation regions. This frame-to-frame subsampling is illustrated in Figure 2.27. The method of interrogation is primarily a statistical approach that probabilistically infers the proper pairings and average displacement of the group of particles in the interrogation cell. The method involves the zero-normalized cross-correlation (or cross-covariance) of the subsampled particle images taken Δt apart.

The discretized cross-covariance for an M by N subdomain is (Westerweel, 1997)

$$C(r,s) = \frac{1}{MN}\sum_{m=1}^{M}\sum_{n=1}^{N}\left[f(m.n) - \overline{f}\right]\left[g(m+r, n+s) - \overline{g}\right], \tag{2.158}$$

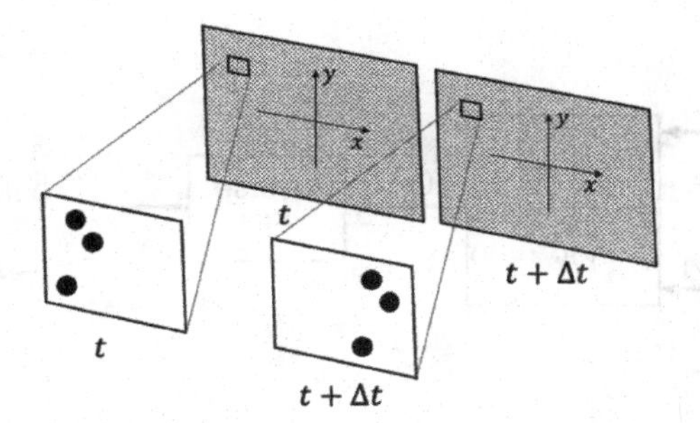

Figure 2.27 Conceptual arrangement of frame-to-frame subsampling associated with digital particle image velocimetry.

where $f(m,n)$ and $g(m,n)$, respectively, correspond to the first and second subsampled images, $\overline{f}$ and $\overline{g}$, respectively, correspond to the average (mean) intensity of f and g over the M by N array, and (r,s) represents the location where the cross-covariance is calculated. The cross-covariance, $C(r,s)$, is computed for all (r,s) to form the cross-correlation function, $C(m,n)$.

Willert (1992) has suggested another discretized cross-covariance formulation that inherently accounts for in-plane loss of particle pairs. This is

$$C(r,s) = \frac{C_{\mathrm{II}}(r,s)}{\sqrt{\sigma_{\mathrm{I}}(r,s)}\sqrt{\sigma_{\mathrm{II}}(r,s)}}, \tag{2.159}$$

where

$$C_{\mathrm{II}}(r,s) = \sum_{m=1}^{M}\sum_{n=1}^{N}[[f(m,n)-\overline{f}]\,[g(m+r,n+s)-\overline{g}(r,s)] \tag{2.160}$$

$$\sigma_{\mathrm{I}}(r,s) = \sum_{m=1}^{M}\sum_{n=1}^{N}[f(m,n)-\overline{f}]^2 \tag{2.161}$$

$$\sigma_{\mathrm{II}}(r,s) = \sum_{m=1}^{M}\sum_{n=1}^{N}[g(m_r,n+s)-\overline{g}(r,s)]^2 \tag{2.162}$$

and average of g being coincident with the interrogation window, f.

The cross-correlation process is more efficiently performed in the frequency domain where the correlation is simply the multiplication of the Fourier transformed arrays. This was first proposed for PIV analysis by Willert and Gharib (1991). In this, a 2-D FFT is performed on the 2-D sub-sampled images. Being a linear operation, the 2-D fast Fourier transform (FFT) can be performed by a series of 1-D FFTs applied separately in each direction. The transformed 2-D array, $\mathcal{F}(f)$, and the complex conjugate of the transformed 2-D array, $\mathcal{F}(g)$, are multiplied together. The resulting array is then inverse Fourier transformed. The process is illustrated in Figure 2.28.

The cross-correlation distribution that results either from the discrete point convolution or the correlation performed in the frequency domain should result in a dominant correlation peak that corresponds to the shift of the particles that occurs between the two subsampled images acquired Δt apart. A sample cross-correlation

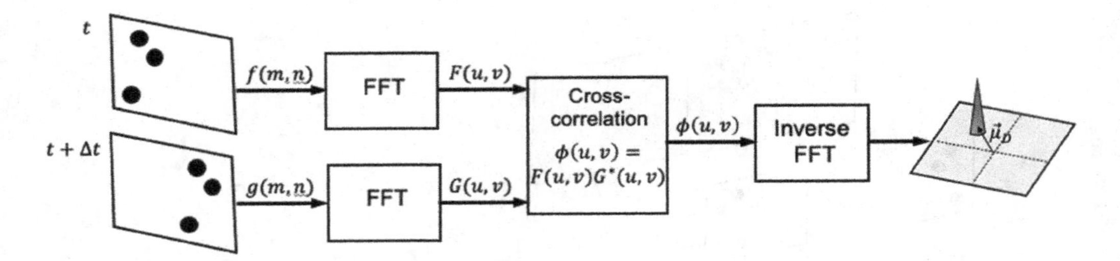

Figure 2.28 Illustration of the process for cross-correlation of subsampled particle images based on 2-D fast Fourier transforms.

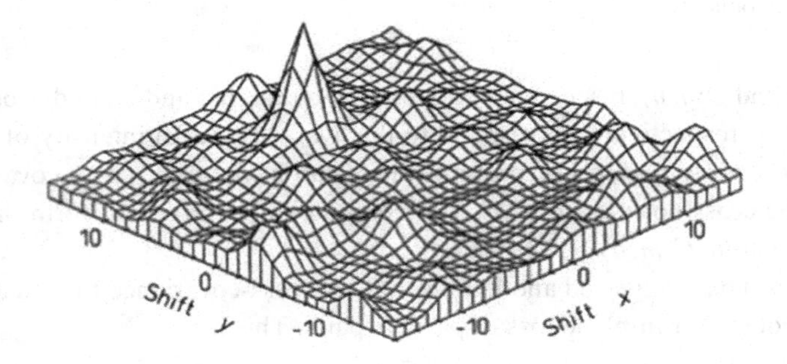

Figure 2.29 Sample cross-correlation distribution showing a single dominant peak representing the direction of particle shift between the two subsampled images. Taken from Willert and Gharib (1991).

distribution (Willert and Gharib, 1991) showing a single dominant peak representing the direction of particle shift between the two subsampled images is shown in Figure 2.29. The discrete shift is converted into a velocity vector, $\vec{u}$, through the calibration parameters, namely

$$\vec{\mu}_1 = \vec{\mu}_D/(M\Delta t), \tag{2.163}$$

where $\vec{\mu}_D$ is the vector distance of the correlation peak from center, and M is the image magnification.

The strength of the correlation peak is proportional to the number of particle-image pairs in the subsample. Large velocities reduce the amplitude of the correlation peak. Additional correlation peaks resulting from random noise are also possible. If the amplitude of one of the noise peaks is comparable to that of the displacement peak, the shift vector could be incorrectly identified. A method to reduce that possibility involves requiring the ratio of the amplitude of the tallest correlation peak to be a factor larger than other correlation peaks. The method is called "thresholding." Another approach is to restrict the search for the tallest correlation peak to a region where it is expected to be based on consistency with neighbor subsamples.

As with all experimental methods, PIV measurements are susceptible to error. There are many parameters that affect the accuracy of PIV measurements: subpixel peak fitting, tracer particle image diameter, tracer particle image intensity distribution, window interrogation size, tracer particle image shift, quantization effects, background noise, displacement gradients within an interrogation window, and out-of-plane tracer particle motion.

A concern with the discrete nature of the images and cross-correlation is that the resolution of the location of the correlation peak is limited by the number of points, m by n, of the subsampled domain. As a result, it has an uncertainty of $\pm 1/2$ pixel. To put this in perspective, for a 32 by 32 subsampled array, with maximum particle shifts of 1/3 of the array, or 10 pixels, the uncertainty in the displacement is, at best, 5 percent. As a result, a number of methods have been developed to obtain sub-pixel accuracy (Westerweel et al., 1996; Westerweel, 1997).

The effect of tracer particle image diameter had been investigated by Raffel et al. (1998). This indicated that a 2-pixel tracer particle image diameter provided the least uncertainty in particle image displacement. They also investigated the effect of tracer particle image density, N_I, given as

$$N_I = C\Delta Z_0 D_I^2/M^2, \tag{2.164}$$

where C is the number density of particle, ΔZ_0 is the width of the 2-D light sheet, D_I^2 is the area associated with the interrogation window, and M is the magnification. This indicated a 95 percent probability for detecting valid displacements when $N_I > 7$.

2.2 Surface Shear Stress Measurement

The wall shear stress in a boundary layer is of fundamental importance to determining its instability state. Further, it can provide an indication of imminent flow separation that, if it occurs, introduces a separated shear layer instability. Direct shear stress measurement requires measuring the shear force on the wall surface. This traditionally involves a moving platen that is linked to a force transducer. The inertial in these devices generally relegates this approach to time-averaged shear stress measurements. In contrast, indirect shear stress measurement approaches are better suited to reactive flow control.

There are a number of indirect methods for measuring boundary layer shear stress. One of the fundamental approaches uses a heated element placed in the wall surface. A schematic of this arrangement is shown in Figure 2.30. Shear stress measurements make use of the resemblance between diffusion of heat and vorticity, where the diffusion of heat equates to heat transfer, and the diffusion of vorticity equates to shear force that follows from Reynolds analogy (Mahulikar and Herwig, 2008).

The basic shear stress gage consists of a heat-generating element located on the surface of the wall in a boundary layer flow that is illustrated in Figure 2.30. The temperature of the heating element is $T = T_w + \Delta T$, where T_w is the surrounding wall temperature. For the boundary layer thickness, δ, free-stream or boundary layer edge velocity, U_∞, and viscosity, μ, the shear stress at the wall is

$$\tau_w \sim \mu \frac{U_\infty}{\delta}, \tag{2.165}$$

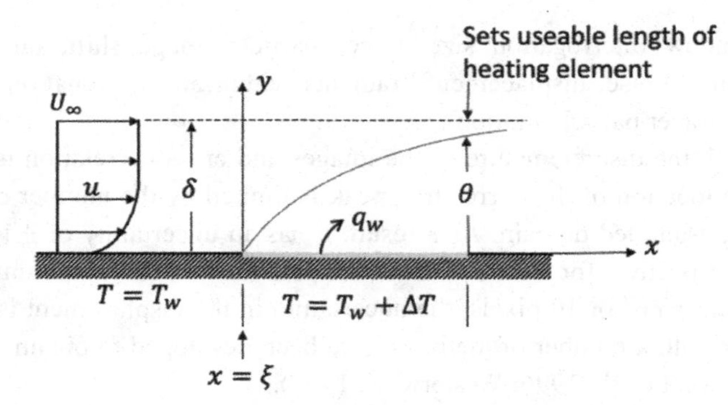

Figure 2.30 Illustration of heated element wall shear stress sensor.

where

$$\delta \sim \sqrt{\frac{\mu}{\rho}\frac{x}{U_\infty}}. \tag{2.166}$$

The heat transfer from the element is

$$q_w \sim \lambda \frac{\Delta T}{\theta}, \tag{2.167}$$

where $\lambda = \mu c_p$ and θ is the thermal boundary layer thickness. Therefore, by analogy,

$$\theta \sim \sqrt{\frac{\lambda}{\rho c_p}\frac{(x-\xi)}{u(\theta)}}, \tag{2.168}$$

where

$$u(\theta) = u \text{ at } y = \theta. \tag{2.169}$$

Now $u(\theta)$ depends on the velocity profile near the wall. To a first approximation, closest to the wall,

$$u(\theta) \simeq \left(\frac{\partial u}{\partial y}\right)_w \theta. \tag{2.170}$$

With

$$\tau_w = \mu \left(\frac{\partial u}{\partial y}\right)_w, \tag{2.171}$$

then

$$u(\theta) \simeq \frac{\tau_w}{\mu}\theta. \tag{2.172}$$

Therefore, substituting for $u(\theta)$ in Eq. (2.168), gives

$$\theta \sim \left(\frac{\lambda}{\rho c_p}\frac{(x-\xi)}{\frac{\tau_w}{\mu}\theta}\right)^{1/2} \tag{2.173}$$

or

$$\theta^3 \sim \frac{\lambda\mu}{\rho c_p}\frac{(x-\xi)}{\tau_w}. \tag{2.174}$$

For the heat flux, q_w,

$$q_w \sim \lambda \frac{\Delta T}{\theta} \sim \lambda \Delta T \left(\frac{\rho c_p}{\mu \lambda}\right)^{1/3} \left(\frac{\tau_w}{x-\xi}\right)^{1/3}. \tag{2.175}$$

Integrating the heat flux for the strip from ξ to $\xi + L$ to obtain the total heat flux, Q_w, gives the basic relation for heat-transfer-based skin friction sensors, namely

$$Q_w(L, \xi) \sim \left(\frac{\rho c_p}{\mu \lambda}\right)^{1/3} \lambda \Delta T \tau_w^{1/3} L^{2/3}. \tag{2.176}$$

Equation (2.176) is valid for laminar and turbulent boundary layers. The linear velocity profile at the wall that is assumed in the analysis is best approximate if the streamwise extent of the element, L, is small.

Previously $u(\theta)$ was approximated to be that closest to the wall. However, more generally,

$$u(\theta) = \left(\frac{\partial u}{\partial y}\right)_w + \frac{1}{2}\left(\frac{\partial^2 u}{\partial y^2}\right)\theta^2 \tag{2.177}$$

or

$$u(\theta) = \frac{t_w}{\mu}\theta + \frac{1}{2}\left(\frac{\partial^2 u}{\partial y^2}\right)\theta^2. \tag{2.178}$$

For a laminar boundary layer, the wall shear stress is a function of the pressure gradient, namely

$$\frac{dP}{dx} = \left(\frac{\partial \tau}{\partial y}\right)_w = \mu \left(\frac{\partial^2 u}{\partial y^2}\right)_w. \tag{2.179}$$

Therefore,

$$u(\theta)_L = \frac{\tau_w \theta}{\mu}\left(1 + \frac{\theta}{2\tau_w}\frac{dP}{dx}\right). \tag{2.180}$$

For a steady, separated flow, $\tau_w = 0$. Therefore,

$$u(\theta) = \frac{\theta^2}{2\mu}\frac{dP}{dx}, \tag{2.181}$$

and as a result,

$$Q_w(L, \xi) \sim \left(\frac{\rho c_p}{\mu \lambda}\right)^{1/4} \lambda \Delta T \left(\frac{dP}{dx}\right)^{1/4} L^{3/4}. \tag{2.182}$$

The streamwise length of the heating element is limited by $\theta < \delta$ or

$$\left(\frac{\lambda \mu}{\rho c_p}\frac{L}{\tau_w}\right)^{1/3} < \mu \frac{U_\infty}{\tau_w}. \tag{2.183}$$

If $\theta < \delta$ is considered an upper limit, a lower limit is $\theta \ll L$ or equivalently

$$\frac{Q_w}{\lambda \Delta T} \gg 1. \tag{2.184}$$

Based on Eq. (2.176), a general form for τ_w is

$$\tau_w^{1/3} = C_1 Q_w + C_2, \tag{2.185}$$

where C_1 and C_2 are calibration coefficients, and C_2 accounts for heat lost into the wall when mounted in the surface.

In constant-current operation, $Q_w = I^2 R_e / \Delta T_0$, where I is constant, R_e is the resistance of the heating element, and ΔT_0 is the temperature difference corresponding to a reference R_{e_0}. The CCA circuit could be the same as that shown in Figure 2.5 where the output voltage is proportional to the change in the element resistance and therefore heat flux, Q.

For constant temperature operation, R_e is constant and $Q^2 = I^2 R_e$. The CTA circuit could be the same as that shown in Figure 2.7. In that case, $Q = CE^2$, where $C = \Delta T_0$ is a calibration constant.

In either case of CCA or CTA operation, a calibration must be performed to determine the coefficients. There are a number of approaches. One is to place the sensor in a fully developed 2-D channel flow in which $\tau_w = \Delta P / 2LH$, and where L is the streamwise distance over which the streamwise pressure gradient, ΔP, is measured, and H is the height of the channel. Another approach is to calibrate the shear stress sensor in a laminar boundary layer where a velocity sensor placed in the linear mean profile region near the wall infers the local shear stress, $\tau_w = \mu(du/dy)$. Finally, one can perform a Clauser fit to the mean velocity profile of a turbulent boundary layer, where

$$\frac{u(y)}{u_\tau} = \frac{1}{\kappa} \log_e \left(\frac{y u_\tau}{\nu} \right) + B \tag{2.186}$$

and

$$u_\tau = \sqrt{\frac{\tau_w}{\rho}}. \tag{2.187}$$

The frequency response of the shear sensor will depend on the thermal mass that includes the resistive element as well as the substrate if it is heated in the process. As with hot-wire sensors, the thermal mass is reduced by reducing the surface area of the element. This favors the use of small diameter wires rather than flat electrodes. Substrate heating can be minimized by using substrates that have low heat conduction coefficients. Another approach is to slightly elevate the element above the surface, while still remaining in the linear profile region. Both of these approaches were used by Corke (1981b).

2.3 Flow Visualization

Flow visualization remains a valuable tool in experimental fluid-mechanical research. Although there have been a number of comprehensive reviews of flow visualization techniques, one of the best remains to be that of Merzkirch (1984). Beautiful examples of visualized flow fields is contained in "An Album of Fluid Motion" (VanDyke, 1982). The advantage of flow visualization is that it can provide a full-field view that can provide immediate insight into the flow physics. This is particularly true with the study of fluid instabilities where flow visualization can be used to determine instability patterns, wavelengths, and turbulence onset. It is also valuable in flow control experiments as an immediate indicator of its effectiveness, for example, in attaching a separated flow or in delaying or accelerating turbulence onset.

Finally, as a general experimental approach, it is advantageous to start with flow visualization as a guide to where in the flow field, quantitative measurements might be performed, as well as on the choice of the most appropriate experimental tools for the task. This is particularly important in high-speed ($M_\infty > 1$) flows where run times can be short, and the number of runs limited.

Flow visualization techniques can be classified roughly into three major groups: surface flow visualization (Maltby and Keating, 1962; Stanbrook, 1962; Kell, 1978), scattering from flow tracers (Corke et al., 1977; Nagib et al., 1978; Mueller, 1980, 1983), and density sensitive flow visualization (Settles and Hargather, 2017). As Merzkirch (1984) demonstrates, there are numerous time-tested approaches to these major groups. Since its publication, the greatest development has been in the visualization of variable density flows. This has particularly drawn importance with experiments on high Mach number flow fields. Therefore, the emphasis here is on such techniques, particularly on the Schlieren flow visualization.

The theory that supports Schlieren and the related shadograph approach is primarily based on geometrical optics with additional consideration of diffraction effects, which can be found in Schardin (1942) and later discussed by Settles (2001), among other sources. This finds its root in the Gladstone–Dale law

$$n - 1 = k\rho, \tag{2.188}$$

where n is the refractive index and ρ is the fluid density, and k is the Gladstone–Dale constant that is specific to a fluid and a function of the light wavelength, λ. As an example for air, $k = 0.2250$ cm^3/g for $\lambda = 0.7034$ μm. For mixtures of gases,

$$k = \sum_{i=1}^{N} k_i \frac{\rho_i}{\rho} = \sum_{i=1}^{N} k_i c_i, \tag{2.189}$$

where c_i is the relative mass concentration of the gas constituents.

Schlieren and shadowgraphy were formally originated by Robert Hooke in the seventeenth century (Settles, 2001). It was later "reinvented" in the mid-nineteenth century by August Toepler, and since has become a standard laboratory tool for observing shock waves. The sensitivity of a Schlieren system is proportional to the *gradient* of the refractive index, $\partial n/\partial \vec{x}$. In contrast, shadowgraphy sensitivity is proportional to $\partial^2 n/\partial \vec{x}^2$, and the sensitivity of optical interferograms is directly proportional to the gas index of refraction, n.

Figure 2.31 shows two examples of classical Schlieren optics setups. Figure 2.31(a) illustrates the Schlieren setup first used by Toepler (Krehl and Engemann, 1995). In that setup, light from a bright source is collimated by lens 1 and brought to a focus by lens 2, beyond which the light beam is projected on a screen (frosted glass) where

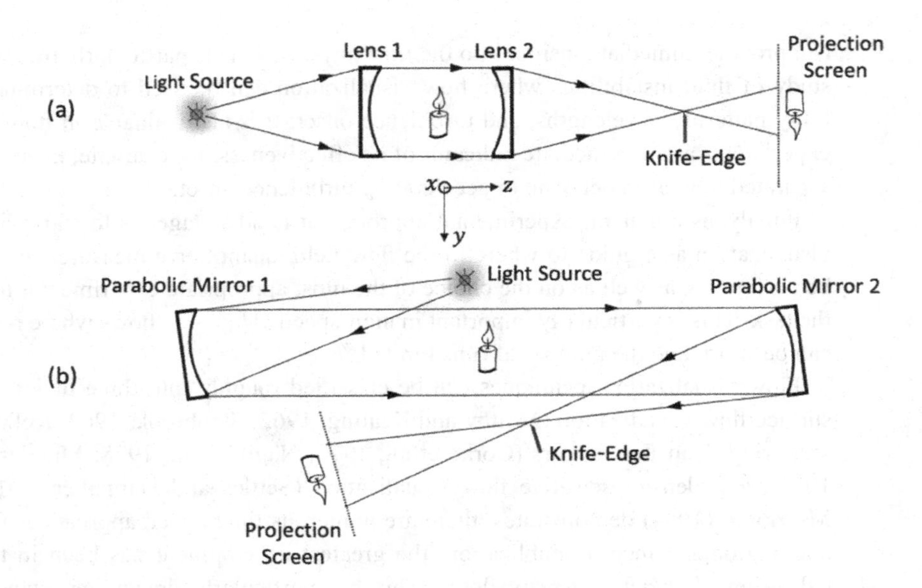

Figure 2.31 Illustrations of two classical Schlieren optic setups where (a) utilizes two lenses in a straight line arrangement, and (b) utilizes parabolic mirrors to allow the light source and projection screen to be off axis.

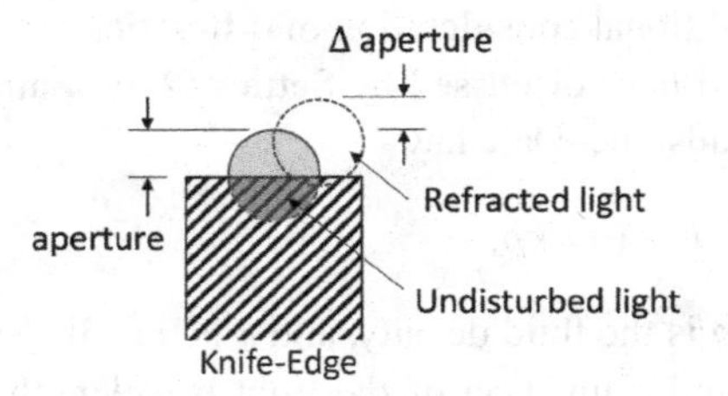

Figure 2.32 Illustration of undisturbed and refracted light apertures relative to a knife-edge in a Schlieren optic setup.

it can be viewed. The image can similarly be projected to the recording plane of a camera. A sharp knife-edge is inserted at the focus, blocking part of the light.

A refractive disturbance in the air between the two lenses, such as a heat source (candle in the illustration), will cause the light in that region to be refracted, for example, upwards. In this illustration, a light beam refracted upwards will avoid the knife-edge and appear bright on the screen. In contrast, a downward refracted light beam will be blocked by the knife-edge and appear dark on the screen. Overall, the traces of every ray in the collimated light source result in a grayscale image that represents the phase disturbance in the field of view.

The knife-edge is adjusted to cut off a certain portion of the image, reducing the intensity at the recording plane. An essential requirement is to use a "point" light source. This is illustrated in Figure 2.32. The image of the point light source is the illuminated circle of which part is masked by the knife-edge. The undisturbed point source that is not blocked by the knife-edge is the aperture, a. The portion of the light that is refracted due to a gradient in the gas density causes the point source to shift relative to the knife-edge, with an amount of Δa.

In the undisturbed, unrefracted image at the recording plane, the light intensity is $I(x,y) =$ constant. With refraction,

$$\Delta a = f \tan \epsilon_y, \tag{2.190}$$

where f is the focal length, y is in the direction perpendicular to the light path between Lens 1 and Lens 2 or Mirrors 1 and 2 in Figure 2.31, and $\epsilon_y = \partial n/\partial y$. The change in light intensity is

$$\frac{\Delta I}{I} = \frac{\Delta a}{a} = \frac{f}{a} \tan \epsilon_y, \tag{2.191}$$

and then

$$\frac{\Delta I}{I} = \frac{\Delta a}{a} \int_{z_1}^{z_2} \frac{1}{n} \frac{\partial n}{\partial y} dz = \frac{kn}{a} \int_{z_1}^{z_2} \frac{\partial \rho}{\partial y} dz, \tag{2.192}$$

where z is in the direction parallel to the light path between the lenses or mirrors. The change in the light intensity by turning the knife-edge 90° is

$$\frac{\Delta I}{I} = \frac{kn}{a} \int_{z_1}^{z_2} \frac{\partial \rho}{\partial x} dz. \tag{2.193}$$

In both orientations of the knife-edge, $\Delta I/I$ increases as f/a increases or as the focal length, f, increases or as the aperture, a, decreases. However, overall, the light intensity decreases as the aperture decreases. It is desirable to measure $\pm\epsilon$ and to limit $-\epsilon$ since a too large $-\epsilon$ produces too large of a $-\Delta I$, reducing the projected light to where the image lacks contrast (too dark). At low light levels, "defraction noise" becomes an issue. Defraction noise is worse for a coherent light source, such as a laser.

Figure 2.31(b) is an example of an off-axis or z-type Schlieren setup. The principles are the same as previously described. Both designs, however, suffer from the fact that they integrate all of the distortions along their optical paths. This can include unwanted effects of refraction/density variations fostered by wind tunnel windows, internal side-wall boundary layers, and unintended flow features (wakes, jets, etc.). These adverse effects motivate the use of a "focused" or the so-called lens-and-grid Schlieren system that is illustrated in Figure 2.33. A comprehensive review of lens-and-grid Schlieren techniques is provided by Weinstein (1993) and Settles (2001). In addition to pure flow visualization, novel applications of the focused Schlieren technique have been demonstrated by VanDercreek et al. (1993) and Hargather et al. (2011) for deflectometry and seedless PIV measurements, respectively.

As illustrated in Figure 2.33, in a focused Schlieren system, the lens or parabolic mirror pairs used in a conventional Schlieren to collimate the light source are replaced by a Fresnel lens. The placement of the light source beyond the focal length of the Fresnel causes the light to converge past the lens. This is followed by a source grid. The source grid consists of equally spaced opaque and translucent horizontal lines that effectively divide the single light source into n discrete sources. A method to fabricate

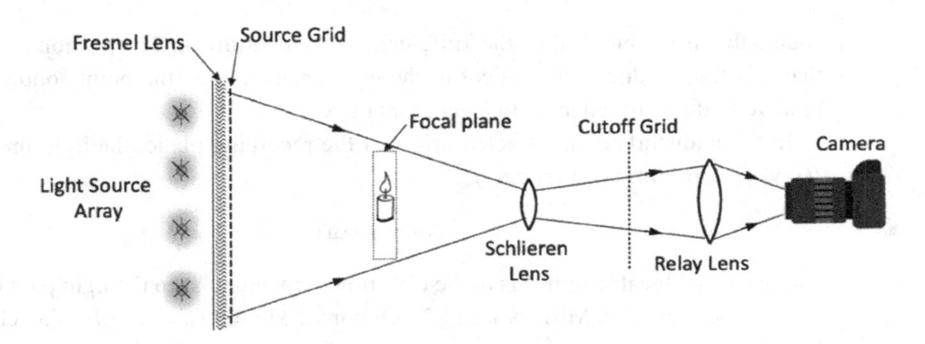

Figure 2.33 Illustration of a focused Schlieren optic setup.

the opaque lines on the source grid involves transferring black vinyl bands onto a polycarbonate sheet.

The placement of the light source beyond the focal point of the Fresnel lens forces the discretized light to converge at various angles with respect to the optical centerline. This causes the Schlieren system to focus on a plane. The disturbances at the focal plane are then reimaged beyond the focal plane. A Schlieren lens is placed near the reimaged light source followed by a cutoff grid. The cutoff grid is a photographic negative of the source grid, thereby creating a knife-edge for each of the n discrete sources.

The cutoff grid is the optical conjugate of the source grid. The method (Bowles, 2012b) to produce this is to place an unexposed film sheet at the reimaged plane of the source grid, exposing the film to the light, and developing the film to produce a perfect reproduction of the reimaged source grid. This is done without any disturbances in the focal plane.

Vertical translation of the cutoff grid blocks light from each discrete source in the same manner as individual knife-edges, limiting the light that reaches the image recording plane and thereby producing the Schlieren effect. In Figure 2.33, the image is recorded by a camera. The selection of the lenses and spacial layout requires tradeoffs in the desired sensitivity, depth-of-focus, and field-of-view. These tradeoffs are best discussed by Settles (2001).

2.4 Flow Actuators

There is a large variety of flow actuators that have been developed over the last 50 or more years. These generally fall into two categories: passive and active, where the latter requires some form of energy input to operate. With regard to flow control based on fluid instabilities, passive flow actuators generally affect the basic (time-averaged) flow. Although active flow actuators can also be used to modify the basic flow, they are generally used to impart an unsteady disturbance to the flow. In cases where the flow control objective is to reduce drag forces, a practical metric of merit is that the power savings resulting from the drag reduction exceed the input power supplied to the flow actuator. This has been a general challenge for active flow actuators.

The following sections describe the various passive and active flow control devices. This provides a preface to the chapters that follow in which these devices have been utilized in various flow fields.

2.4.1 Passive Flow Actuators

Passive flow actuators are defined as ones that require *no* external power input, or control, to operate. Ultimately being simpler to deploy, they are certainly a preferential approach in practical applications. The potential downside to passive actuators is that they are always present and can introduce a parasitic effect (drag) when the condition they are intended to correct (e.g., flow separation) is not occurring. Nevertheless, they represent an important method of flow control.

Surface Roughness

There are a wide range of passive boundary layer "tripping" techniques used to accelerate the streamwise development to a fully turbulent regime. These include distributed roughness particles, dimpled surfaces, and 2-D protrusions. A critical parameter for any form of roughness is the roughness Reynolds number, $Re_k = u^* k_{crit}/\nu$, where k_{crit} is the maximum roughness height that *does not affect* the turbulent transition location. The velocity used in the roughness Reynolds number is the boundary layer friction velocity, $u^* = (\tau_{0k}/\rho)^{1/2}$, where τ_{0k} is the wall shear stress at the location of the roughness.

For the case of a 2-D cylindrical roughness element, the minimum Reynolds number to have an effect on turbulent transition (Tani et al., 1940; Fage and Preston, 1941) is $15 \leq Re_k \leq 20$. In the case of sharp protrusions, the minimum Re_k could be lower.

Dryden (1953) developed a relation between the streamwise transition location, x_{tr}, and the height of a roughness element, k, and its streamwise position, x_k. The correlating factor was the transition Reynolds number, $Re_{tr} = U\delta^*_{tr}/\nu$, where δ^*_{tr} is the boundary layer displacement thickness at the transition location. Dryden (1953) found that

$$\frac{U\delta^*_{tr}}{\nu} = 3.0\frac{k}{\delta^*}\frac{x_k}{k}, \tag{2.194}$$

where x_k is the streamwise location of the roughness element. Kramer (1961a) found that a 2-D cylindrical roughness at an arbitrary position was fully effective in tripping the flow to turbulence if $Uk/\nu \geq 900$.

The criteria for turbulent trips consisting of distributed "sand paper" roughness are not as clear as that for 2-D roughness. For example, in the definition of the roughness height, k, it can refer to the mean height of the particles, the peak height of particles, or the RMS of the roughness height distribution. Assuming a periodic variation (waviness) of the roughness distribution, the three statistics are related. Further details as they relate to exciting instabilities in 2-D and 3-D boundary layers are presented in Chapters 6 and 7.

Finally, a variation on the classical 2-D boundary layer trip that promotes counter-rotating streamwise vortices to accelerate turbulent transition is the so-called zig-zag tape. The parameters include the tape thickness, t, streamwise width, L_2, and zig-zag angle, α. All of these can vary, but most typically $\alpha = 60°$, and t ranges from 1 to 2 mm. Larger thicknesses are obtained by stacking strips. Given the finite thickness of the strips, they can have a tripping effect that is similar to a 2-D cylindrical roughness. However, the streamwise vortices that emanate from the zig-zag pattern are likely to be the overriding factor in their performance.

Vortex Generators

One of the most commonly investigated passive devices are vortex generators (VGs). These are designed to introduce coherent streamwise-oriented circulation patterns that have been shown to be effective in suppressing flow separation by transporting high momentum fluid from the free-stream to the wall region. Vane-type VGs were first introduced in the late 1940s by Taylor (1947). These devices consisted of a row of small plates or airfoils that were mounted on the surface and set at an angle of incidence to the local mean flow direction. Although effective in suppressing flow separation, the Taylor designs produce a relatively large parasitic drag. Nonetheless, Fisher (1954) employed this design to produce attached flows and enhance wing lift.

In early applications of VGs used for flow separation control, their height was on the order of the boundary layer thickness. Kuethe (1972) developed and examined nonconventional "vane-type" VGs with heights that were 27 and 42 percent of the boundary layer thickness. Rao and Kariya (1988) suggested that passive VGs with $h = d/\delta \leq 0.625$ could be just as effective in flow separation control, but with much lower parasitic device drag. Lin (2002) subsequently provided an in-depth review of these so-called low-profile VGs. Illustrations of the different VG designs and parameters are presented in Figure 2.34. Table 2.1 provides a list of most effective VG parameter settings for separation control.

As discussed in Chapter 4, the most effective VG designs for flow separation control devices include the low-profile and conventional VGs. The low-profile VGs are also referred to as "submerged VGs" by Lin et al. (1990a) and "micro VGs" by Lin (1999).

Table 2.1 List of most effective settings for separation control for different VG designs shown in Figure 2.34.

	Most effective settings				
VG type	h/δ	x/h	z/h	Half-angle (°)	$(\Delta x)_{VG}/h$
Doublet	0.1	13	8	25	20
Wishbone	0.2	3	4	23	10
Counter-rotating rectangular vanes	0.2	4	9	25	10
Counter-rotating delta vanes	0.3	10	12	14	52
Corotating trapezoid vanes	0.2	4	4	23	12–19
Forward wedges	0.5	10	–	14	50
Backward wedges	0.5	10	–	14	50

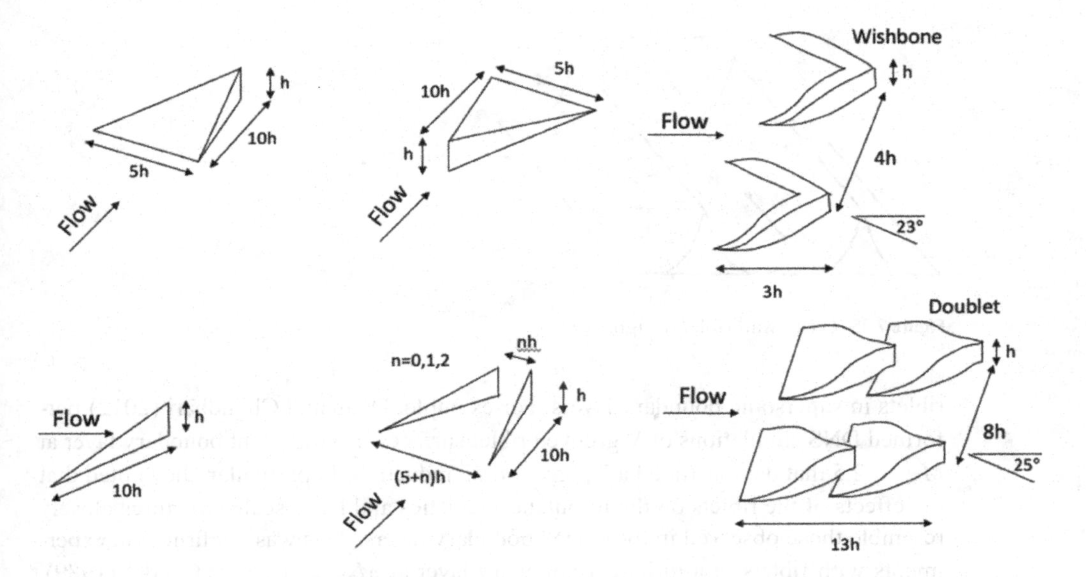

Figure 2.34 Different types of vane-type vortex generators to produce corotating and counter-rotating streamwise vortices. The most effective settings are given in Table 2.1.

Such low-profile VGs having an $h/\delta \simeq 0.2$ were found to be as effective in delaying separation as the conventional VGs with as $h/\delta \simeq 0.8$, while also having a lower device drag.

Wheeler Doublet VGs with an $h/\delta \simeq 0.36$ were found by McCormick (1992) to be effective in suppressing the separation bubble that formed downstream of a shock/boundary layer interaction covered in Chapter 9. The VGs in this application were located 20δ upstream of the shock impingement.

Riblets

Riblet-type surfaces occur in nature, particularly on sharks, and are surmised to be adaptation aimed at reducing the viscous drag (Zhang et al., 2011). Engineered riblets consist of small V-grooves applied to the wall surface. The basic design is shown in Figure 2.35. Experiments have indicated an optimum height (h) and lateral spacing (s) of the grooves for drag reduction that corresponds to $h^+ = 8 - 12$ and $s^+ = 15 - 20$. Such scaling is consistent with those of the low-speed wall streaks observed by Kline et al. (1967a), and suggests an underlying stabilizing mechanism. Relative to the average streak spacing of $100z^+$, the optimum groove height is approximately one-tenth of the streak spacing, and the optimum groove spacing is approximately one-fifth of the streak spacing. This optimum design has resulted in a maximum drag reduction of approximately 8 percent in experiments.

The underlying mechanism for the viscous drag reduction with riblets is the stabilization of the coherent streamwise vorticity associated with the wall "streak" structure. This same mechanism is expected to be valid in turbulent boundary layers at supersonic Mach numbers. The evidence comes from simulations and experiments on

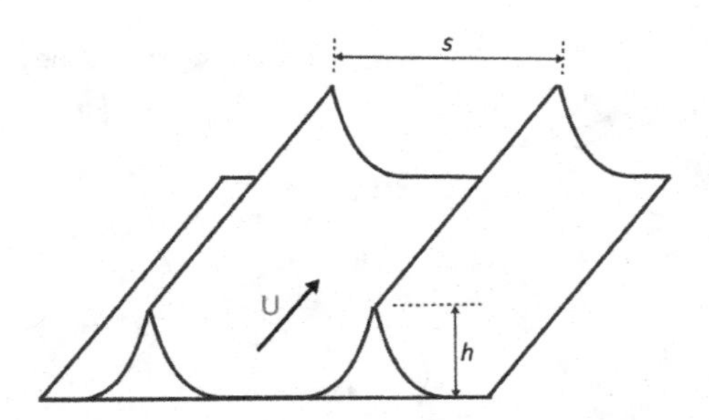

Figure 2.35 Basic wall riblet design.

riblets in supersonic boundary layers. For example, Duan and Choudhari (2012) performed DNS simulations of V-grooved riblet surfaces in a turbulent boundary layer at $M_\infty = 2.5$ that demonstrated a 7 percent drag reduction. In particular, they noted that the effects of the riblets on the turbulence statistics and large-scale structures closely resemble those observed in low-speed boundary layers. This was confirmed in experiments with riblets in a turbulent boundary layer at $M_\infty = 1.25$ by Gaudet (1989). Coustols and Cousteix (1994) extended this to higher Mach numbers of 1.6, 2.0, and 2.5. where they verified that optimum riblet height and spanwise spacing scaled with inner variables in the same way as in subsonic boundary layers.

Compliant Wall Surfaces

Somewhat like riblets, having biological origins from shark skin, the concept of compliant surfaces for drag reduction originated from observations of dolphins (Kramer, 1957, 1960, 1961b, 1962, 1965). Examples of compliant surface designs from Banerjee and Jayakumar (2005) are shown in Figure 2.36. The methodology to selecting an optimum compliant surface is largely attributed to Semenov (1991).

Choi et al. (1997) has indicated that the right combination of material properties is required with at least two requisites. The first of these is that the *dynamic* surface roughness be small enough to be below the value considered to be hydrodynamically smooth. More specifically, the amount of surface deformation of the compliant coating must be much less than the viscous sublayer thickness. The second requirement is that the natural frequency of the compliant surface must be chosen to provide the correct response to the fluctuating pressure forces at the wall. In terms of a nondimensional time, $t_0^+ = f_0^{-1} = t_0 u_\tau^2/\nu$, then $50 \leq f_0^{-1} \leq 150$.

Large-Eddy Break-up Devices

Large-eddy break-up devices or LEBUs have been used for passive flow control to reduce viscous drag in turbulent boundary layers. The theory behind their design is to inhibit large-scale wall-normal velocity fluctuations as well as to add small-scale turbulence.

Their design is an outgrowth of honeycombs used to manage free-stream turbulence (Loehrke and Nagib, 1972, 1976; Tan-atichat et al., 1982). For use in turbulent boundary layers, they consisted of 2-D thin flat plates aligned with the primary flow

Table 2.2 LEBU design from Corke (1981b).

Designation	t/δ	l/δ	h/δ
M-0	–	–	–
M-1	0.002	0.8	0.7, 0.4, 0.2, 0.1
M-2	0.002	0.8	0.6
M-3	0.002	0.8	0.6, 0.2

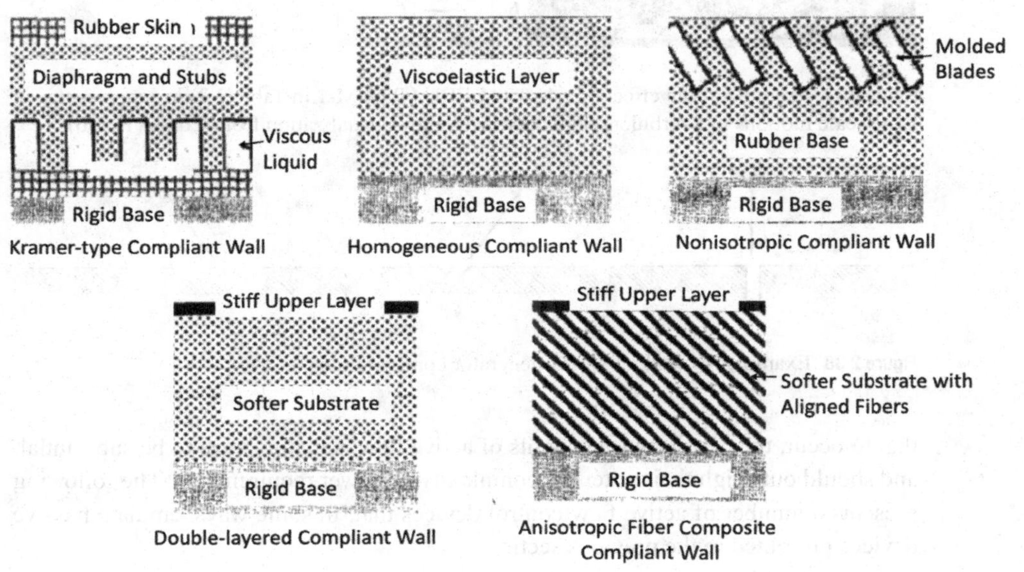

Figure 2.36 Schematic drawing of different designs of compliant walls for turbulent boundary layer viscous drag reduction. Taken from Banerjee and Jayakumar (2005).

direction that are suspended at different heights above the wall. The plates can be used singly, or vertically stacked with different spacings. They can also be placed in tandem in the streamwise direction. Table 2.2 lists the design parameters investigated by Corke (1981b). All the dimensions are normalized by the boundary layer thickness at the upstream edge of the LEBU. The plate thickness should be kept as thin as possible to minimize device drag. In the results of Corke (1981b) and Corke et al. (1982), the normalized plate thickness was $t/\delta = 0.002$. Flow visualization of turbulent boundary layers showing the effect of the four-plate Type M-1 LEBU is presented in Figure 2.37.

2.4.2 Active Flow Actuators

In contrast to passive flow control devices, active flow actuators *require* external power input, or control, to operate. In most cases, active flow control can surpass the effectiveness of passive approaches. However, being more complicated, they seldom get incorporated into practical applications. The challenge is to reverse that trend. For

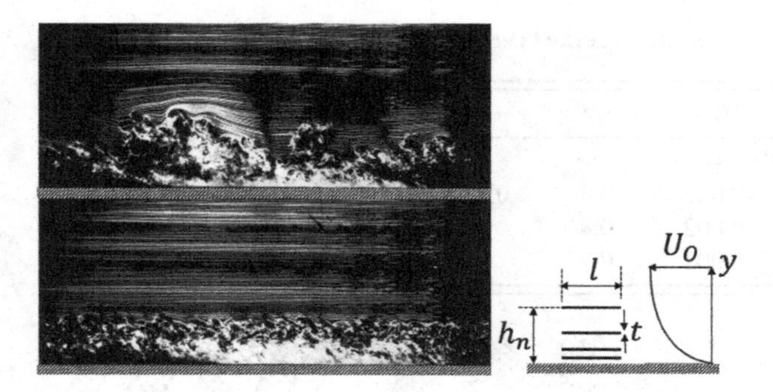

Figure 2.37 Example of the effect of a 4-plate LEBU (Type M-1 in Table 2.2) in suppressing large scale motions in a turbulent boundary layer. Flow visualization from Corke (1981b).

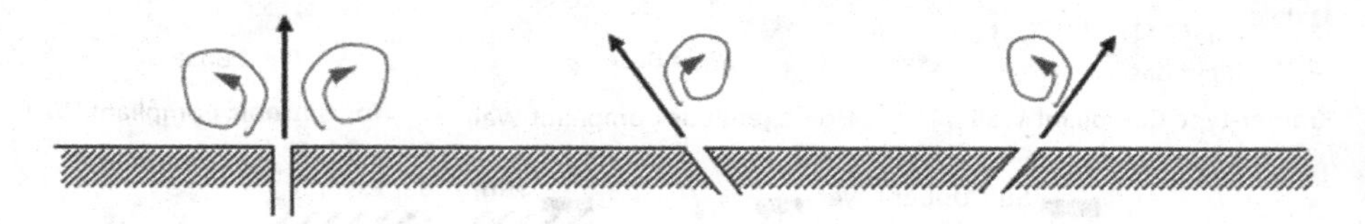

Figure 2.38 Examples of wall jet vortex generator configurations.

that to occur, the performance benefits of active flow actuators have to be substantial, and should outweigh their potential complexity or power requirements. The following presents a number of active flow control devices that, in some ways, emulate passive devices presented in the previous section.

Steady Wall Jets

Many active flow control approaches build upon the characteristics of passive VGs, namely the production of streamwise vorticity that can transport higher momentum fluid toward the wall. Examples illustrated in Figure 2.38 are *steady* wall jets oriented at different angles with respect to the mean flow direction. A vertical jet orientation can generate pairs of counter-rotating vortices. Angling the jets in either cross-stream direction can produce a single streamwise vortex of either circulation direction. Pairs of angled jets can emulate the effect of passive vane-type VGs. Johnston and Nishi (1990) found that jet VGs with velocity ratios of $V_j/U_\infty > 0.8$ were effective in flow separation control.

Synthetic Wall Jets

An unsteady embodiment of the steady wall jet is the so-called "synthetic jet" that was first demonstrated for flow control applications by Smith (1999) and Smith and Glezer (1998).

An isolated synthetic jet is produced by the interaction between a train of vortices that are formed by alternating momentary ejection and suction of fluid across an orifice such that the net mass flux is zero. A unique feature of these jets is that they are formed

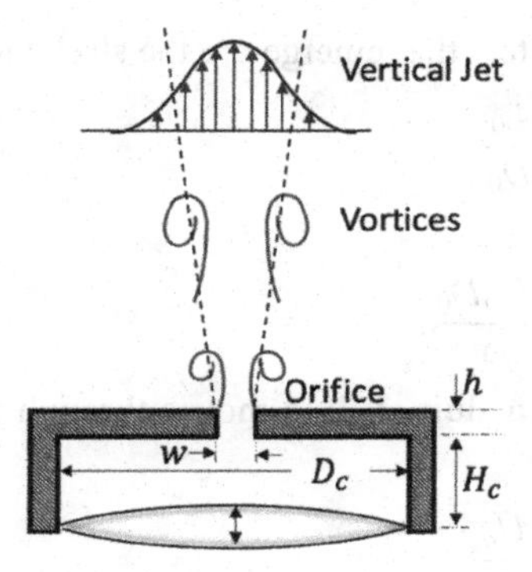

Figure 2.39 Schematic of a synthetic wall jet.

entirely from the working fluid of the flow system and therefore can transfer linear momentum to the flow system without net mass injection across the flow boundary. The approach is illustrated in Figure 2.39.

It is made up of an oscillating diaphragm that is inside a cavity with an orifice opening at one end. The orifice is most typically round in cross section, although there are examples in which the orifice is a 2-D slot. The cavity and orifice form a Helmholtz resonator. As the diaphragm oscillates, fluid is periodically entrained into and expelled from the orifice. During the expulsion portion of the cycle, a vortex ring can form near the orifice and, under certain operating conditions, convect away from the orifice to form a time-averaged jet. This behavior is defined as jet formation and was observed more than 70 years ago by Ingard and Labate (1950). The jet formation is directly related to the flux of vorticity from the synthetic jet.

In place of a moving diaphragm, Ingard and Labate (1950) used an acoustic driver to produce the periodic cavity pressure. In subsequent studies, Mednikov and Novitskii (1975) and Lebedeva (1980) documented the acoustic-driven orifice jet to reach up to 17 m/s. The term "synthetic jet" came from experiments by James et al. (1996) using a piezoelectric element-driven cavity jet that was submerged in water. They observed that a jet of fluid was formed when the excitation level of the actuator exceeded a given threshold. Utilizing bubbles that formed near the orifice as flow field markers, they deduced that the jet was synthesized by a periodic train of vortex rings. This was later confirmed by Smith and Glezer (1998).

The flow characteristics of the synthetic jet are governed by different control parameters that can be classified into three groups: the operating parameters, the geometric parameters, and the fluid parameters. According to Zhong et al. (2007), for a typical synthetic jet actuator with narrow-slot orifice, there are nine independent parameters (see Figure 2.39): the excitation amplitude, A; the excitation frequency, f_e; the slot diameter or width, w; in the case of a slot its length, l, and slot depth, h; the cavity diameter, D_C; the cavity depth, H_C; the fluid kinematic viscosity, ν; and the fluid density, ρ.

From this, the two dimensionless parameters that emerge are the stroke length

$$L = \frac{L_0}{D_0} \tag{2.195}$$

and the Reynolds number

$$\mathrm{Re}_{U_0} = \frac{U_o D_0}{\nu}. \tag{2.196}$$

In these, L_0 is defined as the distance that a slug of fluid moves through the orifice during the blowing cycle. Therefore,

$$L_0 = \frac{U_o}{T}, \tag{2.197}$$

where T is the period of the excitation cycle with frequency, f, or $T = 1/f$. U_0 is the time-averaged blowing velocity during a period of excitation, or

$$U_0 = \frac{1}{T}\int_0^{T/2} u_0(t)dt. \tag{2.198}$$

Smith and Glezer (1998) indicate that the jet stroke length, L, and the Reynolds number, Re_{U_0}, are the most important normalized parameters to characterize the synthetic jet flow. Smith and Swift (2001) argued that the jet formation depends on the stroke length, L. Based on a simple order-of-magnitude analysis, Rampunggoon (2001) showed that the ratio of the Reynolds number to the square of the Stokes number must be greater than some constant to ensure jet formation.

The Strouhal number, St, for the periodic orifice pumping is

$$\mathrm{St} = \frac{f D_0}{U_0} = \frac{D_o}{U_0 T} = \frac{D_0}{L_0} = \frac{1}{L}. \tag{2.199}$$

Another important parameter is the Stokes number, S, which is given as

$$S = \sqrt{\frac{\omega D_0^2}{\nu}} = \sqrt{\frac{\mathrm{Re}_{U_0}}{L}} = \sqrt{\mathrm{Re}_{u_0}\mathrm{St}} \tag{2.200}$$

and where $\omega = 2\pi f$.

The Stokes number determines the thickness of the Stokes layer inside the orifice and therefore the shape of the exit jet velocity prole (Zhou et al., 2009). The shape of the profile ultimately determines the circulation strength of the vortex roll-up, which is important in the jet mixing and spreading. Utilizing PIV measurements of a synthetic jet, Guo and Zhong (2006) suggested that an $S \geq 10$ was required for an appreciable roll-up to occur at an $L = 4$. Zhong et al. (2007) found that a strong roll-up occurred at the orifice with a large Stokes number of 22.

The stroke length, L, represents the distance of the fluid motion during the blowing cycle. It determines the distance from the orifice to the vortex formation, as well as the distance between successive formed vortices (Glezer, 1988). Too large of a spacing between successive vortices reduces their ability to pair, which is an effective mechanism of jet mixing that is discussed in Chapter 5. Too short of a stroke length could cause the vortex to develop too close to the orifice, with the potential of being drawn back into the orifice during the suction portion of the pumping cycle. Holman et al. (2005) suggests that $L > 5$.

The Reynolds number determines the strength of the synthetic jet vortex. However, it does not influence the vortex trajectory (Shuster and Smith, 2007). Based on the slug model, Utturkar (2002) and Holman et al. (2005) developed a condition for the formation of a synthetic jet. They suggested that the formation was governed by the self-induced velocity, V_I, of the vortex formed during the blowing cycle and the mean suction velocity, V_S, formed during the suction cycle. Based on a magnitude analysis, the deduced jet formation criterion is

$$\frac{V_I}{V_S} \sim \frac{1}{\text{St}} = \frac{\text{Re}_{U_0}}{S^2} > K, \tag{2.201}$$

where K is a constant. For a circular orifice cross section that would result in an axisymmetric jet, $K \simeq 0.16$. For a 2-D slot orifice opening, $K \simeq 2$.

With respect to flow control, two important parameters are the momentum coefficient, c_μ, and the velocity ratio, V_R, that are respectively given as

$$c_\mu = \frac{2\overline{U}_0^2 w}{U_\infty^2 D} \tag{2.202}$$

and

$$V_R = \frac{\overline{U}_0}{U_\infty}, \tag{2.203}$$

where w is the width of a 2-D slot orifice. In the case of a circular orifice, the orifice diameter would be the appropriate length scale for c_μ.

Synthetic jet velocities reaching up to 250 m/s were reported by Shaw et al. (2006) using a piezo-ceramic oscillator. However, using a piston-driven oscillator, Gilarranz et al. (2005) reported jet velocities of up to 124 m/s.

Plasma Synthetic Jet

There have been a number of plasma-driven variations on the synthetic jet. One example is shown in Figure 2.40. The concept was discussed in Chapter 6 for the control of 2-D boundary layers at compressible Mach numbers and in Chapter 9 for the control

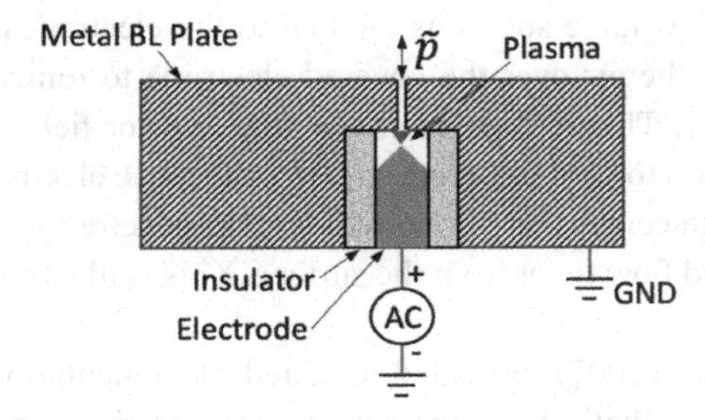

Figure 2.40 Schematic of a subsurface plasma synthetic wall jet.

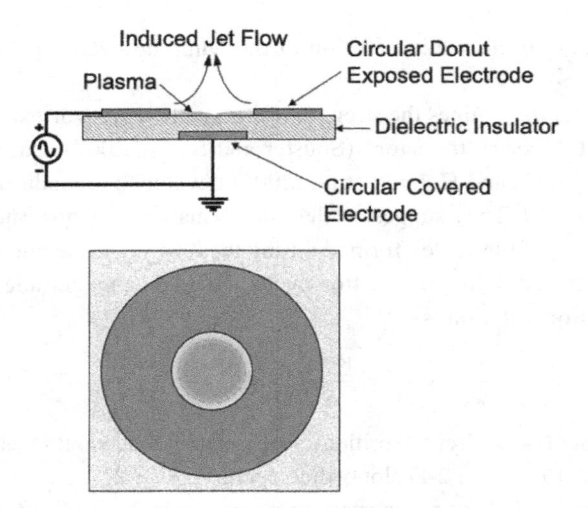

Figure 2.41 Schematic of a surface plasma synthetic wall jet.

of shock–boundary-layer interactions. A schematic of the approach is shown in Figure 2.40. As illustrated, it consists of an electrode that was located in a cavity below the surface of a metal boundary layer plate. The electrode is powered by an alternating current (AC) voltage source. The reference for the voltage source is earth ground which can be the metal plate surface over which the boundary layer develops. A sufficiently large AC voltage will cause a plasma discharge to form between the electrode tip and the inside surface of the cavity. This will produce an unsteady pressure in the cavity that communicates to the flow-side surface of the plate through a small hole. The action of the plasma discharge would replace the oscillating diaphragm in the schematic in Figure 2.39.

A surface mounted single-dielectric barrier discharge (SDBD) plasma actuator design that was motivated by synthetic jets was presented by Santhanakrishnan and Jacob (2007). An illustration of the design is shown in Figure 2.41. It consisted of an exposed annular donut electrode and a smaller diameter covered electrode, leaving an annular gap between the inner radius of the exposed electrode and the outer edge of the inner electrode. The electrodes are separated by a dielectric layer. As with all SDBD plasma actuators, an AC voltage source is applied to the electrodes. A sufficiently high AC voltage causes the air over the covered electrode to ionize. This is illustrated in blue in Figure 2.41. This results in a body force vector field that, with this geometry, acts in the direction toward the center of the concentric electrodes. This results in a flow stagnation at the center that, along with the no-penetration condition at the surface, causes the induced flow to jet from the surface. This is illustrated by the curved arrows in the figure.

Santhanakrishnan and Jacob (2007) provided detailed documentation of the induced jet. As with classical synthetic jet actuators, the plasma ring actuator was periodically pulsed. They found that this resulted in the formation of multiple

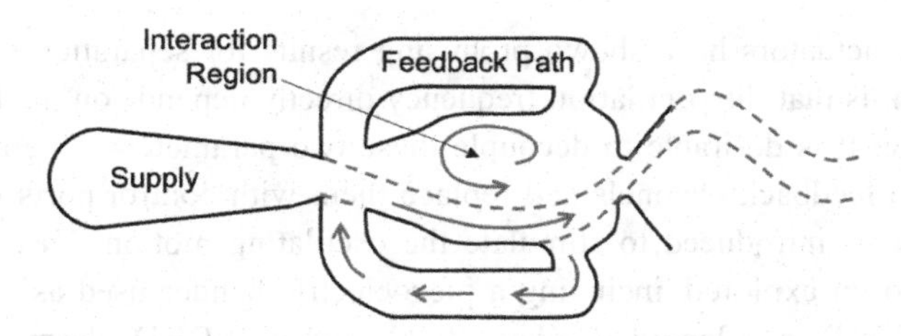

Figure 2.42 Schematic of a fluidic oscillating wall jet.

counter-rotating vortex rings that advected into the flow field in a manner that was similar to cavity-driven synthetic jets.

Oscillating Sweeping Wall Jets

Sweeping wall jets belong to a category of actuators that are based on fluidic oscillators with no moving parts. A sweeping jet actuator produces an oscillating planar sweeping jet at the exit of the device. They have shown good potential in controlling separated flows (Schmidt et al., 2017). In a comparison between synthetic jets and fluidic oscillating jets on the ability to reattach a generic separated flow, Otto et al. (2019) found the oscillating jets to be superior.

The basis for these sweeping wall jets is a bistable fluidic oscillator whose configuration is shown in Figure 2.42. In this design, as a result of the Coanda effect, the primary flow attaches itself to either side of the center channel. This causes the pressure to increase in the feedback leg on the attached side of the center channel. Eventually, the build-up in pressure in the feedback loop drives the primary flow to attach to the opposite side of the center channel. This process repeats in a cyclic fashion. The flow exiting the fluidic oscillator is a pulsed jet that sweeps from one side of the exit nozzle to the other. A similar behavior was exploited in the "blooming jets" that was presented in Chapter 5. In the arrangement shown in Figure 2.42, the flow rate of the supply jet and the length of the feedback loop are the primary parameters that establish the oscillation frequency.

These actuators exploit both fluidics (Kirshner, 1966; Morris, 1973), and fluid interactions (Raghu, 2001). A detailed analysis of the flow physics of this type of fluidic oscillator was performed by Gregory et al. (2007b). Subsequent numerical simulations (Gokoglu et al., 2009) have also endeavored to provide a better understanding of their operation with the objective of optimizing their performance.

The earliest example of fluidic oscillators or the so-called flip-flop nozzles used to produce sweeping jets is attributed to Viets (1975) for use in increasing mixing in turbo-jet ejectors. This was followed by Raman et al. (1994) who tested larger-scale flip-flop nozzles for use as general flow actuators. Such *large-scale* flip-flop nozzles have been extensively studied by Funaki et al. (1999) and Koso et al. (2002). These devices had external feedback loops with a frequency range of a few hertz (for the very large devices) to a maximum of about 300 Hz. Meso-scale devices with nozzle sizes ranging from 200 μm to 1 mm have been used to produce oscillating frequencies from 1–10 kHz (Raghu and Raman, 1999; Gregory et al., 2007b).

Although these actuators have shown promising results for separation control, a potential drawback is that the oscillation frequency directly depends on the flow rate through the device. It is desirable to decouple these two parameters. One approach is to eliminate the feedback channels and replace them with control ports in which fluid pulses could be introduced to stimulate the oscillating motion. Various other approaches have been explored, including a piezoelectric bender used as a diverter (Gregory et al., 2009), a solenoid (Culley, 2006), and an SDBD plasma actuator located in one of the control ports (Gregory et al., 2007a). These have only shown mixed results.

In most of the practical applications involving a large aerodynamic surface such as along the span of a wing or a flap, arrays of the actuators are needed. With the fluidic oscillator actuators, these can share a common plenum air supply. Arrays of such sweeping jet actuators have been used in a number of configurations to control flow separation over airfoils and flaps (Seele et al., 2009; Philips et al., 2010; Tewes et al., 2010; Woszidlo et al., 2010; Woszidlo and Wygnanski, 2011). As an example, Woszidlo et al. (2010) utilized an array of sweeping jets that were embedded in the trailing-edge flap of a symmetric section airfoil. The flap chord length was 60 percent of the airfoil chord. The flow over the large chord-length flap would naturally separate for a flap angle of 20°, reaching a peak lift coeffficient of 1.4. The sweeping jet array was able to extend the useable flap angle to 30° and subsequently increase the maximum lift coefficient to 2.0.

Single-Dielectric Barrier Discharge Plasma Actuators

In a variety of applications, SDBD plasma actuators have been used effectively for flow control purposes. These applications include lift augmentation, boundary layer separation control, jet mixing enhancement, and, most recently, turbulent boundary layer viscous drag reduction. Examples of these applications of SDBD plasma actuators are presented in various chapters in this book, as well as in a number of review articles (Corke et al., 2007a, 2009). Some of the reasons for the popularity of these actuators are their simplicity, their high frequency response, low power consumption, and that they are fully electrical with no moving parts.

The predominant SDBD configuration used for flow control consists of two electrodes, one exposed to the air and the other encapsulated by a dielectric material. The electrodes are generally arranged asymmetrically. An example configuration is shown in Figure 2.43. The SDBD plasma actuator has the electrical properties of a capacitor and is therefore powered by an AC voltage source. A large enough AC voltage causes the air over the encapsulated electrode to weakly ionize. The ionization fraction is typically less than 1 PPM. In the classic description, the ionized air is a “plasma,” which is why these are referred to as “plasma actuators.” The ionized air appears purple, which is a characteristic of the composition of the air as ionized components recombine and de-excite (Davidson and O’Neil, 1964). The emission intensity is extremely low, requiring a darkened space to view by eye. A photograph of the ionized air produced by a plasma actuator is shown in Figure 2.43.

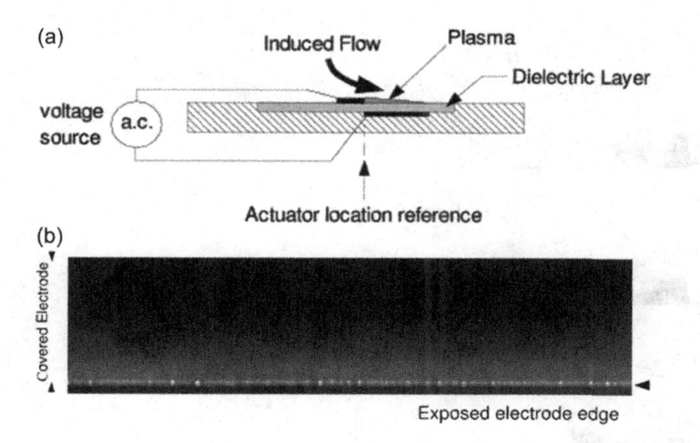

Figure 2.43 Schematic illustration of SDBD plasma actuator (a) and photograph of ionized air at 1 atm. pressure that forms over electrode covered by dielectric layer (b).

The ionized air in the presence of the electric field produced by the arrangement of the electrodes results in a body force vector field that acts on the ambient (nonionized, neutrally charged) air. The body force is the mechanism for active aerodynamic control. In determining the response of the ambient air, the body force appears as a term on the right-hand side of the fluid momentum equation.

For an SDBD, during one-half of the AC cycle, electrons leave the metal electrode and move toward the dielectric where they accumulate locally. In the reverse half of the cycle, electrons are supplied by surface discharges on the dielectric and move toward the metal electrode. The timescale of the process depends on the gas composition, excitation frequency, and other parameters. In air at atmospheric pressure, it occurs within a few tens of nanoseconds (Falkenstein and Coogan, 1997). A 1-D model for the dielectric barrier discharge dynamics was developed by Massines and coworkers (BenGadri et al., 1994, pp. 228–229; Rabehi et al., 1994, pp. 840–845; Massines et al., 1998). It was based on the numerical solution of the electron and ion continuity and momentum transfer equations coupled to Poisson's equation. As is typical in high-pressure discharges, the electrons and ions were assumed to be in equilibrium with the electric field. Their model gave space and time variations in the electric field, and the electron and ion densities. The charge accumulation on the dielectric as the discharge develops was taken into account, and the voltage boundary conditions for dielectrics were derived by considering an equivalent circuit of the gas gap in series with the equivalent capacitor of the dielectric.

Wall-mounted plasma actuators with an asymmetric electrode design like that shown in Figure 2.43 induce a velocity field similar to that of a tangential wall jet. This is illustrated in the measured velocity vector field of an impulsively started SDBD plasma actuator of the design shown in Figure 2.44. The velocity field was measured with a two-component PIV system. There was no external air stream. The PIV laser was triggered following an adjustable time delay from the impulsive start of the actuator. Multiple PIV images were then sampled at different time delays to

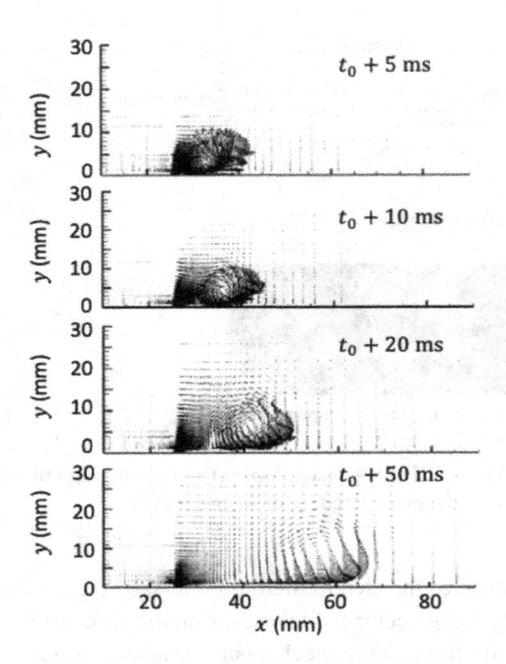

Figure 2.44 Measured velocity vector field at different time delays following an impulsively started single dielectric barrier discharge plasma actuator located on the surface and designed to produce a tangential wall jet. Taken from Post (2004).

obtain phase-averaged spatial velocity fields. The velocity field at time delays of 5, 10, 20, and 50 ms are presented in Figure 2.44. These reveal that the impulsively started plasma actuator produces a "starting vortex" with circulation in the counterclockwise direction that is consistent with a tangential wall jet. As time progressed, the vortex convected away from the plasma actuator (to the right). This same wall-mounted plasma actuator configuration has been extensively used by Thomas et al. (2008) to control the flow around circular cylinders to reduce drag and generated acoustics. It has also been used by Kelley et al. (2014) to control flow separation on the leading edge of airfoils, by Huang et al. (2006a) and Huang et al. (2006b) to control flow separation on low-pressure turbine blades, and by Lombardi et al. (2013) in closed-loop dynamic stall control on periodically pitching airfoils.

Enloe et al. (2004a) and Enloe et al. (2004b) correlated the reaction force (thrust) generated by the induced flow with the actuator AC amplitude. A similar experiment was performed by Thomas et al. (2009). The induced thrust produced by the actuator was found to be proportional to $V_{AC}^{3.5}$. Mertz and Corke (2011) developed a semiempirical model that captured the dynamic nature of the local air ionization and time-dependent body force vector distribution. The mean thrust dependence on voltage was also found to scale with the applied AC voltage as $V_{AC}^{3.5}$.

Wall-mounted plasma actuators whose long axes were aligned with the mean flow direction have been used to generate streamwise vorticity, $\overline{\omega}_x$, similar to that of passive

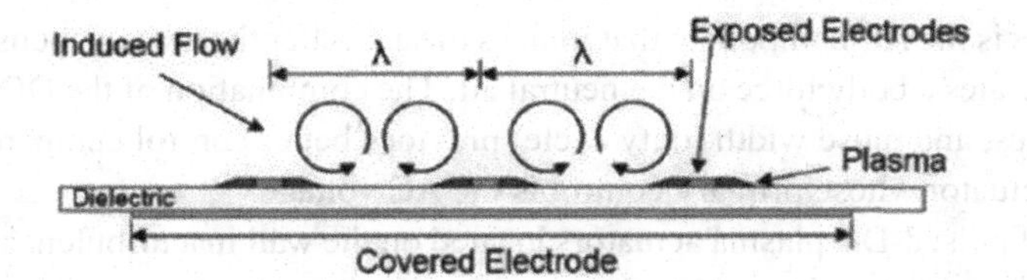

Figure 2.45 Schematic of the basic plasma vortex generator actuator geometry. Taken from Kelley et al. (2016).

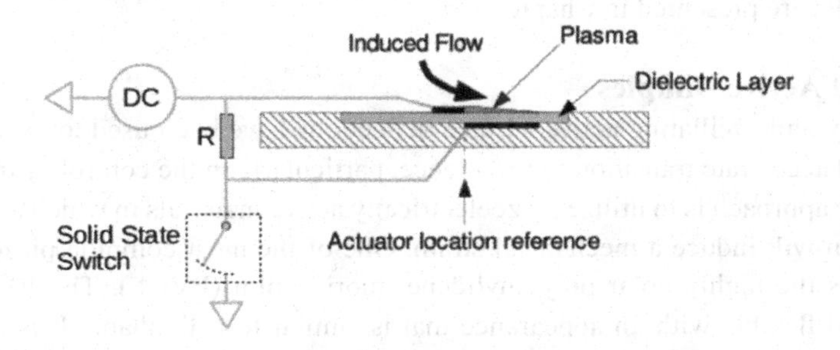

Figure 2.46 Schematic illustration of pulsed-DC SDBD plasma actuator.

vane-type VGs. The electrode configuration of these "plasma streamwise vortex generators" (PSVGs) is shown in Figure 2.45, in which the mean flow is out of the page. When energized, the plasma forms on both sides of each of the exposed electrodes. The body force vector field then gives rise to a series of opposing wall jets in the spanwise direction. These jets collide and subsequently interact with the external mean flow to form pairs of counter-rotating streamwise vortices with a size of $\mathcal{O}(\lambda/2)$. Examples of their use in flow separation control are presented by Schatzman and Thomas (2017) and Kozlov and Thomas (2011).

The details covering the design of PSVGs for the application of flow separation control are presented in Chapter 4. The design parameters are the exposed electrode spanwise spacing, λ, and the electrode streamwise length, L. The criteria for the spanwise spacing should be approximately the same as those for passive opposite-angled vane-VG pairs that produce counter-rotating streamwise vortices. Kelley et al. (2016) found the optimum to be $\Delta z = 1.2\delta$, but generally Δz should be on the order of the boundary layer thickness. With regard to the streamwise length of the plasma VG array, it is based on the timescale of the vorticity, $T_c \geq (\partial \bar{U}/\partial z)^{-1}$, such that $L \geq T_c U_\infty$.

A variation on the AC SDBD configuration that was shown in Figure 2.43 is the pulsed direct current (DC) configuration that is shown in Figure 2.46. The pulsed-DC actuator is a hybrid design that utilizes the best feature of AC, which is more efficient in ionizing the air, and that of DC, which is more efficient in generating a body force. It utilizes a DC voltage source that is supplied to both electrodes, and remains constant in time for the exposed electrode. The DC source for the covered electrode is periodically grounded by means of a solid-state switch for very short instants on the order of 10^{-5}

to 10^{-7} s. This is the AC component that ionizes the air. After the switch opens, the DC potential generates a body force on the neutral air. The combination of the DC voltage, pulse frequency, and pulse width (duty cycle) provides better control compared to the AC plasma actuator whose primary control is the AC voltage.

An array of pulsed-DC plasma actuators located on the wall in a turbulent boundary layer and designed to produce a small spanwise velocity component on the order of the mean shear velocity, u_τ, has produced dramatic reductions on the viscous drag of up to 76 percent (Duong et al., 2019; Duong, 2019). Details of its use and examples of the results are presented in Chapter 8.

Flaps and Active Dimples

Both steady and oscillating active surface deflections have been used to excite instabilities and accelerate transition to turbulence, particularly in the control of free shear layers. One approach is to utilize piezoelectrically active materials in which an electric polarization will induce a mechanical strain. One of the most common piezoelectric materials is the highly polar polyvinylidene fluoride film (PVDF). The PVDF film is clear and flexible with an appearance that is similar to cellophane. It is commercially available under the name Kynar Film. It is available in sheet form in various thicknesses. Electrode elements are formed onto the film by depositing thin layers of conducting metal. A commonly vapor-deposited metal alloy is nickel-aluminum. The metal electrodes overlay each side of the film and will cause the film to mechanically strain in response to an electric potential applied to the electrodes. The motion of the film due to the mechanical strain will depend on the manner in which it is constrained. For example, if constrained on one end, it can produce a cantilever-like deflection. If constrained on both ends, it can produce a dimple-like deflection. The low mass of these films allows them to have a fast response, with motions ranging from static to GHz oscillations. As an example, piezoelectric diaphragms are used as acoustic speakers.

The piezoelectric films exhibit many of the characteristics of a capacitor, one of which is a high input impedance. As a result, there is very little current flow through piezoelectric actuators, and, therefore, they require little power to operate.

A piezoelectric "flap" actuator has been successfully used in a variety of applications, including control of flow separation (Seifert et al., 1998), free shear flows (Wiltse and Glezer, 1993; Cattafesta et al., 2001), and flow-induced cavity oscillations (Cattafesta et al., 1997; Raman and Cain, 2002; Kegerise et al., 2007). Typical free-tip displacements range from 10,100 μm for devices with resonant frequencies in excess of 12 kHz (Kegerise et al., 2007), to approximately 1-mm displacements for designs with resonant frequencies on the order of a few hundred hertz.

An example from Corke et al.(1992a) of a 2-D array of metal electrodes vapor deposited on a PVDF film is shown in Figure 2.47. The array was designed to excite oblique instability waves into the wake of a thin airfoil. The electrodes were all individually controllable in amplitude, frequency, and phase. Each electrode pair was supplied with a periodic voltage series that caused the elements to oscillate in the wall-normal direction. The oblique waves were produced by operating the elements with a spanwise

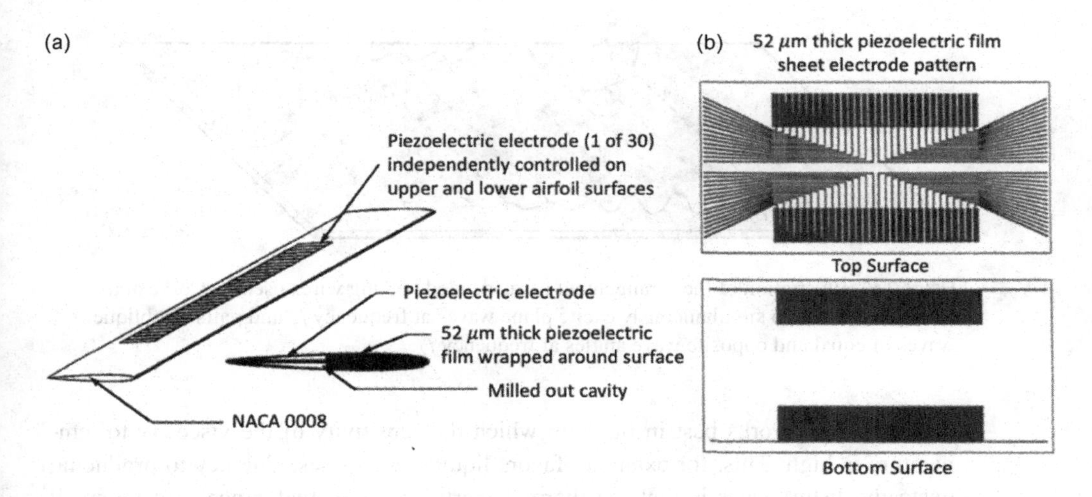

Figure 2.47 Wake-producing airfoil with piezoelectrically active sheet wrapped around (a) and schematic of metallization pattern used to excite 3-D modes (b). Taken from Corke et al. (1992a).

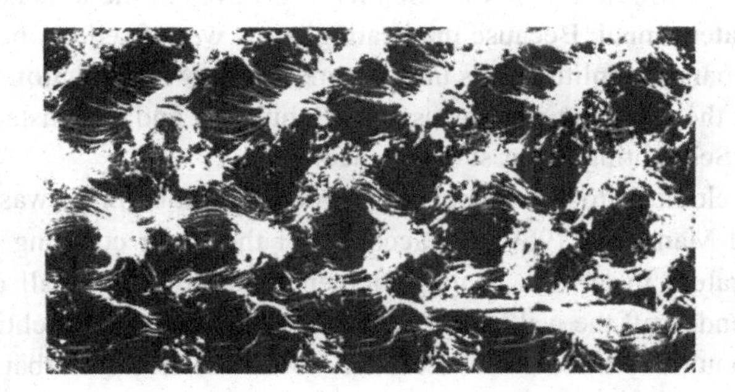

Figure 2.48 Sample phase-averaged smoke-wire flow visualization records showing oblique wave pairs in the wake of a thin airfoil produced by piezoelectrically active sheet design shown in Figure 2.47. Flow is from left to right as viewed in the (x, z)-plane. Taken from Corke et al. (1992a).

phase shift similar to the technique based on that of Corke and Mangano (1989). The spanwise dimension of the elements was chosen so as to be able to produce oblique modes with wave angles in the range from 45° to 70°.

A sample flow visualization image of the oblique wave pairs produced by the piezoelectric actuator array is shown in Figure 2.48. This is made up of multiple realizations that were phase-locked with the actuator array excitation. The sharpness of the smoke pattern is evidence that the wake pattern was correlated with the actuator array excitation.

Unsteady Heating Elements

Heating elements have often been used to introduce controlled disturbances used to excite instabilities. These primarily work by producing a local perturbation in the fluid

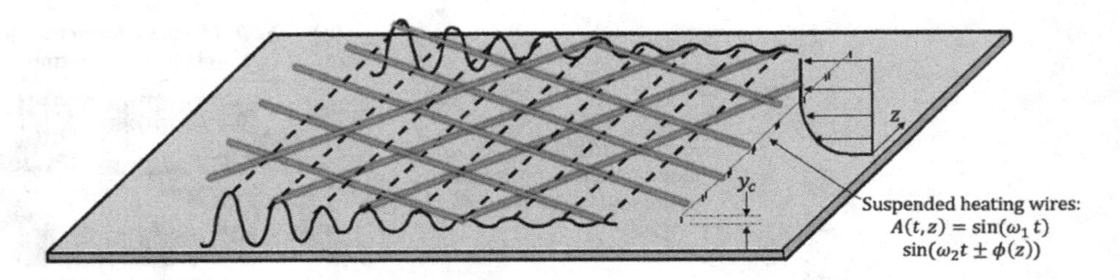

Figure 2.49 Illustration of the arrangement of suspended heating wires used by Corke and Mangano (1989) to simultaneously excite plane waves at frequency f_1 and pairs of oblique waves of equal and opposite wave angles at frequency f_2.

viscosity. This works best in fluids in which the sensitivity of the viscosity to temperature is high. This, for example, favors liquids over gases. The key to producing unsteady disturbances is that the thermal inertia of the actuator needs to be small compared to the timescale (inverse frequency) of the fluid instability of interest.

Liepmann et al. (1982) and Liepmann and Nosenchuck (1982) utilized periodically heated surface strips to introduce Tollmien–Schlichting waves in the boundary layer on a flat plate in a water tunnel. Because the heating strips were fixed on the surface of the wall, the disturbance amplitude was uniform in the spanwise direction which is an improvement over the vibrating ribbons used by Schubauer and Skramstad (1947) to validate Tollmien–Schlichting linear stability theory.

The use of heating elements to excite boundary layer instabilities *in air* was demonstrated by Corke and Mangano (1989). To account for the lower coupling between viscosity and temperature in air, the heating elements consisted of small diameter wires that were suspended off the wall at the height of the Tollmien–Schlichting critical layer, where the boundary layer is most sensitive to disturbance perturbations. An illustration of the arrangement of the suspended heating wires on a flat plate is shown in Figure 2.49.

The experiment was designed to simultaneously excite 2-D and 3-D Tollmien–Schlichting waves. The segments were individually supplied with a time-periodic current of specific amplitude, frequency, and phase, or specifically $A(t,z) = \sin(\omega_1 t) + \sin(\omega_2 \pm \phi(z))$, where z is the spanwise location of a heating wire. In the illustration in Figure 2.49, $\omega_1 = 2\omega_2$, which was designed to set up a subharmonic resonance between plane Tollmien–Schlichting waves at the fundamental frequency and pairs of oblique Tollmien–Schlichting waves at the subharmonic frequency causing the nonlinear growth of the subharmonic oblique wave pairs described in Chapter 6.

2.5 Summary

This chapter provided a background on the types of flow sensors and actuators that are frequently used in establishing the instability of various flow fields, and in the documenting of conditions before and after flow control. The emphasis was to provide a basic overview of these sensors and actuators. Of particular interest was highlighting the conditions where they are best applied. The sensors could be categorized into intrusive, such as probes, and nonintrusive, such as optical approaches. Many of the intrusive sensors could have a frequency response that could capture higher frequency instabilities, including at hypersonic Mach numbers. The frequency bandwidth of the optical techniques were generally more limited, although these continue to improve. Flow visualization has an important role in diagnostics and in guiding point measurements. Thankfully, there are numerous review articles on flow visualization, some of which were cited in this chapter, that are a great resource. Most of these are focused on low-speed flows. With the growing interest in high Mach number flows, this chapter provided more details to the very applicable Schlieren visualization. In that, the benefits of the focused or lens-and-grid Schlieren system were emphasized.

The flow control approaches presented in this chapter are demonstrated in the subsequent chapters that cover the different fundamental flow fields. These approaches have generally been categorized as passive or active. One of the passive techniques is 2-D and 3-D surface roughness that, depending on the placement and height, could trigger an instability or immediately cause turbulence onset. The latter case is shown to be useful in controlling a flow separation. Streamwise VGs are another passive technique that is particularly effective in controlling flow separation. Those discussed in this chapter included physical vane-type devices and less intrusive approaches such as steady wall blowing and plasma actuators.

Some of the passive flow control techniques are specifically designed for turbulent boundary layers. These include riblets and LEBUs, where the latter are derived from devices used to control free-stream turbulence. Both are based on controlling coherent boundary layer vortical structures that are each associated with turbulence production and viscous drag.

With regard to flow field instability control, the passive flow control devices mainly seek to modify the mean flow (basic state). Some of the active flow control devices also have the same effect. An example is the oscillating wall jet that has been shown to be very effective in controlling separated flows. Other active flow control devices produce periodic disturbances that can excite a flow field instability. Examples of these are synthetic jets, plasma jets, unsteady plasma actuators, and oscillating flaps and bumps. Examples of their effectiveness depends on their ability to provide controlled disturbances that can interact with the relevant instability modes. Examples of where this is successful are presented in subsequent chapters.

Problems

2.1 Considering a constant current anemometer (CCA),

1. Derive the static sensitivity in terms of the overheat parameter.
2. What is the static sensitivity at the velocity limits 0 and ∞? What does this say about the operation range?
3. How can the static sensitivity be increased?

2.2 Considering a constant current anemometer (CCA),

1. Derive the first-order time constant given in Eq. (2.44).
2. How can the dynamic response be increased?
3. Can the static and dynamic responses be *simultaneously* maximized?

2.3 Because a hot-wire sensor does not directly measure velocity-only heat transfer, it is sensitive to changes in the ambient temperature.

1. Determine the error in the hot-wire anemometer output voltage as a function of its overheat ratio in the range $1.3 \leq r \leq 1.8$ if the static temperature varies by $10\,°C$. Assume the wire sensor is Tungsten.

2.4 Because of its easy ability to vary the overheat ratio of a constant current anemometer (CCA), it is advantageous for use in mass-flux measurements in supersonic flows. However, the frequency response of the CCA is at best 2 kHz where the frequencies of interest range up to 100 kHz.

1. Design a frequency compensation circuit that will increase the frequency response of the CCA (a first-order system) to 100 kHz. Plot the uncompensated CCA response, the response of the compensation circuit, and the compensated CCA response.
2. The compensated CCA output will be acquired digitally. As a result, to avoid aliasing, the output will need to be low-pass filtered at 100 kHz. Therefore, using the same relation used in the design of the frequency compensation circuit, design a low-pass filter that will pass frequencies up to 100 kHz and then reduce the amplitude at a rate of 20 dB per decade of frequency.

2.5 An LDV is proposed to be used to document the basic flow (mean velocity profile) of a laminar boundary layer to compare it to a Blasius profile.

1. What are the issues regarding signal-to-noise and bias error of the measurements that may come up?
2. How can these issues be resolved?

2.6 Given the parameters that govern a synthetic jet actuator,

1. How might the actuator frequency and jet amplitude be simultaneously increased?
2. How would this affect the Stokes number which is attributed to causing the jet vortex roll-up to occur?
3. If multiple synthetic jet actuators are placed side-by-side, what characteristics might be used to determine their spacing?

2.7 Fluidic oscillating wall jets have been quite successful in maintaining attached boundary layer flows. Separated flows are sensitive to unsteady disturbances at frequencies that scale with the length of the separated region and free-stream speed.

1. In the design of a fluidic oscillator, list the parameters that affect the jet oscillation frequency.
2. How might the frequency be varied to match the optimum for separation control under varying conditions?
3. Could fluidic oscillators also be used as a mass flow meter?

3 Bluff Body Wakes

In aerodynamics, a bluff body can be generally categorized as one in which its length in the flow direction is approximately the same as its height perpendicular to the flow in a two-dimensional (2-D) representation. Such shapes exhibit a wide wake on the scale of the body (2-D) height, and aerodynamic drag that is dominated by the low pressure region that forms in the near wake of the body.

Bluff body wakes are complex, involving boundary layer separation and multiple shear layer interactions. Our understanding of bluff body aerodynamics was once thought of as "entirely empirical" (Roshko, 1993). However, there is now considerable new understanding that has come from decades of innovative experiments, Navier–Stokes simulations, and, possibly most importantly, new insights from hydrodynamic stability theory.

The understanding and control of bluff body aerodynamics has practical importance since it poses technical challenges in numerous applications. Examples include airfoils at high angles of attack, aircraft landing gear (Wicks et al., 2015), ground vehicles (Bearman, 1980), buildings and structures (Cermak, 1976), as well as electric power lines and towed underwater cables. The alternate shedding of vortices from bluff bodies leads to large fluctuating pressures that, at minimum, result in acoustic noise and, at maximum, can resonantly couple with structural vibration modes to produce exponentially growing vibration amplitudes and structural failure.

3.1 Bluff Body Wake Instability

The most widely studied bluff body shape is a large aspect ratio (2-D) circular cylinder. The flow around a circular cylinder can be considered as a prototype of bluff body wakes because of the simplicity of the boundary conditions and the complexity of the physical processes involved. One of the early comprehensive reviews of the flow around circular cylinders was provided by Morkovin (1964). Figure 3.1 provides an illustration of the *steady* viscous flow around a circular cylinder. A boundary layer over the cylinder surface originates at the upstream (windward) stagnation point. The boundary layer eventually separates as it approaches the adverse pressure gradient that exists near the downstream (leeward) side of the cylinder. The separation location will depend on the Reynolds number and surface roughness that determine the state,

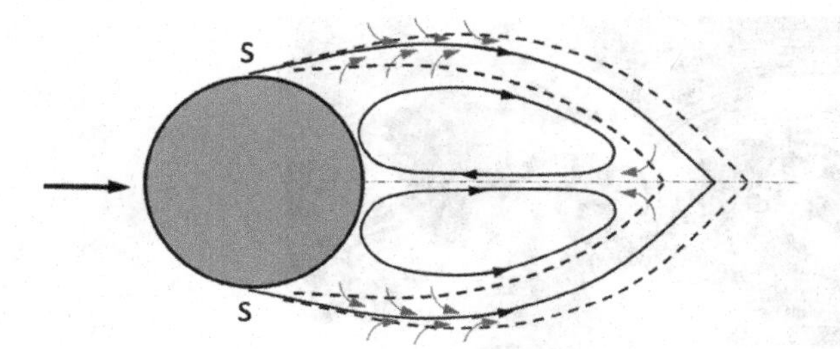

Figure 3.1 Illustration of the mean flow in the wake of a 2-D circular cylinder that notes the origin (S) of the separated shear layers on either side of the cylinder, and the pair of flow recirculation cells.

Figure 3.2 Flow visualization image of the laminar vortex street of a 2-D circular cylinder at $Re_D = 150$. Taken from (Nagib et al., 1978).

laminar or turbulent, of the boundary layer. A low-pressure region is formed between the pair of separated shear layers. This low-pressure region includes a symmetric pair of flow recirculating cells. As will be discussed, the flow recirculation that occurs in the near wake of bluff bodies has implications on the global stability of the wake flow.

Unsteady vortex shedding is observed to occur above a Reynolds number, $Re_D = U_\infty D/\nu \simeq 50$, where D is the cylinder diameter. Figure 3.2 shows a flow visualization image of the wake of a 2-D circular cylinder for $Re_D = 150$ that captures the laminar vortex street. Figure 3.3 shows flow visualization images of the turbulent wake of a 2-D circular for $Re_D = 10{,}000$. In both of these cases, the flow field was visualized by introducing a 2-D sheet of smoke streaklines using a "smoke-wire" (Corke et al., 1977).

In Figure 3.3, the two images correspond to the same phase angle of the periodic vortex shedding. This was accomplished by placing a hot-wire sensor near one of the separated shear layers where it could sense the near-sinusoidal velocity fluctuations that were correlated with the vortex shedding. The hot-wire probe body is visible in the bottom image above the cylinder. The differences in the two images is a result of the smoke-wire placement. In the top image in Figure 3.3, the smoke-wire was located upstream of the cylinder. In that case, the smoke streaklines were unable to penetrate the separated shear layers in order to visualize the flow directly downstream of the cylinder. In order to visualize that region, the smoke-wire was located downstream of the cylinder. With that placement of the smoke-wire, the smoke particles were captured

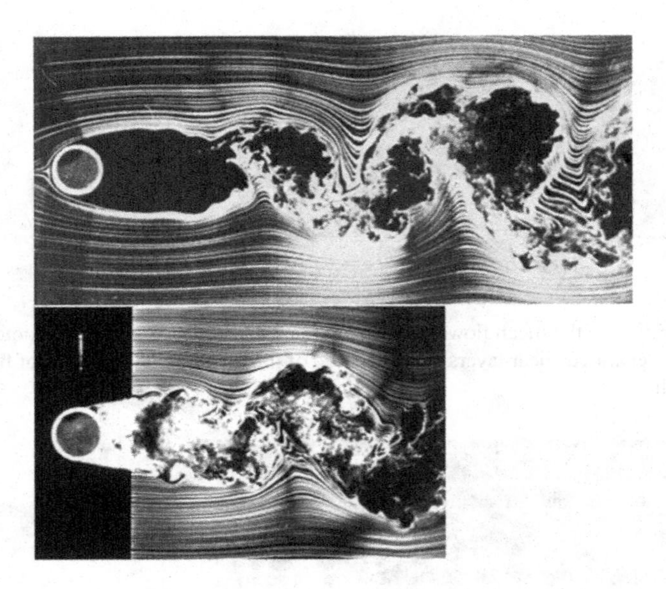

Figure 3.3 Flow visualization images of the turbulent vortex street of a 2-D circular cylinder at $Re_D = 10{,}000$. Taken from (Nagib et al., 1978).

in the flow recirculation cells depicted in Figure 3.1, and transported upstream to the cylinder surface, completely filling the near-wake region. That downstream placement of the smoke-wire also provided better visualization of the waves in the separated shear layer that result from the shear layer instability. The wavelength of the shear layer instability is much shorter than that of the vortex shedding that develops 3–4 cylinder diameters downstream.

Gerrard (1966a) and Perry and Chong (1982) provided some descriptive observations of near-wake vortex formation and shedding. Gerrard (1966a) observed that the formation of a vortex on one side of a cylinder would draw through induction the opposite-side shear layer to move across the wake centerline. This eventually disrupts the growth of the vortex, which would subsequently be shed. The process would then alternately repeat on the opposite side of the cylinder in a quasiperiodic process.

von Kármán (1912) analyzed the stability of vortex street configurations and established a link between the vortex street structure and the aerodynamic drag. He considered the alternate generation of vortices as a natural phenomenon upon which he investigated the linear stability of different point vortex configurations. He determined that two rows of opposite-signed vortices were unstable in both symmetric and antisymmetric arrangements. The exception was an antisymmetric arrangement of vortices in which the ratio of the inter-vortex spacing, b, to the distance between vortex rows, a, was $b/a = 0.28056$. While providing a basis for the commonly observed spacing of the alternate signed shed vortices in the wake of a circular cylinder, it did not address the origin of the vortex shedding. However, based on this pioneering work, the coherent vortex structure in the wake of bluff bodies is referred to as the von Kármán vortex street.

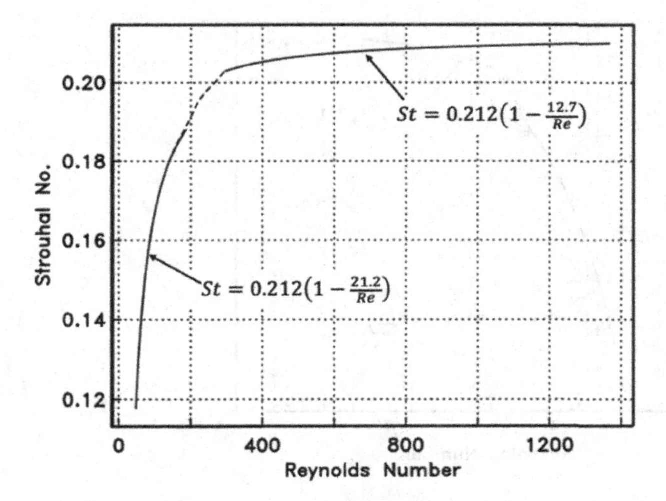

Figure 3.4 Strouhal number dependence on the Reynolds number based on diameter for a circular cylinder. Curves are based on data from Roshko (1954).

The relation between the vortex shedding frequency to the cylinder diameter and external flow conditions is attributed to the classic early experiments of Strouhal (1878). Rayleigh (1915) contributed to the analysis of that data and in the definition of a Strouhal number, $\mathrm{St} = fD/U_\infty$, where f is the physical vortex shedding frequency. Figures 3.4 and 3.5 show the Strouhal number as a function of Reynolds number for a circular cylinder in measurements from Roshko (1954). The solid curves in the two figures are curve fits provided by Roshko (1954). These were in regions of relatively small scatter in the data.

The Re_D range between 150 and 300 exhibits a relatively large amount of scatter, and, therefore, in Figures 3.4 and 3.5, the trend is shown by the short-dashed curve. Figure 3.5 includes data at higher Reynolds numbers, up to approximately $\mathrm{Re}_D = 10{,}000$ that corresponds to the flow visualization images in Figure 3.3. The long-dashed curve signifies the region of Reynolds numbers that extends beyond $\mathrm{Re}_D = 1{,}400$, which was the full extent of the Roshko (1954) curve-fit relation.

Some of the data used in deriving the Strouhal number relation with Reynolds number came from experiments by Kovasznay (1949) utilizing at that time newly developed hot-wire anemometry. He observed that the vortex street only developed at Reynolds numbers above 40, and that it was stable and regular at Reynolds numbers below 160. Kovasznay (1949) went on to characterize three Reynolds number ranges:

Stable range	$40 < \mathrm{Re}_D < 150$
Transition range	$150 < \mathrm{Re}_D < 300$
Irregular range	$300 < \mathrm{Re}_D < 10{,}000+$

It is important to note that wake shedding has been observed well above $\mathrm{Re}_D = 10{,}000$, to Reynolds numbers at least 20 times higher. Over a range of cylinder

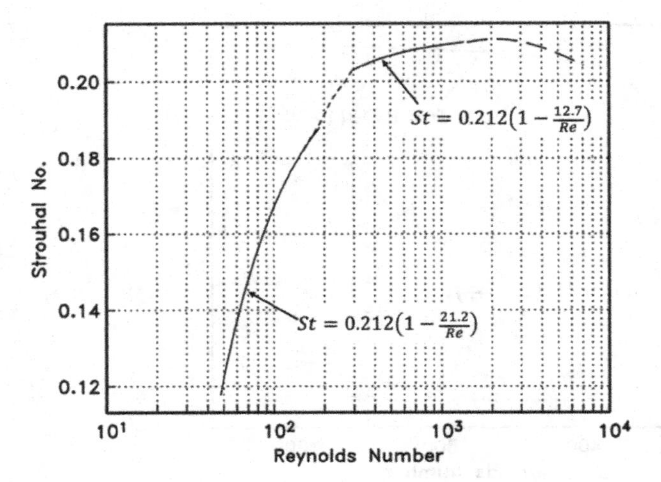

Figure 3.5 Strouhal number dependence on Reynolds number based on the diameter for a circular cylinder. Curves are based on data from Roshko (1954).

Reynolds numbers where the separated shear layers are laminar, the Strouhal number is relatively constant with a value of 0.21. This is referred to as the "subcritical" Reynolds number range.

Morkovin (1964) noticed that the Reynolds number dependence of the *inverse* of the Strouhal number closely followed the dependence of the cylinder drag coefficient on the Reynolds number. This is demonstrated in Figure 3.6. In this format, it appears that $C_D \propto \mathrm{St}^{-1}$, which suggests a link between the vortex shedding and the aerodynamic drag. From a flow control aspect with the intention of reducing aerodynamic drag, this is an important observation.

The drag coefficient corresponds to the total drag that includes the viscous drag due to the boundary layers over the surface, and the form or pressure drag that is produced by the cylinder (or bluff body) wake. A general characteristic of bluff bodies is that the pressure drag is much larger than the viscous drag. A representative composite of C_D values from experiments over a large range of Reynolds numbers is shown in Figure 3.7. The illustrations in the figure denote three regions of the cylinder wake development. In the first region with $40 \leq \mathrm{Re}_D \leq 2000$, the drag coefficient decreases with increasing Reynolds number. This region is denoted by the regular shedding of vortices. Near the onset of vortex shedding at $\mathrm{Re}_D \simeq 100$, the boundary layer separation point is at approximately $\beta = 130°$, where $\beta = 0°$ corresponds to the stagnation point on the cylinder. As the Reynolds number increases to approximately 1000, the angle where the boundary layer separates decreases to about $\beta = 85°$, or just forward of the cylinder apex. The exact separation location and dependence on the Reynolds number can depend on the surface roughness and free-stream disturbances. The boundary layer separation location remains fairly constant over the range $2000 \leq \mathrm{Re}_D \leq 200{,}000$, and, as a result, over this range, the drag coefficient only increases slowly. This is the subcritical Reynolds number range where the Strouhal number is relatively constant.

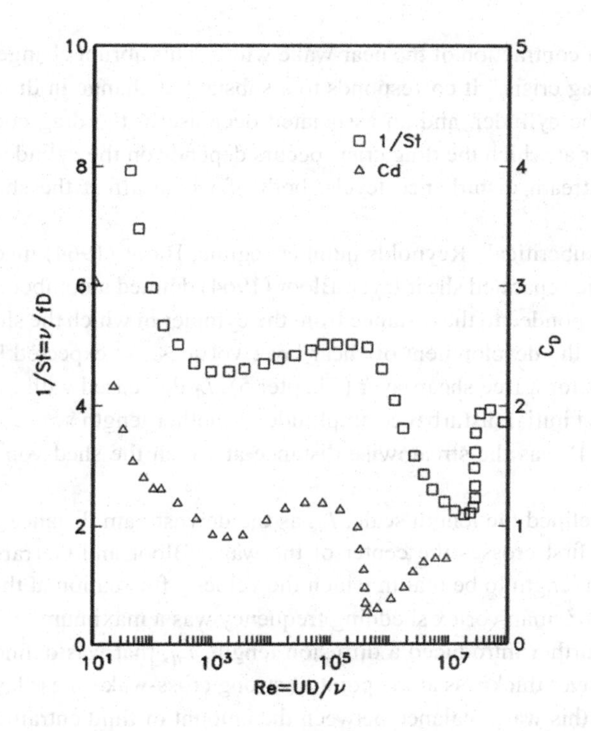

Figure 3.6 Summary of Reynolds dependence of the inverse Strouhal number and drag coefficient for a circular cylinder.

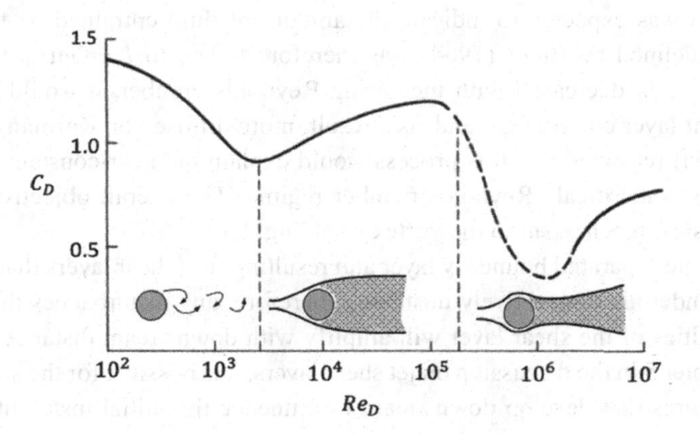

Figure 3.7 Drag coefficient dependence on Reynolds number based on the diameter for a circular cylinder.

An abrupt change in the drag coefficient occurs at $Re_D > 200{,}000$ where turbulent onset occurs for the boundary layers over the cylinder. Since a turbulent boundary layer is more resistant to flow separation than a laminar boundary layer, the turbulent boundary layer remains attached to the sphere surface for a longer distance around the sphere. In this case, the separation location moves to approximately $\beta = 120°$ and, as illustrated, results in a contraction of the near-wake width. This abrupt change is often referred to as the "drag crisis." It corresponds to a substantial change in the pressure distribution around the cylinder, and an associated decrease in the drag coefficient. The Reynolds number at which the drag crisis occurs depends on the cylinder surface roughness and free-stream disturbance levels, both of which affect the shear layer turbulence onset.

Focusing on the "subcritical" Reynolds number regime, Bloor (1964) investigated turbulence onset of the separated shear layer. Bloor (1964) defined a number of length scales. One, L_t, corresponded to the distance from the cylinder in which the shear layer instability resulted in the development of shear layer vortices. As expected based on linear stability theory for a free shear layer (Chapter 5), L_t decreased with increasing Reynolds number, and initial disturbance amplitudes. Another length scale, L_f, introduced by Bloor (1964) was the streamwise distance at which the shed von Kármán vortices first roll up.

Gerrard (1966a) defined the length scale, L_t, as the downstream distance at which the nonvortical fluid first crosses the center of the wake. Bloor and Gerrard (1966) defined the formation length to be that in which the velocity fluctuation at the second harmonic of the von Kármán vortex shedding frequency was a maximum.

Gerrard (1966a) further introduced a diffusion length, L_d, that was defined as the instantaneous shear layer thickness at the point of strong cross-wake shear layer interaction. The basis for this was a balance between the amount of fluid entrained by the free shear layer vortices, and the amount of fluid returned to the formation region that occurred in the process of shedding a Kármán vortex. Thus, the relative thickness of the shear layer was expected to indicate the amount of fluid entrained by the shear layer. The L_t defined by Bloor (1964) was therefore linked to L_d. Gerrard (1966a) surmised that as L_t decreased with increasing Reynolds number, it would result in increased shear layer entrainment and, as a result, more diffuse von Kármán vortices. Gerrard (1966a) reasoned that this process would explain the near-constant Strouhal number in the "subcritical" Reynolds number regime. The overall objective was to provide a physical mechanism to the vortex shedding.

In general, the separated boundary layer and resulting free shear layers that develop off of the cylinder are convectively unstable. Therefore, any disturbances that would excite instabilities of the shear layer will amplify with downstream distance. As presented in Chapter 5 in the discussion on jet shear layers, it is possible for the shear layer vortical structures that develop downstream to influence the initial instability development through acoustic pressure feedback. In a perfect scenario, this can lead to the temporal growth of disturbances, which produces exponentially growing limit-cycle oscillations of discrete frequencies.

In the case of bluff body wakes, the landmark transient experiments of Mathis et al. (1984) and Strykowski (1986) indicate that von Kármán vortex shedding at low Reynolds numbers results from a global instability. As a result, disturbances grow in space and time rather than in space alone. This difference is illustrated in Figure 3.8.

Further evidence of a cylinder wake global instability came from Provansal et al. (1987), Sreenivasan et al. (1987), and Sreenivasan et al. (1989). It was experimentally

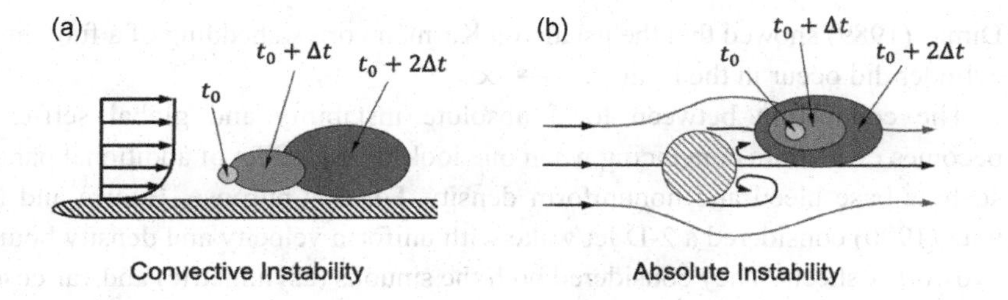

Figure 3.8 Schematic contrasting disturbance growth due to a convective instability that occurs in boundary layers (a) and an absolute instability that occurs in the near-wake region of circular cylinders and other bluff body flows (b).

validated by impulsively increasing the Reynolds number (by a step change in velocity) from just below the critical value ($\text{Re}_D = 46$) to just above where von Kármán vortex shedding initiated ($\text{Re}_D \simeq 49$). The resulting temporal growth of self-sustained oscillations in the linear regime and the final nonlinear amplitude saturation were consistent with a supercritical Hopf bifurcation to a global mode. The experiments also showed that at near-critical conditions, the temporal evolution of the characteristic global-mode amplitude was accurately described by a Landau equation

$$\frac{dB}{dt} = c_1(\text{Re} - \text{Re}_{G_c})B - c_2|B|^2 B, \tag{3.1}$$

where $c_1(\text{Re} - \text{Re}_{G_c})$ is the linear temporal growth rate, and $c_2 > 0$ is for a subcritical bifurcation. The subscript c in Re_{G_c} refers to the minimum critical Reynolds number condition. The saturation amplitude, $|B|_{\text{sat}}$, is obtained by setting $d|B|/dt = 0$ or

$$|B|_{\text{sat}} \propto (\text{Re} - \text{Re}_{G_c})^{1/2} \quad \text{for } \text{Re} \geq \text{Re}_{G_c}. \tag{3.2}$$

The experimental results of Strykowski (1986) clearly documented the exponential growth (in time) of a linear global mode. The concept of a bifurcation to a global mode has been fully confirmed by Zabib (1987) and Jackson (1987) based on global linear stability calculations in which the temporal stability characteristics were determined on the computed nonparallel basic flows. Monkewitz (1988) then firmly tied the local stability properties to the observed global behavior by showing that the sequence of transitions in the cylinder wake as the Reynolds number is raised follows the sequence in the model problem of Chomaz et al. (1987, 1988), namely a transition from stable to convectively unstable, then from convectively unstable to locally absolutely unstable, and then finally a bifurcation to a self-sustained global mode only after a sufficiently large portion of the flow has become absolutely unstable.

Huerre and Monkewitz (1990) point out that one may think that absolute instability is closely related to the existence of reverse mean flow to "carry information upstream." They point out that such an intuitive concept can be misleading. As an example, they cite the "floating wake" simulation by Triantafyllou and Dimas (1989) in which a circular cylinder is immersed half-way in a uniformly flowing stream in which they demonstrate that at a low Froude number, $F = U_\infty(gD/2)^{-1/2}$, the wake is convectively unstable despite the reverse mean flow. However, Triantafyllou and

Dimas (1989) showed that the usual von Kármán vortex shedding of a fully immersed cylinder did occur in the limit as $F \to \infty$.

The connection between local absolute instability and global self-excitation becomes even more convincing when one looks at the effect of additional parameters, such as base bleed and nonuniform density. For this purpose, Huerre and Monkewitz (1990) considered a 2-D jet/wake with uniform velocity and density bounded by two vortex sheets. They considered both the sinuous (asymmetric) and varicose (symmetric) shedding modes for variable velocity and density ratios, which respectively are

$$\Lambda = (U_{\text{wake/jet}} - U_{\infty})/(U_{\text{wake/jet}} + U_{\infty}) \tag{3.3}$$

and

$$S = \rho_{\text{wake/jet}}/\rho_{\infty}. \tag{3.4}$$

They focused on the boundaries between convective and absolute instabilities in (S, Λ) parameter space. This led to the observation that reverse flow, namely $\Lambda > 1$, promotes an absolute instability. In contrast, co-flow for the jet, or base bleed for the wake, promotes a convective instability. The analysis also determined that for the wake flow, the sinuous or von Kármán mode becomes absolutely unstable first, whereas in the jet, the varicose or symmetric mode is absolutely unstable first.

The Huerre and Monkewitz (1990) analysis found that in the wake, lowering the fluid density, for example in air by adding heating, suppressed the absolute instability. This had the opposite effect in a jet.

3.2 Bluff Body Passive Flow Control

The existence of and knowledge about an absolute instability of bluff body wakes offers the possibility of effective wake control, where the suppression of the absolutely unstable region offers a particularly effective method. The following sections explore such flow control approaches.

3.2.1 Splitter Plates

One common method in aerodynamics to suppress the periodically oscillating wake flow is to place a splitter plate in the wake of the cylinder, and so to suppress the absolutely unstable region. The seminal work by Roshko (1954) included cylinders with rigid splitter plates. As illustrated in Figure 3.9(a) and (b), these included splitter plates that were attached to the trailing portion of the cylinder, as well as splitter plates that were displaced from the cylinder in the downstream direction. Figure 3.10 shows the pressure coefficient measured at different distances on the wake centerline with and without a splitter plate whose length was $5D$. Roshko (1954) reported that the periodic vortex formation was completely inhibited by the splitter plate. As evident in the figure, the base pressure increased considerably above that without the splitter plate, which reflects a decrease in the drag coefficient.

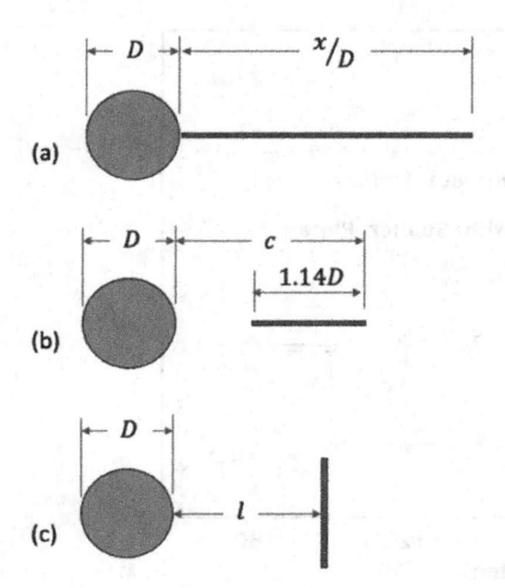

Figure 3.9 Illustration of two types of splitter plate configurations (a and b) investigated by Roshko (1954), and configuration (c) used by Gerrard (1966a).

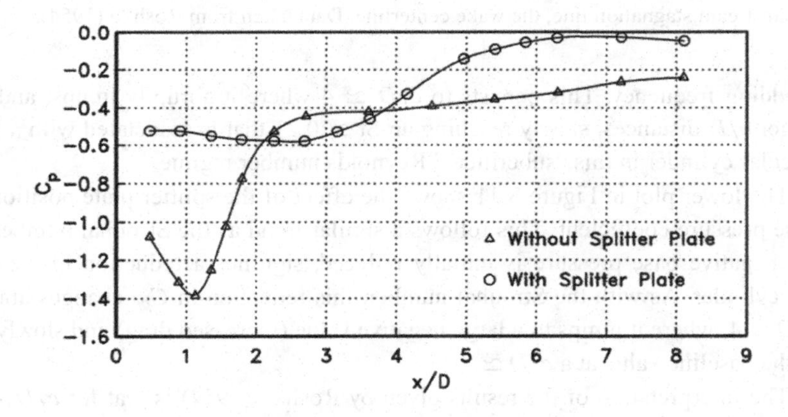

Figure 3.10 Pressure coefficient measured on the wake centerline at $Re_D = 14{,}500$ with and without a splitter plate with a length of $5D$ (see illustration in Figure 3.9(a)). Data taken from Roshko (1954).

As shown in Figure 3.11, the addition of the splitter plate changed the pressure distribution everywhere over the cylinder surface. Roshko (1954) noted that this result indicates that the suppression of the wake shedding produced by the splitter plate had the added effect of changing the outer potential flow as well as that in the wake.

Roshko (1954) also investigated the effect of shorter splitter plates, particularly one whose length was $1.14D$. The arrangement with the shorter splitter plate is shown in the bottom illustration in Figure 3.9. The effect of the splitter plate placement, c/D, on the vortex shedding Strouhal number and the base pressure coefficient, C_{ps}, are shown in Figure 3.12. Focusing on the Strouhal number shown by Figure 3.12(a), the downstream movement of the splitter plate initially has the effect of lowering the

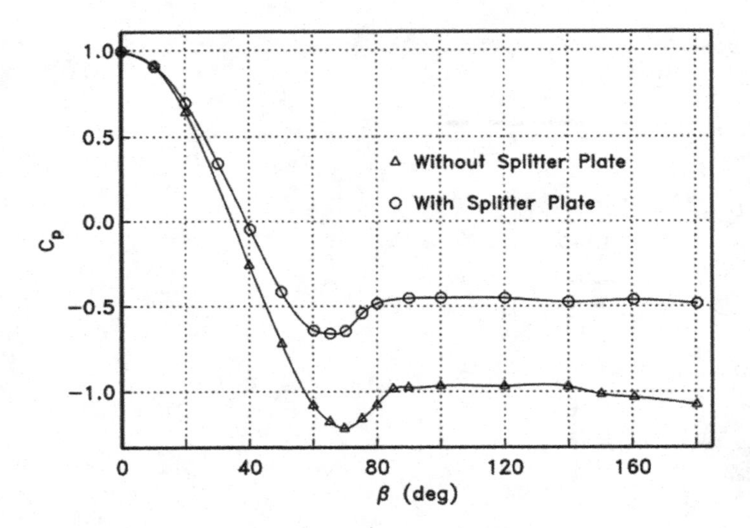

Figure 3.11 Surface pressure coefficient measured around half of the cylinder circumference for the same conditions in Figure 3.10, with and without the splitter plate. $\beta = 0$ corresponds to the upstream stagnation line, the wake centerline. Data taken from Roshko (1954).

shedding frequency. This persists to $c/D \simeq 4$ where it abruptly jumps, and then at larger c/D distances, slowly reaching an St $\simeq 0.21$ that is associated with a baseline circular cylinder in this "subcritical" Reynolds number regime.

The lower plot in Figure 3.11 shows the effect of the splitter plate position on the base pressure coefficient. This follows a similar trend as the Strouhal number, where the negative base pressure is initially reduced, signifies a reduction in the drag on the cylinder. As with the Strouhal number, this reduction in C_{ps} changes abruptly at $c/D \simeq 4$, where it jumps to a large negative value (increased drag) and slowly returns to the baseline value at a $c/D \simeq 7$.

The interpretation of the results given by Roshko (1954) is that for $c/D < 4$, the effective length of the bluff body extends to the trailing edge of the splitter plate. This reflects the respective changes in the wake shedding frequency and base pressure. However, when the gap becomes too large, the splitter plate is isolated with respect to the cylinder, and the effective length becomes that of the cylinder mean wake extent. The abrupt change that occurs at $c/D \geq 4$ suggests the possibility of a bistable state for the wake that might depend on other conditions such as external disturbances that could be receptive to flow control.

Gerrard (1966a) also performed experiments with splitter plates placed behind circular cylinders. These included repeating the Roshko (1954) investigation of splitter plates with a gap between the cylinder and plate. These included different length splitter plates. For smaller plate lengths, when the gap length was small, the reduction in Strouhal number was the same as that with no gap. For larger gaps, however, there was a significant effect of the gap length. Gerrard (1966a) determined that the relevant parameter was the length of the gap rather than the position of the trailing edge of the splitter plate as surmised by Roshko (1954).

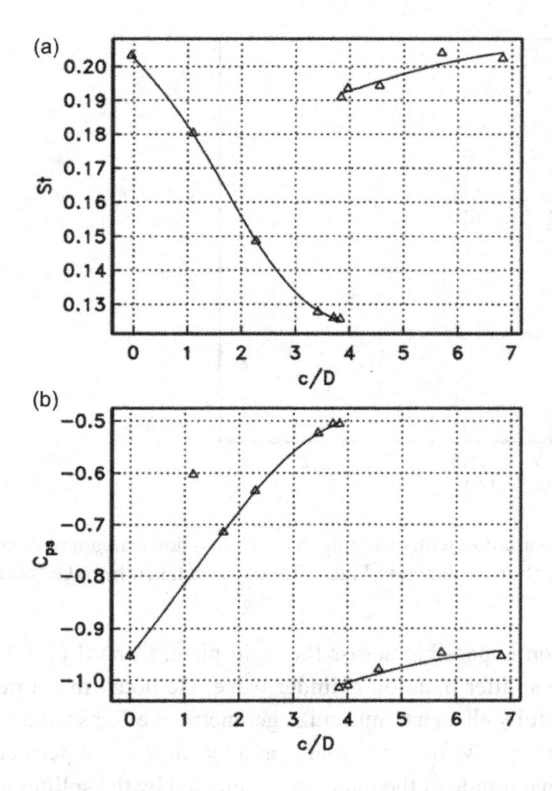

Figure 3.12 Effect on Strouhal number (a) and stagnation pressure coefficient, C_{ps} (b), due to a detached splitter plate with length 1.14D located at different distances from the cylinder as shown in Figure 3.9(b). Data taken from Roshko (1954).

Utilizing flow visualization, Gerrard (1966a) showed that the dramatic discontinuity in the Strouhal number that occurred when the gap between the splitter plate and cylinder increased beyond a critical amount was a transition from a flow regime in which the wake formation region lengthened to include the splitter plate, to one in which shed vortices formed upstream of the splitter plate. The gap lengths where the discontinuities occurred were found to be minimally affected by the lengths of the splitter plates.

Gerrard (1966a) also investigated the effect of a plate placed normal to the mean flow direction. This is illustrated in Figure 3.9(c). The plate height was 0.069D. The effect this had on the vortex shedding Strouhal number is presented in Figure 3.13. The Strouhal number again exhibited an abrupt change when the normalized spacing, l/D, reached a critical value. Gerrard (1966a) further observed that there was a region close to the critical l/D where the shedding intermittently displayed both frequencies. This again suggests a bistable stability state that is likely sensitive to external disturbances and can be exploited by active external forcing.

Based on the previous results, it is clear that fixed–rigid splitter plates provide an effective means of suppressing von Kármán vortex shedding. In its use, no velocity

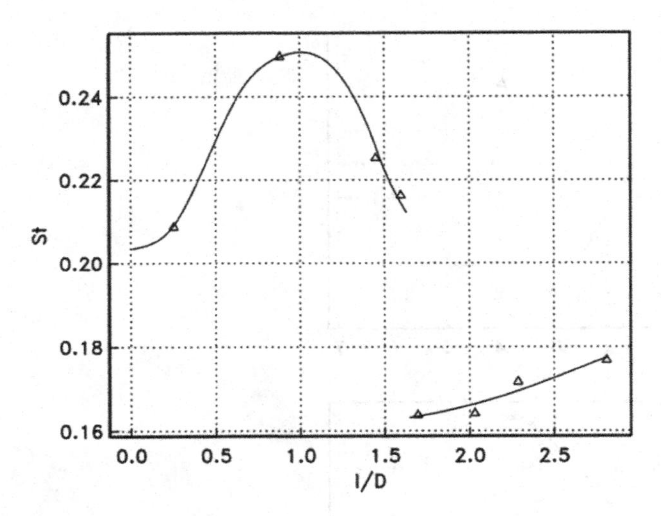

Figure 3.13 Strouhal number variation with changing distance to a downstream plate oriented normal to the mean flow direction as shown in Figure 3.9(c). Data taken from Gerrard (1966a).

or pressure communication is possible across the rigid plate. Cardell (1993) studied the effect of a *permeable* splitter plate on cylinder wake. He noted that a permeable splitter plate that was carefully chosen to minimize geometrical effects would interfere with communication across the wake center plane, and by varying the permeability, it was possible to vary the magnitude of the interference created by the splitter plate. His results showed that by varying the permeability of the splitter plate, the base pressure would smoothly vary from that of the plane cylinder to that of the impermeable rigid splitter plate.

Cylinders with *hinged–rigid* splitter plates have been recently studied by Assi et al. (2009) and Shukla et al. (2009). The control parameter for hinged splitter plates is the damping of the hinge motion. With a relatively large hinge damping, Assi et al. (2009) found that the splitter plates did not oscillate but assumed a stable position at an angle to the flow direction. In this configuration, the vortex-induced vibrations of the cylinder were effectively suppressed along with a significant amount of drag reduction. In contrast, at very low hinge damping levels, Shukla et al. (2009) observed periodic oscillations of the splitter plate with tip motion amplitudes on the order of 0.5 diameters. The periodic oscillations occurred for splitter plates whose lengths were three diameters or lower. With longer hinged splitter plates, Shukla et al. (2009) observed a sharp reduction in the plate oscillation amplitude, with the oscillations also becoming more random.

As somewhat of an outgrowth of the hinged splitter plates, Shukla et al. (2013) considered the effects of a *flexible* splitter plate. In this case, the control parameter was the flexural rigidity (EI) of the splitter plate. The mass per unit length of the splitter plate was another parameter that affects the motion inertia. Time-resolved images of the flexible plate motion indicated that the deformations took the form of a traveling wave with an amplitude that grew approximately linearly with increasing distance from the cylinder.

In the Shukla et al. (2013) experiments, the splitter plate consisted of thin plastic sheets with a thickness of approximately 40 μm. Spanwise stiffeners consisting of 450-μm-thick plastic were attached to the thin sheet at regular intervals along its length. The ratio between the width of each stiffener to the separation distance between adjacent stiffeners was 0.13, so that the flexural rigidity of the composite sheet was essentially the same as that of the thin plastic sheet without stiffeners. The flexural rigidity of the sheets was varied through differences in the sheet thicknesses.

The Shukla et al. (2013) experiments were conducted in water. The mass per unit length of the flexible sheets was small so that the inertial forces due to the sheet were small compared to the fluid inertia. Similarly, the structural damping of the thin sheets was considered small so that the effect of the mass and damping of the flexible splitter plate was considered to be negligible. The Reynolds numbers ranged from 1800 to 10,000. The flexural rigidity per unit width of the splitter plate (EI) for a majority of the cases was 3.96×10^{-6} N-m.

The results suggested that with increasing Reynolds number, there were two regimes of periodic oscillations of the flexible splitter plate. At Reynolds numbers between these two regimes, the splitter plate exhibited aperiodic oscillations and indications of multifrequency competition. Within the two regimes of periodic oscillations, the normalized tip oscillation frequency, $f^*_{\text{tip}} = f_{\text{tip}} D/U_\infty$, was approximately 0.2, or approximately the generally observed cylinder baseline Strouhal number (0.21).

Shukla et al. (2013) observed that increasing the flexural rigidity of the splitter plate would shift the Reynolds number at which the transition between the two regimes would occur. They characterized the stiffness through a nondimensional parameter, K^*, given as

$$K^* = \frac{\text{EI}}{0.5\rho U_\infty^2 L^3}, \tag{3.5}$$

where L is the length of the splitter plate, and $0.5\rho U_\infty^2$ is the fluid dynamic pressure. The parameter K^* was found to collapse the tip oscillation amplitude results for the range of splitter plate stiffness and lengths. This suggests that K^* is the governing dimensionless parameter for this form of wake control.

3.2.2 Control Cylinder

Oertel (1990) reports a private observation of E. Berger that the vortex shedding frequency over the entire span of a cylinder dropped when a hot wire was inserted into one of the shear layers. Such sensitivity was demonstrated by the landmark experiments of Strykowski (1986) who showed that the proper placement of a second smaller *control* cylinder into the near-wake region of a larger cylinder could completely suppress the von Kármán vortex shedding. An example is shown in Figure 3.14 which documents images of flow around a circular cylinder using the hydrogen bubble flow visualization

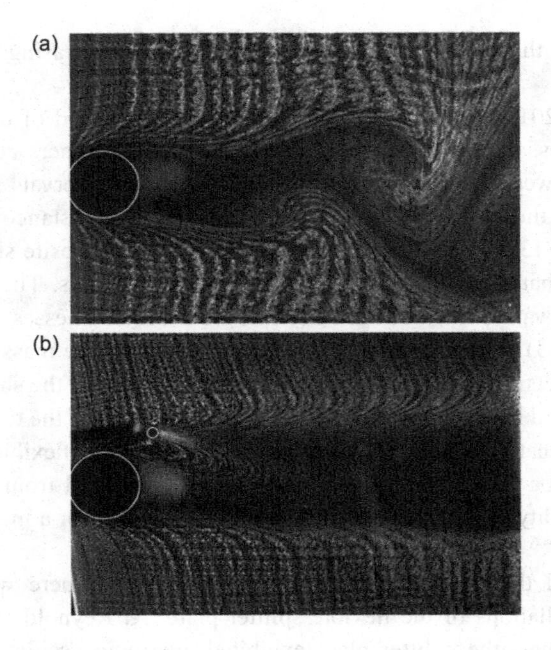

Figure 3.14 Hydrogen bubble flow visualization of vortex shedding behind a cylinder at $Re_D = 90$ alone (a) and with the addition of a small control cylinder located above and downstream of the larger cylinder (b). Taken from Strykowski and Sreenivasan (1990).

technique in water. In both images, $Re_D = 90$, which is twice the critical Reynolds number for von Kármán vortex shedding. The top image is with the larger cylinder, outlined in white, alone in the flow. The bottom image is the visualized flow field that includes the addition of a small control cylinder, also outlined in white. It is clear from the flow visualization that the addition of the control cylinder had completely suppressed the von Kármán vortex shedding.

In the images shown in Figure 3.14, the ratio of the diameter vortex shedding cylinder, D, to the control cylinder diameter, d, was $D/d = 7$. Strykowski and Sreenivasan (1990) point out that for the conditions of the visualized flow, the Reynolds number of the control cylinder was subcritical to vortex shedding, $Re_d < 10$. In addition, special care was taken to ensure that the control cylinder was aligned parallel to the shedding cylinder and was free of vibration.

Strykowski and Sreenivasan (1990) examined regions within which the placement of the control cylinder would suppress the von Kármán vortex shedding. These are shown in Figure 3.15 for three different values of D/d. Although not drawn, identical contours exist on the opposite side of the vortex shedding cylinder so that the control cylinder is equally effective when placed within the boundaries on either side of the wake.

The effect of the control cylinder on the suppression of the von Kármán vortex shedding has been reproduced in a numerical flow simulation by Strykowski and Sreenivasan (1990). However, this did not reveal the underlying mechanism for the

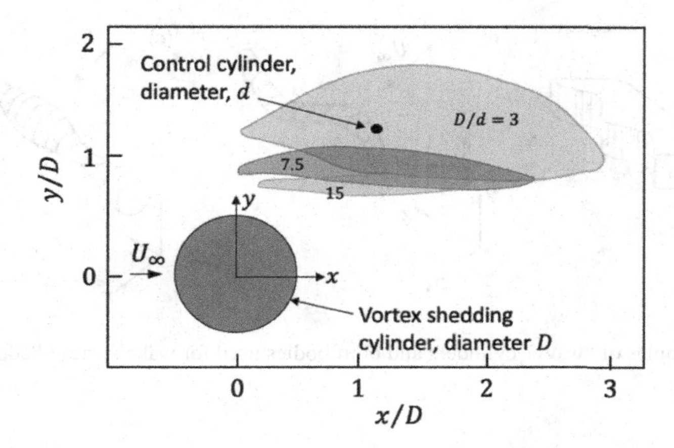

Figure 3.15 Regions of vortex street suppression for three values of the ration of the vortex shedding cylinder diameter, D, to the control cylinder diameter, d. Replotted from Strykowski and Sreenivasan (1990).

suppression of the vortex shedding. The speculation was that the *asymmetry* in the base flow produced by the control cylinder led to a change of local stability properties, particularly those governing an absolute instability.

Strykowski and Sreenivasan (1990) established that the vortex shedding suppression was not limited to circular cylinders. This stemmed from an earlier work of Koch (1985) that pointed out that the important factors are the mean flow (basic state) characteristics of the wake, and not the shape that produced them. Further support of this view came from Triantafyllou and Karniadakis (1990) who showed in numerical flow simulations that a vortex street could be generated without the inclusion of the wake-producing body, as long as the appropriate basic state was utilized.

Strykowski and Sreenivasan (1990) subsequently investigated the use of a control cylinder on the wake of a flat plate with sharp edges. In particular, a control cylinder with a diameter equal to one-seventh of the flat plate height provided suppression of the vortex shedding for Reynolds numbers up to 38 percent higher than the critical value for shedding onset.

In experiments in air, Strykowski and Sreenivasan (1990) found that resistively heating the control cylinder dramatically widened the region of vortex shedding suppression. This might represent a combination of effects of the mean flow modification produced by the control cylinder, and the lowering of the air density that Huerre and Monkewitz (1990) predicted would suppress the absolute instability.

3.2.3 Wavy Cylinders

It has been observed that the nominally 2-D von Kármán vortex shedding eventually exhibits some three-dimensional (3-D) character, particularly at higher Reynolds numbers. One of the first to study this in detail was Gerrard (1966b), who identified a number of sources for the 3-D development. The first of these is oblique shedding in

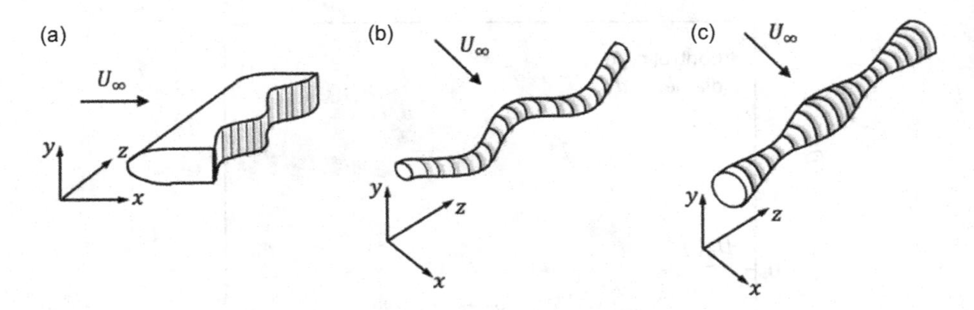

Figure 3.16 Examples of "wavy" cylinders and bluff bodies used for wake vortex shedding and drag control.

which the nominally 2-D vortices are shed at a small angle relative to the cylinder axis. Williamson (1989) demonstrated that 2-D vortex shedding could only be obtained if the cylinder end conditions are very carefully controlled. In his case, this involved suitably angled end plates. Other methods for controlling bluff body end effects are attributed to Eisenlohr and Eckelmann (1989) and Hammache and Gharib (1991).

A second source of 3-D vortex shedding is "vortex dislocation." This is prompted by spanwise variations in the vortex shedding frequency. It can originate from small spanwise variations in the mean wake thickness on which the vortex shedding frequency (Strouhal number) follows. Such vortex dislocations can be promoted by the use of tapered-diameter or stepped-diameter cylinders. As will be presented, this is a method that has been used in bluff body wake control.

There have been a number of cylinder designs that have exploited the effect of spanwise varying surface profiles. Those in which the spanwise variations are sinusoidal are termed "wavy cylinders." Examples of such designs are illustrated in Figure 3.16. The effect of such "wavy cylinder" shapes on bluff body wake development has been investigated both experimentally and in simulations. Examples of these come from Ahmed and Bays-Muchmore (1992), Ahmed et al. (1993), Tombazis and Bearman (1997), Bearman and Owen (1998), Owen et al. (2000), Darekar and Sherwin (2001a), Darekar and Sherwin (2001b), Keser et al. (2001), Lam et al. (2004a, 2004b), Zhang et al. (2005), Dobre et al. (2006), Lee and Nguyen (2007) and Lam and Lin (2007, 2008).

Tombazis and Bearman (1997) investigated the effect of a spanwise wavy surface added to the trailing face of the cylindrical bodies. The effect of a wavy leading edge on cylinder bodies was investigated by Bearman and Owen (1998) and Dobre et al. (2006). Bearman and Owen (1998) obtained a 30 percent drag reduction at Re_D = 40,000.

Owen et al. (2000) studied flow past a sinuous bluff body, shown by the center illustration in Figure 3.16. An example of the visualized wake of the sinuous bluff body at $Re_D = 100$ from Owen et al. (2000) is shown in Figure 3.17. The wake generating body is a constant diameter cylinder having a spanwise sinuous wavelength of $7.5D$ and an amplitude of $0.166D$. For reference, a cross section of the visualized wake of a straight cylinder is shown in image (a). Images (b) and (c) show cross sections of the

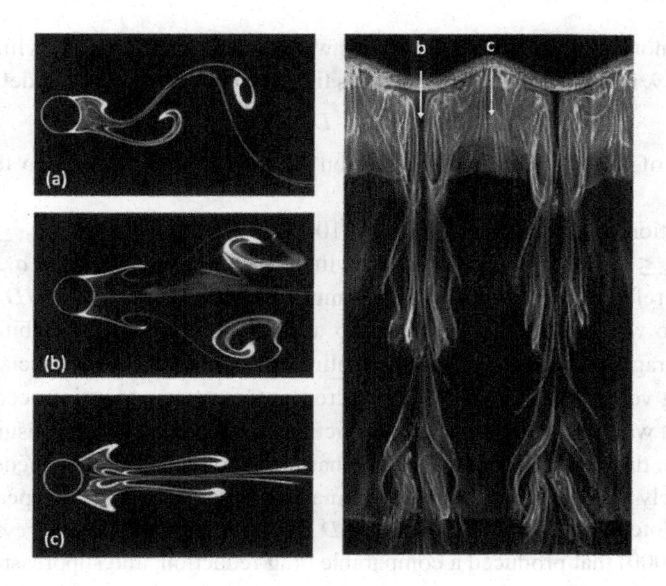

Figure 3.17 Visualized wake of a constant diameter cylinder having a spanwise sinuous axis with a wavelength of $7.5D$ and an amplitude of $0.166D$. $\mathrm{Re}_D = 100$. Image (a) shows a cross-sectional view of the wake of a straight cylinder. Images (b) and (c) show cross sections of the wake of the sinuous cylinder at the spanwise locations correspondingly labeled in the top-view image at the right. Taken from Owen et al. (2000).

visualized wake at two spanwise locations of the sinuous cylinder corresponding to a valley (b) and a peak (c) of the cylinder shape. These spanwise locations are similarly labeled in the top-view of the visualized wake flow field. The key features are that the alternate von Kármán vortex shedding observed in image (a) is suppressed. Instead the wake structure consists of a spanwise periodic variation in the wake width. A wide wake occurs where the cylinder body extends furthest in the downstream direction (b). The narrow wake occurs where the body protrudes furthest in the upstream direction (c). The section views show these two regions to be dramatically different, with large symmetric vortex loops forming at spanwise location (b). The top-view flow visualization image suggest that section (b) also includes coherent streamwise vortical motion that is not evident in the thinner wake region of section (c).

In flow field simulations, Darekar and Sherwin (2001a) and Darekar and Sherwin (2001b) observed that a sinuous leading edge on a bluff body could suppress the natural von Kármán vortex shedding. With this, they obtained a 16 percent drag reduction at $\mathrm{Re}_D = 100$. They expected the drag reduction would increase at higher Reynolds numbers.

Lam and Lin (2009) performed flow simulations of a cylinder having a sinusoidally varying diameter in the spanwise direction like that illustrated in Fig. 3.16(c). The spanwise diameter of the cylinder corresponded to

$$D(z) = D_m + 2a\cos(2\pi z/\lambda), \tag{3.6}$$

where $D(z)$ denotes the local diameter at spanwise location, z, along the cylinder, a is the amplitude, λ is the wavelength, and D_m is the mean cylinder diameter defined as

$$D_m = (D_{\min} + D_{\max})/2. \tag{3.7}$$

The diameter of the spanwise middle portion of the cylinder is equal to the mean diameter, D_m.

The simulations corresponded to $\text{Re}_D = 100$ and examined $1 \leq \lambda/D_m \leq 10$ and $0.02 \leq a/D_m \leq 0.30$. Constant-diameter cylinders with $\lambda/D = \infty$ and $a/D_m = 0$ were used for reference comparisons. The simulations revealed that for $a/D_m > 0.1$, there were two wavelengths of $\lambda/D_m \simeq 2.5$ and $\lambda/D_m \simeq 6.0$ that exhibited sharp drops in the drag coefficient, and the fluctuating lift coefficient, C_L', associated with the alternating vortex shedding. The larger drop in these two quantities occurred for $\lambda/D_m \simeq 6.0$ at which an 18 percent drag reduction was recorded. Commensurate with the decrease in drag was a zero level of C_L' that implied that the vortex shedding had been completely suppressed. Thus, the optimal spanwise wavelength appears to be $\lambda/D_M \simeq 6$. Note that this is close to the $7.5D$ wavelength of the sinuous cylinder of Owen et al. (2000) that produced a comparable drag reduction, and suppression of the von Kármán vortex shedding.

A variation on the wavy cylinder concept for bluff body wake control is the use of spanwise spaced vertical tabs that are placed along the trailing edge. An example of this is illustrated in Figure 3.18. Such tabs produce a spanwise variation in the wake thickness that is similar to the sinuous variations in the bluff body thickness that were illustrated in Figure 3.16. Such tabs could equally be substituted with streamwise vortex generators.

A parametric study of the effect of trailing-edge tabs on the wake of a bluff body was performed by Park et al. (2006). This included the effect of the tab height, l_y, width, l_z, and wavelength, λ. The shape of the bluff body was the same as that used by Tombazis and Bearman (1997), which was a flat plate with an half-elliptic leading edge having a major-to-minor axis ratio of 8. Tabs were placed on both the upper and lower surfaces of the flat plate. As illustrated in Figure 3.18, the tabs on the upper and lower surfaces were either aligned or staggered by $\lambda/2$.

Park et al. (2006) investigated a tab design that would minimize the overall drag on the bluff body. This included the parasitic drag on the tabs. Considering a tab alone,

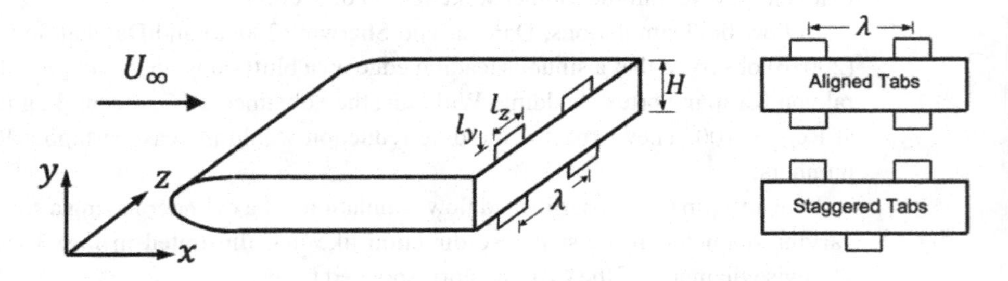

Figure 3.18 Schematic showing a variation on the wavy cylinder concept involving spanwise spaced vertical tabs along the trailing edge of a blunt body.

they estimated that for $(l_y/H, l_z/H) = (0.2, 0.2)$ the drag increase was approximately 2 percent, which was found to be small compared to the drag reduction on the bluff body produced by the addition of the tabs, which was on the order of 23 percent.

Utilizing tabs with dimensions $(l_y/H, l_z/H) = (0.067, 0.2)$, Park et al. (2006) determined that the optimum spacing between *aligned* tabs was $\lambda/H = 1.7$. If one assumes that H in this case is equivalent to D for a circular cylinder, then this spacing is on the order of $\lambda/D_m \simeq 2.5$, which is where the simulations of Lam and Lin (2009) observed the first drop in drag for a sinusoidally varying diameter cylinder.

Park et al. (2006) observed that the tabs caused the vortex formation length behind the bluff body to significantly increase. In a fashion that is similar to a cylinder with a sinusoidally varying diameter, the wake width was observed to become larger near a tab, and smaller away from a tab. As a result, the vortices shed from the upper and lower trailing-edge surfaces lost their 2-D nature and vortex dislocation occurred. As a result, the von Kármán vortex shedding was subsequently suppressed, along with a commensurate increase in the base pressure.

The staggered tabs with the same l_y/H, l_z/H and λ/H were found to be less effective for drag reduction than the aligned tabs. They did however lower the drag by approximately 13 percent compared to the baseline bluff body.

3.3 Bluff Body Active Control

This section focuses on *active* flow control methods for controlling bluff body wakes. These are generally designed to exploit the same wake instabilities as the previous passive approaches. However, the active approaches have the potential to manipulate the wake in ways that are not possible with passive methods alone.

3.3.1 Base Bleed

One of the earliest methods of active bluff body wake control is "base bleed," in which some fraction of fluid is injected at the aft part of the body. In an early demonstration of the effect of base bleed, Wood (1964) documented that a sufficiently large base bleed produced a suppression of von Kármán vortex shedding behind a blunt airfoil trailing edge, as well as a substantial drag reduction. His critical bleed coefficient was found to be in good agreement with stability calculations of Monkewitz and Nguyen (1987) that showed the wake to be everywhere convectively unstable for the previously given velocity ratio, $\Lambda > -0.84$. Castro (1971) demonstrated a similar effect by examining porous wake bodies made up of perforated plates and screens with different open areas. For this, an open area ratio between 0.37 and 0.39 produced an abrupt change in the von Kármán vortex shedding in which the shedding was completely suppressed. Such a result is consistent with the concept of a global instability, where the nonlinear saturation amplitude of limit cycle oscillations, $|B|_{sat}$ in Eq. (3.2), abruptly increases when the control parameter (open area ratio) increases beyond a critical value.

Bearman (1967) documented the effects of base bleed on the flow around a 2-D model with a blunt trailing edge for Reynolds numbers, based on model base height, between 1.3×10^4 and 4.1×10^4. It was found that a sufficiently large bleed quantity would suppress the regular vortex street, and when that occurred, the base drag of the section was reduced by one-third. Bearman (1967) further reported that those results agreed closely with a previous splitter plate investigation.

In numerical simulations, Oertel (1990) investigated the effect of base bleed on the wake development of a flat plate. They defined a bleed coefficient of the form

$$c_q = \frac{m}{U_\infty D}, \tag{3.8}$$

where m was the bleed mass flow rate, and D was the height of the wake generating body, similar to a cylinder diameter. The simulation indicated a total suppression of the von Kármán vortex street at a critical base bleed of $c_q = 0.22$.

3.3.2 Vibrating Body

Experiments with a vibrating cylinder in order to influence the vortex shedding have been carried out by Taneda (1963) and Bishop and Hansan (1963). The application of forced oscillations to suppress the von Kármán vortex street has been reported by Wehrmann (1965) and Berger (1965). Wehrmann (1967) utilized a cylinder consisting of a piezoelectrically active material (Clevite PZT Bimorphe). The cylinder had an oblong shape, 0.154-cm long by 0.0675-cm wide cross-sectional dimensions. It was arranged in the flow so that the longer side was parallel to the flow direction.

The piezoelectric cylinder was vibrated by applying an AC voltage in a frequency range from 20 Hz to 500 kHz. The experiments focused on small Reynolds numbers within the "stable range" (Kovasznay, 1949), $40 < Re_D < 150$. The cylinder vibration was set to match the natural shedding frequency at the given Reynolds number. The experiments indicated that the first stage of turbulence in the wake could be delayed by the mechanical vibrations. Specifically, the fluctuation energy was concentrated to a single frequency that remained stable to higher Reynolds numbers compared to the case without the cylinder vibration.

3.3.3 Unsteady Blowing

A number of active bluff body flow control approaches involve the introduction of unsteady disturbances on the body. An example is Huang (1996) who incorporated a spanwise slit through the wall of a hollow cylinder in which to introduce surface-normal 2-D jet. An illustration of the setup is shown in Figure 3.19. The flow out of the slit was driven by a pair of acoustic speakers located at either end of the hollow cylinder. The cylinder could also be rotated about its long axis in order to orient the 2-D jet relative to the mean flow direction. The Reynolds numbers in the Huang (1996) experiment ranged from 4×10^3 to 1.3×10^4. Over this range, the shedding frequency ranged from 94 to 98 Hz.

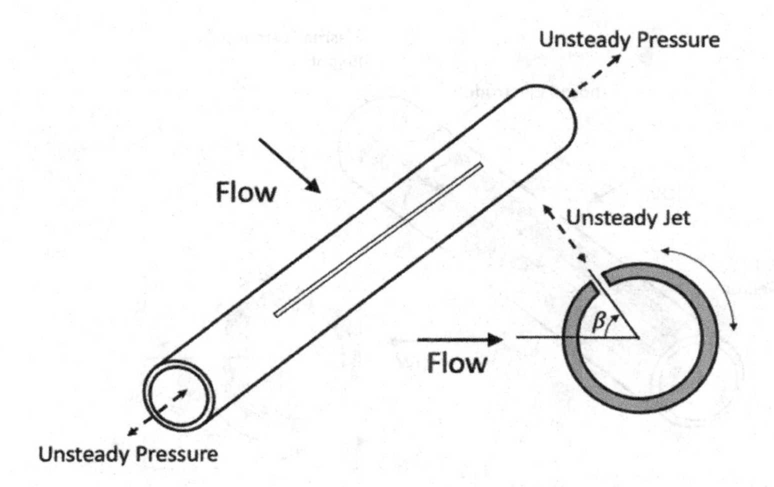

Figure 3.19 Illustration of hollow cylinder with spanwise slit through the cylinder wall used by Huang (1996) to introduce unsteady surface-normal jet.

The angle of the slit in the Huang (1996) experiment was predominately at $\beta = 80°$, which placed it near the flow separation location on the cylinder. This is a location that is highly receptive to controlled unsteady disturbances. As one would expect, the unsteady forcing from the slit in the cylinder wall enforced spanwise uniformity in the vortex shedding. Unsteady forcing at frequencies that were shifted from the natural vortex shedding frequency was found to suppress the natural shedding. The forcing was not as effective for slot angles of $\beta > 90°$, which placed the slot downstream of the flow separation location for the range of cylinder Reynolds numbers of the experiment.

3.3.4 Plasma Actuators

A more general approach to unsteady blowing for bluff body wake control involves the use of single-dielectric barrier discharge plasma actuators. An example of a plasma actuator array applied to a circular cylinder is shown in Figure 3.20. This design was used as a proof of concept by Thomas et al. (2008) to create a "plasma fairing" that would effectively streamline a circular cylinder in a cross-flow.

As illustrated in Figure 3.20, the cylinder had four plasma actuators located at angles measured from the upstream stagnation line ($\beta = 0$) of $\beta = \pm 90°$ and $\beta = \pm 135°$. Each of the plasma actuators consisted of two electrodes that were separated by a dielectric layer, which in that case was the hollow cylinder wall made of glass. The electrodes in each pair were staggered so that the ionized air (plasma) forms on the cylinder surface over the area of the covered (inside the cylinder) electrode. As indicated, each of the electrode pairs were connected to separate AC voltage sources. When the AC voltage amplitude source was large enough, the air over the covered electrode ionized. This is illustrated by the blue coloring. The ionized air in

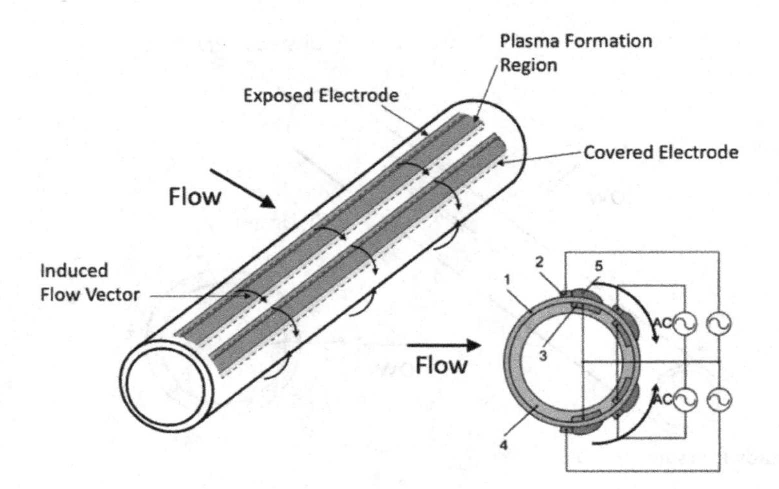

Figure 3.20 Illustration of four single-dielectric barrier discharge plasma actuators used for active cylinder wake control. (1) is the wall of the cylinder which is made of glass. (2) and (3) are plasma actuator respective exposed and covered electrodes. (4) is insulation over the internal electrodes. (5) is the ionized air (plasma) that forms over the covered electrode in the staggered electrode arrangement. Taken from Thomas et al. (2008).

the presence of the electric vector field created by the staggered electrodes results in a body force vector field that acts on the neutral air. Further details on plasma actuators are contained in the review article by Corke et al. (2010b).

In the design shown in Figure 3.20, the body force acts in the direction from the outer, exposed electrode toward the covered electrode, or from left to right in the image. This is indicated by the curved arrows in the illustration. In the application by Thomas et al. (2008), the body force vectors were in the direction of the mean flow approaching the cylinder. They were designed to produce a velocity component along the cylinder surface toward the wake spanwise centerline, in a manner similar to base bleed.

The plasma actuators that utilize a dielectric layer between the electrodes are electrically equivalent to capacitors, and therefore only pass an AC current. Thus, the actuators are powered by a sinusoidal voltage waveform. In Thomas et al. (2008), the AC frequency was 10 kHz, which was approximately 1000 times higher than the 9 Hz natural shedding frequency corresponding to a St $=$ 0.21 in the experiment. Figure 3.21 illustrates the different AC voltage time series applied to the plasma actuators. The actuators were generally operated in pairs that covered the top-half (positive β) and bottom-half (negative β) surfaces of the cylinder. For "steady" operation, the actuators are operated with a continuous running 10-kHz AC voltage time series that had the same amplitude and phase for all four actuators. As illustrated in Figure 3.21, an "unsteady" operation involved switching the 10-kHz AC ON and OFF. This was defined by two timescales, T_1 that corresponded to the time between the leading edges of the AC ON part of the waveform and T_2 that corresponded to the time over which the AC waveform was ON. Based on these, the unsteady frequency was $1/T_1$, and

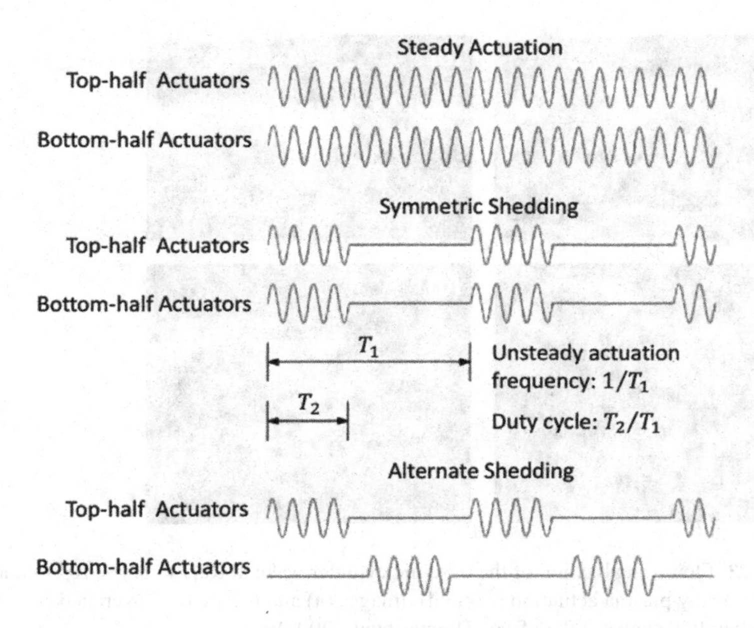

Figure 3.21 Plasma actuator AC voltage time series for steady actuation and unsteady symmetric and alternating shedding actuation. Taken from Thomas et al. (2008).

the waveform duty cycle is T_2/T_1. For "symmetric shedding" the waveforms to the actuators on the top and bottom halves of the cylinder were in phase. For "alternate shedding," the actuators on the top and bottom halves of the cylinder were 180° out of phase.

Figure 3.22 documents the global structure of the flow around the cylinder with the plasma actuators off (a and b) and with the plasma actuators in the "steady" actuation condition (c and d). In Figure 3.22(a) and (c), the flow visualization involved introducing a continuous smoke stream upstream of the cylinder on the spanwise centerline, and illuminating the smoke with a high-intensity photo floodlight. The images were recorded with a single reflex digital camera with a long exposure setting to obtain a time average view of the flow field.

Figure 3.22(b) and (d) were obtained by seeding the air upstream of the cylinders with nominally 1-μm diameter olive oil droplets introduced with a pressurized air atomizer. The oil droplets were illuminated by an Nd:Yag laser in a narrow sheet that bisected the cylinder spanwise centerline. This setup was used to perform PIV measurements of the wake flow. For that, the laser was double-pulsed with a 50-μs interval between pulses. The images were recorded using a digital PIV camera synchronized to the laser. The images in Figure 3.22(b) and (d) are single short-time exposures taken from the PIV measurements.

As evident in Figure 3.22, the "steady" plasma actuation had a profound influence on the global structure of the cylinder wake flow. In particular, the steady plasma actuation had substantially reduced the extent of the separated flow region, and completely

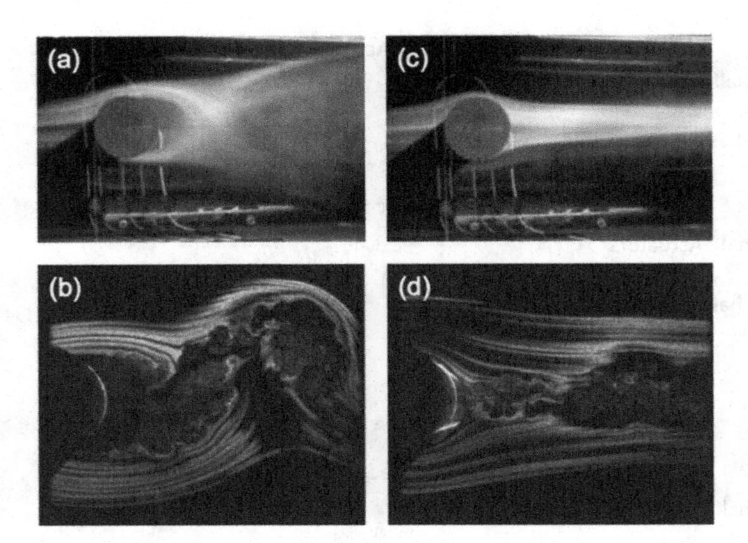

Figure 3.22 Flow visualization of the baseline cylinder wake at $Re_D = 3.3 \times 10^4$ (a and b) and with the steady plasma actuation (c and d). Images (a) and (c) are time averaged over many vortex shedding cycles. Taken from Thomas et al. (2008).

suppressed the von Kármán vortex shedding. With the plasma actuation, the flow separation from the cylinder has been greatly reduced, producing flow streaklines that possessed a strong top-to-bottom symmetry. The steady operation of the four symmetrically placed plasma actuators was effective in maintaining an attached flow over most of the aft portion of the cylinder. This effectively channeled fluid into the near wake, in a manner that is similar to base bleed. As indicated by the flow visualization, this produced a dramatic suppression of the vortex shedding.

Flow visualization was also used by Thomas et al. (2008) to characterize the effect of unsteady plasma actuation. For this, the duty cycle was fixed at 25 percent. Exploring a range of unsteady frequencies, they found that an optimum to suppress vortex shedding and subsequently to maximum drag reduction occurred for an unsteady actuation frequency corresponding to St = 1, or five times the natural shedding Strouhal number.

Examples of the wake structure with unsteady plasma actuation at St = 1 are shown in Figure 3.23. Figure 3.23(a) corresponds to the symmetric shedding condition where the top-half and bottom-half plasma actuators are operating in phase. Figure 3.23(b) corresponds to the alternate shedding condition where the top-half and bottom-half plasma actuators are operating with a 180° phase shift. In both cases, the flow visualization indicates that the von Kármán vortex street has been completely suppressed. This was confirmed by spectra, shown in Figure 3.24, of the streamwise velocity fluctuations measured in the wake at $x/D = 2$, and just off of the wake centerline at $y/D = 1$. The plasma OFF case shows the spectral peak at 9 Hz that corresponded to the natural vortex shedding at St = 0.21. That spectral peak was completely suppressed for all three of the plasma actuation cases. Spectral peaks at 50 Hz

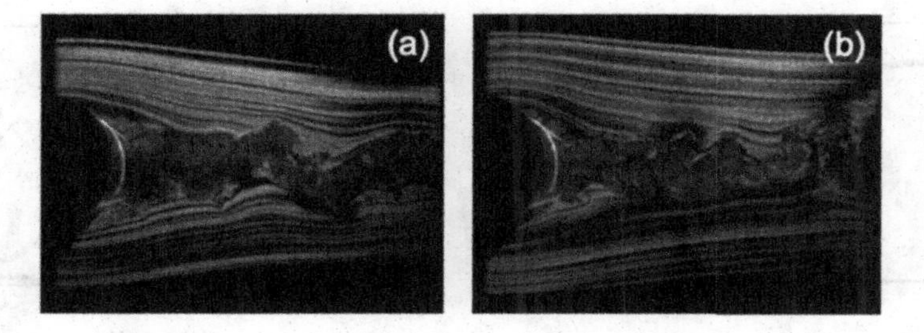

Figure 3.23 Flow visualization of the cylinder wake at $Re_D = 3.3 \times 10^4$ with unsteady plasma actuations to excite symmetric shedding (a) and alternate shedding (b) at five-times the natural Strouhal number, $St = 1$. Taken from Thomas et al. (2008).

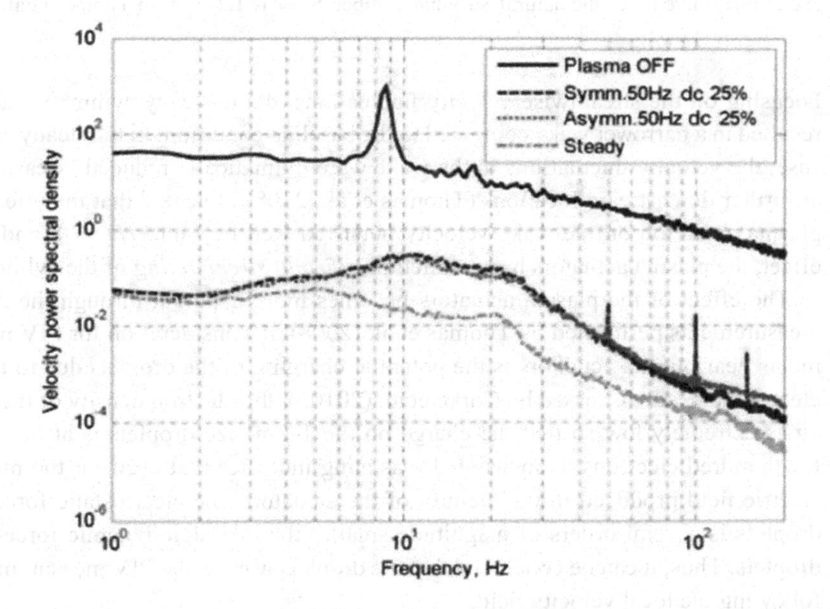

Figure 3.24 Spectra of streamwise velocity fluctuations measured at $x/D = 4$ and $y/D = 1$ for the baseline cylinder (plasma OFF), steady plasma actuation, and unsteady symmetric and alternate shedding excitation at five-times the natural Strouhal number, $St = 1$. Taken from Thomas et al. (2008).

corresponding to the $St = 1$ excitation are evident for the two unsteady actuation cases.

The effect of the plasma actuation on the cross-stream profiles of the mean velocity and root mean square (RMS) of streamwise velocity fluctuations are documented in Figure 3.25. These correspond to $x/D = 2.0$ and include the baseline (plasma OFF), steady plasma actuation, and unsteady symmetric excitation at $St = 1$.

It is apparent that the wake defect was greatly reduced by the steady plasma actuation. The unsteady symmetric plasma actuation produced a reduction in the wake defect that fell between the steady actuation and baseline (plasma OFF) cases.

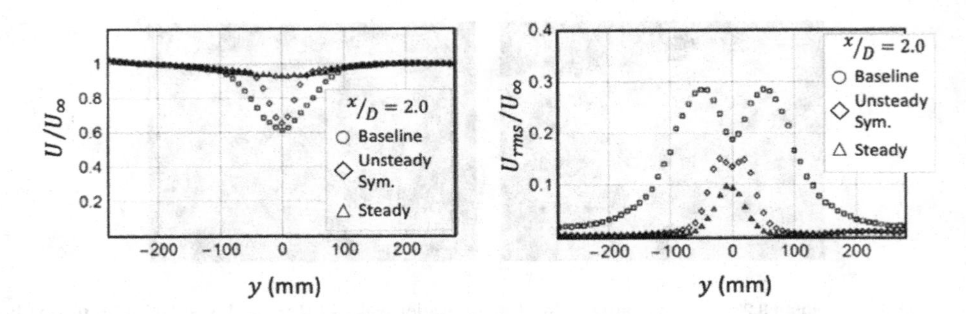

Figure 3.25 Mean and RMS velocity profiles across the wake measured at $x/D = 2$ for the baseline cylinder (plasma OFF), steady plasma actuation, and unsteady symmetric shedding excitation at five times the natural Strouhal number, St = 1. Taken from Thomas et al. (2008).

Focusing on the streamwise velocity fluctuations, the unsteady symmetric actuation resulted in a narrower wake compared to the baseline condition. In the steady actuation case, the velocity fluctuations in the wake were dramatically reduced. Measurements at further downstream locations (Thomas et al., 2008) indicated that the effect of the plasma actuation on the wake velocity persisted well beyond $x/D = 8$ and that, in effect, the plasma actuators had resulted in a *virtual streamlining* of the cylinder.

The effect of the plasma actuators becomes more apparent through the 2-D PIV measurements performed by Thomas et al. (2008). A consideration for PIV measurements near plasma actuators is the potential charging of the droplets due to the local electric field. As discussed by Corke et al. (2010b), the electron density of the ionized air is extremely low so that the charge on the 1-μm size droplets is at most only a few hundred electrons. Dimensional reasoning indicates that even for the maximum electric field produced in the vicinity of the actuators, the electrostatic force on the droplets is several orders of magnitude smaller than the aerodynamic forces on the droplets. Thus, it can be concluded that the droplets used in the PIV measurements are following the local velocity field.

With that in mind, Figure 3.26 shows ensemble-averaged velocity vectors, velocity magnitude (m/s), and spanwise vorticity (s^{-1}) for steady plasma actuation. The PIV measurements confirm that the steady induced blowing on the aft surfaces of the cylinder produced by the plasma actuator array caused the flow around the cylinder to remain attached. The resulting velocity field looks very much like an Oseen cylinder flow. The vorticity field reveals the narrow wake that occurs with the steady plasma actuation. The length scales of the vortical fluctuations are much smaller than the cylinder diameter, signifying a complete suppression of the von Kármán vortex street.

The PIV images that follow involved ensemble averages in which the images were acquired phase-locked to the unsteady plasma actuator frequency. This was accomplished by synchronizing the PIV laser pulses with the plasma actuator control circuit. This included a phase delay that allowed ensemble images at selected phases of the actuator cycle. For these, each data ensemble consisted of 150 images.

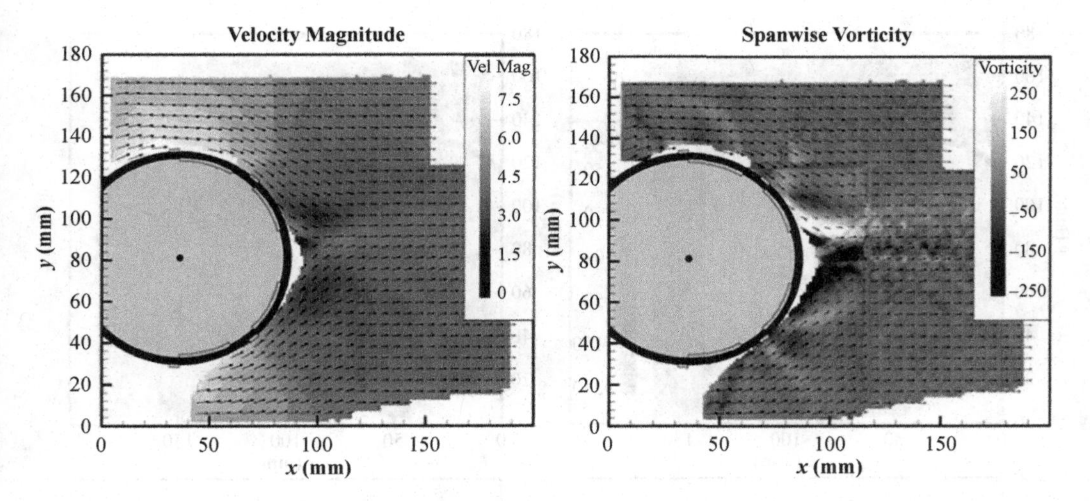

Figure 3.26 Two-dimensional PIV ensemble-averaged velocity vectors, velocity magnitude (m/s), and spanwise vorticity (s^{-1}) for steady plasma actuation. Taken from Thomas et al. (2008).

Figure 3.27 presents phase-averaged velocity vectors and spanwise vorticity (s^{-1}) for four phase angles of unsteady symmetric shedding excitation at four times the natural Strouhal number, St = 1. These images reveal that the actuator pulsing produced a series of compact discrete vortices of opposite sign that formed symmetrically on either side of the wake centerline. The vortices emerged from the cylinder surface near the plasma actuators located on the aft side of the cylinder. The discrete vortices were shed at the unsteady plasma actuation frequency, St = 1.

The vortices are observed to propagate toward the wake centerline. A significant cancelation of the vortex strength occurs when the vortices of opposite sign meet. As a result, the spanwise vorticity emanating from the near wake is quite small, and any remaining coherent vortical structures appear to be uncorrelated with any natural wake shedding. Thus again, the natural von Kármán vortex street has been completely suppressed by the symmetric forcing at the five-times higher Strouhal number.

Figure 3.28 presents phase-averaged velocity vectors and spanwise vorticity (s^{-1}) for four phase angles of unsteady alternate shedding excitation at five times the natural Strouhal number, St = 1. As illustrated in Figure 3.21, the alternate shedding is accomplished by introducing a 180° phase shift in the actuator time series of the top and bottom pairs of plasma actuators. As with the symmetric forcing, the PIV data reveal that the actuator pulsing produced a series of compact discrete vortices. However, in this case, the sign of the vorticity is staggered across the wake centerline. As with the previous symmetric forcing, the vorticity field indicates that the von Kármán vortex street has been completely suppressed by the alternate shedding forcing at the four-times higher Strouhal number.

The plasma actuators can also be used to vector the cylinder wake. An example is shown in the flow visualization image in Figure 3.29 from Thomas et al. (2005)

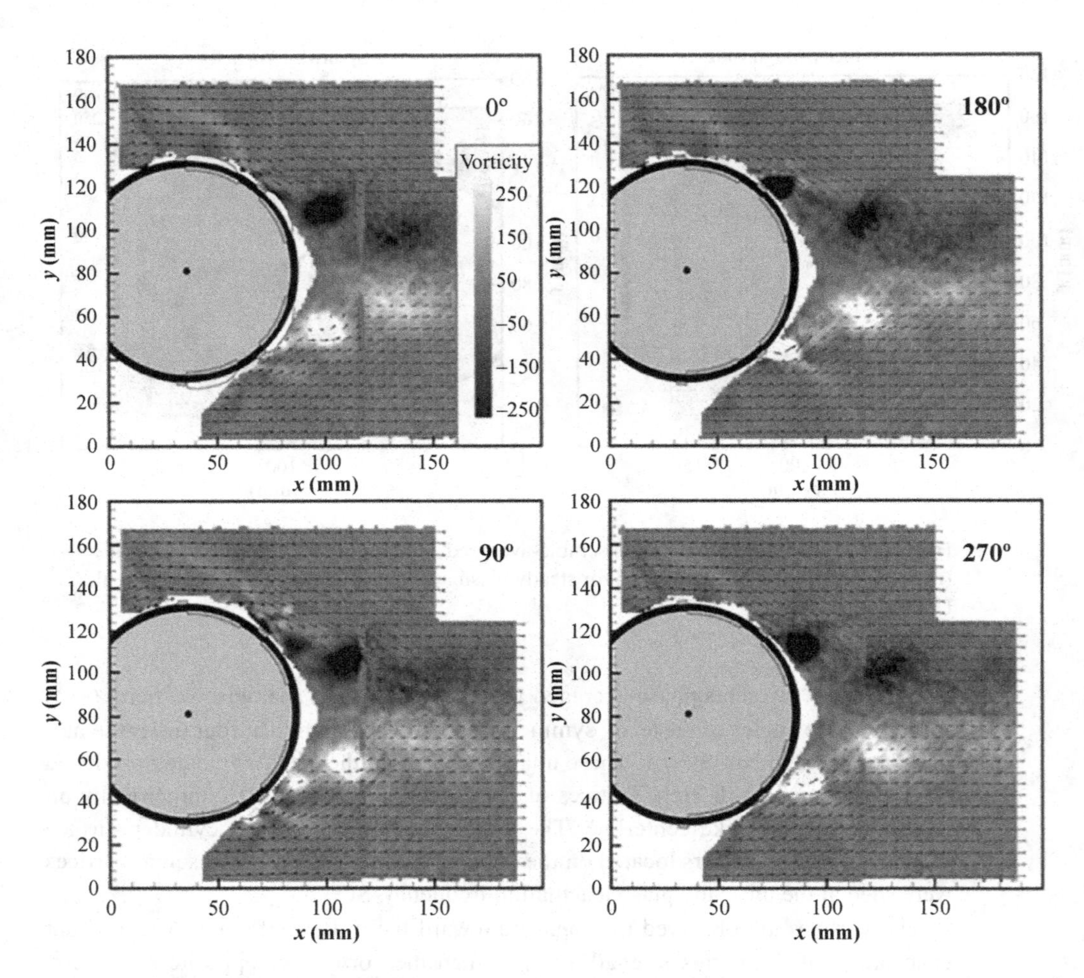

Figure 3.27 Two-dimensional PIV phase-averaged velocity vectors and spanwise vorticity (s^{-1}) for unsteady symmetric shedding excitation at four times the natural Strouhal number, St = 1. Taken from Thomas et al. (2008).

where only the top pair of plasma actuators were operated. The result is to maintain an attached flow on the upper aft surface of the cylinder. The cylinder upper to lower asymmetry in the flow separation results in the flow leaving the cylinder at a downward vector angle. The effect is similar to that in which the cylinder was rotating in the clockwise direction in which case a lift force acting on the cylinder would be generated.

The previous examples of plasma actuators for cylinder wake control aligned the actuators in the spanwise direction, where they would induce fluid momentum in the mean flow direction. The result was to maintain attached flow over the aft portion of the cylinder, significantly reducing the wake width and pressure drag. Another approach used by Thomas and Koslov (2010) was to align the plasma actuators in

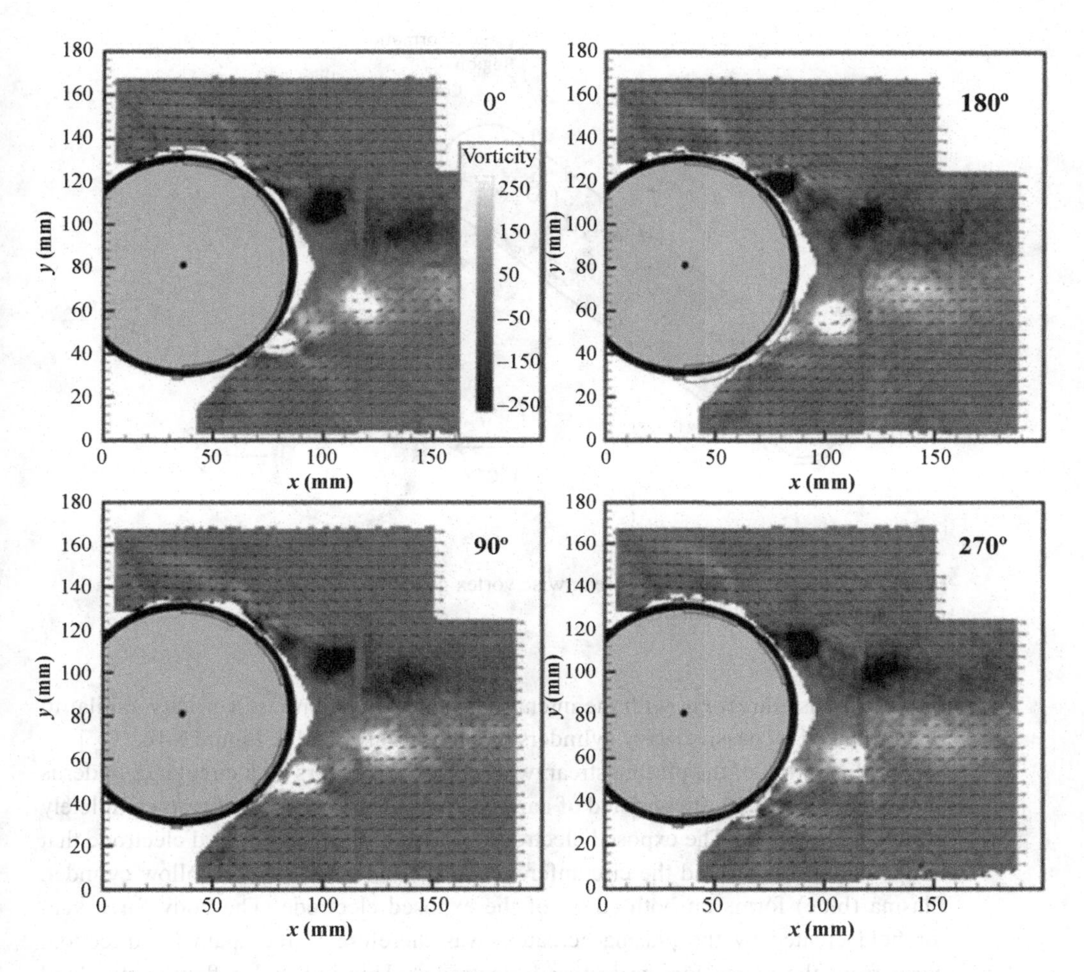

Figure 3.28 Two-dimensional PIV phase-averaged velocity vectors and spanwise vorticity (s^{-1}) for unsteady alternate shedding excitation at four times the natural Strouhal number, St = 1. Taken from Thomas et al. (2008).

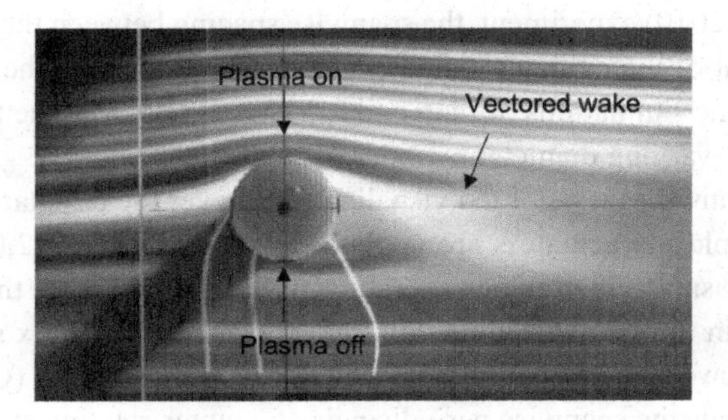

Figure 3.29 Example of wake vectoring produced by a plasma actuator located on the upper surface of the cylinder. Taken from Thomas et al. (2005).

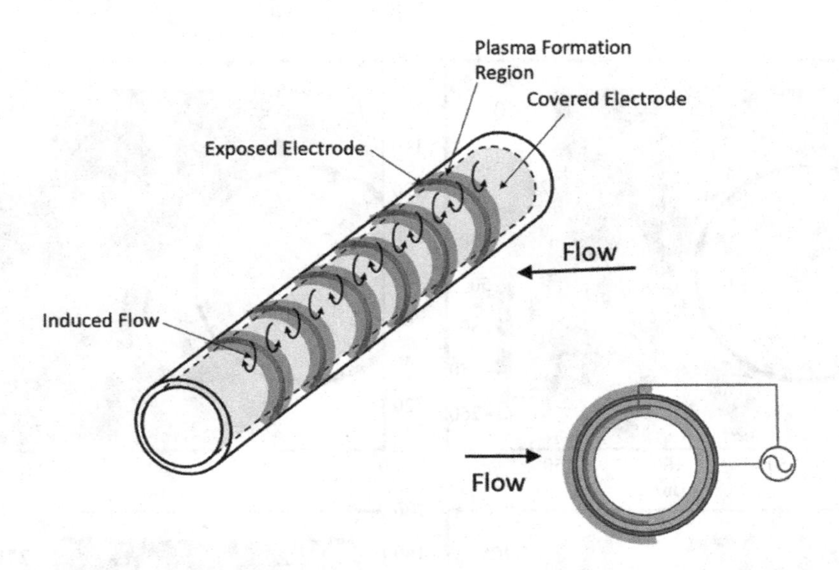

Figure 3.30 Illustration of plasma streamwise vortex generators applied to the surface of a cylinder for active wake control.

the streamwise direction, with the intent of producing streamwise vorticity similar to that produced by passive wavy cylinders like those illustrated in Figure 3.16.

An illustration of the plasma streamwise vortex generators on a circular cylinder is shown in Figure 3.30. It consisted of narrow exposed electrodes that wrap completely around the cylinder. The exposed electrodes shared a common covered electrode that wrapped halfway around the circumference of the inner wall of the hollow cylinder. Plasma (blue) forms on both edges of the exposed electrode. The body force vector field created by the plasma actuators was therefore in the spanwise direction, away from the edges for the exposed electrodes. This induced a flow vector field that is illustrated by the curved arrows drawn over each of the exposed electrodes. This arrangement creates opposing wall jets that collide in the space between adjacent exposed electrodes and subsequently give rise to pairs of counter-rotating vortices. In the Thomas and Koslov (2010) experiment, the spanwise spacing between the exposed electrodes was twice that of the cylinder diameter. This spacing was on the order of the $\lambda/D_m \simeq 2.5$, where the simulations of Lam and Lin (2009) observed the first drop in drag for a sinusoidally varying diameter cylinder.

Figure 3.31 shows spanwise views of the visualized flow over the circular cylinder over which streamwise plasma actuators are applied. For this, $\mathrm{Re}_D = 22{,}000$. The top image is with the plasma actuators not operating (OFF). In this case, the smoke streaklines are uniform in the spanwise direction, reflecting the 2-D vortex shedding from the cylinder. The lower image is with the plasma actuators operating (ON). The smoke visualization reveals the spanwise periodic pair of counter-rotating streamwise vortices produced by the plasma actuators.

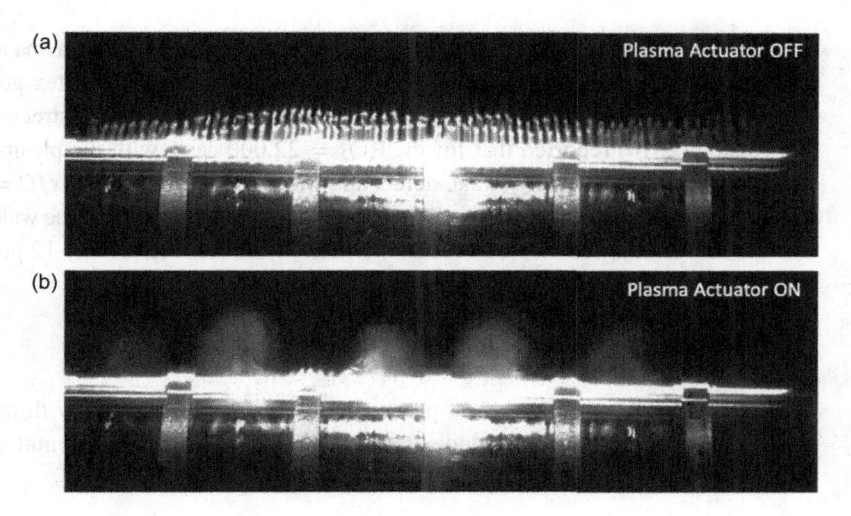

Figure 3.31 Flow visualization showing spanwise view of circular cylinder with plasma streamwise vortex generators OFF (a) and ON (b). Taken from Thomas and Koslov (2010).

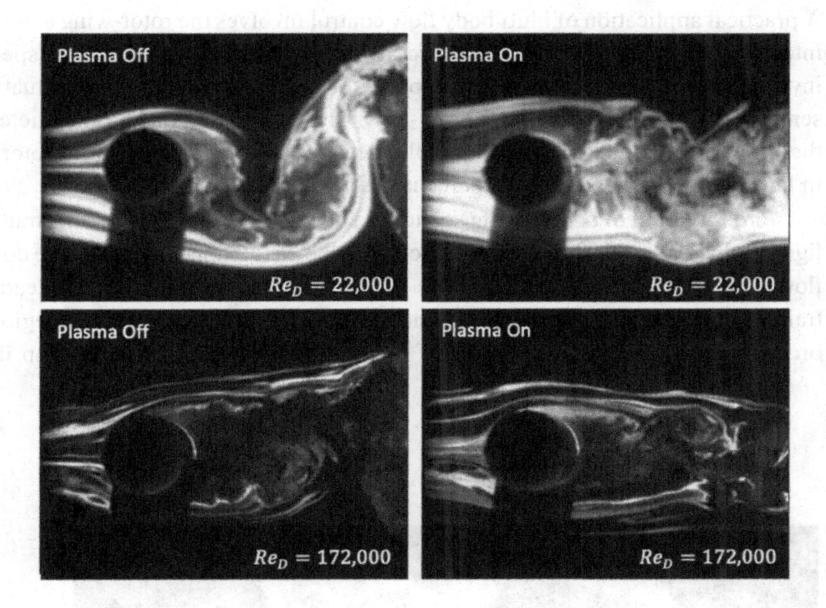

Figure 3.32 Flow visualization showing the effect of plasma streamwise vortex generators on the near wake of a circular at $\mathrm{Re}_D = 22{,}000$ and 172,000. Taken from Thomas and Koslov (2010).

Figure 3.32 presents flow visualization images of cylinder wake with the plasma streamwise vortex generators that are viewed in a plane that cuts across the wake at a spanwise location between a pair of plasma electrodes. These images correspond to $\mathrm{Re}_D = 22{,}000$ and 172,000. The left images at each Reynolds number are for the basic cylinder with the plasma actuators OFF. The right images are with the plasma actuators operating. In both Reynolds number cases, the plasma streamwise vortex generating arrangement on the cylinder fully suppressed the von Kármán vortex street. Thomas and Koslov (2010) reported that for the $Re_D = 22{,}000$ case, with the plasma vortex generators, the mean velocity wake defect and wake width measured at $x/D = 2$ were respectively reduced by 24 and 20 percent. In the $Re_D = 172{,}000$ case, the wake deficit was not substantially reduced, although the wake width was reduced by 12 percent.

3.4 Practical Wake Control Applications

This section presents a number of practical applications of bluff body flow control. These make use of all of the understanding gained through fundamental research, particularly that involving circular cylinders.

3.4.1 Hover Flight Rotor–Wing Interaction

A practical application of bluff body flow control involves the rotor-wing aerodynamic interaction that occurs on tilt-rotor aircraft such as the V-22 Osprey. This specifically involved the downwash from the propeller on the wing which in this situation presented itself as a bluff body in a cross-flow. This interaction had a dramatic effect on the hover performance. For example, the download on the wing of a tilt rotor aircraft in hover could be as large as 15 percent of the total rotor thrust.

A cross section of a V-22 wing section is shown in Figure 3.33. As illustrated in the figure in hover, the air flow is perpendicular to the airfoil upper surface. The downwash flow over the wing results in a large flow separation at the otherwise leading and trailing edges of the airfoil. This generates a large flow recirculation region and a broad wake that is a characteristic of bluff bodies. The flow visualization image in

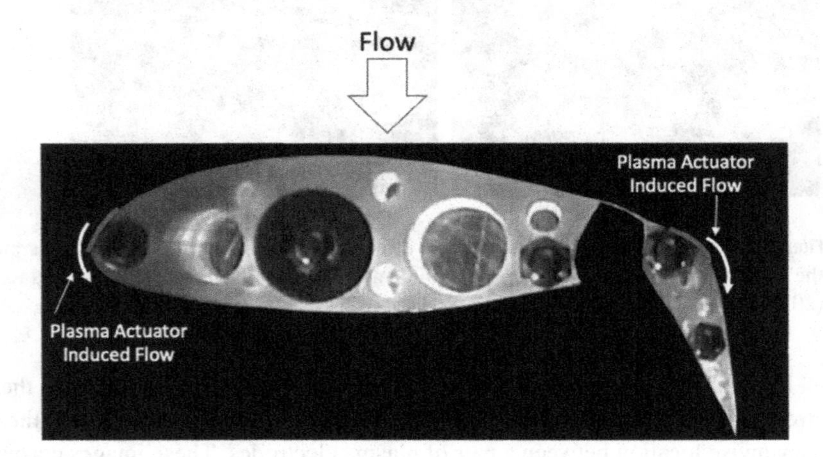

Figure 3.33 Photograph of a scaled version of a V-22 wing section used in wind tunnel tests of flow separation control in a hover configuration.

Figure 3.34(a) provides an illustration of the broad wake produced by the wing in the hover configuration. In an attempt to minimize the download aerodynamic force in hover, the trailing-edge flap is extended to its maximum deflection angle in order to project the minimum cross-sectional area.

There have been a number of attempts to reduce the aerodynamic drag produced by the wing during hover. These have concentrated on controlling the flow separation that occurs at the leading- and trailing-edge surfaces. These regions are indicated by the white curved arrows in Figure 3.33. The most prevalent of these have involved active control, since passive devices that modify the shape of the wing could be detrimental in the horizontal flight configuration. One of the most reported approaches is from Steele et al. (2009), who utilized unsteady surface jets located near the flow separation locations. As with any separation control approach, to be most effective, the flow control device needs to be placed as close as possible to the flow separation location. With the V-22 wing, one flow separation location was on the trailing-edge flap. Unfortunately, Steele et al. (2009) were unable to fit the unsteady jet hardware within the V-22 flap. Therefore, they placed it at the closest possible location in the wing body, upstream of the extended flap gap.

An alternate approach that was investigated was to utilize plasma actuators for flow separation control on the V-22 wing section. This had the specific advantage of being able to place a plasma actuator on the surface of the trailing edge, just upstream of the flow separation location. The locations of the plasma actuators at the leading-edge and trailing-edge flaps of the V-22 wing section are illustrated by the blue curves in Figure 3.33. The plasma actuators were aligned parallel to the leading and trailing edges of the airfoil, similar to how they were placed on the circular cylinder that was shown in Figure 3.20. As with the circular cylinder, the plasma actuators were designed to produce the effect of a tangential wall jet, in this case directed toward the leading and trailing edges. The tangential velocity components generated by the plasma actuators are illustrated by the white arrows in Figure 3.33.

The plasma actuators were operated to produce a periodic disturbance that corresponded to a reduced frequency, $F^+ = fc/U_\infty = 1$, that is shown (see Chapter 4) to be optimum for controlling separated flows. Here, c is the airfoil chord length and U_∞ is the free-stream downwash velocity. In the experiment, a duty cycle of 10 percent (defined in Figure 3.21) was found to be sufficient to minimize the drag on the airfoil section in the downwash configuration. Increasing the duty cycle had minimal effect and otherwise increased the power to the actuator.

Figure 3.34 shows flow visualization images that document the effect of the bluff body wake control. For these, the airfoil section was placed in a wind tunnel and oriented with the upper surface normal to the mean velocity vector. Flow visualization was performed by introducing smoke streaklines at the wind tunnel entrance. The smoke streaklines were oriented in the vertical direction at the airfoil section spanwise centerline. Circular end plates were used to minimize end effects on the finite span airfoil section. The end plates were clear to allow optical access to the visualized wake. Finally, the smoke streaks were illuminated by a high-intensity light source that entered through a glass-covered slot in the test section floor. A mirror on the ceiling

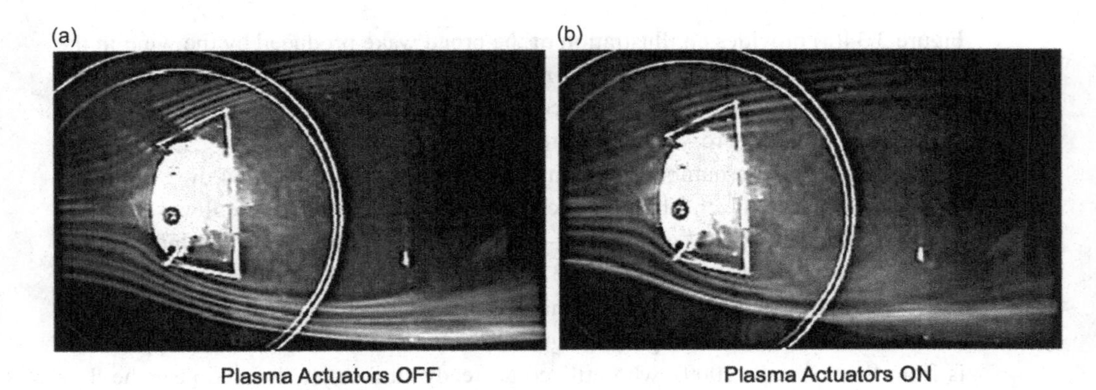

Figure 3.34 Visualized flow around V-22 airfoil in hover configuration for 20-m/s free-stream velocity with actuators OFF (a), and leading- and trailing-edge actuators operating with unsteady frequency with $F^+ = 1$ (b).

of the test section reflected some of the light back to help to redirect the light more uniformly.

Figure 3.34 provides a comparison of the visualized flow around the V-22 airfoil section in the hover configuration with the plasma actuators OFF (left image) and operating (right image). The airfoil section was oriented with the trailing-edge flap at the top and the leading edge at the bottom. In order to provide a scale, the airfoil chord length with the trailing edge deflected at 70° was 15.24 cm. The free-stream velocity was 20 m/s. The flow visualization indicates a noticeable reduction in the width of the wake with the plasma actuators operating (ON). Quantitative results of the effect of the actuators come from mean velocity measurements in the wake of the wing section.

Figure 3.35(a) provides an example of the mean velocity profiles measured across the wake of the V-22 airfoil section. These were measured on the spanwise centerline of the wing section. As in the flow visualization image, the airfoil trailing-edge flap is at the top and therefore its effect corresponds to the upper part of the profile. The measurements were taken approximately three chord lengths (45.73 cm) downstream of the airfoil. This was downstream of the flow recirculation region which allowed the mean velocity profiles to be used to determine the effect of the separation control on the aerodynamic drag.

The effect of the plasma actuators was to noticeably reduce the velocity defect and width of the wake. The baseline wake profile (plasma actuator OFF) was noticeably asymmetric, with a larger velocity deficit occurring on the side of the trailing-edge flap. The flow separation control with the plasma actuator operating both at the leading and trailing edges resulted in a more symmetric wake, similar to what would be expected from a symmetric section shape.

The pressure drag was determined from mean velocity profiles like those in Figure 3.35(a). The drag normalized by the chord length, c, is

$$\frac{D}{c} = \int_0^h (P_0 + \rho U_0^2) dy - \int_0^h (P_1 + \rho U_1^2) dy, \tag{3.9}$$

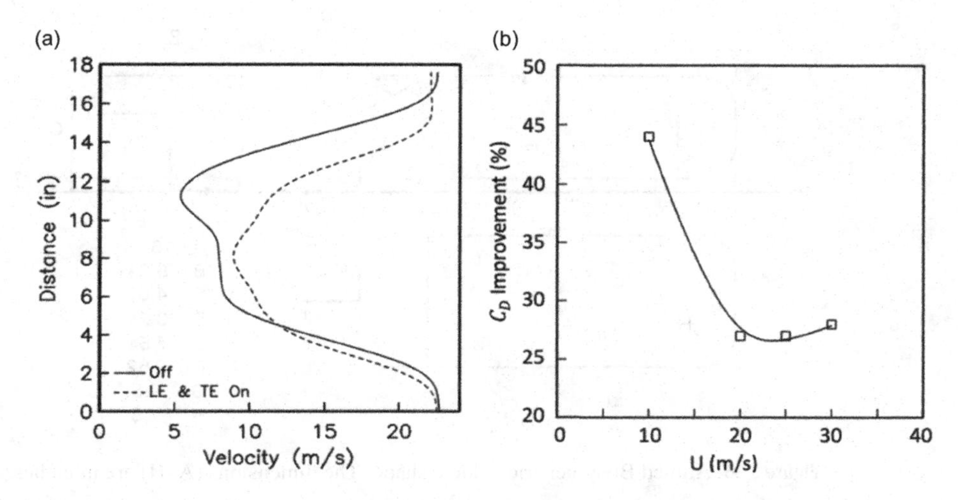

Figure 3.35 Mean velocity profiles in the wake of V-22 section in the hover configuration for a 20-m/s velocity with the plasma actuators OFF and operating (a), and percent of drag coefficient improvement as a function of free-stream velocity (b).

where the subscripts 0 and 1 refer to the respective measurement stations upstream and downstream of the airfoil, P is the static pressure, U is the mean velocity, and y is the cross-stream direction with a total length of h that completely encompasses the width of the wake of the airfoil. Upstream of the airfoil, the mean velocity profile is uniform. In addition, the static pressure is not a function of the cross-stream location. As a result, the equation for drag reduces to

$$\frac{D}{c} = h(P_0 - P_1) + h\rho U_0^2 - \int_0^h (\rho U_1^2)dy. \tag{3.10}$$

Figure 3.35(b) documents the amount of drag reduction produced by the wake control as a function of the free-stream velocity. At the lower free-stream velocity, a nearly 45 percent drag reduction is achieved. In this case, the boundary layers at the flow separation location are laminar and therefore the separation control is more effective. At the higher free-stream velocities, the boundary layers at the separation location are turbulent, and the drag reduction is less, but still a respectable 27 percent. When considering that the download on the wing of a tilt rotor aircraft in hover can be as large as 15 percent of the total rotor thrust, a 27 percent decrease can reduce the download to 11 percent which is directly translatable to extra payload or fuel that can extend the range.

3.4.2 Surface Vehicle Aerodynamics

Another practical application of bluff body wake control involves surface vehicles, specifically automobiles and trucks. A comprehensive review of surface vehicle aerodynamics was given by Hucho and Sovran (1993). The problem of reducing

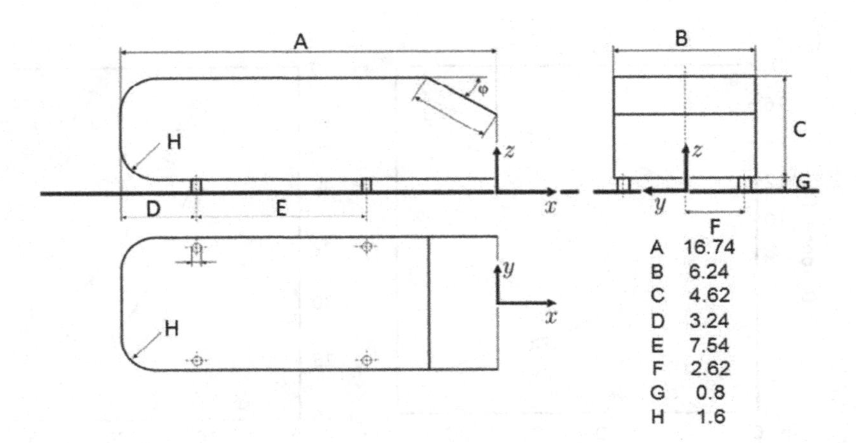

Figure 3.36 Ahmed Body generic vehicle shape. The dimensions (A–H) are in inches.

aerodynamic drag on surface vehicles has been approached in many different ways. As with all bluff bodies, the predominant component is form or pressure drag.

Passive flow control devices have included flaps (Beaudoin, 2008) and vortex generators (Aider, 2009). Buchheim et al. (1976) suggested that a roof ramp angle between 10° and 15° would produce the optimum drag reduction. Krishnani and Zhou (2010) found a similar result with the lowest drag occurring with a wake edge flow angle of approximately 10° (toward the ground plane). As significant, they observed a dramatic *increase* in the aerodynamic drag if the flow turning angle was larger than the optimum.

The flow over the upper-rear surface has been of particular interest in the effort to reduce aerodynamic drag on surface vehicles. The influence of the slant angle of the upper-rear surface has been extensively investigated by Ahmed (1984a) and Hucho and Sovran (1993). The so-called Ahmed Body (Ahmed, 1984b) has been a generic vehicle shape used in these investigations. An Ahmed Body is shown in Figure 3.36. The dimensions (A–H) in this representation are in inches. The relative dimensions are specific to the Ahmed Body. Therefore, a linear scale factor that preserves the shape is used to define an Ahmed Body of any size.

The effect of the upper-surface downward slant angle, ϕ, on the aerodynamic drag coefficient of an Ahmed Body is presented in Figure 3.37. Initially, a downward deflection of the upper-rear surface results in a decrease in the drag coefficient, C_D. The drag coefficient reaches a minimum at $\phi \simeq 12°$. Surface flow visualization by Ahmed (1984a) indicated that up to that deflection angle, the flow was attached to the ramp surface. Based on this, one can assume that the downward vector angle of the mean flow matched that of the ramp angle, ϕ.

Ahmed (1984a) observed that larger upper-surface downward slant angles of $\phi >$ 12° resulted in a rapid increase in the drag coefficient. Flow visualization (Ahmed, 1984a) revealed that the larger slant angles resulted in the development of a pair of streamwise vortices that formed at the side-corners of the ramp. This is illustrated in

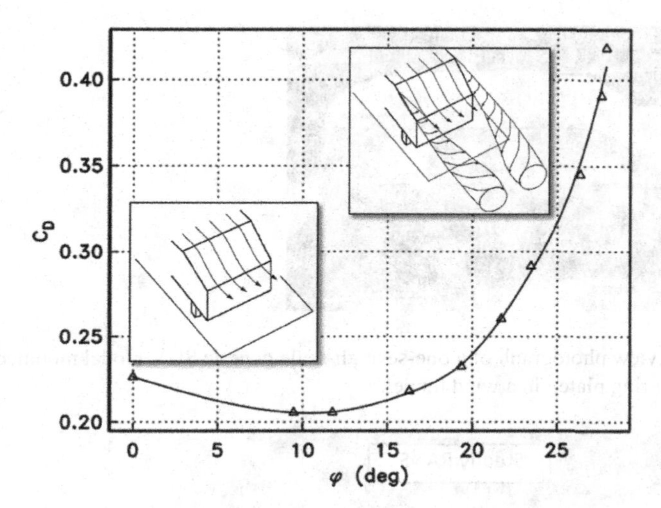

Figure 3.37 Effect of the upper-rear angle on aerodynamic drag coefficient of an Ahmed Body. C_d data is replotted from Ahmed (1984a).

the upper right inset to the plot in Figure 3.37. Streamwise vortices are known to add significant drag. A case in point are wing-tip vortices that contribute a majority of the drag on an aircraft wing, and prompt the development of passive control winglets. Thus, if the objective is drag reduction, these larger slant angles are to be avoided.

There have been a number of *active* flow control approaches applied to the slanted upper-rear surface of an Ahmed Body, with the objective of reducing the aerodynamic drag. These have included steady blowing (Lehugeur and Gillieron, 2006; Brunn et al., 2007), steady suction (Roumeas et al., 2009), pulsed blowing (Krentel et al., 2010), synthetic jets (Kourta and Leclerc, 2013), and plasma actuators (Boucinha et al., 2011; Shadmani et al., 2018; Kim et al., 2020). Collectively, these approaches resulted in drag reduction that ranged from 5 to 13 percent.

The following presents an active flow control *example* on a less generic vehicle shape consisting of a one-seventh-scale model of a sport utility vehicle (SUV) that illustrates an active flow approach. A side view of the SUV model is shown in Figure 3.38, where it is mounted on a drag force measuring platen in a wind tunnel.

As a first step in designing an active approach for drag reduction, steady Reynolds-averaged Navier–Stokes (RANS) flow simulations were performed to characterize the flow in the vicinity of the aft upper surface of the SUV. As a result of the dependence of the drag coefficient on the aft flow turning angle that was presented in Figure 3.37, the focus of the simulations was on the angle of the flow entering the wake of the SUV. As shown in Figure 3.39, based on a line where the local time-averaged velocity was 70 percent of the free-stream velocity, the simulation predicts a mean wake angle of $\psi = 4°$. This flow vector angle is less than the 12° where the drag is expected to be a minimum. Therefore, the *objective* of the active flow control was to vector the flow toward the optimum angle.

Figure 3.38 Side view photograph of a one-seventh-scale generic SUV model mounted on a drag force measuring platen in a wind tunnel.

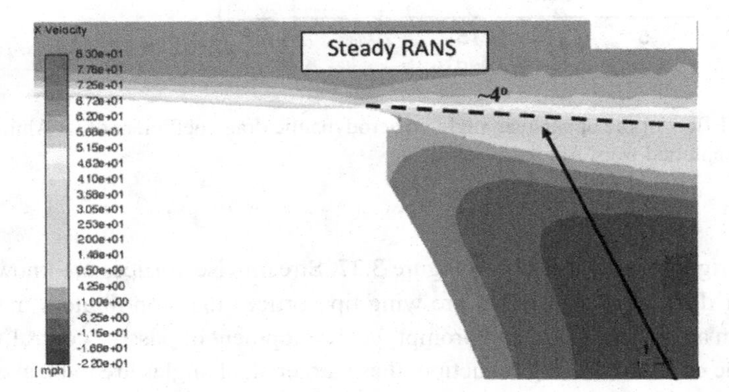

Figure 3.39 Contours of constant mean velocity at the center-span of the wake of the one-seventh-scale SUV based on a steady RANS simulation.

The active flow control involved placing a plasma actuator on the aft portion of the SUV upper surface. This is illustrated in Figure 3.40. The plasma actuator was designed to produce a body force vector field that would accelerate the flow over the curved edge of the SUV upper surface. This was intended to move the separation location downstream toward the trailing edge of the SUV upper surface. The unsteady RANS simulations that included the plasma actuator body force indicated that this would vector the flow in the wake to the optimum 12° angle.

The plasma actuator was operated to produce a periodic disturbance. The drag on the SUV was measured for a range of disturbance frequencies. Figure 3.41 shows the results for the frequency that produced the minimum drag (largest drag reduction) for two free-stream velocities. For this, the optimum disturbance frequency was represented by its reduced frequency, $F^+ = fL/U_\infty$, where U is the local free-stream velocity, and L is the streamwise extent of the separated flow region on the aft portion of the SUV. As presented in Chapter 4, $F^+ = 1$ is generally regarded as optimum for

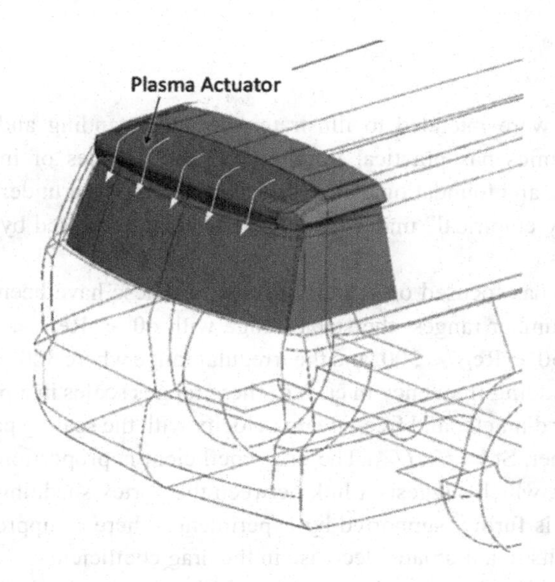

Figure 3.40 Illustration of the placement of a plasma actuator on the aft upper-surface of the SUV that is designed to vector the flow to the optimum $\psi \simeq 12°$ angle that minimizes the drag coefficient.

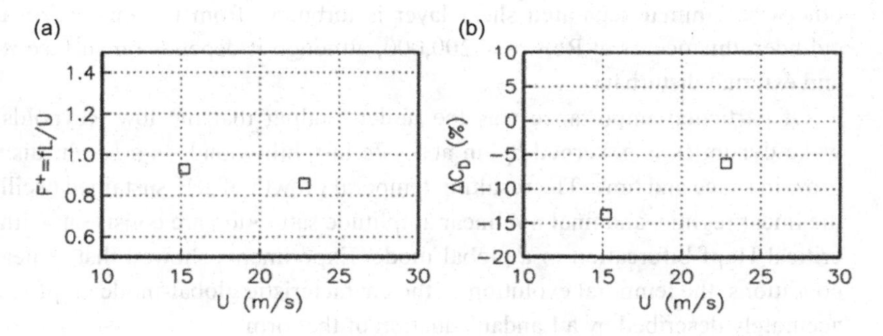

Figure 3.41 Strouhal number of unsteady plasma actuator frequency to maximize drag reduction (a) and percent change in drag coefficient (b) for two free-stream speeds for one-seventh-scale SUV model.

the control of separated flows. In the case with the SUV, for both free-stream speeds, the maximum drag reduction occurred for controlled periodic disturbances at $F^+ \simeq 1$.

Figure 3.41(a) documents the percent change in the drag from the baseline (actuator OFF) condition at the two velocities. At the lower velocity, $\Delta C_D = 14$ percent. At the higher velocity, the drag reduction is somewhat less at $\Delta C_D = 6$ percent. Both these values are within the range (5–13 percent) of other active flow control approaches applied to ground vehicles.

3.5 Summary

As these last examples were intended to illustrate, the understanding and control of bluff body aerodynamics has practical importance. The decades of innovative experiments, simulations, and foundational analysis have expanded our understanding well beyond the "entirely empirical" understanding previously expressed by Roshko (1993).

Much of the research has focused on circular cylinders. These have been used to define three Reynolds number ranges: the stable range with $40 < \mathrm{Re}_D < 150$, the transition range where $150 < \mathrm{Re}_D < 300$, and the irregular range where $300 < \mathrm{Re}_D <$ 10,000+. The vortex shedding frequency in each of these ranges scales in a particular manner with the cylinder diameter and free-stream velocity with the scaling parameter being the Strouhal number, $\mathrm{St} = fD/U_\infty$. The drag coefficient is proportional to the inverse Strouhal number, which suggests a link between the vortex shedding and the aerodynamic drag. This is further supported by experiments where a suppression of the vortex shedding results in a dramatic decrease in the drag coefficient.

The boundary layer separation location remains fairly constant over the range $2000 \leq \mathrm{Re}_D \leq 200{,}000$, and, as a result, over this range, the drag coefficient only increases slowly. This is the subcritical Reynolds number range where the Strouhal number is relatively constant. The drag on a bluff body abruptly drops when the otherwise laminar separated shear layer is turbulent from the onset. For a circular cylinder, this occurs at $\mathrm{Re}_D \sim> 200{,}000$, although it depends on surface roughness and external disturbances.

Of particular importance was the understanding that the low Reynolds number wake dynamics is governed by an absolute instability, and, as a result, disturbances grow in space and time. The resulting temporal growth of self-sustained oscillations in the linear regime, and final nonlinear amplitude saturation are consistent with a super-critical Hopf bifurcation to a global mode. Experiments showed that at near-critical conditions, the temporal evolution of the characteristic global-mode amplitude can be accurately described by a Landau equation of the form

$$\frac{dB}{dt} = c_1(\mathrm{Re} - \mathrm{Re}_{G_c})B - c_2|B|^2 B, \tag{3.11}$$

where $c_1(\mathrm{Re} - \mathrm{Re}_{G_c})$ is the linear temporal growth rate, and $c_2 > 0$ is for a subcritical bifurcation.

An intrinsic feature of the bluff body wakes is a mean flow recirculation in which downstream disturbances can travel upstream. With this in mind, Huerre and Monkewitz (1990) considered a 2-D jet/wake having variable velocity and density ratios given as

$$\Lambda = (U_{\mathrm{wake/jet}} - U_\infty)/(U_{\mathrm{wake/jet}} + U_\infty) \tag{3.12}$$

and

$$S = \rho_{\mathrm{wake/jet}}/\rho_\infty, \tag{3.13}$$

on which they focused on the boundaries between convective and absolute instability in the (S, Λ) parameter space. With this, they observed that reverse flow, $\Lambda > 1$, promotes an absolute instability. In contrast, co-flow for the jet, or base bleed for the wake, promotes a convective instability. The analysis also determined that for the wake, the sinuous or von Kármán mode that is commonly observed becomes absolutely unstable first, compared to the varicose or symmetric shedding mode. Lowering the fluid density, for example in air through heating, suppressed the wake absolute instability.

The absolute instability governing the wake dynamics makes it sensitive to even small changes in the mean flow. As an example, the addition of a small "control cylinder" in the near-wake region of a larger circular cylinder could fully suppress vortex shedding (Strykowski, 1986; Strykowski and Sreenivasan, 1990). Furthermore, the sensitivity to the control cylinder placement was dramatically reduced when it was heated, supporting the Huerre and Monkewitz (1990) predictions on the effect of density.

Other approaches that modified the near-wake and fully suppressed vortex shedding were rigid and flexible splitter plates, base bleed, and surface mounted actuators such as unsteady blown slots and plasma actuators. Considerable attention has also been given to bluff bodies that incorporated different forms of spanwise waviness, or variations in thickness or diameter in the case of cylinders. These were intended to introduce a 3-D character in the otherwise 2-D vortex shedding. In these, there appeared to be an optimum spanwise wavelengths at which the vortex shedding was suppressed and the drag was minimized. These spanwise wavelengths corresponded to $\lambda/D \simeq 2.5$ and $\lambda/D \simeq 6.0$, with the larger wavelength producing the greater drag reduction. The effect could also be produced by spanwise spaced tabs and vortex generators located on the surface of the bluff body. An active approach using plasma vortex generators was equally effective.

The knowledge gained from such fundamental investigations has benefited numerous practical applications involving bluff body flows. This chapter highlighted just two of these to demonstrate the flow control aspects. In each, the flow control approach exploited some aspect of the wake instability to achieve the desired effect.

Problems

3.1 A method sometimes used to calibrate hot-wire sensors is to place them in the wake of a circular cylinder where they can detect the vortex shedding. The shedding frequency is related to the velocity through the relation $\mathrm{St} = 0.212(1 - 21.2/\mathrm{Re})$ that was shown in Figure 3.4.

1. Based on this approach, determine the number of cylinder diameters that might be needed to allow hot-wire calibration for velocities ranging from 5 to 60 m/s.
2. How might the calibration be performed?
3. What difficulties might arise in calibrating at the highest velocities?

3.2 Hot-wire sensors are simply small diameter cylinders that are heated. Velocity is inferred by measuring the heat transfer from the wire, where the governing parameter is the Nusselt number, Nu. An empirical relation used in the calibration of hot-wire sensors is $\mathrm{Nu} = C_1 + C_2\sqrt{\mathrm{Re}_D}$, where C_1 and C_2 are functions of the Prandtl number. This relation is generally valid for $1 \leq \mathrm{Re}_D \leq 1000$.

1. We expect there to be a relation between the heat transfer and the drag coefficient. Use the information in Figure 3.6 to generate a plot of Nu versus C_D for a circular cylinder with $1 \leq \mathrm{Re}_D \leq 1000$.

3.3 Circular cylinders undergo a dramatic drop in drag ("drag crisis"), at $\mathrm{Re}_D >$ 200,000. The drop in drag is associated with the turbulence onset of the boundary layer which caused the separation location to move further aft along the cylinder surface with the result being a contraction of the wake.

1. Describe three passive methods that could be used to lower the drag on a circular cylinder.
2. Based on Problem 2.2, what effect would the "drag crisis" have on the heat transfer from the "cylinder." Is there a practical application for this?
3. Spheres also undergo a drag crisis at higher Reynolds numbers. List three passive methods that apply to spheres to lower the drag on spheres. Provide practical examples.

3.4 It has been established that at low Reynolds numbers, the wake undergoes an absolute instability to disturbances with a temporal amplitude that can be described by a Landau equation of the form given in Eq. (3.1). This applies near criticality where $\mathrm{Re} \simeq \mathrm{Re}_{G_c}$.

1. Based on Eq. (3.1), plot the amplitude, $B(t)$, as a function of time for $c_1 = c_2$ where (a) both are positive, and (b) where c_1 is positive and c_2 is negative.
2. For the case in which c_2 is positive, determine the saturation amplitude.

3.5 A concept to lower the drag on an undersea towed cable is to add flexible trailing filaments as shown in the following image.

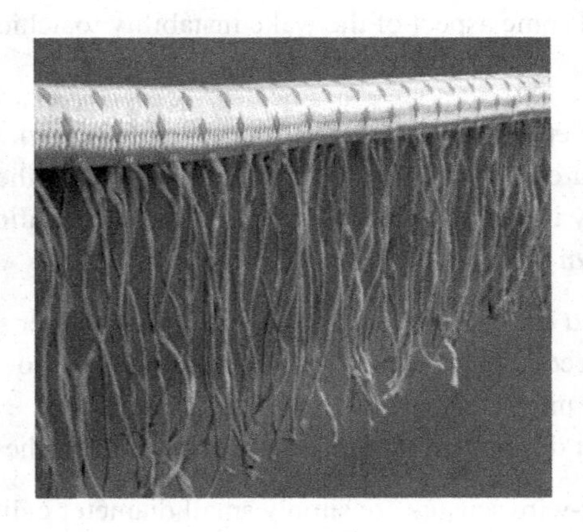

1. Describe how filaments might reduce the drag on the cable. What governs the length of the filaments?
2. What is the closest analog to the passive bluff body wake control devices presented in this chapter?
3. The producer of the cable shown in the image also claims that the acoustics generated by the towed cable are lowered. What does that apply about the wake dynamics?
4. Can you think of other approaches that can be applied to the towed cable to reduce the drag?

4 Separated Flows

Flow separation can occur in a variety of applications involving external flows, particularly related to aircraft, and internal flows such as within turbo-machines. In most applications, flow separation has adverse effects on performance and is to be avoided. However, with helicopters, it is an asset to its performance. This chapter discusses the sources and characteristics of a separated flow field and methods to either prevent flow separation or to reattach a separated flow.

4.1 Introduction

Flow separation results when the flow does not have sufficient momentum to overcome an adverse pressure gradient, or when viscous dissipation occurs along the flow path. It is almost always associated with some form of aerodynamic penalty, including a loss of lift, an increase in drag, a loss of pressure recovery, and an increase in entropy. Although flow separation is often viewed to be two-dimensional (2-D) and *steady*, experiments have shown the process to be strongly three-dimensional (3-D) and highly unsteady, involving separated shear layer instabilities that are highly receptive to external excitation. This is particularly the case in turbo-machines that exhibit strong unsteadiness and high turbulence levels.

Thick airfoil sections have a tendency for the flow to separate in the adverse pressure gradient that forms past the airfoil maximum thickness point. As depicted in the illustration in Figure 4.1(a), this can result in a separation "bubble" or a large separation zone that extends past the airfoil trailing edge. On an airfoil, this depends on the state of the approaching boundary layer (laminar or turbulent) and the streamwise pressure gradient which is a function of the angle of attack. Trailing-edge separation is more likely to occur if the approaching boundary layer is laminar. In either case, flow separation results in a loss of lift and an increase in drag. The angle of attack where this occurs is defined as the "steady stall angle of attack," α_{ss}. Methods to increase α_{ss} directly impact aircraft performance, including maneuverability, and takeoff and landing distances.

Having too large of an airfoil camber will cause the flow to separate at the leading edge, forming a leading-edge separation bubble depicted in the drawing in Figure 4.1(a). A flow visualization image capturing leading-edge separation on a NASA

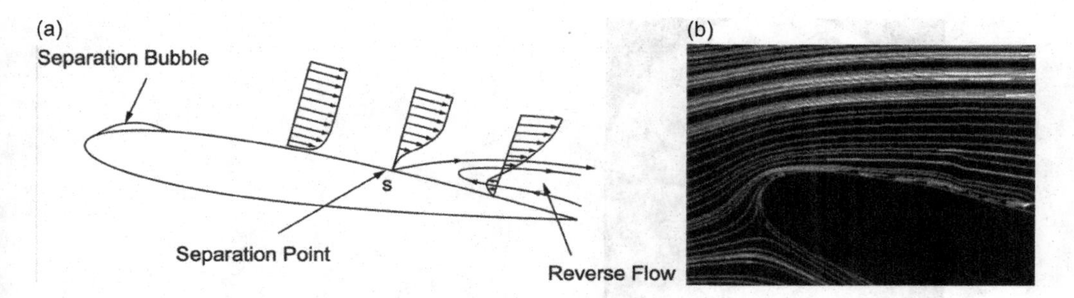

Figure 4.1 Example of flow separation that can occur on airfoils at moderate or large angles of attack (a) and flow visualization of the flow over the leading edge of a NASA EET airfoil section at a 14° angle of attack at $M = 0.30$ (b). Taken from Kelley et al. (2014).

Energy Efficient Transport (EET) wing section shape is shown in Figure 4.1(b). This can occur on aircraft with deflected trailing-edge flaps whose effect is equivalent to increasing wing camber. Since the leading edge of an airfoil produces the largest percentage of lift, suppressing leading-edge separation is important. Passive approaches aim to increase the leading-edge radius through leading-edge flap mechanisms. Such approaches add weight and complexity to the wing, which has prompted interest in active flow control methods.

Flow separation on wings can also result in a shock wave that forms over the surface of an airfoil. Figure 4.2 shows a visual example of a trailing-edge flow separation resulting from a standing shock wave obtained by Pearcey and Holder (1954). This is just one example of a shock–boundary-layer interaction in which the pressure drop across the shock, coupled with the local adverse pressure gradient, causes the flow to separate. Other examples include supersonic and hypersonic inlet ducts to propulsion systems in which shock waves are used to lower the local Mach number and build up pressure prior to combustion. Boundary-layer separation that occurs downstream of a shock wave introduces unsteadiness, and in the case of laminar boundary layers, causes turbulent transition, both of which are important issues seeking flow control measures. Chapter 9 is dedicated to this topic.

Flow separation is not limited to aircraft wings. It also can occur in the boundary layer around the fuselage. An example is presented in Figure 4.3 which shows visualization of the flow on the underside of a prototype fuselage model. Of particular interest is the aft portion of the fuselage in which the flow is clearly separated. A cross-cut through the separated flow region in Figure 4.3(b) shows it to be highly 3-D, with streamwise-oriented flow recirculation on either side of the center separation zone. These recirculation zones detach as streamwise vortices, similar to wing tip vortices, and significantly increase the fuselage drag. Flow control that can suppress the separated flow that leads to the formation of these vortices can have a significant impact on the aircraft performance.

Helicopters in forward flight require the rotor to periodically pitch to increase and decrease the rotor angle of attack as it respectively precesses through the retreating and advancing portions of the rotor revolution cycle (Corke and Thomas, 2015). It

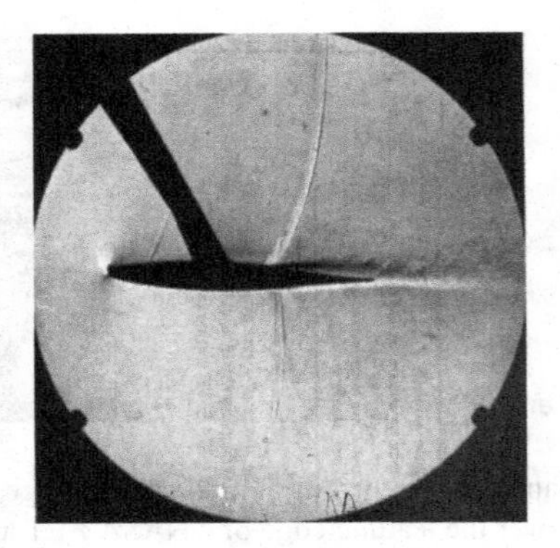

Figure 4.2 Schlieren photograph of the flow past an airfoil with 2° angle of attack at Mach 0.83 that reveals a shock-induced flow separation. Taken from Pearcey and Holder (1954).

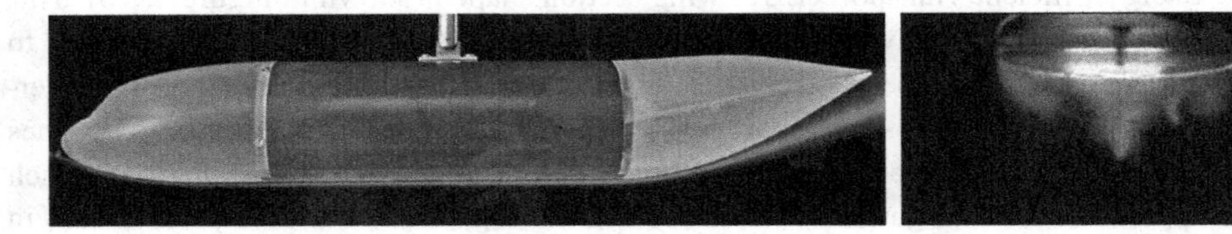

Figure 4.3 Example of flow separation in the aft portion of a fuselage where the cross-cut image on the right reveals the 3-D nature of the separated flow field.

is common for the rotor blade to undergo *dynamic stall* during the retreating portion of the rotor cycle. Dynamic stall is an incredibly rich fluid dynamics problem that manifests itself on an airfoil during rapid, transient motion in which the angle of incidence approaches or surpasses the static stall limit, α_{SS}. It is an important element of many man-made and natural flyers, including helicopters and super-maneuverable aircraft, and low Reynolds number flapping-wing birds and insects. The fluid dynamic attributes that accompany dynamic stall include an eruption of vorticity that organizes into a well defined Dynamic Stall Vortex (DSV) and massive excursions in aerodynamic loads that can couple with the airfoil structural dynamics. The dynamic stall process is highly sensitive to surface roughness that can influence turbulent transition and to local compressibility effects that occur at free-stream Mach numbers that are otherwise incompressible. Under some conditions, dynamic stall can result in negative aerodynamic damping that leads to limit-cycle growth of structural vibrations and rapid mechanical failure.

The dynamic stall process that occurs on an airfoil oscillating about its quarter-chord location is illustrated in Figure 4.4. Figure 4.4(a) shows the aerodynamic loads and pitch moment during the pitching cycle. Figure 4.4(b) shows the flow field structure during the pitching cycle that was visualized with particle streak lines introduced at an upstream location. Characteristic *stages* of the flow development in the pitching cycle are denoted by the numbers next to specific flow visualization images and aerodynamic load and moment cycle maps. The conditions of the visualized flow exemplify "deep" dynamic stall which occurs when the maximum angle of attack of

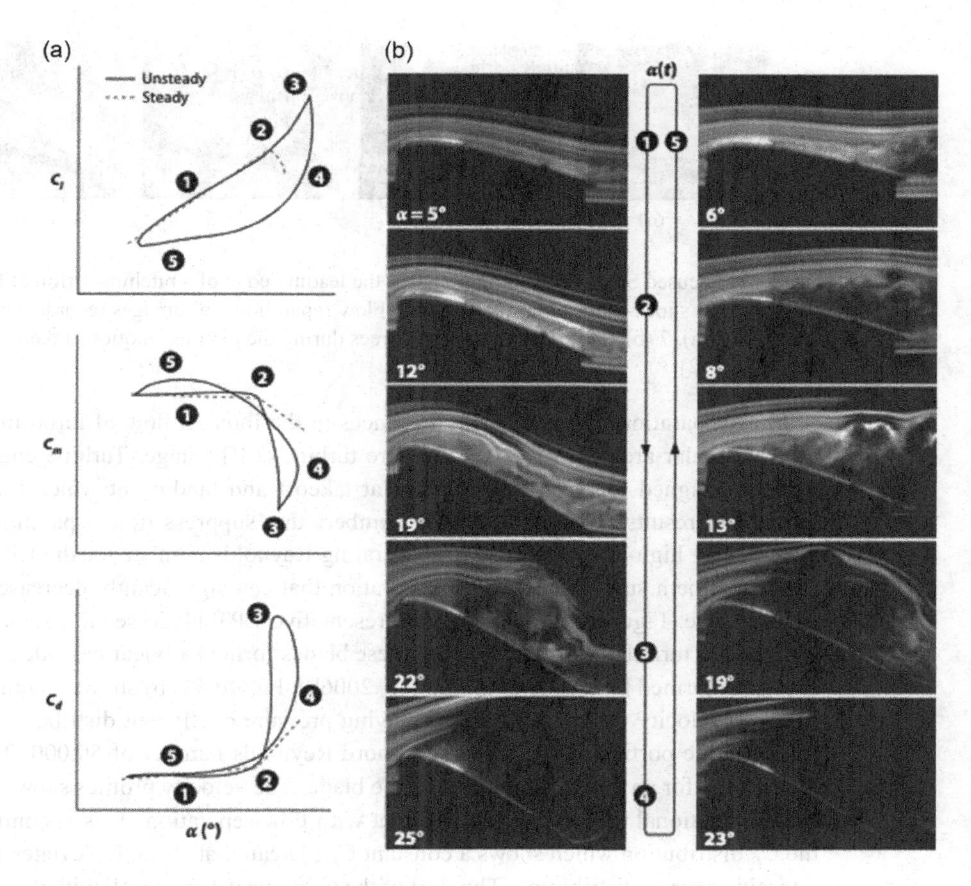

Figure 4.4 Illustration of dynamic stall events based on air loads and pitch moment cycle adapted from McCroskey (1982) (a) and visualized flow about a pitching airfoil undergoing deep dynamic stall (b). Taken from Corke and Thomas (2015).

the pitching motion exceeded the steady airfoil stall angle of attack, α_{SS}. This results in a "fully developed" DSV shedding phenomenon that produces large peak aerodynamic loads and severe cycle hysteresis. The DSV formation and development has been found to be sensitive to leading-edge turbulent trips, which suggests that passive and active flow control could be an effective approach toward mitigating the detrimental effects of dynamic stall. Such approaches will be explored later in this chapter.

The airfoil pitching motion illustrated in Figure 4.4 can lead to weak shock formation near the leading edge at otherwise subsonic free-stream Mach numbers. An example is shown in Figure 4.5 which presents focused Schlieren images of the flow near the leading edge of the airfoil in a sequence of the pitch-up motion for a free-stream Mach number of 0.6. The weak shocks that form at the lower pitch angle initially proceed aft as the airfoil pitches up but eventually coalesce into a λ-shock. The λ-shock eventually becomes unstable, and a "stall vortex" emanates immediately downstream of the shock.

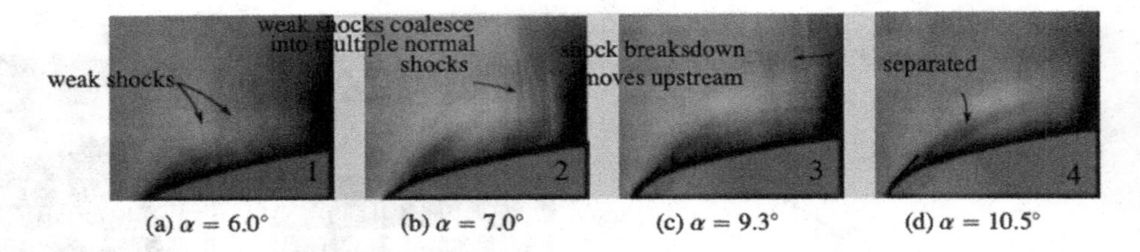

(a) $\alpha = 6.0°$ (b) $\alpha = 7.0°$ (c) $\alpha = 9.3°$ (d) $\alpha = 10.5°$

Figure 4.5 Focused Schlieren photograph near the leading edge of a pitching airfoil at Mach 0.6 that reveals a shock-induced flow separation. Flow separation with images recorded at pitch angles of 6 (a), 7 (b), 9.3 (c) and 10.5 (d) degrees during the pitch-up motion. Taken from Bowles (2012).

Flow separation occurs in many instances in the internal flow of turbo-machines. One particular area is in the low-pressure turbine (LPT) stage. Turbine engines are usually designed for peak performance at takeoff and landing at which the higher air density results in larger Reynolds numbers that suppress flow separation. However, during high-altitude cruise, the operating Reynolds number for the LPT blades can make them susceptible to flow separation that can significantly decrease turbine performance. Figure 4.6(a) shows a representative LPT blade section shape that is designed to turn the flow 95°. Nine of these blades formed a linear cascade in experiments performed by Huang et al. (2006a, 2006b). Figure 4.6(b) shows a combination of mean velocity profiles and accompanying pressure coefficient distribution for the trailing edge portion of the blade at a chord Reynolds number of 50,000. The solid C_p curve is for an inviscid flow around the blade. The velocity profiles show the onset of an inflectional profile that is consistent with flow separation. This is confirmed in the C_p distribution which shows a constant C_p plateau that strongly deviates from the inviscid pressure distribution. The start of the C_p plateau agrees well with the zero wall shear stress velocity profile that identifies the point of flow separation, S_p . The flow reattachment location on the blade, R_p, is identified in the C_p distribution as the point where the measured value returns to match the inviscid value. This is downstream of the last velocity profile, which is still highly inflected and includes a small amount of reverse flow. Control of flow separation in this case will be presented later in the chapter.

Another example of flow separation that can occur in the LPT stage of turbo-machines involves flow through the space (gap) between the rotor blade and the side wall. This is illustrated in Figure 4.7(a). The gap is obviously necessary to prevent the rotating turbine blade from contacting the casing surface. However, the pressure gradient that occurs around the blade can drive flow through the gap. The flow passing through the gap can lead to a flow separation on the underside of the blade. This was documented by Stephens et al. (2007, 2008) using surface visualization dots applied to the underside of representative blade shapes in a linear cascade. Examples of the results are shown in Figure 4.7(b). The tracks of the dots indicate the surface shear velocity magnitude and direction. The arrows have been added for clarity. These clearly reveal the separated flow pattern that existed on the underside of the blade. This has serious implications to turbine blades because the backflow recirculation

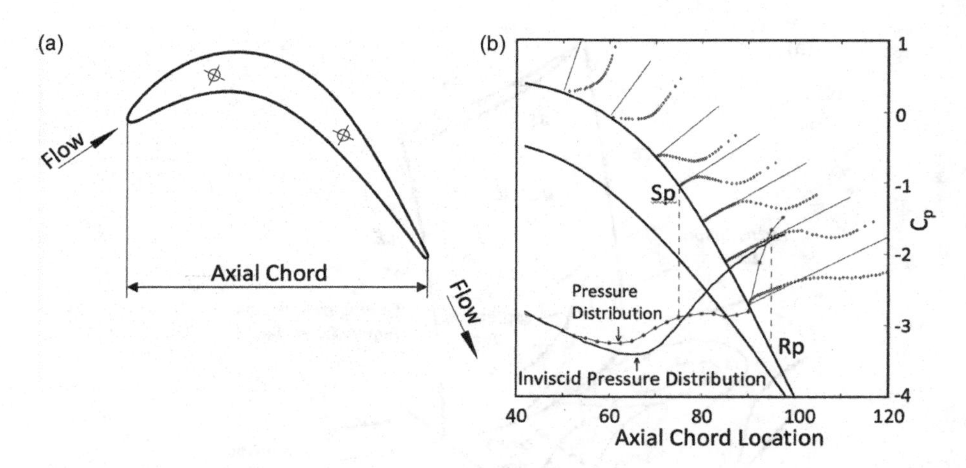

Figure 4.6 Low-pressure Pak-B turbine blade geometry (a) and experimental upper surface pressure distribution and corresponding mean velocity profiles revealing a trailing-edge separation bubble (b). Taken from Huang et al. (2006a).

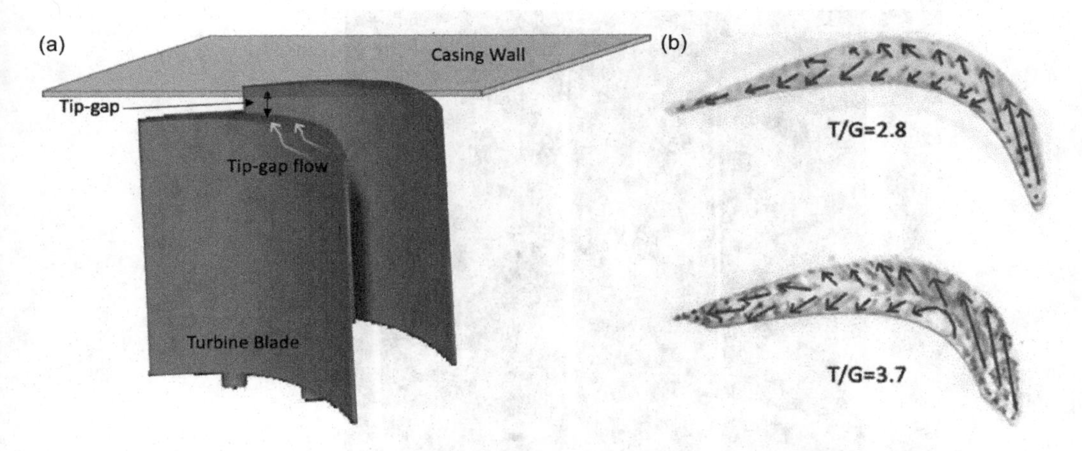

Figure 4.7 Schematic illustrating tip-gap used in turbine blades (a) and surface visualization of the separated flow field on the underside of the turbine blades (b) caused by the flow through the gap for two blade thickness-to-gap, T/G, ratios. Taken from Stephens et al. (2007).

associated with separated flows can trap hot gases that can degrade the blade tip. Flow control approaches to counter this are presented later in the chapter.

The previous two examples referred to rotating components of turbo-machines. The following refers to stator blades used in the compressor stage of turbo-machines. Of particular interest is a 3-D flow separation that is intrinsic to the hub and suction surface corner region of compressor stator blades, which is referred to as "corner stall." The surface flow topology of corner stall is illustrated in Figure 4.8(a). Figure 4.8(b) shows an image of the visualized flow on the suction-side surface of a representative pair of stator blades in a linear cascade that included an adverse pressure

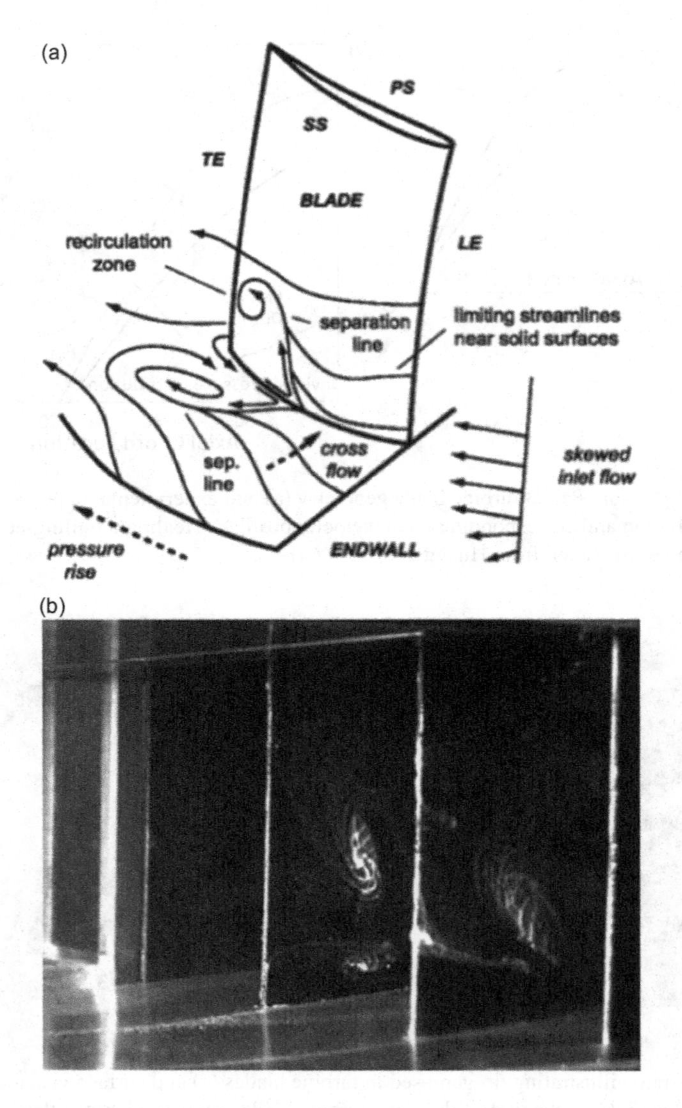

Figure 4.8 Schematic representation (a) and surface flow visualization of the corner stall separated flow that can occur in compressor stator stages (b). Flow visualization from Klevin (2016).

gradient Klevin (2016). Corner stall is responsible for the majority of the flow blockage and entropy generation in the compressor stage and, as a result, sets the limit on compressor performance. This makes it an important application for flow control.

Finally, many practical applications involve turning a flow though a duct. Figure 4.9 is an example from Samper (2019) of turning the flow in a 2-D duct by an angle of 120°. This shows contours of mean velocity and cross-stream vorticity, ω_z, in the region downstream of the bend. Both types of measurements show characteristics of a large flow separation region that originates at the inside surface of the bend. Flow

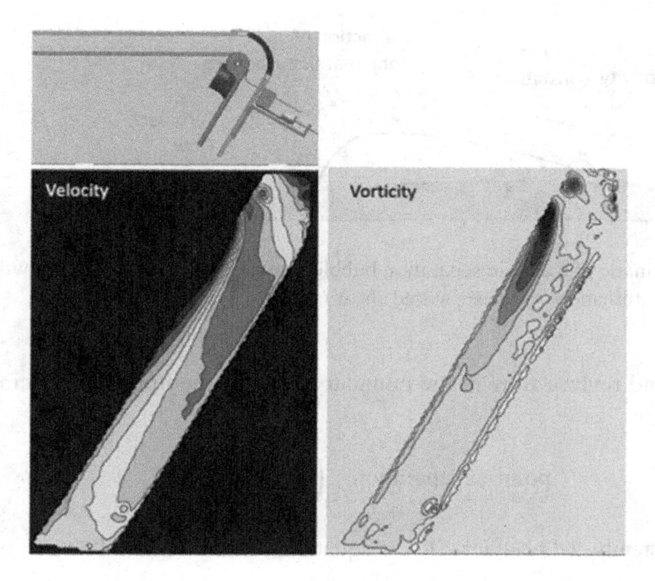

Figure 4.9 Velocity and vorticity measurements in a rectangular duct with a 120° bend. Taken from Samper (2019).

fields like this can be used as a test case to investigate different passive and active flow control techniques. Examples of these techniques will be presented later in the chapter.

The previous examples are a sampling of practical systems where flow separation impacts their performance and subsequently motivates methods of flow separation control. As a preface to a discussion on separation control techniques, and in keeping with the theme of this book, the following section first addresses the fluid instability aspects of separated flows.

4.2 Stability of Separated Flows

A feature of laminar separated flows is that they become unstable even at relatively low Reynolds numbers. As a result, such flows usually involve unsteadiness and/or a transition to turbulence. The mean-flow pattern and unsteady behavior of a separated flow is then largely governed by fluid instability and transition to turbulence. This section aims to establish the main *stages* of instability, turbulent transition, and reattachment of a separated flow. This includes the transformation of external disturbances into small amplitude oscillations of the separated flow, the linear stability of the separated shear layer, interactions and nonlinear growth, amplitude saturation and spectral broadening, and turbulent transition. An illustration of these stages is shown in Figure 4.10.

In boundary layers, separation in steady flows occurs only in regions of adverse pressure gradient, $dp/dx > 0$. The point of separation is defined as the limit between

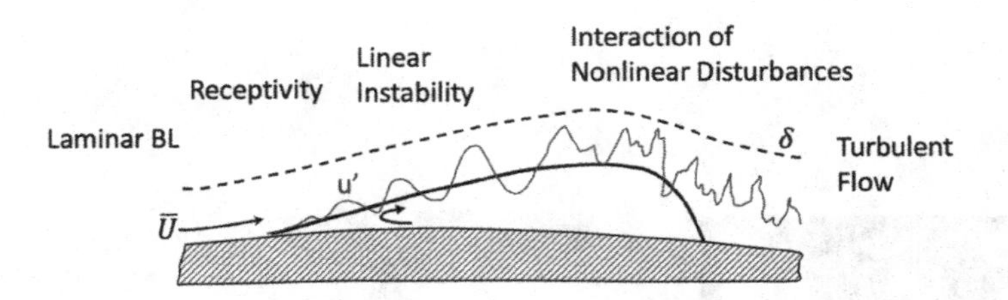

Figure 4.10 Schematic of a classic separation bubble showing the stages of the growth and breakdown of disturbances of the separated shear layer.

downstream and reverse flow in the boundary layer in the immediate vicinity of the wall ($y = 0$), or

$$\text{point of separation:} \quad \left(\frac{\partial u}{\partial y}\right)_{y=0} = 0. \tag{4.1}$$

Based on the steady, 2-D Navier–Stokes equations, with $u = v = 0$ at $y = 0$,

$$\mu \left(\frac{\partial^2 u}{\partial y^2}\right)_{y=0} = \frac{dp}{dx}. \tag{4.2}$$

Differentiating Eq. (4.2) with respect to y gives

$$\left(\frac{\partial^3 u}{\partial y^3}\right)_{y=0} = 0. \tag{4.3}$$

In a decelerated flow where $dp/dx > 0$, based on Eq. (4.2), $(\partial^2 u/\partial y^2)_{\text{wall}} > 0$. Regardless of pressure gradient, the mean profile far from the wall requires that $\partial^2 u/\partial y^2 < 0$. Therefore, under adverse pressure gradient conditions, a point of inflection must exist somewhere in the boundary-layer profile.

Even prior to separation, an adverse pressure gradient significantly affects the stability to linear disturbances of the attached boundary layer. For example, for a 2-D laminar Falkner–Scan–Hartree boundary layer with streamwise pressure gradient parameter, β, the minimum critical Reynolds number for disturbances to amplify varies from $\text{Re}_{\text{crit}} = 420$ for zero pressure gradient ($\beta = 0$) to approximately $\text{Re}_{\text{crit}} = 20$ for an adverse pressure gradient at the point of separation ($\beta = -0.1988$). Therefore, one can expect disturbances to amplify in the boundary layer approaching flow separation. This is illustrated in Figure 4.11, which shows a representative pressure gradient over an airfoil. Near the leading edge, the convex curvature results in a favorable pressure gradient. However, past the maximum thickness point, an adverse pressure gradient is required to satisfy the Kutta condition at the trailing edge. In this region indicated by the left vertical dashed line, the adverse pressure gradient results in an inflected mean profile that promotes boundary-layer instability, with the resulting growth in amplitude of velocity fluctuations, u'_f. At high angles of attack, the adverse pressure gradient can result in flow separation. The onset of flow separation is marked by the right vertical dashed line. The disturbances that emanate in

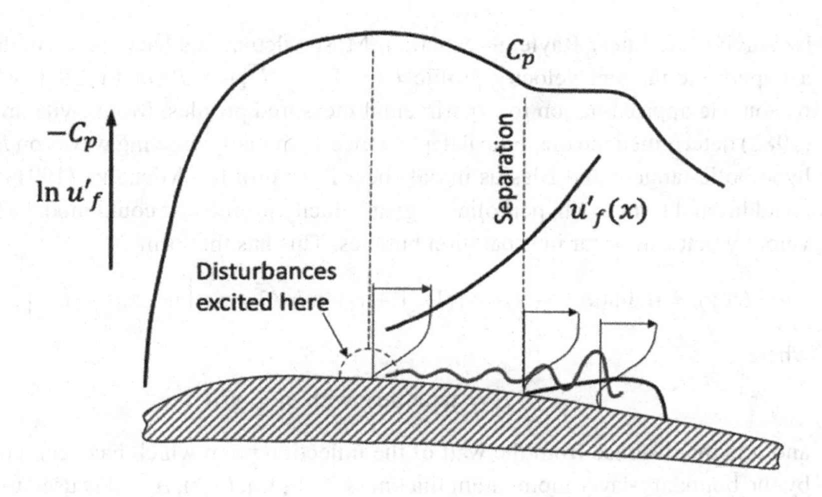

Figure 4.11 Schematic relating to the effect of the local streamwise pressure gradient on the mean velocity profile and to the growth of disturbances in the boundary layer that precede flow separation.

the upstream attached boundary layer can subsequently influence the separated flow development and particularly drive the instability of the separated shear layer.

The initial disturbances that are amplified by the separated shear layer instability can be a combination of vortical and acoustic (pressure) fluctuations. A source for the former can be linear instability waves in the upstream (laminar) boundary layer, or through upstream turbulence fluctuations. The sensitivity of separated shear layers to acoustic disturbances has been known for more than 150 years, following the dramatic examples of acoustic control of gas jets and buoyant smoke plumes by Tyndall (1864). This however requires a mechanism to convert long-wavelength acoustic pressure fluctuations into shorter-wavelength vortical fluctuations that can be amplified by the shear layer instability. That mechanism is "acoustic receptivity" (Morkovin and Paranjape, 1971; Morkovin, 1988). As illustrated in Figure 4.10, the acoustic receptivity site in a separated flow occurs at the interface between the separated shear layer and the wall surface.

The separated shear layer is inviscidly unstable to small disturbances through the Kelvin–Helmholtz instability mechanism. The same instability mechanism generally applies to free shear layers, which is covered in Chapter 5. At incompressible Mach numbers, this results in the exponential growth of 2-D waves. These waves eventually roll up into discrete vortices. If the basic flow $U(y)$ is taken to be strictly parallel, and the disturbances are infinitesimally small (satisfying the linear stability assumptions), the perturbation stream function can be cast into the form

$$\psi(x,y,t) = \psi(y)\exp[i(\alpha x - \omega t)] + c.c., \tag{4.4}$$

where $\omega = 2\pi f$ is the angular frequency, α is the wave number, and $c.c.$ denotes the complex conjugate. In the inviscid limit, $\mathrm{Re} = \infty$, the eigenfunction $\psi(y)$ is obtained by solving the linear Rayleigh equation. Most calculations have been conducted for a hyperbolic-tangent velocity profile $U(y;R) = \overline{U}[1 + R\tanh(y/2\theta)]$, which is a reasonable approximation to experimental measured profiles. Monkewitz and Huerre (1982) determined the theoretical dependence of spatially growing waves on R for both hyperbolic-tangent and Blasius initial shear layer profiles. Michalke (1991) included an additional term to a hyperbolic-tangent velocity profile that could model a backflow velocity that can occur in separation bubbles. This has the form

$$U(y) = [\tanh(a(y-d)) + t_1]/(1+t_1) + b\sqrt{3}\eta \exp\left[-1.5\eta^2 + 0.5\right]. \tag{4.5}$$

where

$$t_1 = \tanh(ad) \text{ with } \eta = y/d \tag{4.6}$$

and d is the distance from the wall of the inflection point which has been normalized by the boundary-layer momentum thickness, θ. In Eq. (4.5), $b < 0$ is used to generate a backflow component. The constant, a, is determined so that when normalizing y by the momentum thickness of each profile,

$$\int_o^\infty U(y)[1 - U(y)]dy = 1. \tag{4.7}$$

Note that for $b = a = 0$, the traditional $U(y) = \tanh(ay)$ is recovered.

Figure 4.12 shows examples of four profiles derived from Eq. (4.5). These correspond to combinations of d/θ and b. For reference, the Falkner–Skan–Hartree velocity profile at separation $\beta = -0.1988$ has $d/\theta = 4.0026$ and $b = 0$. This profile closely matches the dashed curve designated by the circle symbol that is shown in Figure 4.12. Therefore, this curve is representative of the velocity profile for incipient separation. The short-dashed curve designated by the star symbol in which $d/\theta = 0$ and $b = 0$ approximates a Blasius boundary-layer velocity profile. The longer dashed curve designated by the diamond symbol represents a velocity profile that is intermediate between the Blasius profile and that at the onset of flow separation. Finally, the solid curve designated by the triangle symbol represents a fully separated shear layer with backflow as it would occur over a separation bubble. The symbol notation used in Figure 4.12 to denote the different basic flow profiles is carried to the following figures.

The normalized spatial growth rates, $-\alpha_i\theta$, and phase velocities, c_r/U_∞, for the four profiles in Figure 4.12 (Dogval et al., 1994) are presented in Figure 4.13. Both are plotted as functions of the dimensionless frequency, $\omega\theta/U_\infty$. In the three cases of inflectional profiles in which $d/\theta > 0$, the condition is $\mathrm{Re} = \infty$. In the one profile in which $d/\theta = 0$, representing a Blasius layer, $\mathrm{Re} = 7000$. Also included are experimental results from Michalke (1991) that are shown by the filled square symbols. Regarding the spatial growth rates, it is clear that in each case there is a band of frequencies over which disturbances are amplified. This of course is typical of convective shear layer instabilities. It is also evident that the growth rates and the span of unstable frequencies increase rapidly as the profile inflection point, d/θ, moves away from the wall. The effect of the backflow is to reduce the span of unstable frequencies. It also

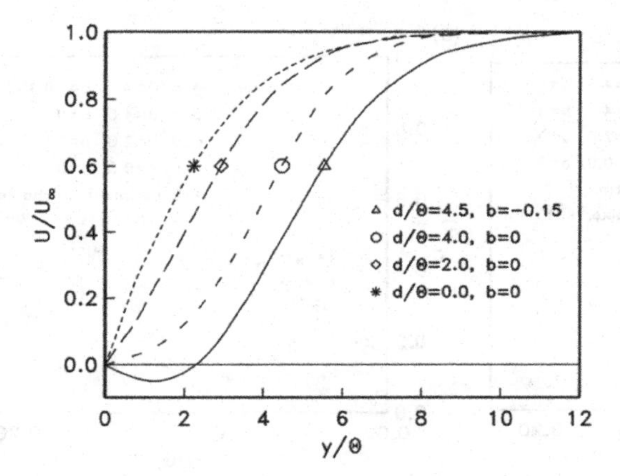

Figure 4.12 Normalized tanh mean velocity profiles with different inflection points, d/Θ, and backflow parameter, b.

appears to increase the growth rate, although that effect can also be the result of the larger d/θ in that case. Comparing the $d/\theta = 0$ case to that of the case with backflow, the growth rate is 20 times larger, and the most unstable frequency is approximately 5 times higher. Therefore, the separated flow is much more likely to result in turbulence transition than a Blasius layer. *Appreciating this is critical to methods used in flow separation control.*

With regard to the phase velocities shown in Figure 4.13(b), the phase speed increases with increasing frequency, ω, from $c_r/U_\infty = 0$ at $\omega = 0$ to where the phase velocity c_r matches the local mean velocity at the height of the inflection in the mean profile, d. This location where $c_r = \overline{U}(d)$ corresponds to the "critical layer." The difference in the heights of the inflection point from the wall in the four cases separates four phase velocity curves. For the profile in which $d/\theta = 4.0$ and $b = 0$ representing incipient separation, the phase velocity is approximately twice that of the Blasius profile. There is a slight lowering of the phase velocity for higher frequencies with the addition of the backflow that is more reminiscent of a free shear layer (Michalke, 1991).

At incompressible Mach numbers in 2-D shear layers, 2-D waves are most amplified; however, 3-D (oblique) waves are amplified and can impact the development of separated flows. In order to include the growth of 3-D disturbances, the linear stability exponential growth term that was given in Eq. (4.4) takes the form $\exp[i(\alpha x + \gamma z - \omega t)]$, where γ is the spanwise (z) wave number. For oblique waves, the wave angle corresponds to $\psi = \arctan(\gamma/\alpha_r)$. Figure 4.14 presents the normalized growth rates and phase velocities for different spanwise wave number oblique waves as a function of the dimensionless frequency for the profile in Figure 4.12, where $d/\theta = 4$ and $b = 0$ and $\mathrm{Re} = \infty$. The case with $\gamma\theta = 0$ corresponds to the 2-D wave case that was shown in Figure 4.13.

As expected for linear amplitude disturbances, oblique waves are less amplified than 2-D waves. In addition to having a lower growth rate, the range of

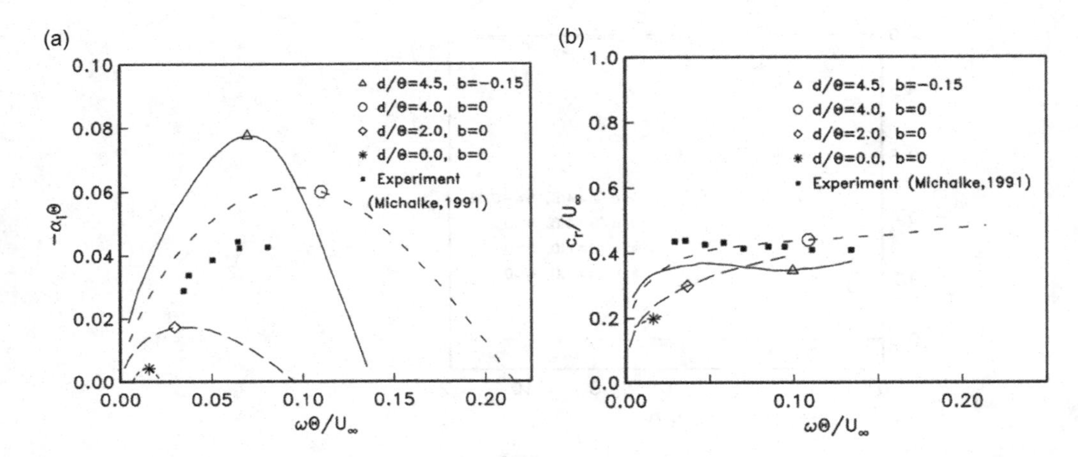

Figure 4.13 Normalized spatial growth rate (a) and (b) phase velocity (b) as a function of normalized frequency for the tanh profiles shown in Figure 4.12. For $d/\Theta > 0$, $\mathrm{Re} = \infty$. For $d/\Theta = 0$, $\mathrm{Re} = 7000$. Data from Dogval et al. (1994). Experimental results from Michalke (1991).

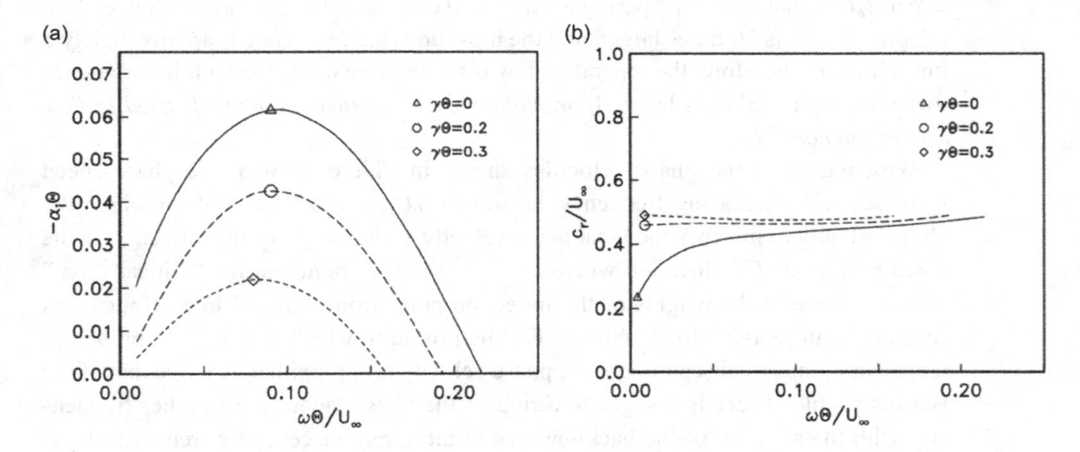

Figure 4.14 Normalized spatial growth rate (a) and phase velocity (b) as a function of normalized frequency of 3-D disturbances for the tanh profile with $d/\Theta = 4$ and $b = 0$ that was shown in Figure 4.12 and $\mathrm{Re} = \infty$. Data from Dogval et al. (1994).

frequencies where amplification occurs decreases with increasing spanwise wave number. Although this is shown for $\mathrm{Re} = \infty$, the trend is the same for lower Reynolds numbers (Dogval et al., 1994). In contrast to the 2-D waves, where at lower frequencies the phase velocity varied with frequency, the phase velocity of the 3-D wave is independent of the frequency. This trend is also independent of Reynolds number (Dogval et al., 1994).

As mentioned, the 3-D wave angle is related to the spanwise and streamwise wave numbers, γ and α_r, as $\psi = \arctan(\gamma/\alpha_r)$. The streamwise wave number is a function of the wave frequency, ω. Figure 4.15 provides an example of the effect of 3-D wave

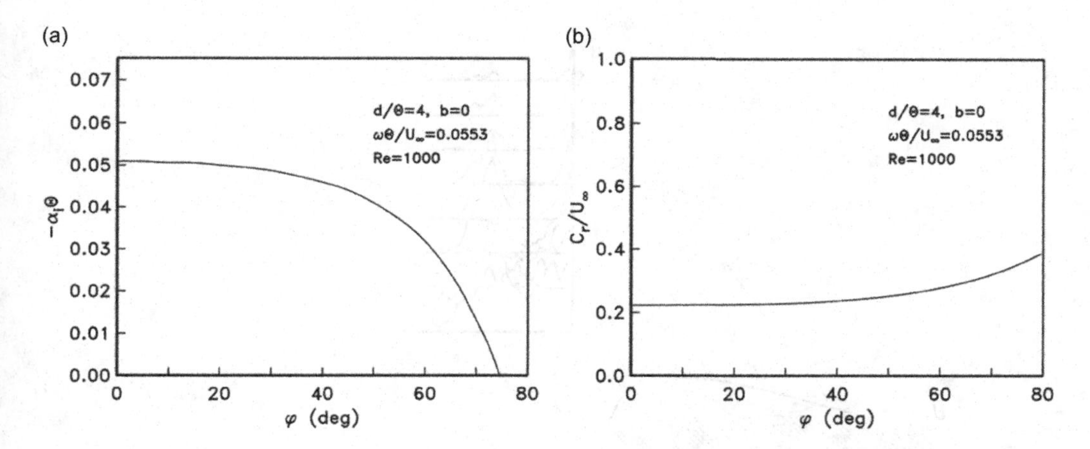

Figure 4.15 Normalized spatial growth rate (a) and phase velocity (b) as function wave angle of 3-D oblique wave for the tanh profile with $d/\Theta = 4$ and $b = 0$ that was shown in Figure 4.12 for $\omega\Theta/U_\infty = 0.05553$ and Re = 1000. Data from Dogval et al. (1994).

angle on the normalized spatial growth rate and phase velocity. This corresponds to the tanh profile with $d/\Theta = 4$ and $b = 0$ that was shown in Figure 4.12, a normalized frequency of $\omega\Theta/U_\infty = 0.05553$, and an Re = 1000. Figure 4.12 clearly shows the monotonic decrease in the growth rate that occurs with increasing 3-D wave angle. The corresponding phase velocity is observed to increases with increasing wave angle.

The variation in phase velocity with wave angle offers interesting implications for seeking parametric interactions between 2-D and 3-D waves that could accelerate turbulent transition of separated flows. Such interactions were initially proposed by Craik (1985) in the context of boundary-layer transition but are likely to apply to shear layers as well. This would involve a pair of 3-D modes with equal but opposite spanwise wave numbers in which their streamwise wave number is one in which the 3-D mode phase speed is the same as that of a 2-D mode. This often happens when the 2-D wave number is twice that of the 3-D streamwise wave number and thereby results in a subharmonic resonance.

The previously discussed linear stability analysis provides an insight into the initial growth of disturbances in the separated shear layer. Eventually, these disturbances grow sufficiently in amplitude where nonlinear effects can occur, with the generation of broad frequency content and eventual transition to turbulence. These stages of development are illustrated in spectra of streamwise velocity fluctuations measured in the separated shear layer of a separation bubble by Boiko et al. (1989) and shown in Figure 4.16. The velocity spectra document the evolution of the separated shear layer. Upstream near the separation location, the spectra indicate two closely spaced peaks at the higher frequency. From linear theory, such as that represented in Figure 4.13, we expect amplification of a *band* of frequencies. The two spectral peaks that appear in the initial development of the separated shear layer fall within the amplified band but appear likely distinct as a result of external excitation. There is also a large peak

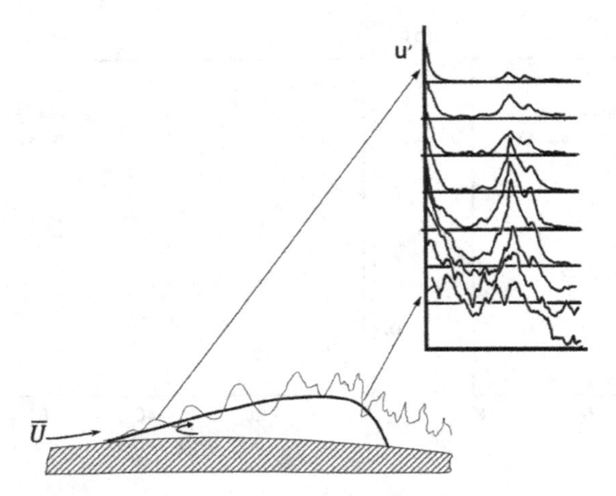

Figure 4.16 Spectra of streamwise velocity fluctuations measured in the separated shear layer at axial different stations along the separation bubble. Taken from Dogval et al. (1994).

at very low frequencies that is not expected from linear theory but is common to separation bubbles and referred to as "flapping" or "breathing." In some cases, this low-frequency unsteadiness can interact with the higher frequency instability causing spectral broadening through sum and difference interactions.

The streamwise evolution of the spectra in Figure 4.16 indicates that the dominant frequency in this case was the left-most of the two initial spectral peaks. We would presume that this was the most amplified of the two or that it was driven by an external source. As it grows to large amplitude, the central peak broadens, with spectral peaks emerging in frequency sidebands. Such sideband development has been documented to occur in shear layers as a result of 3-D effects (Miksad, 1972). Spectral broadening and the lack of a dominant frequency is a tenet of turbulent flows. This appears to be nearly achieved at the last measurement location.

A feature of shear layers in separation bubbles is a relatively small dispersion of phase velocities of small amplitude disturbances. It applies to 2-D waves whose phase velocity is almost independent of frequency, as well as to 3-D waves of a fixed frequency whose streamwise component of the wave vector, α_r, varies only slightly with wave angle. This is illustrated in Figure 4.17 which plots the ratio of the streamwise wave number of oblique waves at different wave angles to that of a 2-D wave for two different experiments. The curve marked (1) is experimental results for a separation bubble on an airfoil (Gilev et al., 1988). The curve marked (2) is for a separation bubble formed downstream of a step in a Blasius boundary layer (Boiko et al., 1989). These indicate that at 3-D wave angles of less than 50°, the convection velocity, $c_r = \alpha_r \omega$, is nearly constant. Such a small dispersion of linear disturbances in separation bubbles means a constancy of phase correlation between different frequency components of both 2-D and 3-D waves. This aspect is extremely important as it points to a mechanism of resonant interactions of the type first postulated in boundary layers by Craik (1985).

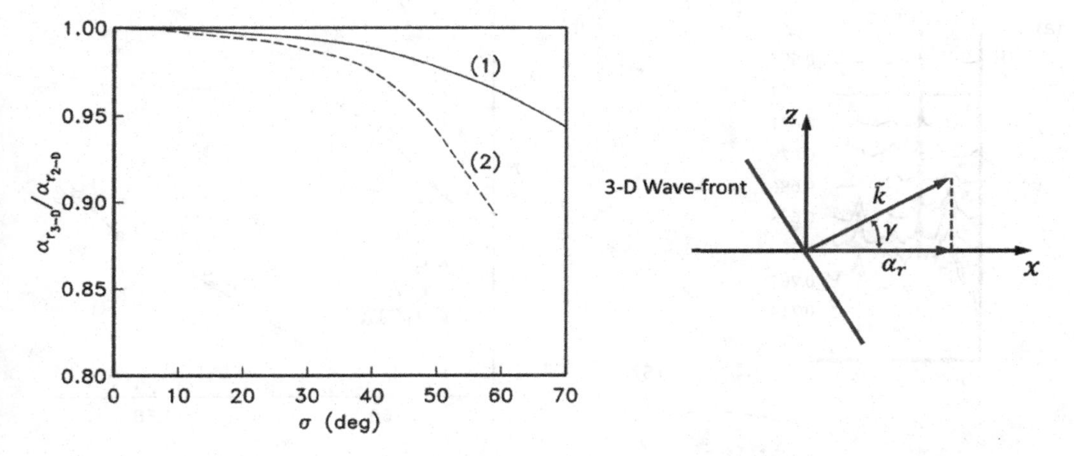

Figure 4.17 Streamwise component of the wave vector of an oblique wave normalized by that of a 2-D wave for the separation bubble on an airfoil (1) (Gilev et al., 1988) and at a hump in a flat plate boundary layer (2) (Boiko et al., 1989).

As discussed in Chapter 5, one of the most striking features in shear layer dynamics is the emergence of a subharmonic frequency component (Sato, 1959; Browand, 1966; Miksad, 1972; Ho and Huang, 1982). In order to understand the conditions that could lead to the growth of the subharmonic, Monkewitz (1982) examined the linear *temporal* (secondary) instability of a 2-D periodic shear flow with 2-D wave amplitude, A, to a perturbation consisting of a 2-D subharmonic wave with amplitude, B. This resulted in an amplitude function for the subharmonic wave

$$\frac{dB}{dT} = c\exp(-i\beta)AB^*, \tag{4.8}$$

where c is a constant, β is the phase difference between fundamental and subharmonic waves, and $*$ denotes the complex conjugate. The initial total subharmonic growth rate

$$\omega_{i_{1/2}} + [d(\ln|B|/dT] \tag{4.9}$$

is then proportional to the periodic amplitude, A, and varies continuously from a maximum when both waves are in phase, $\beta = 0$, to a minimum when they are out of phase, $\beta = \pi$. In a more elaborate analysis of subharmonic resonance (Kelly, 1968; Maslowe, 1977), the subharmonic amplitude, B, is allowed to react back nonlinearly on the fundamental amplitude, A, through a second coupled equation

$$\frac{dA}{dT} = gB^2\exp(i\beta). \tag{4.10}$$

Evidence of the growth of a subharmonic of the initial 2-D instability wave in a separation bubble is contained in results from Boiko et al. (1989) that are shown in Figure 4.18. Figure 4.18(a) shows frequency spectra of streamwise velocity fluctuations in the separation bubble on an airfoil. In this, a vibrator located upstream of the separation bubble introduced a 2-D disturbance at a frequency f_0. This excitation

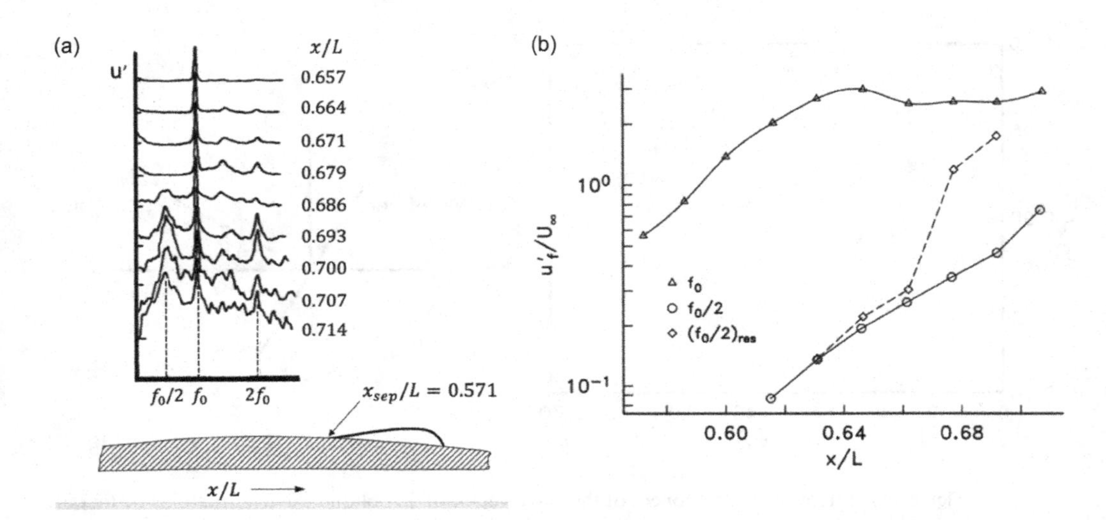

Figure 4.18 Spectra of streamwise velocity fluctuations (a) and corresponding growth in amplitude (b) of fundamental and subharmonic disturbances measured in the separated shear layer at axial different stations along the separation bubble. Taken from Boiko et al. (1989).

resulted in a sharp peak in the spectrum at the furthest upstream location. A spectral peak at the subharmonic frequency, $f_0/2$, is observed to emerge further downstream. Also apparent is a spectral peak at the first harmonic, $2f_0$.

Based on the Monkewitz (1982) secondary instability model, the growth in the subharmonic mode is tied to the amplitude of the fundamental mode. Evidence of this comes by following the streamwise growth in the amplitude of the fundamental and subharmonic waves. These were taken from the spectra in Figure 4.18(a), and shown in Figure 4.18(b). In this, a log scale is used on the ordinate to highlight exponential growth. This plot documents the initial linear growth of the fundamental wave up until nonlinear effects cause its amplitude to saturate. When the subharmonic wave emerges from the background, its growth rate matches that of the fundamental in agreement with the Monkewitz (1982) model. However, the subharmonic takes on a more rapid growth when the fundamental amplitude reaches a nonlinear level. This is likely evidence of a nonlinear resonance of the type proposed by Kelly (1968) and Maslowe (1977). Both mechanisms point to a powerful approach to the control of flow separation bubbles.

4.3 Flow Separation Control

As previously discussed, flow separation occurs when the mean flow lacks sufficient momentum to overcome an adverse pressure gradient, or when viscous dissipation occurs along the flow path. Boundary-layer separation control methods are generally designed to increase the momentum in the flow either through mixing with high-momentum fluid or direct injection of high-momentum fluid. Both passive and active approaches have been developed. As always, each has its positive and negative aspects.

Table 4.1 Methods of separation control.

Control method	Type	Mechanism
Trips	P	Boundary-layer turbulence onset
Roughness	P	Boundary transition to turbulence
Vortex generators	P	Boundary layer mixing
Bumps and dimples	P	Boundary layer mixing
Wall suction	A	Boundary layer thinning
Wall blowing	A	Boundary layer thickening
Tangential blowing	A	Momentum addition

For example, passive techniques require no power to operate; however, they can produce parasitic losses when not required. As an example, passive separation control devices needed for leading-edge separation control on aircraft are only required during takeoff and landing; otherwise, if deployed, they add unnecessary drag. Although active approaches require power to operate, their operation can be tailored to the immediate conditions, including turning them off when not needed. In this, their optimum operation would require sensing and feedback control. To be most effective, active separation control approaches should account for the instability characteristics of the separated shear layer, which can amplify the unsteady actuator output and thereby reduce the necessary power input to the actuator. A general list of passive (P) and active (A) boundary-layer separation control methods are provided in Table 4.1. The active methods can be either steady or unsteady.

4.3.1 Passive Separation Control

Passive control of flow separation has been extensively studied for nearly a century. One of the most commonly investigated passive devices are vortex generators (VGs). These are designed to introduce coherent streamwise-oriented circulation patterns that are intended to suppress flow separation by transporting high-momentum fluid from the free-stream to the wall region. Vane-type VGs were first introduced in the late 1940s by Taylor (1947). These devices consisted of a row of small plates or airfoils that were mounted on the surface and set at an angle of incidence to the local mean flow direction. Although effective in suppressing flow separation, the Taylor designs produce a relatively large parasitic drag. Nonetheless, Fisher (1954) employed this design to produce attached flows and enhance wing lift.

Bearman and Harvey (1993) used dimples to control the flow separation that occurs over a circular cylinder. The ratio of the depth of the dimples to the diameter was approximately 0.001. The dimples were intended to simulate surface roughness which, when placed upstream of a laminar flow separation region, is effective in causing the attached boundary layer to become turbulent, increasing the shear stress at the wall and thereby maintaining attached flow further on the cylinder. The result was to reduce the wake deficit and therefore to reduce the aerodynamic drag. Bearman and Harvey

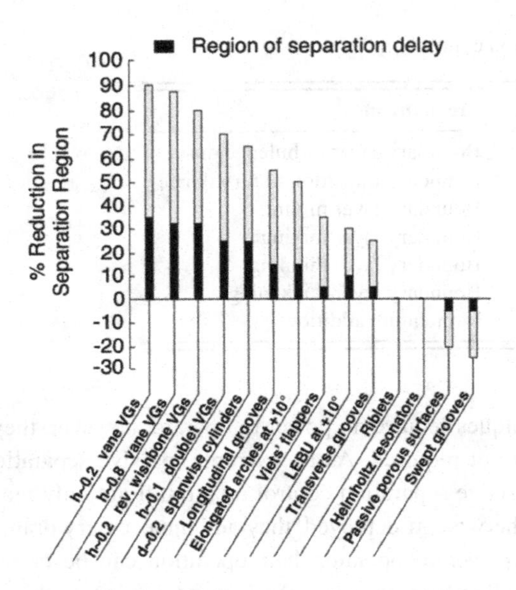

Figure 4.19 Plot of flow separation and reattachment performance for types of passive vortex generator designs listed in Table 4.2.

(1993) found that dimples with comparable characteristics (depth to diameter) to random roughness performed better, leading to lower drag. This is the motivation for dimpled surfaces on golf balls which have been shown to be similarly effective in lowering drag (Kimura and Tsutahara, 1991).

Turbulent transition caused by surface roughness is one technique to add momentum to the near-wall flow. However, the most commonly used passive approaches are variations of the Taylor (1947) streamwise VGs.

In early applications of VGs used for flow separation control, their height was on the order of the boundary-layer thickness. Kuethe (1972) developed and examined nonconventional "wave-type" VGs with heights that were 27 and 42 percent of the boundary-layer thickness. Rao and Kariya (1988) suggested that passive VGs with $h/\delta \leq 0.625$ could be just as effective in flow separation control but with much lower parasitic device drag. Lin (2002) subsequently provided an in-depth review of these so-called low-profile VGs. Figure 4.19 provides a general summary of the effectiveness of a number of vane VG designs in terms of their ability to suppress flow separation and/or cause flow reattachment. The different VG designs and parameters are presented in Figure 4.20. The numbers on the ordinate in Figure 4.19 refer to the 12 types of vane-type VGs that are listed in Table 4.2.

Figure 4.19 and the accompanying Table 4.2 indicate that the most effective groups of flow separation control devices were those that generated streamwise vortices. These included the low-profile and conventional VGs, and the large longitudinal surface grooves. The low-profile VGs are also referred to as "submerged VGs" by Lin et al. (1990a) and "micro VGs" by Lin (1999). Such low-profile VGs having an

Table 4.2 List of passive vortex generator designs corresponding to the numbers on the abscissa of Figure 4.19.

VG type	Description
1	Vane VG, $h/\delta \simeq 0.2$
2	Vane VG, $h/\delta \simeq 0.8$
3	Reverse wishbone VG, $h/\delta \simeq 0.1$
4	Doublet VG, $h/\delta = 0.1$
5	Spanwise cylinders, $d/\delta \simeq 0.2$
6	Longitudinal grooves
7	Elongated arches at $+10°$ angle
8	Viets' flappers
9	LEBU at $+10°$ angle
10	Transverse grooves
11	Passive porous surface
12	Swept grooves

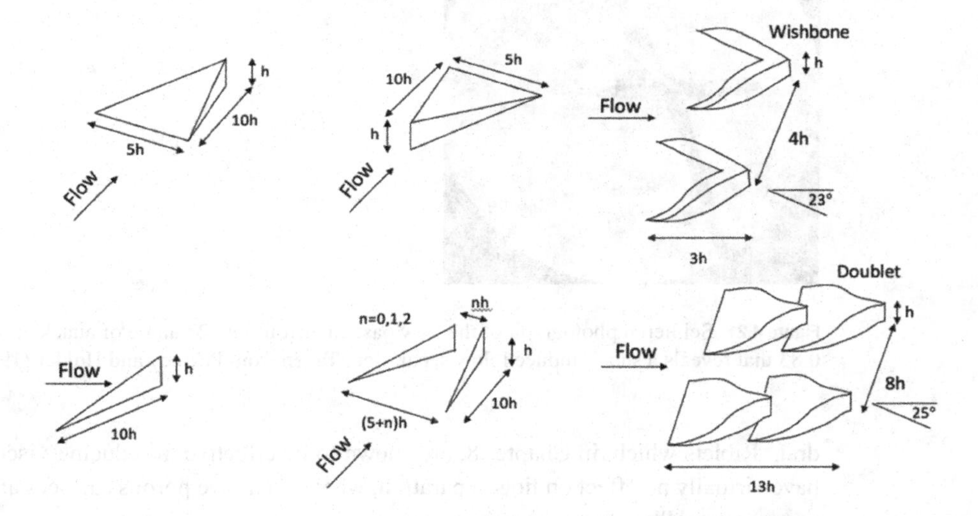

Figure 4.20 Different types of vane-type vortex generators to produce corotating and counter-rotating streamwise vortices. The most effective settings are given in Table 4.3.

$h/\delta \simeq 0.2$ were found to be as effective in delaying separation as the conventional VGs with a $h/\delta \simeq 0.8$, while also having a lower device drag.

The second most effective VG group in Figure 4.19 includes those devices that generate transverse vortices, such as the spanwise vortex sheet produced by suspended spanwise cylinders or thin flat plates used in large-eddy breakup devices (LEBUs). Other transverse vortex generating devices that were examined include elongated arches, flow-driven "Viets flappers," and transverse grooves. As indicated, these devices were generally less effective than the low-profile VGs. In addition, because they require more complete spanwise coverage, they generally had a higher device

Table 4.3 List of most effective settings for separation control for different VG designs shown in Figure 4.20.

	Most effective settings				
VG type	h/δ	x/h	z/h	Half-angle (°)	$(\Delta x)_{VG}/h$
Doublet	0.1	13	8	25	20
Wishbone	0.2	3	4	23	10
Counter-rotating rectangular vanes	0.2	4	9	25	10
Counter-rotating delta vanes	0.3	10	12	14	52
Corotating trapezoid vanes	0.2	4	4	23	12–19
Forward wedges	0.5	10	–	14	50
Backward wedges	0.5	10	–	14	50

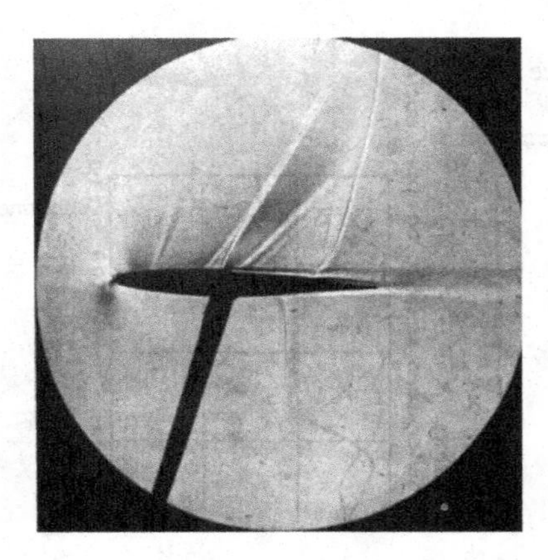

Figure 4.21 Schlieren photograph of the flow past an airfoil with 2° angle of attack at Mach 0.83 that reveals a shock-induced flow separation. Taken from Pearcey and Holder (1954).

drag. Riblets which, in Chapter 8, are shown to be effective in reducing viscous drag have virtually no effect on flow separation, whereas passive porous surfaces and swept grooves actually enhance separation.

Most of the studies investigating the performance of different types of VGs for separation control have been performed at subsonic and incompressible Mach numbers. However, McCormick (1992) had demonstrated that low-profile VGs could significantly suppress a shock-induced boundary-layer separation bubble, which improved the boundary-layer characteristics downstream of the shock. The experiment utilized a wheeler doublet with $h/\delta \simeq 0.36$. The VGs were located approximately 20δ upstream of the shock location. Shock–boundary-layer interaction is discussed in detail in Chapter 9. Figure 4.21 provides an example of the use of passive VGs to control shock–boundary-layer-induced flow separation on the same airfoil and Mach number that was shown in Figure 4.2. The VGs were located at the 41 percent chord location. Comparing this image to that in Figure 4.2 indicates that the VGs were able to maintain attached flow.

4.3.2 Active Separation Control

As in any active flow control method, they require some form of external energy input. This can be an air source in the case of jets or an electric power source in the case of voltage-driven actuators. In addition, the excitation can be steady or unsteady, with each acting on some aspect of the basic flow or fluid instability. The following sections describe the different approaches used in controlling separated flows.

Steady Streamwise Blowing

One of the simplest active approaches to flow separation control involves steady tangential surface blowing. This is applied just upstream of the separation location to add momentum to the boundary layer at the wall to overcome an adverse pressure gradient. A widely used example of this approach is split trailing-edge flaps on aircraft wings that allow high-pressure air from the underside of the wing to pass to the upper side to add momentum to the boundary layer over the flap surface. This allows larger flap angles to be used without flow separation. The result is to increase the wing maximum lift.

Another approach used to add momentum to the flow near the wall involves a plasma actuator. An example of this was performed by He and Corke (2007), which was also reported by He (2008). This was demonstrated on a "generic" separated flow that occurred over a "Glauert Hump." The separated flow characteristics of the Glauert Hump have been well-documented and utilized as a test case for a number of separated flow control experiments. Details of the Glauert Hump and flow separation characteristics have been compiled by Greenblatt et al. (2005).

A photograph of the Glauert Hump with a spanwise-oriented plasma actuator for flow separation control is shown in Figure 4.22. The flow direction in the figure is from right to left. The pressure coefficient, $-C_p$, distribution for the Glauert Hump (He, 2008) is shown in Figure 4.23(a). The symbols correspond to the measured distribution. The various dashed curves correspond to the results from computational fluid dynamics simulations using different turbulence models. Flow separation is evident by the $-C_p$ plateau that begins at $x/c \simeq 0.625$.

The plasma actuator shown in Figure 4.22 consists of two electrodes that are separated by a dielectric material. One of the electrodes is exposed to the air. The other electrode is fully covered by the dielectric material. The positions of the two electrodes are staggered, with the covered electrode shifted in the streamwise direction leaving a small amount of overlap with the exposed electrode. A high-voltage AC source supplied across the electrodes caused the air over the covered electrode to ionize. The ionized air in the presence of the vector electric field produced by the staggered electrode geometry results in a body force vector field that acts on the neutral air. In this case, the body force vector is directed in the downstream direction. The result is to produce the equivalent of a tangential wall jet that is directed toward the flow separation location. The wall jet adds momentum to the separated flow region causing the flow to remain attached. The effect on the $-C_p$ distribution is shown in Figure 4.23(b). With the plasma actuator operating (spanwise plasma actuator On), the plateau in the

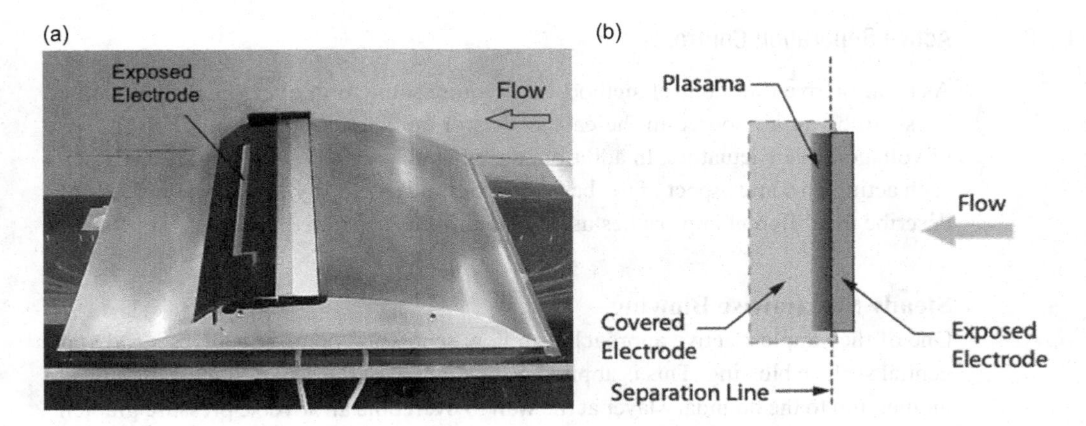

Figure 4.22 Photograph of spanwise plasma actuator on the Glauert Hump model (a) and schematic of plasma actuator (b) indicating its location relative to the flow separation line. Taken from He (2008).

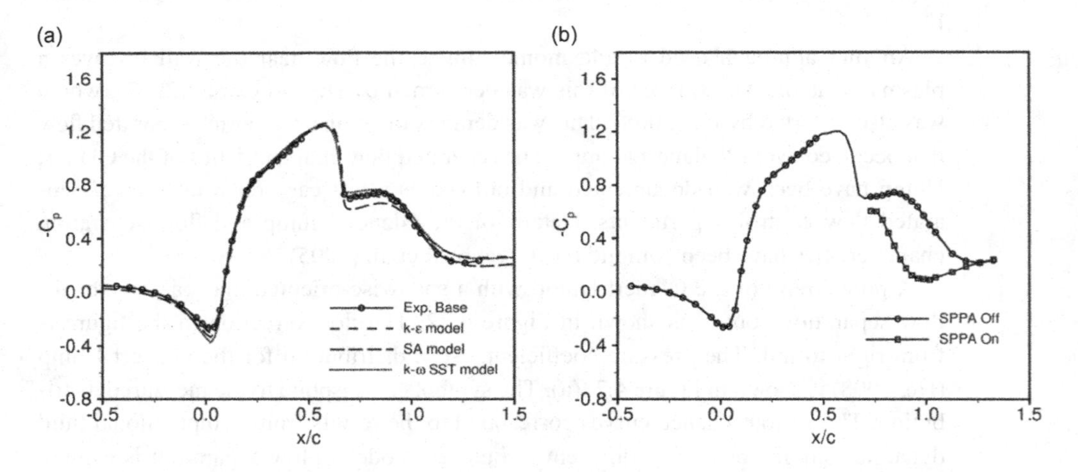

Figure 4.23 Baseline pressure coefficient distribution for Glauert Hump (a) and comparison of the pressure coefficient distribution with spanwise plasma actuator operating (On) (b). Taken from He (2008).

pressure coefficient is eliminated, along with a significant drop in the suction pressure associated with a separation bubble. Note that when the plasma actuator is not operating, there is no effect on the baseline flow.

Steady Spanwise Blowing

Many active separation control approaches build upon the characteristics of passive VGs, namely the production of streamwise vorticity that can transport higher momentum fluid toward the wall. One example from Johnston and Nishi (1990) involves the use of an array of wall jets oriented at different angles with respect to the mean

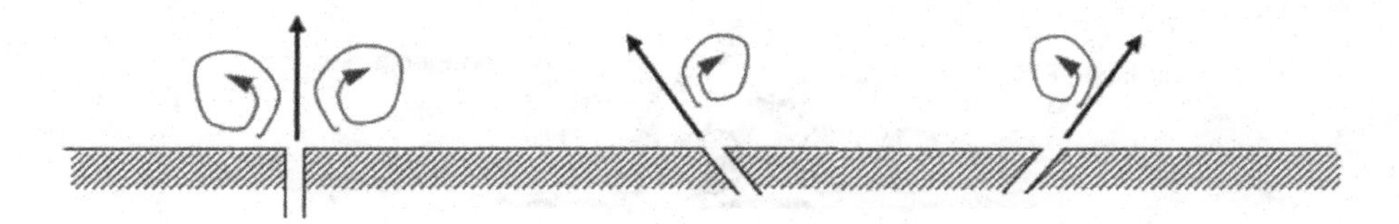

Figure 4.24 Examples of wall jet vortex generator configurations.

flow direction in order to produce localized streamwise vortices. As illustrated in Figure 4.24, a vertical orientation of the wall jet can result in pairs of counter-rotating vortices, while a single streamwise vortex of either circulation direction can be generated by angling the jet in either cross-stream direction. Pairs of angled jets could emulate the effect of passive vane-type VGs illustrated in Figure 4.20. In separation control experiments using jet VGs, Johnston and Nishi (1990) found that velocity ratios, V_j/U_∞, above 0.8 were effective.

Similarly, Rixon and Johari (2003) studied the development of a steady VG-jet in a turbulent boundary layer. The jet was pitched upward from the wall at 90° and skewed at 45°. They observed that the jet created a pair of streamwise vortices. One of the vortices was stronger and dominated the flow field. The circulation, peak vorticity, and wall-normal position of the primary vortex increased linearly with the jet velocity.

Another approach that is designed to replicate passive streamwise VGs involves plasma actuators. In this application, the plasma actuators consist of electrodes aligned in the flow direction that are designed to produce a body force in the *spanwise* direction. A basic geometry is shown in Figure 4.25(a). It consists of a series of exposed electrodes that are aligned with the mean flow direction, and that overlie a common covered electrode. A thin dielectric barrier material separates the exposed and common electrodes. In this example, the multiple exposed electrodes are equally spaced in the spanwise direction by an amount, λ. When energized, the plasma forms on both sides of each of the exposed electrodes. The body force vector field then gives rise to a series of opposing wall jets in the spanwise direction. These jets collide and subsequently interact with the external mean flow to form pairs of counter-rotating streamwise vortices with a size of $\mathcal{O}(\lambda/2)$ (Kozlov and Thomas, 2011; Schatzman and Thomas, 2017). The effect of the pairs of counter-rotating vortices are evident in the flow visualization photograph shown in Figure 4.25(b). This was produced by introducing a uniform distribution of marker particles into the boundary layer upstream of the plasma VG array, and illuminating the particles with a 2-D laser sheet located at the downstream edge of the array. This image documents the upwelling of fluid that is caused by the pumping action of the pair of counter-rotating vortices that form between the exposed electrodes.

The generation of streamwise vorticity with a plasma VG involves (1) the reorientation of spanwise vorticity present in the approaching boundary layer and (2) the generation of vorticity along the length of the exposed electrodes. The ability to generate mean streamwise vorticity, $\bar{\omega}_x$, can be viewed by considering the x-component of the mean vorticity transport equation given in Eq. (4.11).

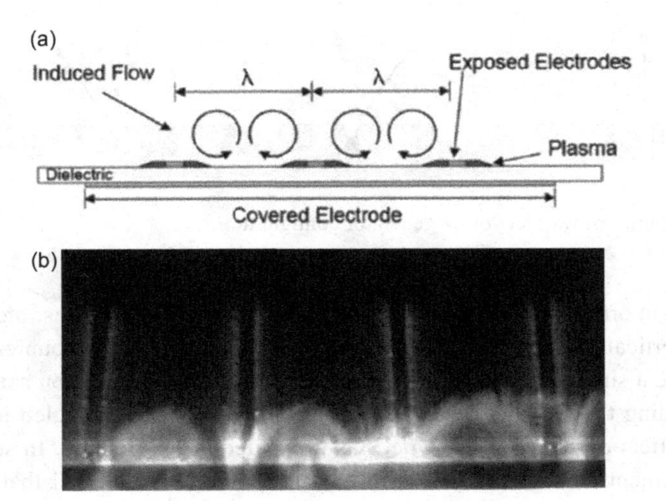

Figure 4.25 Schematic of the basic plasma vortex generator actuator geometry (a) and photograph of visualized counter-rotating flow patterns produced by plasma vortex generator array (b). Taken from Kelley et al. (2016).

$$\frac{D\bar{\omega}_x}{Dt} = \bar{\omega}_j \bar{S}_{xj} + \overline{\omega_j' s_{xj}} - \overline{u_j \frac{\partial \omega_x'}{\partial x_j}} + \nu \frac{\partial^2 \bar{\omega}_x}{\partial x_j \partial x_j}. \tag{4.11}$$

In this, the primary streamwise vorticity generation occurs through the mean vorticity–strain-rate interaction terms, $\bar{\omega}_j \bar{S}_{xj}$. In Cartesian coordinates, this consists of the three terms on the right-hand side of the equality in Eq. (4.12).

$$\bar{\omega}_j \bar{S}_{xj} = \bar{\omega}_x \bar{S}_{xx} + \bar{\omega}_y \bar{S}_{xy} + \bar{\omega}_z \bar{S}_{xz}. \tag{4.12}$$

By applying a thin shear layer approximation, Eq. (4.12) can be further written as

$$\bar{\omega}_j \bar{S}_{xj} = \bar{\omega}_x \frac{\partial \bar{U}}{\partial x} + \bar{\omega}_y \frac{\partial \bar{U}}{\partial y} + \bar{\omega}_z \frac{\partial \bar{U}}{\partial z}. \tag{4.13}$$

The first term involves streamwise dilatation (vortex stretching/compression). The second term represents a reorientation of the wall-normal component of vorticity into the streamwise direction. The third term is similar but involves the streamwise reorientation of initially spanwise boundary-layer vorticity.

Upstream of the plasma VG, the boundary layer is expected to be relatively 2-D so that $\bar{\omega}_y$ is initially quite small. This is especially true in the favorable pressure gradient region for example, near the leading edge of an airfoil. However, over the plasma VG array, the induced counter-rotating flow pattern shown in Figure 4.25(b) results from a wall-normal flow directed toward the surface electrodes and away from the wall at spanwise locations centered between electrodes. The result is a nonzero $\partial \bar{U}/\partial z$ that generates wall-normal vorticity. Therefore, the second term in Eq. (4.13) simplifies to

$$\bar{\omega}_y \bar{S}_{xy} \approx \frac{\partial \bar{U}}{\partial z} \frac{\partial \bar{U}}{\partial y}. \tag{4.14}$$

Following similar arguments, the last term in Eq. (4.13) simplifies to

$$\bar{\omega}_z \bar{S}_{xz} \approx \frac{\partial \bar{U}}{\partial y} \frac{\partial \bar{U}}{\partial z}. \tag{4.15}$$

For an approaching 2-D boundary layer, $\partial \bar{U}/\partial z$ is negligible.

A key function of a plasma VG array is to generate a necessary spanwise mean velocity gradient $\partial \bar{U}/\partial z \approx \mathcal{O}(\Delta U/\lambda)$, where λ is the spanwise spacing between electrodes and ΔU is the change in streamwise mean velocity induced by the actuator. Substituting this into Eq. (4.11) gives

$$\frac{D\bar{\omega}_x}{Dt} \approx \mathcal{O}\left(\frac{\Delta \bar{\omega}_x}{L/U_c}\right) \approx \bar{\omega}_x \frac{\partial \bar{U}}{\partial x} + 2\mathcal{O}\left(\frac{U_p}{\delta^*} \frac{\Delta U}{\lambda}\right), \tag{4.16}$$

where L is the streamwise length of the plasma VG electrodes and U_c is a characteristic mean velocity of the flow over the actuator array such that the convective timescale of a fluid particle over the array is given by $T_c \equiv L/U_c$. Reordering terms and simplifying, one obtains the following relation for the change in mean streamwise vorticity due to the plasma VG array:

$$\Delta \bar{\omega}_x \approx \bar{\omega}_x \mathcal{O}\left(\frac{\partial \bar{U}}{\partial x} \frac{L}{U_c}\right) + 2\mathcal{O}\left(\frac{L}{\delta^*} \frac{U_p}{U_c} \frac{\Delta U}{\lambda}\right). \tag{4.17}$$

The second term in Eq. (4.17) embodies the effect of the plasma VG to create streamwise vorticity by reorienting both wall-normal and spanwise mean components, $\bar{\omega}_y$ and $\bar{\omega}_z$, respectively, into the streamwise direction. To be effective, the convective timescale, $T_c = L/U_c$, must be greater than both S_{xy}^{-1} and S_{xz}^{-1}. The former is set by the approach boundary layer, and the latter is set by the plasma VG array design. In most flow control applications, $S_{xz}^{-1} \gg S_{xy}^{-1}$ so that the criterion for the minimum plasma VG length should be based upon $T_c \geq S_{xz}^{-1}$.

The magnitude and scaling of the terms in Eq. (4.17) has been investigated by Kelley et al. (2016). This involved placing an array of plasma VGs on the upper surface of a V-22 wing model that provided a generic flow field that undergoes trailing-edge flow separation as a precursor to lift stall. Different streamwise pressure gradients were examined by changing the airfoil angle of attack. Results from that investigation are shown in Figures 4.26 and 4.27. Figure 4.26(a) compares the vorticity timescale, $S_{xz} = (\partial \bar{U}/\partial z)^{-1}$, against the convective timescale, $T_c = L/U_\infty$, for the different cases produced by changing the airfoil angle of attach, free-stream velocity, and streamwise length, L, of the plasma VG array. As stated, the scaling criterion for the minimum L is that $T_c \geq S_{xz}^{-1}$. This was observed to occur in a majority of the cases. This is affirmed in Figure 4.27(b) where it is clear that the vorticity, $\partial \bar{U}/\partial z$, scales linearly with the convective timescale for the full range of conditions in the experiment.

The effect of the convective timescale on the $\omega_x \simeq (\partial \bar{V}/\partial z)_{p-p}$ that is generated by the plasma VG is shown in Figure 4.27. This indicates that for the range of conditions, a majority of the $(\partial \bar{V}/\partial z)$ values scale linearly with the convection timescale. Those that deviated from that trend corresponded to the lowest Mach number and therefore the longest timescale. This may be the effect of the vorticity dilatation term

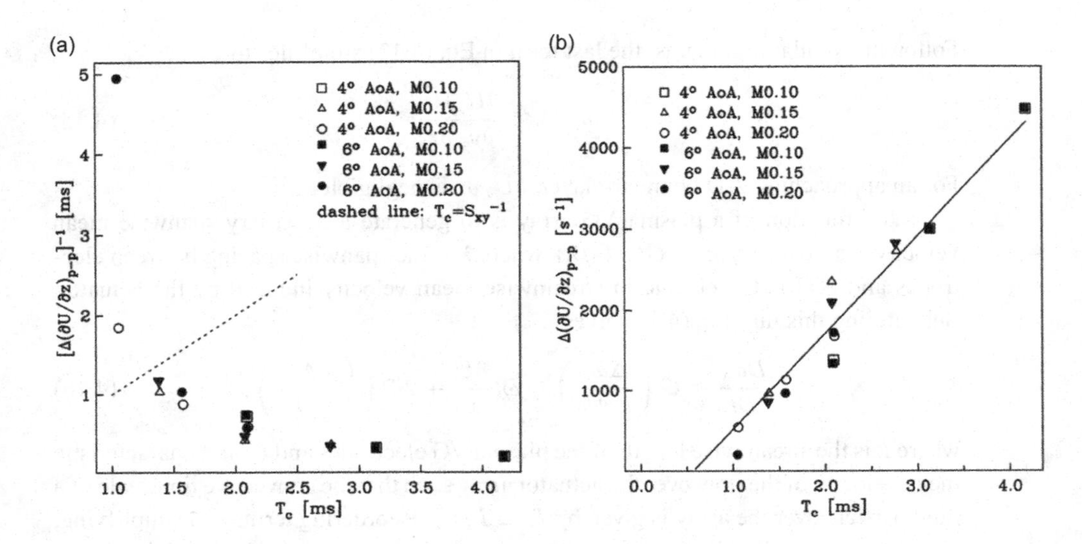

Figure 4.26 Vorticity timescale $(\partial \bar{U}/\partial z)^{-1}$ (a) and vorticity $\partial \bar{U}/\partial z$ (b) versus the convective timescale. Taken from Kelley et al. (2016).

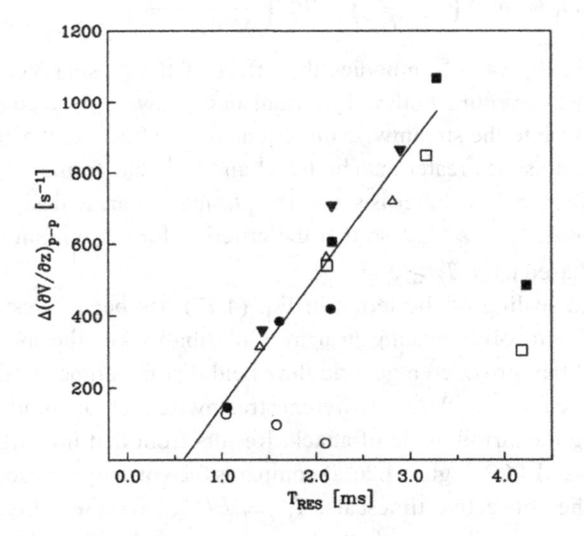

Figure 4.27 Effect of vorticity residence time on $\partial \bar{V}/\partial z$. The symbol convention is the same as in Figure 4.26. Taken from Kelley et al. (2016).

in Eq. (4.15), which scales with $\Delta \bar{U}/U_\infty$. Therefore, its effect is the largest with the longest L, and the lowest free-stream speed (Mach number), which result in the largest T_c values. In contrast, those cases with the shorter electrode lengths at the lowest Mach numbers follow the linear trend.

From a plasma VG design aspect, the parameters are (1) the exposed electrode spanwise spacing, λ, and (2) the electrode streamwise length, L. The criteria for the spanwise spacing should be approximately the same as for passive opposite-angled vane-VG pairs that produce counter-rotating streamwise vortices. Based on Table 4.3, $\Delta z = 0.8\delta$. In their plasma VG application, Kelley et al. (2016) found the optimum to be $\Delta z = 1.2\delta$, but generally Δz should be on the order of the boundary-layer thickness. With regard to the streamwise length of the plasma VG array, it is based on the timescale of the vorticity, $T_c \geq (\partial \bar{U}/\partial z)^{-1}$, such that $L \geq T_c U_\infty$.

The plasma VGs produced the equivalent of a spanwise-directed tangential wall jet. Therefore, the same scaling criterion should equally apply to a pressure-driven tangential wall jet that extends in the streamwise direction.

Unsteady Excitation

The previous examples focused on *steady* streamwise blowing to energize the boundary layer, and *steady* spanwise blowing designed to promote streamwise vorticity to transport higher momentum fluid toward the wall as a means to prevent flow separation or hasten reattachment. However, none of these approaches make use of the natural instability of the separated shear layer that can be exploited with *unsteady* excitation.

Seifert et al. (1996) conducted experiments on four different airfoils to examine the separation control effect using oscillatory blowing. It was found that oscillatory blowing could delay boundary-layer separation more effectively than steady blowing. An important observation was that the most effective excitation frequency was the one having a reduced frequency of $F^+ = \omega L/U_\infty = 1$, where ω is the excitation frequency with units of s^{-1}, U_∞ is the local free-stream velocity, and L is the streamwise length of the separated region, which in the case of Seifert et al. (1996) corresponded to the distance between the actuator and the trailing edge of the airfoil.

The observation by Seifert et al. (1996) of an optimum excitation frequency to reattach a separated flow suggests that it is acting on the separated shear layer instability mechanism. If one considers the instability of a hyperbolic-tangent free-shear layer profile like that presented in Figure 4.14(b), the average phase speed is approximately $c_r = \omega\lambda \simeq 0.45U_\infty$. Therefore, for the optimum reduced frequency, $\omega L/U_\infty = 1$, in which L is the length of the separated region,

$$L = \frac{U_\infty}{\omega}. \tag{4.18}$$

Substituting for ω based on $c_r = \lambda\omega = 0.45U_\infty$, where λ is the instability wave length, then

$$L \simeq \frac{\lambda}{0.45} \simeq 2\lambda. \tag{4.19}$$

Therefore, the optimum excitation frequency is the one in which there are approximately *two* instability wavelengths spanning the separated flow region.

Nishioka et al. (1990) utilized external acoustic excitation to control leading-edge flow separation on an airfoil. In their case, an acoustic speaker was located in the wind tunnel wall, directly under the leading edge of the test article that was located in the center of the test section. In the same theme, Nishizawa et al. (2003) utilized miniature acoustic speakers located inside an airfoil as a disturbance actuator for leading-edge separation control. Passage holes from the speaker cavity to the leading edge were used to communicate the pressure disturbances. The speakers were operated to produce a pure tone at a *fixed* amplitude and frequency. It was observed that the separated flow reattached when the free-stream velocity was above a certain threshold. Based on Seifert et al. (1996), one would presume that attachment occurred when the combination of the excitation frequency and the free-stream velocity resulted in the optimum $F^+ = 1$ condition. Possibly more notable from this work was the development of a flow separation sensor that consisted of a micro-electromechanical system cantilever beam that could sense flow direction. It was used to detected boundary-layer separation based on sensing a reverse-flow component associated with a separation bubble.

Greenblatt et al. (2005) examined the effect of unsteady blowing on the separated flow over the Glauert Hump. The unsteady blowing emanated from a 2-D slot in the surface of the model, which was located at the separation location. The optimum frequency of excitation for reattaching the flow was observed to fall in the range $1.05 \leq F^+ \leq 1.35$, which is consistent with the earlier observations of Seifert et al. (1996). Here again, the characteristic length, L, used in the normalized frequency was the length of the separated region.

Kelley et al. (2014) utilized a pulsing plasma actuator to control leading-edge flow separation on a NASA EET airfoil section shape. The plasma actuator was placed directly along the leading edge of the airfoil. It utilized a staggered geometry between the exposed and covered electrodes, like that previously shown in Figure 4.22. The electrodes were oriented so that the body force vector field produced by the actuator was directed toward the suction side of the airfoil at an angle of attack. The plasma actuator was operated to produce unsteady pulsing at a prescribed frequency.

The voltage time series to the plasma actuator to produce unsteady pulsing is shown in Figure 4.28(a). It consisted of an AC carrier at a frequency that is most efficient to ionize the air, which was switched on and off to produce a periodic duty cycle. The frequency of the duty cycle was $1/(2\pi P)$, where P was the time period between leading edges of the AC carrier duty segments. In this approach, the time of the AC carrier duty needs to be at least 2–3 AC cycles to cause the air to ionize (Corke et al., 2010b). Figure 4.28(b) shows a sequence of flow visualization images over a NACA 0015 airfoil at a post-stall angle of attack that documents the temporal development of spanwise vortical structures that are produced by the pulsing actuator. The pulsing frequency corresponds to an $F^+ = 1$. The three images correspond to dimensionless times, tU_∞/c, of 1.0, 4.9, and 5.9, corresponding to labels (a)–(c), respectively. As previously mentioned, the $F^+ = 1$ should produce two vortices within the flow separation region. This is clearly supported by image (b).

Kelley et al. (2014) documented the lift, drag, and pitch moment of the EET airfoil for free-stream Mach numbers from 0.1 to 0.4. An example shown in Figure 4.29 corresponds to near- and post-stall angles of attack at Mach 0.10. This compares both steady and unsteady actuation. The unsteady actuation is at $F^+ = 1.0$, which was found to be most effective. The comparison shows that although the steady actuation

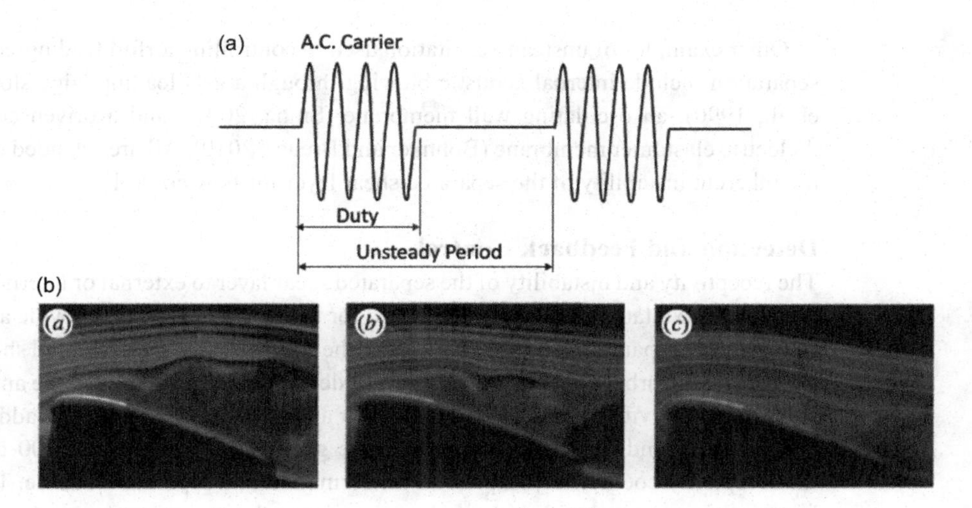

Figure 4.28 AC voltage time series used to produce unsteady pulsing of plasma actuators (a) and sequence of flow visualization images over a NACA 0015 airfoil at a post-stall angle of attack (b) that shows the initial response of the separated flow to a leading-edge plasma actuator pulsing at a frequency corresponding to $F^+ = 1$. $tU_\infty/c = 1.0, 4.9$, and 5.9 from (a) to (c).

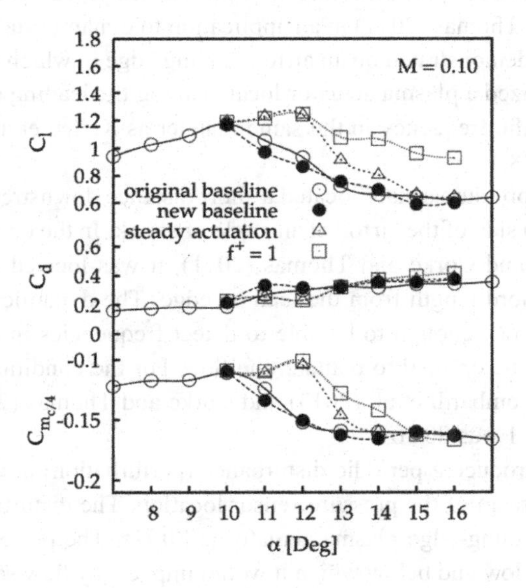

Figure 4.29 Lift, drag, and pitch moment coefficients for NASA EET airfoil section at near- and post-stall angles of attack showing baseline condition, and that with steady and unsteady at $F^+ = 1$, leading-edge plasma actuator for separation control. Taken from Kelley et al. (2014).

was effective in increasing the post-stall lift coefficient, the unsteady actuation was more effective. If you account for the difference in actuator power between the steady (a duty cycle of 1) versus the unsteady (a duty cycle of 0.2), the benefits of the unsteady actuation are even more pronounced.

Other examples of unsteady excitation used in controlling airfoil leading-edge flow separation include internal acoustic blowing through a 2-D leading-edge slot (Hsiao et al., 1990), an oscillating wall membrane (Sinha, 2001), and a driven compliant dielectric elastomer membrane (Bohnker and Breuer, 2019). All are intended to utilize the inherent instability of the separated shear layer for flow control.

Detection and Feedback Control

The receptivity and instability of the separated shear layer to external or internal disturbances and excitation that can be exploited for flow control can also provide a method for detecting separation onset. In particular, the receptivity of the separated shear layer to acoustic disturbances provides a means of detection. For example, in the analysis of acoustic receptivity of a boundary layer over a parabolic leading edge, Haddad et al. (2005) had found that there was a dramatic growth, on the order of 100 times, in the receptivity coefficient just prior to the formation of a separation bubble. Thus, the boundary layer was significantly more responsive to the acoustic pressure disturbances when the flow was on the verge of separation.

In its first application, He (2008) utilized the extreme receptivity of the separated flow shear layer to devise a flow separation detection that became an integral part of a closed-loop separation control approach. This was further described by Lombardi et al. (2013) and Corke and Thomas (2011) in an application to dynamic stall control. The approach was initially demonstrated on an airfoil leading edge in which the open-loop separation control utilized a plasma actuator located along the leading edge, and operated to pulse at a specific frequency in the same manner as Kelley et al. (2014), and illustrated in Figure 4.28.

The approach utilized a pressure sensor located a short distance downstream of the leading edge, on the suction side of the airfoil at an angle of attack. In the embodiment of Lombardi et al. (2013) and Corke and Thomas (2011), it was located at $x/c = 0.05$, or 5 percent of the chord length from the leading edge. The dynamic response of the pressure sensor was fast enough to be able to detect frequencies in excess of $F^+ = 1$ that were shown to be optimal to reattach the flow. For the conditions of the experiments described by Lombardi et al. (2013) and Corke and Thomas (2011), the physical frequency at $F^+ = 1$ was 80 Hz.

The approach was to introduce a periodic disturbance (perturbation) at a location on the airfoil that was upstream of the pressure sensor location. The disturbance was produced by pulsing the leading-edge plasma actuator at 80 Hz. The power level of the plasma actuator was set low and below which it would impact any flow separation. Figure 4.30 shows spectra of the time series from the pressure transducer located at $x/c = 0.05$ for two airfoil angles of attack of 6° and 14°. The steady stall angle of attack for the airfoil was $\alpha_{ss} = 15°$. At stall, the flow over the airfoil becomes fully separated and produced a dramatic drop in lift and an increase in drag. Both the spectra shown in Figure 4.30 are at pre-stall angles of attack.

At the 6° angle of attack, which is well below the angle where leading-edge separation occurs, the pressure spectra shows no indication of the 80-Hz disturbance frequency. However, at the 14° angle of attack, which is 1° prior to airfoil stall

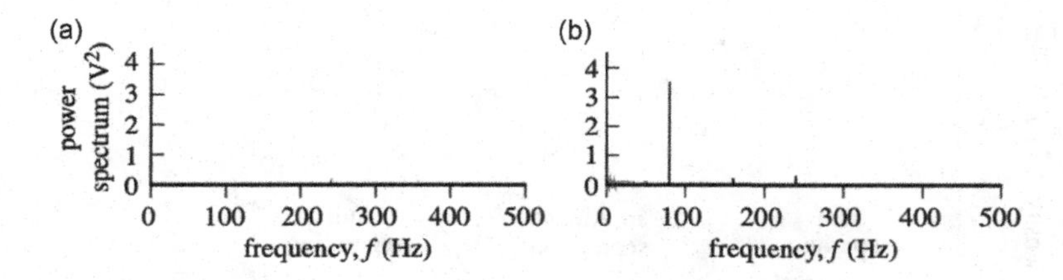

Figure 4.30 Spectra of the output from a pressure transducer located at $x/c = 0.05$ from the leading edge of an airfoil that is responding to low amplitude 80-Hz pulsing from a plasma actuator located at the leading edge for two angles of attack: $\alpha = 6°$ (a) and $\alpha = 14°$ (b). Taken from Lombardi et al. (2013) and Corke and Thomas (2011).

and complete flow separation, the spectra reveal a sharp peak at the 80-Hz disturbance frequency. Flow visualization images (Corke and Thomas, 2011) correlated the appearance of the spectral peak with the formation of a small separation bubble that extended over the location of the pressure port.

Lombardi et al. (2013) observed that the magnitude of the spectral peak detected in the static pressure fluctuations increased with the angle of attack. This observation, along with accompanying flow visualization, led to criteria based on the spectral peak amplitude of flow separation that was used in feedback separation control. The criteria defined three regions based on the amplitude of the spectral peak. These were "no flow separation," "imminent flow separation," and "fully separated" flow. These are shown in Figure 4.31(a). The feedback separation control flowchart that was based on this detection approach is shown in Figure 4.31(b).

The control loop starts with a low-power pulsing at the frequency that corresponds to $F^+ = 1$, which will be optimum at higher actuator power to attach the flow. Spectra of the time series from the detection pressure transducer are monitored to determine the amplitude, if any, of a spectral peak at the $F^+ = 1$ frequency. Based on the amplitude of the spectral peak, and its correlation with the separation criteria, the separation control actuator either continues to remain at low power, where the state of the flow is continually monitored, or switches to high power in order to maintain attached flow. The need to remain at high power is continually assessed in the control loop. If, based on the amplitude of the spectral peak, it indicates that the flow will naturally attach without actuation, the actuator power will be returned to the low "sense" state.

To illustrate feedback control approach, it was applied to a periodically pitching NACA 0015 airfoil. The airfoil was pitched about its quarter-chord location to produce a time-dependent angle of attack, $\alpha(t) = \alpha_{\text{mean}} + \alpha_{\text{amp}} \sin(2\pi f t)$. As with Kelley et al. (2014), a plasma actuator was located along the leading edge and pulsed at a specific frequency. As in the previous discussion, a pressure sensor was located in the surface of the airfoil at $x/c = 0.05$.

Figure 4.32 shows an example of the short-time spectral analysis of the pressure sensor output for the periodically pitching airfoil having a mean angle of attack of 10° and a pitching amplitude of 10° that leads to a condition of "light" dynamic stall (Corke and Thomas, 2011). The airfoil pitching frequency was 4 Hz, which resulted

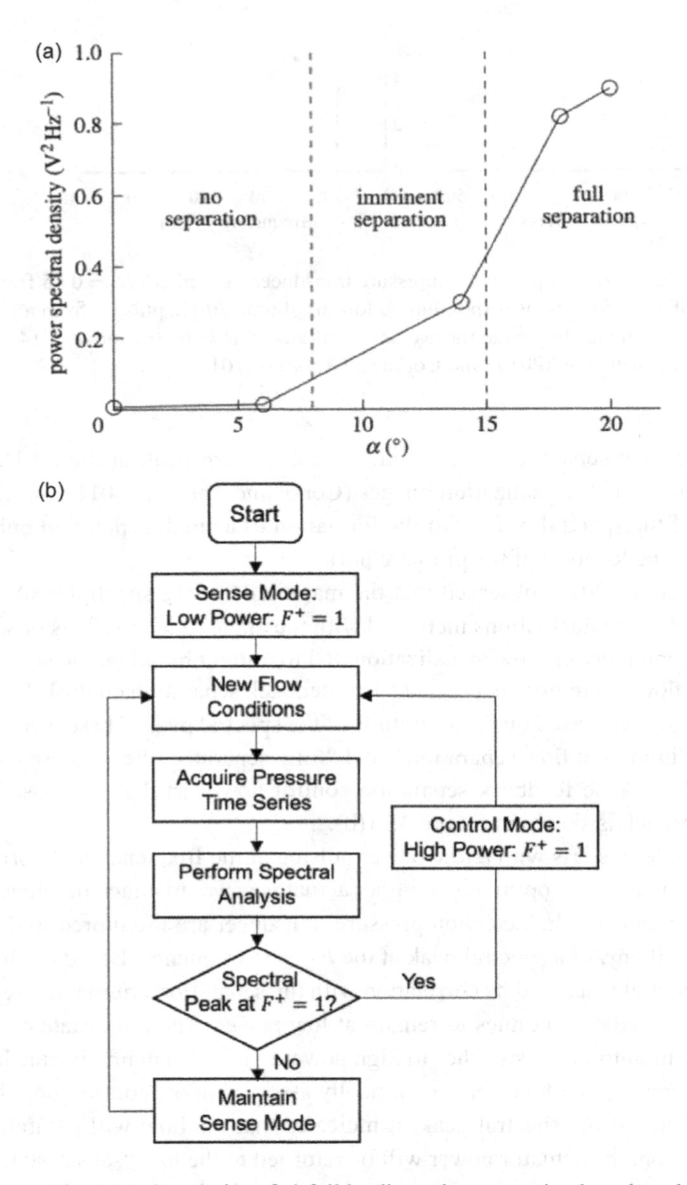

Figure 4.31 Categories of airfoil leading-edge separation based on detected pressure sensor 80-Hz amplitude (a) and associated flowchart for closed-loop control (b). Taken from Lombardi et al. (2013) and Corke and Thomas (2011).

in a reduced frequency, $k = \pi fc/U_\infty = 0.16$. As in the previous example, the plasma actuator was pulsing at a frequency of 80 Hz, which in this case was again the optimum $F^+ = 1$. The magnitudes of the short-time spectra are presented as constant level contours. The horizontal axis is time (ms), where the 4-Hz period of the airfoil pitching cycle is 250 ms. The vertical axis is frequency, and for reference, a horizontal dashed line has been drawn at the 80-Hz actuator pulsing frequency. Figure 4.32a shows time

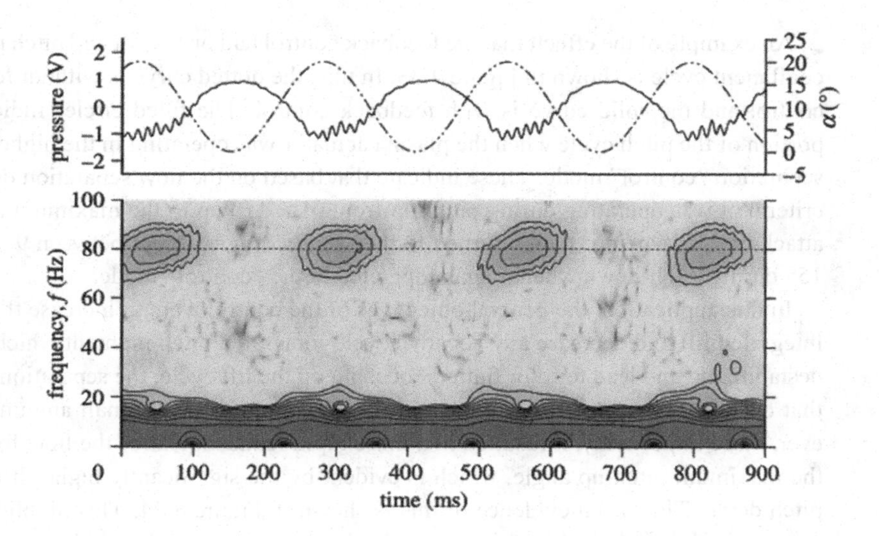

Figure 4.32 Contours of short-time spectral analysis of the separation detection pressure sensor time series subject to low-amplitude 80-Hz pulsing of leading-edge plasma actuator on a NACA 0015 airfoil pitching at conditions to produce light dynamic stall. (b) Corresponding angle of attack (dashed curve) and pressure (solid curve) time series. Taken from Corke and Thomas (2011).

series of the airfoil angle of attack, $\alpha(t)$, and the voltage proportional to the mean removed static pressure, $p(t)$, used for detection at $x/c = 0.05$.

Throughout the pitching cycle, the leading-edge plasma actuator was pulsing at the lower "sense" amplitude level. During the pitch-up portion of the cycle where the angles of attack are lower, the pressure time series is smooth. Flow visualization during this portion of the cycle indicated that the boundary layer over the airfoil was fully attached. However, during pitch-up when the airfoil reaches larger angles, higher-frequency fluctuations appear in the pressure time series starting at $\alpha \simeq 16°$ and continuing until $\alpha \simeq 5°$ during airfoil pitch-down motion. The short-time spectral analysis indicates that the pressure fluctuations were centered at a frequency of 80 Hz. This repeats from one pitching cycle to the next over which the detection of the leading-edge actuator pulsing frequency signifies that the boundary layer was periodically separating and reattaching. This was again confirmed by flow visualization (Corke and Thomas, 2011).

Lombardi (2011) and Lombardi et al. (2013) performed experiments to determine the impact of the feedback control on the lift and moment cycle of a periodically pitching airfoil undergoing dynamic stall. The experimental setup was the same as that used to describe Figure 4.32. The airfoil pitching frequency was the same 4 Hz. However, the mean angle of attack was higher, at 15°, with a pitching amplitude of 10° that lead to a more challenging "deep" dynamic stall condition (Corke and Thomas, 2011). Flow visualization and accompanying lift and pitch moment cycles were previously shown in Figure 4.4.

An example of the effectt that the feedback control had on the lift and pitch moment coefficient cycle is shown in Figure 4.33. In this, the dotted curve is without feedback control and the solid curve is with feedback control. The filled circles indicate the portion of the pitch cycle when the plasma actuator was operating in the higher power separation "control" mode. These indicate that based on the flow separation detection criteria, it was operating during pitch-up from $\alpha \simeq 15°$ up to the maximum angle of attack of 25°, where it then returned to the "sense" mode except between $9° \leq \alpha \leq 15°$ during pitch-down when it again operated in the "control" mode.

In this application, the general objectives of the control were to increase the cycle-integrated lift and decrease any negative cycle-integrated pitch moment which can be destabilizing and lead to rotor flutter. Focusing on the lift cycle, the separation control that occurred during pitch-up increased the maximum lift by a small amount. However, more importantly, it resulted in a more rapid reattachment of the flow following the maximum pitch-up angle, which is evident by the significantly higher lift during pitch down. The visual evidence of this is shown in Figure 4.34. This highlights two instants in the pitch cycle, one at $\alpha = 22°$ during pitch-up and the other at $\alpha = 19°$ during pitch-down, and contrasts the change that occurred with the separation control. Without control. the pitch-up of the airfoil results in the formation of a DSV that is evident in the no-control image at $\alpha = 22°$. The DSV results in a larger maximum lift coefficient compared to static airfoils. However, as the airfoil pitches up to its maximum angle, the DSV convects off the airfoil, resulting in a dramatic drop in lift that persists well into the pitch-down portion of the cycle (see Figure 4.4). In contrast, as evident from the flow visualization image, the feedback control is able to maintain a nearly attached flow. As was shown in Figure 4.33, the separation control during pitch-up did not result in a loss in the maximum lift.

Without separation control, the massive flow separation that occurs following the maximum pitch-up angle takes a substantial portion of the pitch-down cycle for the flow to reattach. This delay results in a significant loss of lift. Based on Figure 4.33, which utilized the separation criteria given in Figure 4.31, the actuator was not operating in the "control" mode during pitch-down until $9° \leq \alpha \leq 15°$. Thus, the evident improvement in flow attachment during pitch-down is the residue of the strategic control during pitch-up that resulted from the feedback control separation criterion.

Considering the global effect of the feedback control, the overall lift corresponds to the area under the cycle curve. In the presented case, the feedback separation control increased the cycle-integrated lift by 10 percent (Lombardi et al., 2013). With regard to the pitch moment, the flow control increased the cycle-integrated pitch moment from a negative value to a preferable near-neutral value (Lombardi et al., 2013). Finally, these results were comparable to those produced by open-loop separation control that was applied over the full pitch cycle (Lombardi et al., 2013). Therefore, in this application, the further benefit of the closed-loop control was to minimize the power supplied to the actuator.

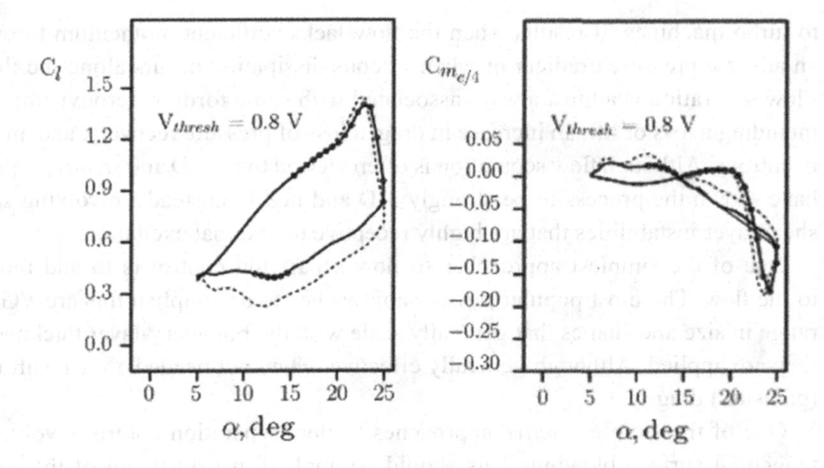

Figure 4.33 Lift and pitch moment cycles for conditions resulting from closed-loop control outlined in Figure 4.31. Dotted curve is no control. Solid curve is with feedback control. Filled circles indicate portion of pitch cycle where $F^+ = 1$ pulsing was activated by feedback control. Taken from Lombardi et al. (2013).

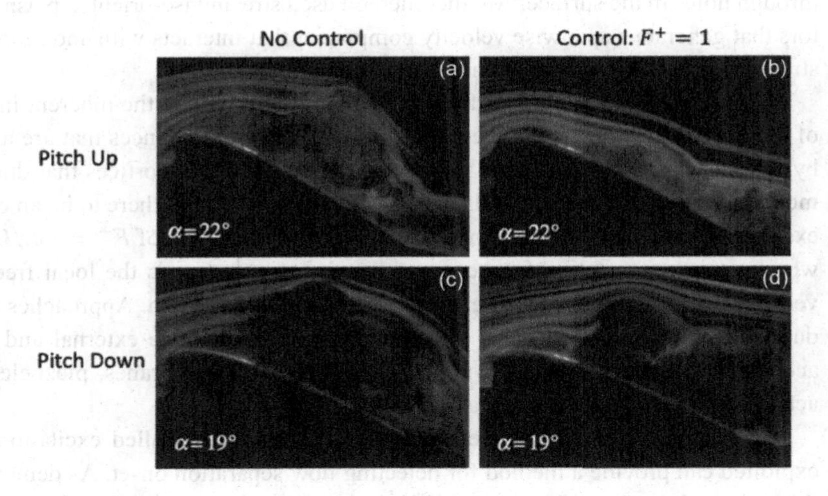

Figure 4.34 Samples from a flow visualization sequence of a periodically pitching NACA 0015 airfoil with a mean angle of attack of 15°, and a pitching amplitude of 10° that is producing deep dynamic stall. The top row corresponds to snapshots during pitch-up when the angle of attack was 22°. The bottom row corresponds to snapshots during pitch-down when the angle of attack was 19°. The columns correspond to conditions with no control (a) and control with periodic leading-edge pulsing at $F^+ = 1$ (b).

4.4 Summary

Flow separation can occur in a variety of important applications including external flows particularly related to aircraft and internal flows, with particular applications to turbo-machines. It results when the flow lacks sufficient momentum to overcome an adverse pressure gradient or when viscous dissipation occurs along the flow path. Flow separation is almost always associated with some form of aerodynamic penalty, including a loss of lift, an increase in drag, a loss of pressure recovery, and an increase in entropy. Although flow separation is often viewed to be 2-D and *steady*, experiments have shown the process to be strongly 3-D and highly unsteady, involving separated shear layer instabilities that are highly receptive to external excitation.

One of the simplest approaches to flow separation control is to add momentum to the flow. The most popular *passive* approaches to accomplish this are VGs. These range in size and shapes that generally scale with the boundary-layer thickness where they are applied. Although generally effective, when not needed, they result in added (parasitic) drag.

One of the simplest *active* approaches to flow separation control involves steady tangential surface blowing. This should be applied just upstream of the separation location to add momentum to the boundary layer. An example of this approach is split trailing-edge flaps on aircraft wings which allow high-pressure air from the lower side of the wing to pass to the upper side to add momentum to the boundary layer over the flap surface. Active VGs can be produced by blowing air at shallow angles through holes in the surface. Another method used streamwise-oriented plasma actuators that generate a spanwise velocity component that interacts with and reorients the streamwise flow to produce streamwise vorticity.

Approaches that require less flow actuator authority utilize the inherent instability of the separated shear layer. These generate unsteady disturbances that are amplified by the shear layer to cause it to form coherent cross-stream vortices that draw high-momentum fluid toward the wall. Experiments have indicated there to be an optimum excitation frequency that corresponds to a reduced frequency of $F^+ = \omega L/U_\infty = 1$, where ω is the excitation frequency with units of s^{-1}, U_∞ is the local free-stream velocity, and L is the streamwise length of the separated region. Approaches to introduce the unsteady disturbances for separation control include external and internal acoustic speakers, internal blowing, wall jets, flexible membranes, piezoelectrically active surfaces, and plasma actuators.

Finally, the receptivity the separated shear layer to controlled excitation can be exploited can provide a method for detecting flow separation onset. As demonstrated, this can result in closed-loop separation control that can minimize flow actuator power compared to open-loop control.

Problems

4.1 Experiments have indicated that a reduced frequency of $F^+ = fL/U_\infty$ is optimum to reattach a separation bubble whose streamwise length is L. Figure 4.14 indicates that the maximum amplification of a flow separation with reverse flow $(d/\theta = 4.5)$ occurs at $\omega\theta/U_\infty = 0.075$.

1. Based on this, determine the ratio of θ/L such that $F^+ = 1$ occurs where the spatial growth rate is a maximum. Note that $\omega = 2\pi f$.
2. Assuming that the ratio of the boundary layer thickness to momentum thickness is $\delta/\theta \simeq 10$, what is the ratio δ/L such that $F^+ = 1$ occurs where the spatial growth rate is a maximum.

4.2 Both wall-normal suction and blowing can reduce the tendency for flow separation. Can this be explained from the point of view of the separated shear layer instability? Refer to Figures 4.12 to 4.14.

1. Describe how d/θ of the profile is affected by wall-normal suction or wall-normal blowing.
2. How does the profile d/θ affect the shear layer instability properties? Are the effects the same for suction and blowing?
3. Based on these observations, list the effects that reduce the tendency for flow separation in the cases.

4.3 Spectra of velocity fluctuations at different streamwise locations in a flow separation bubble shown in Figure 4.18 identify dominant spectral peaks at f_0 and $f_0/2$. The growth in amplitude of fluctuations at the two frequencies is shown in Figure 4.18(b). Of particular interest is the enhanced growth at $f_0/2$ for $x/L > 0.63$ that is evidence of a "subharmonic resonance."

1. How is this relevant to the reducing the x-extent of the separation bubble?
2. Based on the temporal model for the effect of the growth of f_0 on $f_0/2$ given in Equations 4.8–4.10, how can this interaction be enhanced by unsteady excitation?
3. If fluctuations at both frequencies are excited, would the phase difference between them be important? If so, what would be the optimum phase difference based on the temporal model.
4. In the figure, the disturbance growth occurs in space (x). How could the temporal model be converted to a spatial model?

4.4 Figure 4.15 displays the normalized growth rates and phase velocities as a function of wave angle in a Falkner–Skan boundary layer just prior to separation. Considering this,

1. What is the most important instability characteristic for a triad resonance of the type described by Craik (1985) to occur between 2-D and 3-D waves?
2. Based on Figure 4.15, describe conditions that could lead to a triad resonance between a 2-D wave and a pair of 3-D waves of equal but opposite wave angles.
3. What effect do you expect this would have on the amplification of the 3-D waves?
4. Could this have some benefit in controlling flow separation? Describe how.

4.5 Table 4.3 indicates that the optimum spanwise spacing of vortex generators to suppress a flow separation is 0.8δ, where δ is the boundary-layer thickness.

1. Considering a Falkner–Skan boundary layer just prior to separation, $d/\theta = 4.0$ in Figure 4.15, the most amplified dimensionless frequency is $\omega\theta/U_\infty = 0.12$,

and the streamwise wavelength of the instability waves is $\lambda_x = 2\pi/\omega$, write λ_x in terms of θ/U_{infty}.

2. Assuming that $\theta/\delta \simeq 0.1$, determine the effective wave angle, $\phi = \tan^{-1}(\lambda_z/\lambda_x)$ produced by the vortex generators separated by $\lambda_x = 0.8\delta$ for a range of free-stream velocities of 5, 10, 15, and 20 m/s.
3. Can you draw a conclusion on a connection between the optimum spacing of vortex generators and the boundary-layer instability to 3-D disturbances?

4.6 Based on plasma vortex generators, Kelley et al. (2016) found that the streamwise length of the actuator, L, was based on the timescale of the vorticity, $T_c \geq (\partial\bar{U}/\partial z)^{-1}$, such that $L \geq T_c U_\infty$. The plasma vortex generators produced the equivalent of a spanwise-oriented tangential wall jet, and therefore the same scaling criterion should equally apply to a pressure-driven tangential wall jet that extends in the streamwise direction.

1. A typical spanwise velocity of a plasma actuator is about 3–5 m/s. Based on Figure 4.26 for free-stream Mach numbers of 0.15–0.20 (51–68 m/s), a $T_c \simeq 1.5$ ms met the criteria that $T_c \geq (\partial\bar{U}/\partial z)^{-1}$. Based on this, what is the required actuator length, L?
2. Using wall jet vortex generators, Johnston and Nishi (1990) found that $V_j/U_\infty > 0.8$ was effective. Based on this and assuming that T_c scales linearly with the actuator velocity, what is the required actuator length?
3. Does this justify the typical size (diameter) of wall jet vortex generators?

5 Free Shear Layers and Jets

Free shear layers of the kind that are formed by the merging of two streams initially separated by a thin surface is a simple flow configuration that arises in numerous natural phenomena, as well as in engineered devices such as combustors and gas lasers. This generic flow field results in intensive mixing that occurs in the velocity gradient region between the two streams. As a result, the flow field is often referred to a mixing layer. A representative example of a mixing layer is shown in Figure 5.1.

In many practical cases, there is a motivation to manipulate the downstream evolution of shear flows to enhance or retard mixing, or to alter the acoustic field generated by the mixing layer. This can be accomplished through a variety of passive and active flow control approaches. In keeping with the theme of this book of an instability approach to flow control, the mixing layer is viewed as a fundamental class of inviscidly unstable free shear flows. It is therefore important to understand the stability characteristics of these flows.

5.1 Stability of Free Shear Flows

Crow and Champagne (1971) discovered that the shear layer of a jet can support orderly vortical structures and operate as a finely tuned amplifier of upstream disturbances. Brown and Roshko (1974) subsequently confirmed that large-scale coherent structures are indeed intrinsic features of turbulent mixing layers at high Reynolds numbers. Furthermore, sequential merging of vortices provide the primary mechanism for the spreading of the layer in the downstream direction, as underscored by the experiments of Winant and Browand (1974).

The approach of this book is to rely on classical hydrodynamic stability theory where the unsteady mixing layer is conceptualized as a superposition of interacting instability waves that propagate and amplify in the downstream direction. Linear stability analysis has been shown to very satisfactorily describe the initial development of the mixing layer up to the first roll-up of the fundamental mode vortices. In addition, appropriate phenomenological models can determine the main length scales of the flow much farther downstream. Furthermore, analytical methods are available to account for observed wave interactions that can be attributed to weakly nonlinear effects.

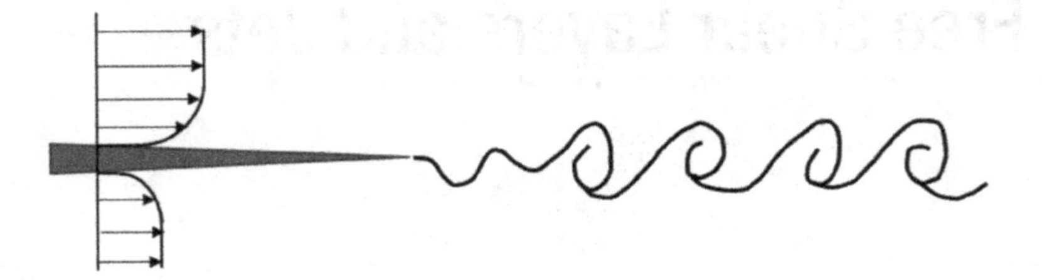

Figure 5.1 Schematic of a classic 2-D mixing layer produced by the merging of two streams separated by a thin splitter plate.

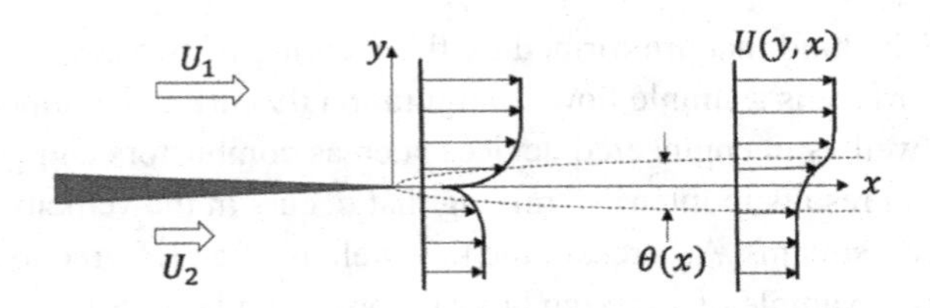

Figure 5.2 Schematic of a classic 2-D free shear flow formed by two streams with velocities U_1 and U_2 that are separated by a thin splitter plate.

It has long been realized that the evolution of mixing layers can be highly susceptible to very low-amplitude disturbances. Thus, artificially generated perturbations have been used to both deepen our basic understanding of shear layer dynamics, as well as to significantly alter the downstream development. Such low-amplitude disturbances or perturbations occur in the natural environment. These can have a broad frequency content, or even be monochromatic. Linear stability analysis determines the response of the basic (time-averaged) flow to such perturbations. The incentive behind such analysis is to guide approaches for efficiently interacting with and controlling the initial mixing layer development.

The classic free shear layer basic flow consists of two laminar streams with free-stream velocities U_1 and U_2. A schematic of this arrangement is shown in Figure 5.2, where $U_1 > U_2$, and as shown, x is the streamwise direction and y is the cross-stream direction. The other variables are the spanwise direction, z, and time, t.

In practical applications, the two streams are initially separated by a thin splitter plate over which a boundary-layer flow develops. The free shear layer basic flow then consists of the mean profile that results from a merging of the two boundary layers at the trailing edge of the splitter plate. The trailing edge of the splitter plate represents the virtual origin of the free shear layer, at which point the momentum thickness is a minimum and denoted as θ_0. The mixing of the two streams that occurs through the downstream development of the shear layer causes the momentum thickness to increase. Close to the trailing-edge virtual origin, $\theta(x)$ increases as $\sqrt{x}$.

The basic flow can be conveniently characterized by two nondimensional parameters. One is a velocity ratio, $R = \Delta U/(2\overline{U})$, where $\Delta U = U_1 - U_2$, and $\overline{U} = (U_1 + U_2)/2$. In this, $0 \leq R \leq 1$, where $R = 0$ is a wake flow, and $R = 1$ has only one stream present. The other nondimensional parameter is the trailing-edge Reynolds number, $\mathrm{Re}_0 = \overline{U}\theta_0/\nu$, where θ_0 is the initial momentum thickness at the trailing edge of the splitter plate, and ν is the kinematic viscosity.

The development of mixing layers downstream of a splitter plate is inviscidly unstable to small perturbations via the Kelvin–Helmholtz instability mechanism. Thus, two-dimensional (2-D) waves grow exponentially with downstream distance. These waves eventually roll up into discrete vortices. If the basic flow $U(y)$ is taken to be strictly parallel, and disturbances are infinitesimally small (satisfying the linear stability assumptions), the perturbation stream function can be cast into the form

$$\psi(x,y,t) = \psi(y) \exp i(\alpha x - \omega t) + c.c, \tag{5.1}$$

where $\omega = 2\pi f$ is the angular frequency, α is the wave number, and $c.c$ denotes the complex conjugate.

In the inviscid limit $\mathrm{Re} = \infty$, the eigenfunction $\psi(y)$ is obtained by solving the linear Rayleigh equation. Most calculations have been conducted for a hyperbolic-tangent velocity profile $U(y;R) = \overline{U}[1 + R\tanh(y/2\theta)]$, which is a reasonable approximation to experimental measured profiles. Monkewitz and Huerre (1982) determined the theoretical dependence of spatially growing waves on R for both hyperbolic-tangent and Blasius initial shear layer profiles.

Monkewitz and Huerre (1982) found that the nondimensional growth rate, $-\alpha\theta/R$ varied with Strouhal number, $\mathrm{St} = f\theta/\overline{U}$. This is shown in Figure 5.3 for two cases with $R = 0.5$ and 1. Excellent agreement was found in experiments by Freymuth (1966) which is represented by the symbols in the figure. The stability analysis determined that the largest amplification occurred at $\mathrm{St} = 0.032$. The analysis found that the most-amplified Strouhal number varied by only 5 percent over the full range of $0 \leq R \leq 1$. Monkewitz and Huerre (1982) also found that the maximum amplification rate, $(-\alpha_i\theta)_{\max}$, increased approximately linearly with R. Additionally, the phase speed of the most amplified wave was found to equal the average velocity, $\overline{U}$, of the two streams. These results have a clear impact on shear layer control that involves unsteady disturbance excitation.

According to linear theory, the most amplified instability wave with frequency, f_n, grows exponentially in amplitude with downstream distance, and travels at an average velocity of $\overline{U}$. As depicted in Figure 5.1, these waves eventually evolve into a periodic array of compact spanwise vortices that are spaced in the streamwise direction by a wavelength of $\lambda_n = \overline{U}/f_n$. Experimental observations indicate that the roll-up process is predominantly 2-D and is completed at the downstream station where the fundamental component at frequency, f_n, reaches its maximum amplitude. It is also accompanied by the generation of a subharmonic component, $f_n/2$.

5.1.1 Subharmonic Mode and Vortex Pairing

The emergence of a subharmonic component, $f_n/2$, constitutes one of the most striking features in mixing-layer dynamics, as underscored by the experiments of

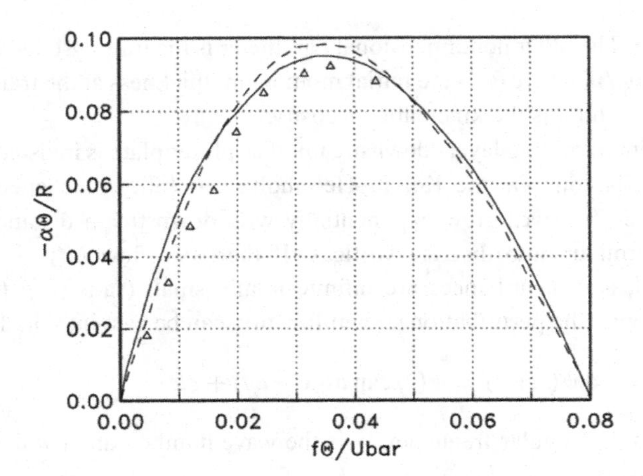

Figure 5.3 Normalized amplification rate as a function of Strouhal number based on linear stability theory. Solid line for $R = 0.5$ and dashed for $R = 1$ from Monkewitz and Huerre (1982). Symbol from experiments of Freymuth (1966).

Sato (1959), Browand (1966), Miksad (1972), and Ho and Huang (1982). To understand conditions that could lead to the growth of the subharmonic mode, $f_n/2$, Monkewitz (1982) examined the linear temporal (secondary) instability of the periodic flow

$$U(y) + \epsilon \left(A\phi_1(y) \exp\left[\frac{i}{2\theta}(x - \overline{U}t) - i\beta\right] + c.c.\right) \tag{5.2}$$

to a subharmonic wave of the form

$$\psi(x, y, t) \sim \epsilon B(T)\phi_{1/2}(y) \exp\left[\frac{i}{4\theta}(x - \overline{U}t) + \omega_{1_{1/2}}t\right] + c.c., \tag{5.3}$$

where $\omega_{i_{1/2}}$ and $\phi_{1/2}(y)$ are the respective subharmonic temporal amplification rate and eigenfunction, β is the phase difference between the fundamental and subharmonic waves, and $T = \epsilon t$ is a slow timescale. Based on this formulation, subharmonic resonance required that both waves have approximately the same phase speed, namely $\overline{U}$.

The amplitude function $B(T)$ is found (Monkewitz, 1982) to satisfy

$$\frac{dB}{dT} = c \exp(-i\beta)AB^*, \tag{5.4}$$

where c is a constant and * denotes the complex conjugate. The initial total subharmonic growth rate

$$\omega_{i_{1/2}} + [d(\ln |B|/dT] \tag{5.5}$$

is proportional to the periodic amplitude A and varies continuously from a maximum when both waves are in phase, $\beta = 0$, to a minimum when they are out of phase, $\beta = \pi$. This dependence on phase suggests an active flow control approach.

In a more elaborate analysis of subharmonic resonance (Kelly, 1968; Maslowe, 1977), the subharmonic amplitude, B, is allowed to react back nonlinearly on the fundamental amplitude, A, through a second coupled equation

$$\frac{dA}{dT} = gB^2 \exp(i\beta). \tag{5.6}$$

This again has relevance to shear layer control approaches in which fundamental and subharmonic instabilities are simultaneously excited.

It is now well established (Wille, 1963) that the selective growth of the subharmonic component leads to the pairing of neighboring vortices further downstream. Winant and Browand (1974) were the first to demonstrate that successive mergings of vortices is the primary process governing the streamwise growth of the mixing layer. In this process, the vorticity initially contained in the basic flow velocity profile is being constantly redistributed into larger and larger vortices, with their wavelength and strength being doubled after each interaction. The passage frequency is similarly halved after each coalescence, and the spectrum exhibits a shift toward lower frequencies with increasing distance from the trailing edge (Kibens, 1980). Under controlled conditions, the first few pairings are accompanied by an approximate doubling in momentum thickness, θ. The variations of θ between mergings are otherwise negligible.

The interacting spanwise vortices observed in the early laminar stages of the flow evolution continue to exist farther downstream in the turbulent region, where they coexist with a fine-scale motion. These features persist up to Reynolds numbers at least as high as 10^7, and well into the self-preserving region of classical theory (Dimotakis and Brown, 1976). Conditional measurements in turbulent shear flows that document the coherent vortical structure evolution (Browand and Weidman, 1976) have documented sequential mergings similar to those in laminar shear layers. However, the random locations of pairing locations resulted in a smooth linear spreading rate without the jumps in momentum thickness.

Based on these observations, it is increasingly clear that the evolution of vortical structures in both laminar and turbulent shear layers is governed by essentially the same dynamical processes. This has prompted the application of hydrodynamic stability concepts to *turbulent* shear layers. Within this framework, the large-scale structures are then viewed as instability waves propagating on the pseudo-laminar flow defined by the time-averaged velocity field. This approach was born out by Fiedler et al. (1981), who found that a quasi-parallel inviscid linear approximation satisfactorily predicted the local amplification rate of large-scale structures generated by low-frequency forcing.

5.1.2 3-D Mode Development

To this point, the focus has been on the development of a 2-D vortex sheet. The occurrence of significant three-dimensional (3-D) effects was first discovered by (Miksad, 1972). This revealed spanwise-periodic features consisting of counter-rotating

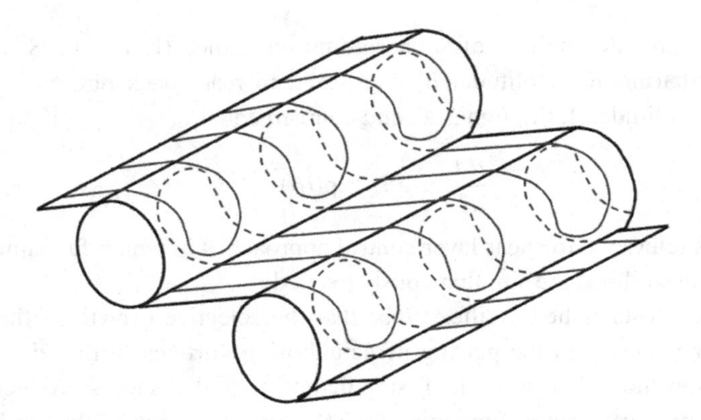

Figure 5.4 Schematic representation of secondary streamwise vortices superposed on primary 2-D vortices in an initially 2-D shear flow based on flow visualization records from Bernal (1981).

streamwise vortices superposed onto the primary 2-D vortices. The morphology of the secondary motion derived from observations of Bernal (1981) is shown in Figure 5.4. The spanwise wavelength was insensitive to the small irregularities in the boundary layer on the splitter plate, indicating that the spanwise features were the result of a fluid instability.

The earliest theoretical analysis of streamwise structures in a mixing layer was by Benney and Lin (1960). Their approach involved modeling weakly nonlinear interactions between a 2-D wave of wave number α, and a 3-D wave with an *identical streamwise wave number* and a spanwise wave number, γ. The nonlinear growth of the 3-D wave results in a mean streamwise-independent, spanwise distortion of the basic tanh(y) profile that takes the form of two horizontal layers of counter-rotating streamwise vortices. If A and B denote the respective amplitudes of the 2-D and 3-D components, the spanwise wave number associated with the counter-rotating eddies varies from γ to 2γ as the ratio $|B|/|A|$ increases from zero to infinity. According to Stuart (1962), A and B are coupled via evolution equations of the form

$$\frac{dA}{dT} = \sigma_A A + b_a|A|^2 A + c_A|B|^2 A + d_A A^* B^2 \tag{5.7}$$

$$\frac{dB}{dT} = \sigma_B B + b_B|A|^2 B + c_B|B|^2 B + d_B B^* A^2, \tag{5.8}$$

where σ_A and σ_B are the respective linear stability amplification rates of the 2-D and 3-D modes, and T is a slow timescale. The streamwise structure is therefore determined by the superposition of the steady mean spanwise distortion and the finite-amplitude 3-D wave of amplitude $B(T)$.

Another interaction model initially proposed by Craik (1985) in the context of boundary-layer transition is likely to apply to mixing layers as well. This would involve a pair of 3-D modes with equal but opposite spanwise wave numbers in which their streamwise wave number is one in which the 3-D mode phase speed is the same as that of a 2-D mode. This often happens when the 2-D wave number is twice that of the 3-D streamwise wave number and thereby results in a subharmonic resonance.

5.1.3 Effect of Initial Conditions

Given the dependence of the initial shear layer instability on the initial momentum thickness, θ_0, it is not surprising that the downstream development of a shear layer is sensitive to the state (laminar or turbulent) of the boundary layer on the splitter plate. In a mixing layer with laminar upstream conditions, the peak turbulence intensity $u'/\Delta U$ increases in the streamwise direction and reaches a maximum around the first vortex-pairing location. The turbulence intensity eventually settles farther downstream to an asymptotic value characteristic of a "fully developed" turbulent mixing layer. In contrast, starting as a turbulent boundary layer at the trailing edge of the splitter plate ($x = 0$), the turbulence intensity relaxes monotonically to approximately the same constant value as in the laminar boundary-layer case, with no overshoot as in the laminar case (Drubka, 1981).

As previously discussed, sequential mergings of coherent spanwise rollers are believed to be responsible for most of the entrainment into the shear layer and for the occurrence of small-scale transition. Hence, in order to manipulate most effectively the development of the flow, imposed excitations should, as a first requirement, be spatially coherent along the span. One approach is through acoustic excitation. An issue at low Mach numbers is the large mismatch between the speed of sound and the phase speed of instability waves. As a result, a mechanism is required to convert long wavelength acoustic pressure fluctuations into shorter wavelength vortical fluctuations that can be amplified by the shear layer instability. That mechanism is "acoustic receptivity" and it occurs at the splitter plate trailing edge (Morkovin and Paranjape, 1971; Morkovin, 1988). The efficiency of the trailing-edge conversion mechanism has been studied extensively in connection with the jet-noise problem. Bechert (1980), among others, has investigated the coupling effects arising between an incident acoustic wave traveling down a circular nozzle and instability waves on the jet shear layers. This has similarly been studied in a low Mach number axisymmetric jet by Corke et al. (1991), where pressure fluctuations associated with vortex roll-up and pairing receptivity feed upstream and reenforce the linear stability of the jet shear layer.

Either as a result of self-excitation or external excitation, the spatial evolution of the shear layer depends on the frequency, amplitude, and phase of each excited spectral component. For example, Ho and Huang (1982) observed that exciting a higher initial level fundamental shear layer frequency, f_n, can temporarily suppress the growth of the subharmonic, $f_n/2$, and therefore delay vortex pairing. Similarly when exciting a frequency, f_e, such that $f_n/3 < f_e < f_n/2$, the frequency that emerged in the shear layer jumped to the first harmonic, $2f_e$, that came closest to the natural fundamental shear layer frequency, f_n. In this case, the excitation became the subharmonic of the initial vortex passage frequency, and in contrast with the previous fundamental excitation, vortex pairing was promoted.

Ho and Huang (1982) documented that further reductions in the excitation frequency, f_e, could lead to successive frequency-locking stages in which the resulting shear layer frequency became the second and third subharmonics of f_e, resulting in the coalescence of as many as three or four vortices. As a result of such excitation, the shear layer spreading rate could be either enhanced by promoting multiple-vortex amalgamations or reduced by delaying them. Provided that the excitation is at the *proper* frequency, the results of Ho and Huang (1982) indicate that very low forcing levels on the order of 0.01–0.1 percent of $\overline{U}$ are sufficient.

5.1.4 Feedback Effect

Dimotakis and Brown (1976) were the first to suggest that pressure disturbances occurring in the formation and pairing of vortices can be felt at the splitter plate trailing edge, and thereby influence the initial unsteady vorticity through the receptivity mechanism. Following this, Laufer and Monkewitz (1980) considered a global feedback mechanism that was associated with the sudden change of circulation that occurs during merging interactions. Based on this, each vortex-pairing event was assumed to be linked to the trailing edge via a feedback loop that consisted of a downstream-propagating subharmonic instability wave and an upstream-propagating acoustic wave. According to Ho and Nosseir (1981), the phase difference between the two waves is required to be of the form $2N\pi$, where N is an integer. Thereby, if x_j denotes the location of the jth vortex merging, λ_j to the corresponding subharmonic wavelength, and λ_a to the acoustic wavelength, the following integer wavelength constraint must be satisfied:

$$\frac{x_j}{\lambda_j} + \frac{x_j}{\lambda_a} = N. \tag{5.9}$$

Evidence of the shear layer feedback mechanism has come from axisymmetric jet experiments in which the exit shear layer is thin compared to the jet nozzle exit radius of curvature satisfying a 2-D shear layer approximation. The most sited evidence of the feedback mechanism comes from Kibens (1980) in which the shear layer of a low Mach number axisymmetric jet was acoustically excited. The excitation frequency, f_e, corresponded to the naturally most amplified shear layer frequency, f_n. The response of the shear layer to the excitation was measured using a hot-wire velocity sensor that was sequentially moved downstream from the jet exit. The results from Kibens (1980) are reproduced in Figure 5.5. The ordinate corresponds to the shear layer frequency observed at different x/D locations in the shear layer which is plotted on the abscissa The shear layer frequency, f, is presented as a Strouhal number, fD/U_j, where U_j is the centerline jet velocity.

The measurements documented a stair-step-like response with regions of spatial frequency "lock-in" at discrete frequencies corresponding to f_e, $f_e/2$, $f_e/4$, and $f_e/8$. Such frequency lock-in is indicative of a feedback resonance. Kibens (1980) indicated that the frequency jumps corresponded to vortex-pairing events, which is consistent with the factor of two frequency changes at each jump. Acoustic measurements

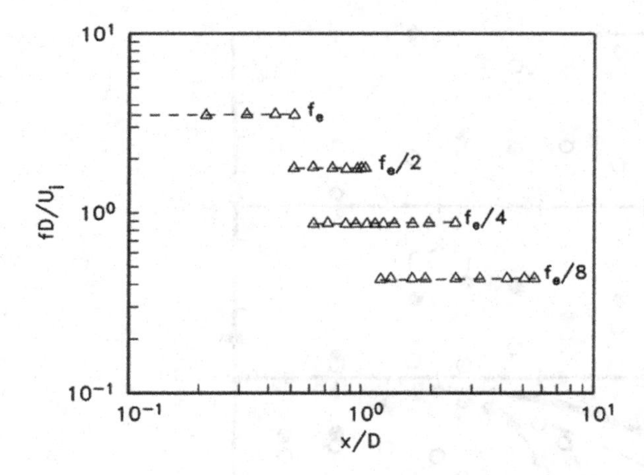

Figure 5.5 Shear layer frequency measurements at different downstream distances when acoustically excited at frequency, f_e, that reveals frequency lock-in indicative of feedback resonance involving a subharmonic cascade. Results from Kibens (1980).

confirmed a correspondence between the flow field measurements and the near-field pressure spectrum.

Evidence of the role of acoustic feedback in a natural, unforced shear layer is offered by Corke et al. (1991). As with Kibens (1980), this was performed in an axisymmetric jet, but, in contrast, the shear layer was not artificially excited but rather responded to natural disturbances. A summary of the results is presented in Figure 5.6. This includes the streamwise development of the streamwise velocity fluctuation amplitude, $u'/U_\infty(f)$, of the shear layer fundamental and subharmonic frequencies, linear coherence, $\Gamma(f)$, between velocity fluctuations at downstream locations and pressure fluctuations at the jet exit for the fundamental and subharmonic frequencies, and the cross-bicoherence, $\beta(f_1,f_2)$, between downstream velocity fluctuations and pressure fluctuations at the jet exit. The linear coherence is a measure of linear phase locking, which ranges from 0 to 1, with 1 indicating perfect linear phase locking. The cross-bicoherence is a measure of the nonlinear phase locking between two frequencies and their sum and difference frequencies.

As expected, the most amplified fundamental shear layer mode grows exponentially in amplitude with downstream distance. This appears as a linear growth with the slope corresponding to the amplification rate on the log-based ordinate. Linear theory predicts a lower amplification rate for the subharmonic mode, which is evident by its lower initial growth in amplitude. The fundamental mode amplitude peaks at $x/D \simeq 0.3$. Based on the literature (Ho and Huerre, 1984), this corresponds to the roll-up of the fundamental mode wave into a vortex. The linear coherence between the fundamental mode velocity and jet exit pressure fluctuations indicates a degree of phase locking that would suggest a measurable degree of acoustic feedback. Because of the relatively low linear coherence, this is shown by a thin arrow indicating a weak effect on the fundamental mode initial amplitude.

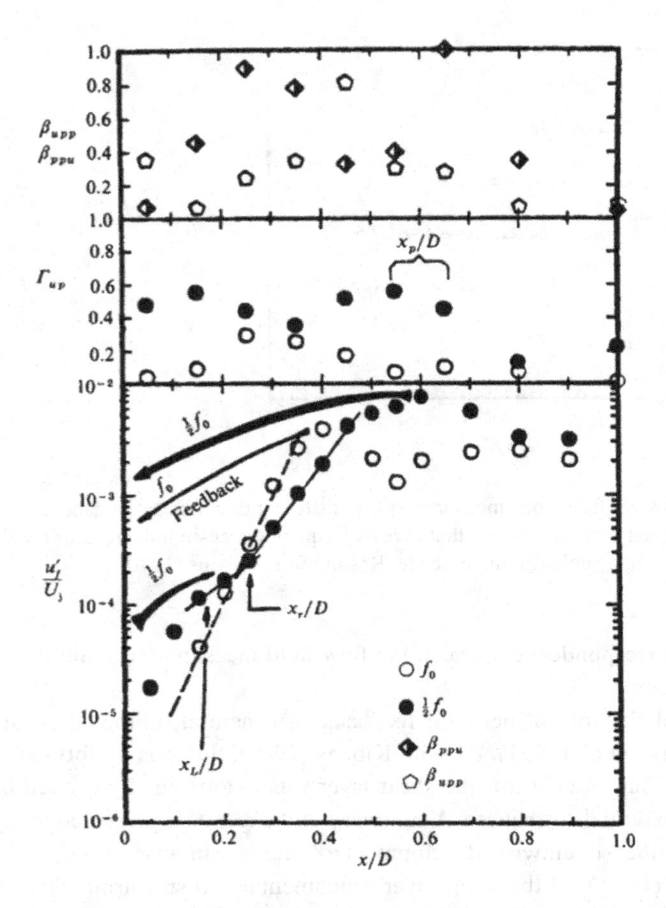

Figure 5.6 Streamwise development of amplitude, linear coherence and cross-bicoherence of the initial axisymmetric mode and its subharmonic in the shear layer of an axisymmetric jet with natural excitation. Taken from Corke et al. (1991).

In contrast to the fundamental mode, when the subharmonic mode amplitude reaches a peak, signifying pairing of the fundamental mode vortices (Ho and Huerre, 1984), there is a more significant linear phase locking between the subharmonic mode velocity and jet exit pressure fluctuations, which indicates a strong degree of phase locking that suggests a strong degree of acoustic feedback. This is signified by the thicker arrow to indicate a strong effect on the subharmonic mode development. This is evident in the larger initial amplitude of the subharmonic mode velocity fluctuations compared to that of the fundamental mode.

These observations have focused on linear interaction and pressure feedback. However, a feature of the subharmonic mode growth is the change in the growth rate that occurs at $x/D \simeq 0.25$. This happens to coincide with a near perfect (1) cross-bicoherence, β_{ppu}, which is a measure of the nonlinear interaction between pressure fluctuations at the fundamental and subharmonic frequencies at the jet exit with that of velocity fluctuations at the subharmonic frequency in the form $p'(f) - p'(f/2) = p'(f/2)$. This interaction resulted in faster than linear stability growth of the subharmonic mode, and a strong self-forced phase locking between the fundamental and subharmonic shear layer modes.

5.1.5 Jet Instabilities

For many years, investigations have been conducted to understand the flow processes that occur in jets. One practical purpose of these was to determine the relationship between the observed characteristic velocity fluctuations in the jet and the generation of measured far-field acoustic disturbances associated with these flows. Previous investigations in naturally and artificially excited jets have determined the importance of two instability length scales: one associated with the initial shear layer thickness at the exit of the nozzle, and the other associated with the jet diameter which governs the shape of the mean velocity profile at the end of the potential core. The former shares common features with free shear layers illustrated in Figure 5.1. As with free shear layer, the instability modes in the first region develop through continuous and gradual frequency and phase adjustments to produce a smooth merging with the jet potential core region. This process makes this problem fundamentally interesting, and for that reason, it has received a great deal of attention.

In experiments in low-disturbance axisymmetric jet flows, most of the emphasis has been placed on the growth of axisymmetric disturbances close to the nozzle exit. However, the analysis of Michalke (1971) has pointed out that the first helical mode has amplification characteristics in this region which are nearly identical to those of the fundamental axisymmetric mode. This analysis further suggests that as the shear layer thickens or grows, the amplification of the helical mode becomes more dominant over the axisymmetric mode. Mattingly and Chang (1974) performed a similar analysis with a different prescribed mean velocity distribution that showed that the amplification of the axisymmetric mode was only approximately 12 percent larger than that for the helical mode, and that the streamwise frequency of the most amplified helical disturbance in that case was approximately 20 percent higher than that of the most amplified axisymmetric mode. Drubka (1981) experimentally verified these results.

Flow visualization in an axisymmetric jet by Corke and Kusek (1993) captured images of the axisymmetric and helical shear layer modes. Examples of these are shown in Figure 5.7. The linear stability analysis wave number representation of the helical mode is $(\alpha, \pm\beta)$, where α is the streamwise wave number, and β is the azimuthal wave number, where the local wave angle is $\psi = \arctan(\beta/\alpha)$. Note that the linear stability theory does not distinguish between positive or negative azimuthal wave numbers so that either, or both simultaneously, are possible and likely. Figure 5.8 is a flow visualization image from Kusek et al. (1989) that captures the simultaneous occurrence of helical mode pairs of equal–opposite wave angles.

Jet Column Mode

Although the initial shear layer in jets shares common features with classical free shear flows, a fundamental difference is the second instability length scale associated with

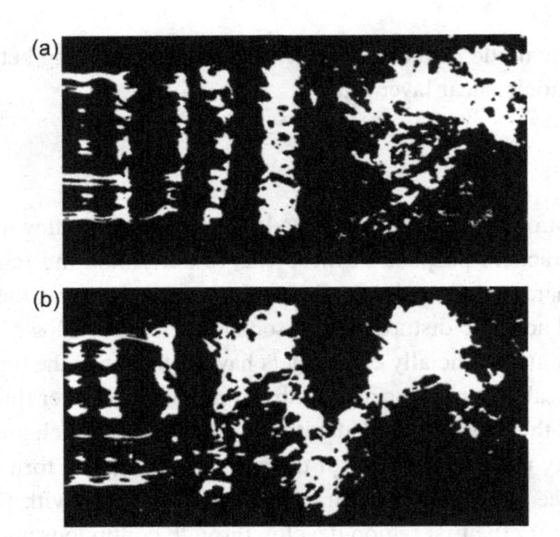

Figure 5.7 Flow visualization images in an axisymmetric jet showing axisymmetric (a) and helical shear layer modes (b). Taken from Corke and Kusek (1993).

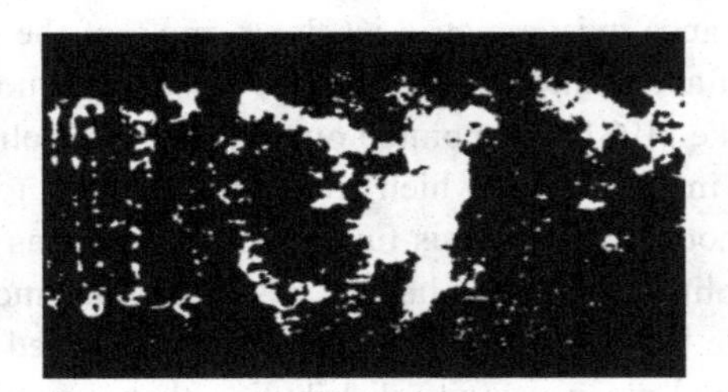

Figure 5.8 Flow visualization images in an axisymmetric jet showing pairs of helical shear layer modes of equal–opposite wave angles. Taken from Kusek et al. (1989).

the jet diameter, D, in an axisymmetric jet, or height, H, in the case of a 2-D jet. In axisymmetric jets, a potential flow of velocity U_j issuing from the exit plane comes to an end approximately $5D$ downstream. This marks the end of the "potential core." In a 2-D jet, the end of the potential core occurs approximately $3H$ downstream of the nozzle exit. In both types of jets, orderly vortical structures associated with the shear layer instability can be clearly identified to surround the potential core.

In experiments involving monochromatic excitation of an axisymmetric jet flow, Crow and Champagne (1971) documented a natural instability of the jet column mode with a frequency that scaled with the jet exit velocity and diameter. The maximum amplification was found to occur at a Strouhal number of $f_c D/U_j \simeq 0.3$. Experiments indicate that the column mode frequency can vary between 0.25 and 0.5, with the difference in frequencies likely due to uncontrolled facility factors. Note that the column mode Strouhal number is approximately an order of magnitude lower than that of the most amplified shear layer instability, which occurs at $\mathrm{St} = 0.032$.

As previously discussed, the initial shear layer instability frequency scales with θ_0 which varies as $(\mathrm{Re}_D)^{-1/2}$. In contrast, the jet column instability scales as $(\mathrm{Re}_D)^{-1}$. As a result, there is a possibility in a jet where, at a particular Reynolds number, the initial shear layer frequency can develop through a succession of vortex pairings to reach a passage frequency that matches the jet column mode frequency at the end of the potential core. Such a jet would be highly organized with strong feedback coupling.

For this to occur, the initial shear layer Strouhal number, based on the jet nozzle diameter and jet velocity, namely $\mathrm{St}_n \equiv f_n D/U_j$, would need to be an integer power of 2 of the jet column mode Strouhal number, $\mathrm{St}_c \equiv f_c D/U_j$, or

$$\mathrm{St}_n = 2^N \mathrm{St}_c. \tag{5.10}$$

The Reynolds number at which this resonance coupling would occur would then be

$$\mathrm{Re}_{\mathrm{resonance}} = C(2^N \mathrm{St}_c)^2, \tag{5.11}$$

where N refers to the number of shear layer vortex pairings, and C is a constant that is unique to each jet.

In experiments at different jet Reynolds numbers and jet core turbulence intensity ($\mathrm{TI} = u'/U_j$) levels, Drubka (1981) had observed that at $\mathrm{Re}_D = 42{,}000$ with the lowest TI, there resulted a sharp increase in the amplitude of pressure fluctuations at the jet exit that were at the subharmonic shear layer frequency. That data has been reproduced in Figure 5.9. Such pressure fluctuations were associated with fundamental mode vortex pairing. Under those conditions, Drubka (1981) had determined the frequency of velocity fluctuations at the end of the potential core that corresponded to a column mode Strouhal number of $\mathrm{St}_c = 0.42$. Under the same conditions, the Strouhal number for the initial shear layer frequency based on the jet diameter was $\mathrm{St}_n = 3.33$. Based on these, following Eq. (5.10),

$$N = \frac{\ln\left(\frac{\mathrm{St}_n}{\mathrm{St}_c}\right)}{\ln(2)} = \frac{\ln\left(\frac{3.33}{0.42}\right)}{\ln(2)} = 2.98 \simeq 3. \tag{5.12}$$

Thus, this Reynolds number set up a condition in which three pairings of the initial fundamental shear layer mode matched the jet column mode frequency at the end of the potential core, setting up a feedback resonance condition. Furthermore, based on Eq. (5.11) for $\mathrm{Re}_D = 42{,}000$, the coefficient, C, was 3720. With that, Eq. (5.11) could be used to predict other jet Reynolds numbers where similar resonance would occur based on $n = 1, 2$, and 4. These Reynolds numbers were 2600, 10,500, and 16,800, respectively (Drubka, 1981; Drubka et al., 1989). From a flow control aspect, the concept of a resonant coupling between the initial shear layer mode and the jet column mode is a powerful mechanism. However, based on Figure 5.9, the resonance was suppressed with the higher jet exit core turbulence levels. This observation offers an approach if this resonance mechanism is to be avoided.

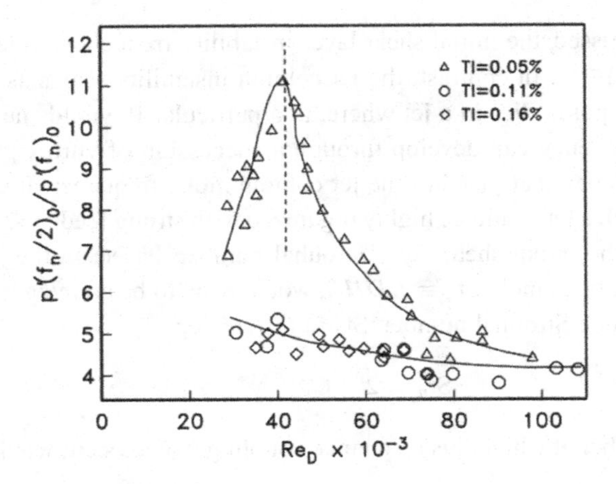

Figure 5.9 Normalized shear layer subharmonic mode pressure fluctuations at the jet exit indicating a feedback resonance at Re = 42,000 for the lowest core exit turbulence intensity condition. Data from Drubka (1981).

5.2 Free Shear Layer Control

There are quite a few spectacular examples of dynamic flow control of free shear layers and jets dating back more than 150 years. Notable examples from Tyndall (1864), include the effect of an acoustic source on gas jets and buoyant smoke plumes that are illustrated in Figures 5.10 and 5.11. With reference to the gas flame, Tyndall (1864) recounts that "a barely perceptible motion of the interior of the flame may be noticed when a shrill whistle is blown close to it. But by turning the cock more fully on, the flame is brought to the verge of flaring. And now, when the whistle is blown, the flame thrusts suddenly out seven quivering tongues. The moment the sound ceases, the tongues disappear, and the flame becomes quiescent." Similarly, Tyndall (1864) documented dramatic changes in the extent and pattern of the smoke "break-up" produced by different types of acoustic excitation. In one case shown in the far right image in Figure 5.11, there is evidence of a flow bifurcation that was later demonstrated by Parekh et al. (1989) and Reynolds et al. (2003), and discussed in Section 5.2.

Passive control can come through geometric changes in the development path of the boundary layer over the splitter plate or jet nozzle wall. The shape of the trailing edge of the splitter plate or jet exit is also a parameter. Specifically, the trailing-edge sharpness controls its acoustic receptivity which is the principal factor in converting long wavelength acoustic disturbances into short wavelength vortical disturbances. This impacts acoustic feedback resonance as well as shear layer control using acoustic sources. Finally, another passive method of shear layer control is through control of the free-stream turbulence levels, such as with the addition of turbulence generating screens, grids, or honeycombs. As previously presented in Section 5.1.5, higher turbulence levels eliminated the resonant coupling between the jet shear layer and jet column mode in two of the three cases of Drubka (1981).

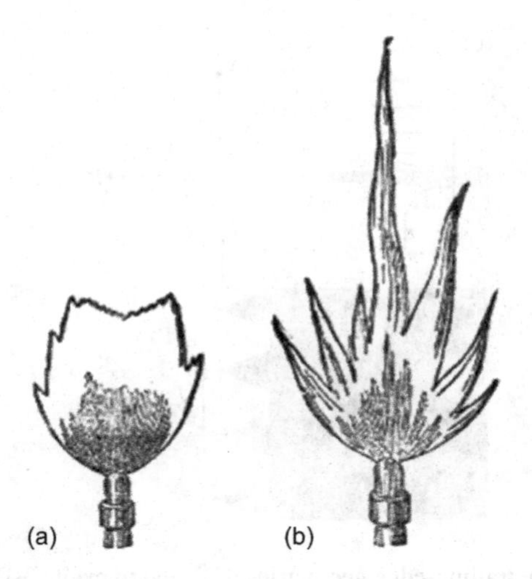

Figure 5.10 Images of a gas flame in a quiescent state (a) and when excited by monochromatic sound "whistle." (b) Taken from Tyndall (1864).

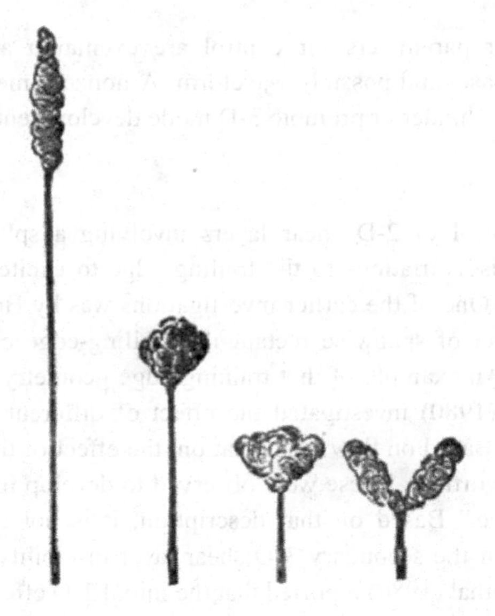

Figure 5.11 Images of smoke columns from a gas burner demonstrating the effect of acoustic excitation on the column length and shape. Taken from Tyndall (1864).

Dynamic control of a turbulent flow is possible by periodic excitation at one or more frequencies that fall in the amplified band based on the shear layer instability characteristics. In the shear layer of a round jet in which the shear layer thickness is small compared to the round jet radius of curvature, typically $\theta/r < 0.01$, the linear stability amplification rates of helical modes are comparable to that of the axisymmetric mode making it easy to excite either with low initial amplitude disturbances.

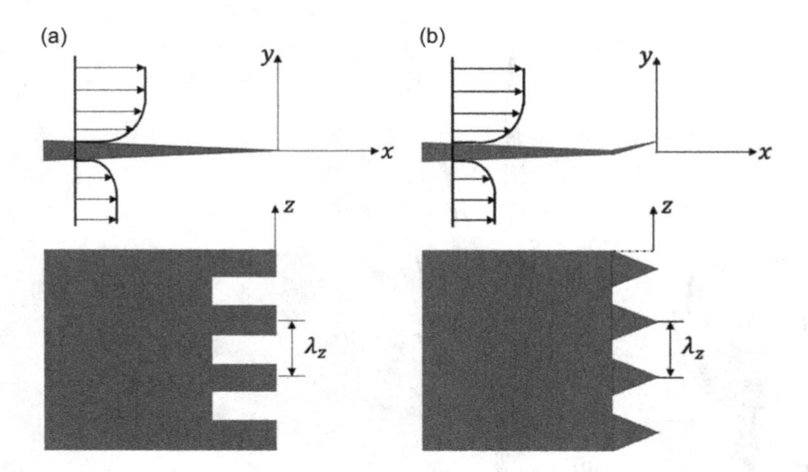

Figure 5.12 Examples of splitter plate trailing-edge geometries designed to excite 3-D shear layer instabilities like that illustrated in Figure 5.4 where (a) involves planar notches and (b) are involves deflected notches.

Besides frequency, the other parameters for control are excitation amplitude, streamwise and cross-stream phase, and possibly waveform. A nonzero mean cross-stream or azimuthal variation can hinder or promote 3-D mode development.

Passive Control

A majority of the passive control of 2-D shear layers involving a splitter plate have focused on adding spanwise variations to the trailing edge to excite 3-D primary or secondary instabilities. One of the earlier investigations was by Breidenthal (1980), who examined the effect of spanwise rectangular trailing-edge cutouts on the shear layer spreading rate. An example of that training edge geometry is shown in Figure 5.12(a). Breidenthal (1980) investigated the effect of different spanwise wavelengths, λ_z, of the cutouts. Based on flow visualization, the effect of the cutouts introduced pairs of streamwise vortices. These were observed to develop into "interconnected loops like a necklace." Based on that description, it is not unlike the observations of Bernal (1981) of the secondary 3-D shear layer instability that was illustrated in Figure 5.4. Breidenthal (1980) reported that the initial 3-D effect quickly relaxed and that further downstream the shear layer assumed its characteristic 2-D vortex structure.

If the motivation is to introduce streamwise vorticity at the trailing edge of the splitter plate, a better geometry is triangular cutouts that are deflected into one or both of the flow streams. An example of this is shown in Figure 5.12(b). A similar geometry is commonly used as streamwise vortex generators to control boundary-layer separation (Lin, 2002). Lasheras and Choi (1988) and Kit et al. (2007) examined the passive effect that this type of modification of the trailing edge had on the 3-D instability development of plane mixing layers. Lasheras and Choi (1988) observed that the characteristic time of growth of the 2-D shear instability was much shorter

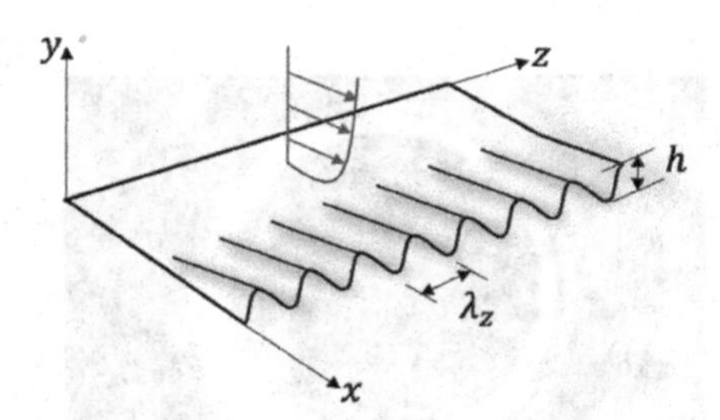

Figure 5.13 Splitter plate trailing edge with surface-normal sinusoidal deflections designed to excite 3-D shear layer instabilities to enhance mixing.

than that of the 3-D instability. As expected from linear theory, the primary Kelvin–Helmholtz instability developed first, leading to the formation of an almost 2-D array of spanwise vortex tubes. They observed that the strain field created by the serrated trailing edge resulted in the formation of streamwise vortex tubes. As observed by Breidenthal (1980), during the formation of the streamwise vortex tubes, the spanwise vortices maintained their two-dimensionality, suggesting an almost uncoupled development of both instabilities. The vortex tubes that formed through the 3-D instability further underwent nonlinear interactions with the spanwise vortices that were observed to induce on their cores a wavy undulation. The authors also investigated the influence of the amplitude and wavelength of the 3-D perturbation introduced by the serrated trailing edge.

In applications where induced drag is important, triangular vortex generators introduce a significant parasitic drag. A more refined approach to introduce streamwise vorticity at trailing edges with less added drag is through the addition of a surface-normal sinusoidal deflection of the trailing edge. An example of this geometry is illustrated in Figure 5.13. The effect of such a trailing-edge geometry was the subject of numerical simulations by Valtchanov et al. (2020). They observed that the sinusoidal deflected surface introduced streamwise vorticity into the attached boundary layer that developed over the splitter plate. The added streamwise vorticity in the boundary layer subsequently convected downstream into the free shear layer where it acted as a 3-D perturbation of the nominally 2-D shear layer. The 3-D disturbance was observed to introduce two distinct and mutually counter-rotating streamwise-oriented vortex structures into the braid region and the region surrounding the 2-D Kelvin–Helmholtz spanwise vortices. The relative strength of the 3-D structure is determined by the parameters of the trailing-edge waviness. For example, according to the linear stability theory (Klaassen and Peltier, 1985), the 3-D disturbance yielding the highest amplification rate has a spanwise wavelength of approximately half the shear layer thickness.

The application of shear layer flow control has clearly been applied in turbo-jet acoustic control. This is evident in the photographs in Figure 5.14 of engine exit nozzle "mixer" designs that employ triangular "chevron" cutouts and sinusoidal edge deflections that are designed to enhance mixing and subsequently lower the acoustic levels.

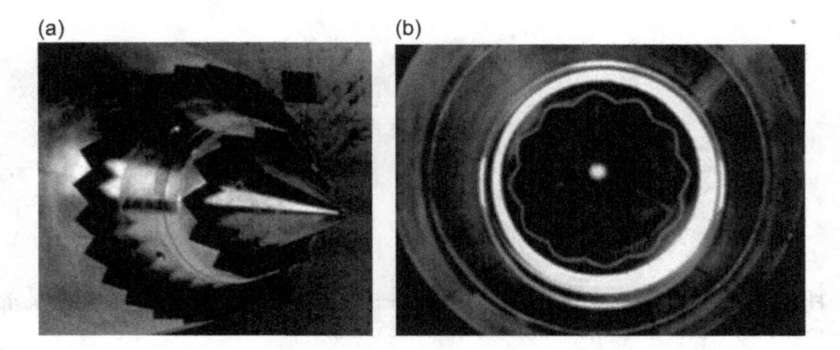

Figure 5.14 Application of shear layer flow control on commercial jet engines consisting of chevron-shaped cutouts (a) and surface-normal sinusoidal trailing edge (b). NASA Glenn Research Center photograph.

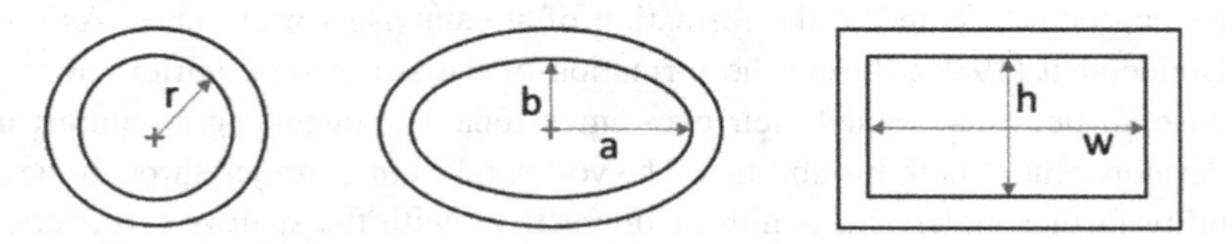

Figure 5.15 Examples of different jet exit geometries investigated by Gutmark and Ho (1983).

With regard to jet flows, a well-documented flow control technique involves modification of the nozzle exit geometry by adding azimuthal asymmetry. Examples of this include elliptic nozzles extensively investigated by Ho and Gutmark (1987), and rectangular nozzles that have been investigated by Krothapalli et al. (1981). Examples of these along with a reference circular nozzle are shown in Figure 5.15.

With elliptic nozzles, the spreading of the jet flow has an interesting dynamic in which it alternately switches from the major to minor nozzle axes with downstream distance. In an unforced elliptic jet with a small aspect ratio (2:l) (Ho and Gutmark, 1987), the mass entrainment ratio was several times higher than that in a circular or a 2-D jet. Ho and Gutmark (1987) attribute this to an increase in the azimuthal distortion of the elliptic vortex caused by self-induction. Most of the entrainment occurs around the minor-axis plane where the vortex core moves away from the jet axis and the external fluid is entrained into the jet stream. Owing to the self-induction of the vortex, axis switching occurred three times in the range of $x/a < 40$, where a is the major axis radius. Because of the axis switching, the flow properties, mean velocity, turbulence intensities, and Reynolds stress, were very different in the major- and minor-axis planes. In addition, self-similarity was not observed out to $x/a = 40$ (Ho and Gutmark, 1987).

An important metric for jet flow control is the entrainment ratio. This is defined as the difference between the mass flow rate, Q, at a downstream station and that at the nozzle exit, Q_0. A comparison of the entrainment ratios, $(Q-Q_0)/Q_0$) for a 2:1 elliptic jet, a circular jet, and the 2-D jet with a 24:1 aspect ratio is shown in Figure 5.16. This

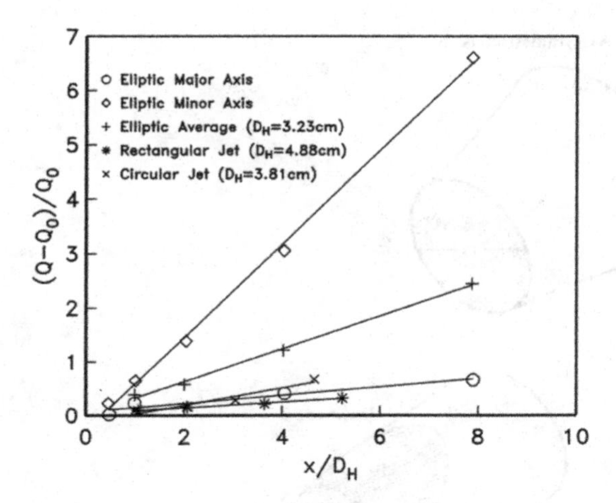

Figure 5.16 Entrainment ratio for an elliptic jet with hydraulic diameter, $D_H = 3.23$ cm; circular jet, $D_H = 3.81$ cm; and a 2-D jet, $D_H = 4.88$ cm. Data from Ho and Gutmark (1987).

is based on data from Ho and Gutmark (1987). For the elliptic jet, the entrainment ratios of the major and minor axes are included along with the average entrainment rate. The lowest entrainment ratio occurred for the rectangular jet. The circular jet performed better than the entrainment along the major axis of the elliptic jet. However, the tremendously enhanced entrainment along the minor axis of the elliptic jet more than compensated so that the average entrainment ratio of the elliptic jet significantly exceeded that of the circular jet.

As a variation on elliptic nozzles, Wlezien and Kibens (1988) investigated what they termed as "indeterminate-origin" (IO) nozzles in which the axial position of the nozzle lip varied around the circumference of the nozzle. Examples of these are shown in Figure 5.17. As presented in Section 5.1.5, the shear layer instability characteristics scale with the initial shear layer momentum thickness. For a round nozzle, one can expect the mean velocity profile and the shear layer momentum thickness to be azimuthally invariant. As a result, the instability characteristics of the shear layer at any axial location up to the end of the potential core are largely expected to be independent of the azimuthal position. In contrast, the IO nozzles like those shown in Figure 5.17 have a nozzle lip in which the boundary-layer development length, and subsequently the initial shear layer momentum thickness, vary azimuthally. As a result, the instability characteristics and streamwise development length vary azimuthally.

An example of the effect of the IO nozzles with three inclined exit angles is shown in Figure 5.18. As illustrated in Figure 5.18(a), the exit angles are defined by a rotation of the exit by an amount that displaces the top- and bottom-dead-center positions by a fraction of the nozzle exit diameter, D. In Figure 5.18, the angles correspond to those produced by displacements of $D/4$ (b), $D/2$ (c), and $D/1$ (d).

The effect of these IO nozzles is presented by smoke visualization of the jet shear layer. For the smallest nozzle exit angle in Figure 5.18(b), the shear layer is observed to develop similar to that of an axisymmetric nozzle, with the shear layer vortices aligned

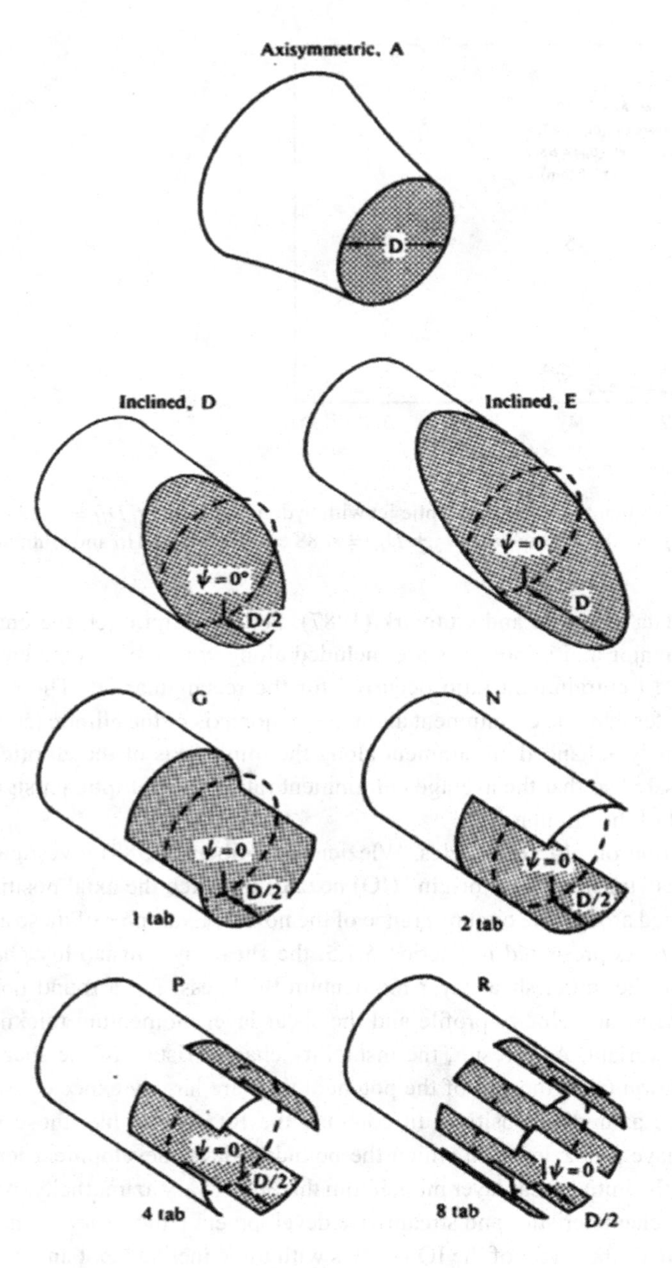

Figure 5.17 Examples of different jet nozzle azimuthal asymmetric geometries consisting of baseline axisymmetric (A), two inclined exits (D and E), and four notched exits made up of one (G), two (N), four (P) and eight (R) tabs that have been investigated Wlezien and Kibens (1986, 1988).

with the nozzle exit. However, in this case, in order to maintain the same streamwise development length, the resulting azimuthal vortical structures are slanted at an angle that mirrors that of the nozzle exit.

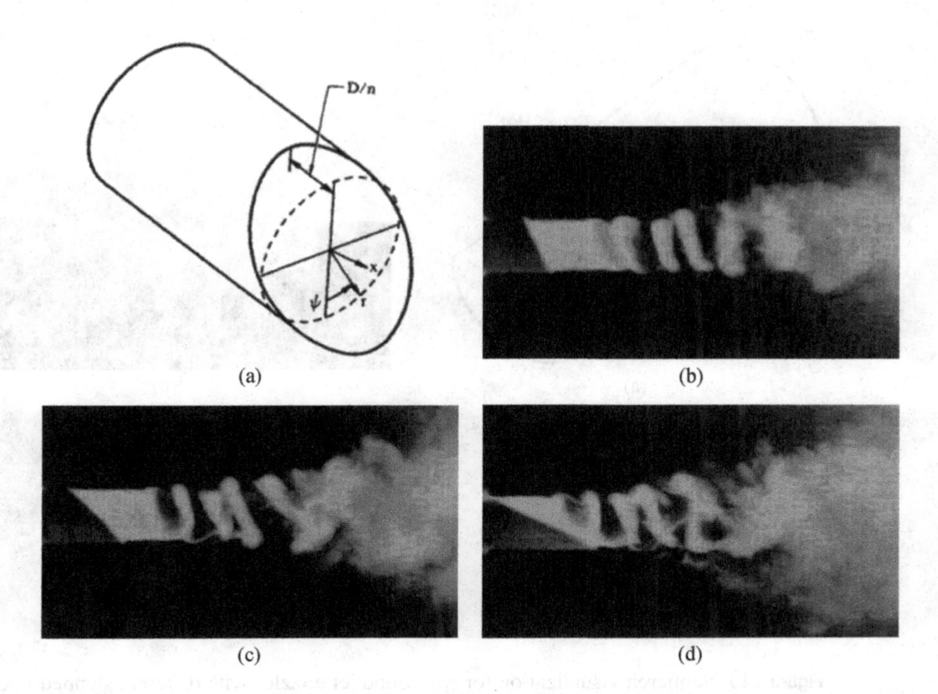

Figure 5.18 Smoke flow visualization for three round jet nozzles with different inclined exit angles corresponding to $D/4$ (b), $D/2$ (c), and $D/1$ (d). Taken from Wlezien and Kibens (1986).

In contrast to the smaller nozzle angle, for the conditions with the next larger angle shown in Figure 5.18(c), the azimuthal vortex structure that develops is no longer parallel to the nozzle exit. Instead, the vortical structures appear to occur at the same angle as with the smaller exit angle in Figure 5.18(b). However, there appears to be an acceleration of downstream asymmetry of the vortical structures with the larger nozzle angle. This is further exaggerated in the flow visualization with the largest nozzle exit angle shown in Figure 5.18(d). There also appears to be larger spreading of the jet at the furthest downstream location.

A more extreme IO nozzle is the stepped-back design illustrated in Figure 5.19(a). In this design, half of the circumference of the round nozzle has a shorter boundary-layer development length which would reduce the initial shear layer momentum thickness in that portion of the nozzle and subsequently its instability characteristics. Two cases are presented that correspond to step-back distances of $D/4$ and $D/2$. In these two cases, the shear layer is visualized by replacing the previously used smoke tracer with a Helium tracer gas, and visualizing the gas density using a Schlieren optical method.

In the flow visualization for the smaller step-back shown in Figure 5.19(b), it is apparent that the shear layers from the upper and lower sections of the nozzle initially develop independently. With the smaller step-back, the difference in the initial momentum thicknesses and subsequently initial shear layer frequency is small so that

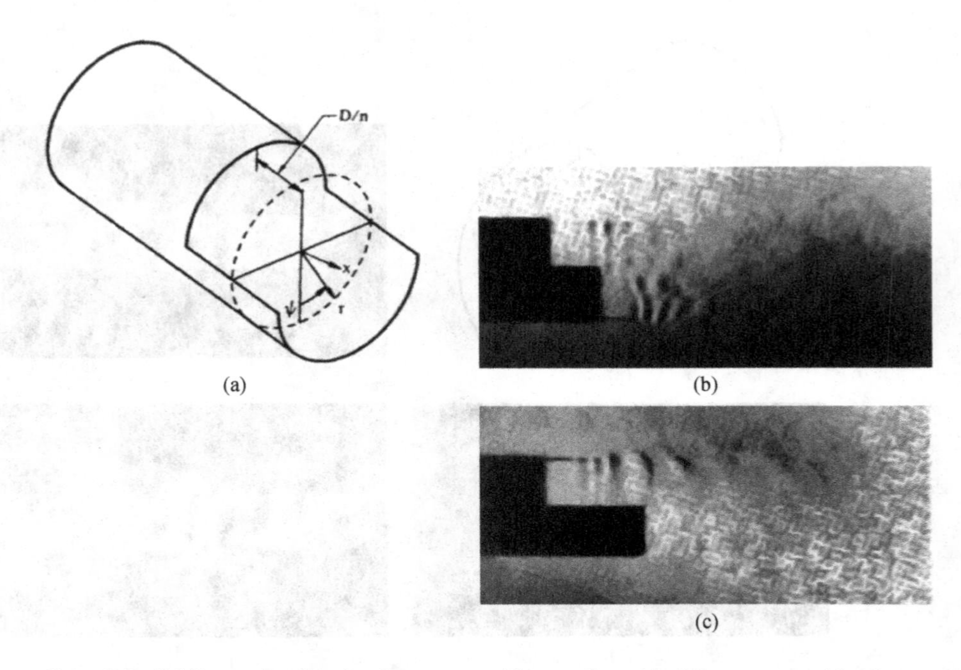

Figure 5.19 Schlieren visualization for two round jet nozzles with different stepped-back sections corresponding to $D/4$ (b) and $D/2$ (c). Taken from Wlezien and Kibens (1986).

the wavelengths of the vortical structures that form in the upper and lower sections are visually the same. However, having different axial origins, the periodic phase between the two shear layers is different. This results in a phase adjustment at the azimuthal locations of the steps, which is evident in the flow visualization by a tilting and merging of the vortical structures. With the larger step-back, the visualization shown in Figure 5.19(c) is not as conclusive. The shear layer for the upper half of the nozzle is clearly developed and undergoes one vortex pairing before reaching the end of the lower half of the nozzle. This will produce a large difference in the wavelengths of the vortical structures between the upper and lower shear layers that will be difficult to merge at the interfaces. Velocity measurements (Wlezien and Kibens, 1986) documented that the upper and lower shear layers did develop independently leading to asymmetry in the jet mean velocity distributions.

The effect that the "intermediate origin" nozzles had on the jet spreading is shown in Figure 5.20. This shows the azimuthal distributions of the shear layer momentum thickness at $x/D = 4$ for a baseline circular nozzle, for the $D/4$ inclined exit nozzle with flow visualization shown in Figure 5.18(b), and for the $D/2$ step-back nozzle with flow visualization shown in Figure 5.19(b). As expected, the baseline round nozzle results in an axisymmetric distribution in the momentum thickness. The two IO nozzles result in an asymmetric spreading in the jet that is reflected in the asymmetric distribution in the momentum thickness. In both IO nozzle cases, the spreading is the largest in the direction of the nozzle exit asymmetry. Although the azimuthally integrated spreading of the IO nozzle jets may not be significantly different from the

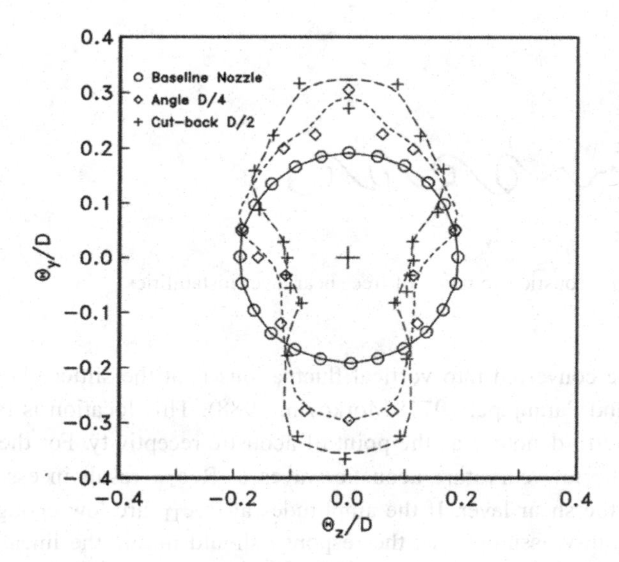

Figure 5.20 Comparison of azimuthal distribution of the shear layer momentum thickness measured at $x/D = 4$ for the baseline circular nozzle, $D/4$ inclined exit nozzle, and $D/2$ step-back nozzle. Data reproduced from Wlezien and Kibens (1986).

round jet, use of such IO nozzles offers a method for selective shaping of the jet spreading that can be important in jet noise control and combustion mixing.

Active Control

As previously presented, passive shear layer control involves static geometric modifications that influence the basic flow, and thereby the resulting instability characteristics. In contrast to this, active control seeks to control the initial amplitude and spectrum of perturbations to the existing basic state in order to feed energy into selected fluid instabilities. The dramatic examples of acoustic control of gas jets and buoyant smoke plumes that were shown in Figures 5.10 and 5.11 from Tyndall (1864), and our gained understanding of acoustic receptivity (Morkovin, 1988), have made acoustic excitation a popular approach to free shear layer control.

Acoustic excitation for shear layer control involves an acoustic generator (speaker) that is driven by a function generator with control of waveform, amplitude, and frequency. In most cases, the waveform is a sine wave or possibly a combination of sine waves at different frequencies, amplitudes, or phase angles. The location of the acoustic source is also a factor. The acoustic waves from the source are assumed to expand radially and decrease in amplitude as the inverse of the squared distance from the source. A far-field acoustic source is one in which the curvature of the acoustic waves at the point of acoustic receptivity is small and can be considered as 2-D. An example of a far-field acoustic source used in exciting free shear layer instabilities is shown in Figure 5.21. Because of the mismatch between the acoustic wavelength and the instability wavelength, the only location in the flow field where the acoustic pressure

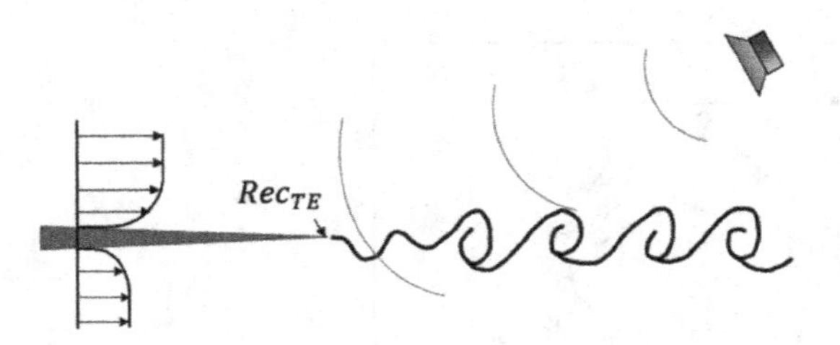

Figure 5.21 Method of acoustic excitation of free shear layer instabilities.

fluctuations can be converted into vortical fluctuations is at the slitter plate trailing edge (Morkovin and Paranjape, 1971; Morkovin, 1988). This location is labeled as Rec_{TE} in the figure to denote it as the point of acoustic receptivity. For the far-field acoustic source, the low curvature acoustic waves at Rec_{TE} result in essentially a 2-D excitation of the shear layer. If the amplitudes at Rec_{TE} are low enough to satisfy the linear stability assumptions, the response should mimic the linear stability amplification dependence on frequency that is shown in Figure 5.3. It is possible to excite multiple frequencies and phase angles that, depending on the application, can enhance or suppress shear layer growth and spreading. Combinations of conditions that illustrate these initial conditions and effects on the shear layer development were documented by Ho and Huang (1982) and discussed in Section 5.1.1.

It is possible to involve near-field acoustic excitation of plane shear layers, particularly in the motivation to excite 3-D instabilities. However, the approach is more generally through in situ active control devices such as those depicted in Figure 5.22. These consist of a periodically pitching trailing edge (a), periodic wall-normal deflection (b), and a periodic streamwise body force (c). The pitching trailing edge will generate spanwise vorticity of the same sign as that of the time-averaged boundary-layer profile on each surface of the splitter plate. Therefore, the pitching action will reenforce the natural instability mechanism. This type of motion can be accomplished in a number of ways ranging from a motor-driven hinged trailing edge to electromagnetic (Suzuki et al., 2004) or piezo electric (Cattafesta et al., 2001) strips. Because the most unstable shear layer frequency scales with velocity, mechanical approaches may not be feasible for high-speed flows.

Electromagnetic or piezoelectric strips can be used to provide periodic wall-normal motion. This can be viewed as imparting a wall-normal velocity component, dv/dt, that will produce perturbations of the spanwise vorticity, $d\omega_z/dt$, that could directly drive the 2-D shear layer instability. Alternatively, if the wall-normal motion amplitude is small, the motion could primarily generate pressure fluctuations, $p(t)$, that could excite the shear layer instability through the acoustic receptivity mechanism at the trailing edge.

A third method for introducing controlled perturbations of a free shear layer is a dielectric-barrier discharge (DBD) plasma actuator. Details about DBD plasma actuators are presented in Chapter 2 and in a general review by Corke et al. (2010a). This involves two electrodes separated by a dielectric insulator that are connected to an

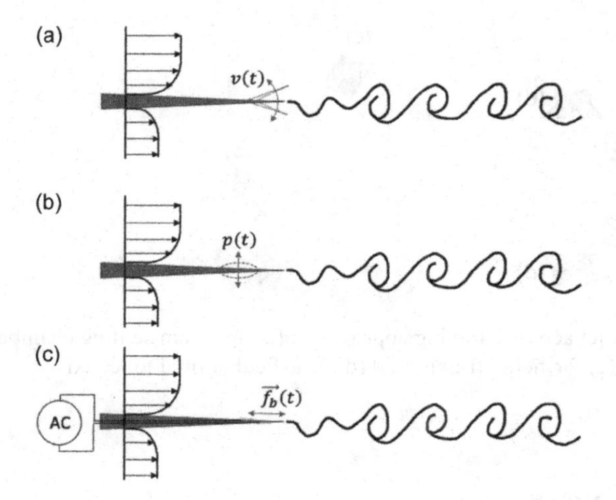

Figure 5.22 Examples of active in situ approaches used to excite shear layer instabilities consisting of a periodic vibrating trailing edge (a), a periodic wall-normal pressure disturbance (b) and a periodic axial body force produced by a plasma actuator (c).

AC voltage source. One electrode is exposed to the air. The other is encased in an electrically insulating substrate. The electrodes are generally overlapping by a small amount, as shown in the bottom illustration in Figure 5.22(c). A large enough AC voltage causes the air over the covered electrode to ionize. The ionized air in the presence of the electric field gradient results in a vector body force, $\vec{f_b}(t)$. The unsteady body force acts as an acoustic dipole with a frequency that is twice the AC source frequency (Mertz and Corke, 2011c). Being a dipole, the sound amplitude is directional. As before, the unsteady acoustics can be converted into unsteady vorticity at the trailing edge.

Any of the three approached depicted in Figure 5.22 can be individually applied in spanwise segments in order to produce 3-D disturbances to the shear layer. These can be phase shifted in the spanwise direction in order to excite oblique waves at different wave angles. Finally, the time series used to drive the disturbance actuators can be made up of multiple frequencies such as a most amplified fundamental and its subharmonic for both 2-D and 3-D disturbances. A similar technique was used by Corke et al. (1992b) to excite a 3-D mode resonance in far wakes.

Any of the controlled disturbance techniques demonstrated in Figure 5.22 for 2-D shear layers can be applied to flow control in jets. However, the dramatic examples of acoustic control of gas jets and buoyant smoke plumes from Tyndall (1864) which are shown in Figures 5.10 and 5.11, have motivated the multivaried use of acoustic forcing for active flow control of jet flows. Examples of the types of acoustic excitation, illustrated by the placement of acoustic drivers, are shown in Figure 5.23(a)–(d). Recall that jet flow development involves two instability mechanisms, one scaling with the exiting shear layer momentum thickness, and the other scaling with the jet exit diameter. An approach that is effective to excite the latter mechanism, the jet column instability, is to place an acoustic driver in the plenum chamber, upstream of the

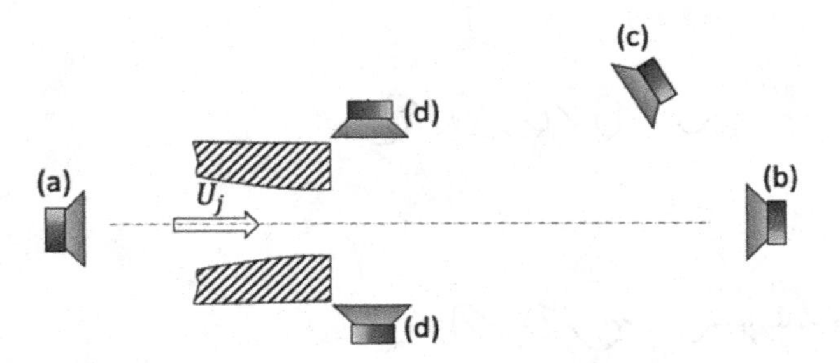

Figure 5.23 Examples of jet acoustic forcing approaches: (a) upstream settling chamber, (b) far-field along jet axis, (c) far-field off-axis, and (d) near-field normal to jet axis.

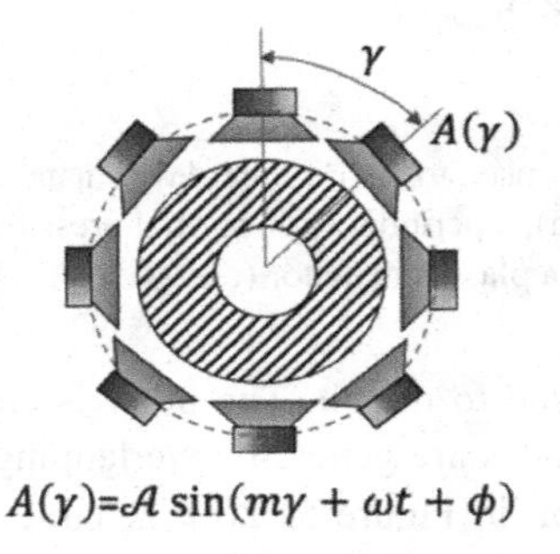

Figure 5.24 Jet forcing approach using jet exit near-field azimuthal acoustic excitation.

jet exit. This can also potentially couple with a Helmholtz instability of the plenum chamber, thereby amplifying the effect. This approach was used by Cohen and Wygnanski (1987) for producing large-amplitude, axisymmetric excitation of the shear layer and jet column instabilities.

Examples of far-field location for acoustic drivers are shown in Figure 5.23(b) and (c). Such far-field placement will produce constant azimuthal phase disturbances around the jet nozzle exit that will excite axisymmetric instability modes. The placement (b) on the jet axis was used by Drubka et al. (1989), where the acoustic driver was located 70 jet diameters downstream of the jet exit. When placed off-axis (c) at a similar distance from the jet exit, the acoustic driver will also only excite axisymmetric instability modes at the nozzle exit. Such off-axis placement can be more practical with high-speed jet flows.

Azimuthal phase variation around the nozzle exit needed to excite helical shear layer instabilities requires that acoustic drivers be placed near to the nozzle exit. This is illustrated in Figure 5.23(d). The number of individually controlled acoustic drivers around the jet azimuth determines the maximum azimuthal mode number, m, of a helical instability wave that can be excited. For example, as illustrated in Figure 5.24, to excite a helical instability wave with positive wave number m (clockwise moving) and frequency $\omega = 2\pi f$, the periodic time series supplied to any individual acoustic drivers located at an azimuthal angle, γ, would be

$$A(\gamma) = \mathcal{A}\sin(m\gamma + \omega t + \phi_1), \tag{5.13}$$

where ϕ_1 is an arbitrary phase shift, and $\mathcal{A}$ is the maximum amplitude to the acoustic drivers. Because of the 2π circular phase limit, the largest azimuthal phase difference between neighbor acoustic drivers is 2π. Therefore, the number of acoustic drivers around the nozzle determines the largest helical mode number that can be excited. For example, the lowest azimuthal mode number, $m = 1$, helical mode requires a minimum of two acoustic drivers spaced $\gamma = \pi$ apart around the nozzle exit. Figure 5.25 shows the arrangement from Corke and Kusek (1993) of 12 acoustic drivers located around the exit of a circular jet. The plot in Figure 5.25(b) shows the azimuthal amplitude distribution, $A(\gamma)$, that was used to simultaneously excite an axisymmetric ($m = 0$) instability wave at one frequency, and pairs of clockwise and counterclockwise helical waves with azimuthal wave numbers of $m = \pm 1$. In this case, the counterclockwise helical wave would have an azimuthal amplitude distribution of

$$A(\gamma) = \mathcal{A}\sin(-m\gamma + \omega t + \phi_2). \tag{5.14}$$

The addition of these two periodic functions for helical waves at the same frequency but equal–opposite mode numbers yields

$$A(\gamma) = 2\mathcal{A}\sin\left(\omega t + \frac{1}{2}(\phi_1 + \phi_2)\right)\cos\left(m\gamma + \frac{1}{2}(\phi_1 - \phi_2)\right). \tag{5.15}$$

Therefore, to produce equal and opposite helical modes with azimuthal wave number $\pm m$ and frequency ω, each acoustic driver operates with a periodic input, $\sin(\omega t + \frac{1}{2}(\phi_1 + \phi_2))$, and amplitude determined by its azimuthal position according to $2\mathcal{A}\cos(m\gamma + \frac{1}{2}(\phi_1 - \phi_2))$. With the 12 speakers used by Corke and Kusek (1993), it was possible to excite helical waves with azimuthal mode numbers up to $m = \pm 6$. The flow visualization image in Figure 5.8 that showed pairs of $m = 1$ helical waves with equal–opposite wave angles was produced by the azimuthal amplitude distribution presented in Figure 5.25. In that case, the axisymmetric excitation was at the most amplified shear layer frequency, and that of the helical wave excitation was at the subharmonic of the axisymmetric excitation frequency. Corke and Kusek (1993) observed that this combination of instability waves resulted in a parametric resonance of the type described by Craik (1985) and previously discussed in Section 5.1.2, which resulted in an enhanced growth in amplitude of the helical waves given by

$$\frac{dA_H}{A_H} = \left(-\alpha_{i_H} + \frac{b}{C_r}A_{A_0}\exp(-\alpha_{i_A}x)\right)dx, \tag{5.16}$$

where H and A refer to the helical wave and axisymmetric waves, respectively, α_i refers to the linear spatial amplification rate, A is an initial amplitude, C_r is the phase velocity of the waves which, for resonance, is the same between the three instability waves, and b is a coupling coefficient. Integrating this equation gives

$$A_H = A_{H_0}\exp\left(-\alpha_{i_H}x + b'\exp(-\alpha_{iu_A}x)\right), \tag{5.17}$$

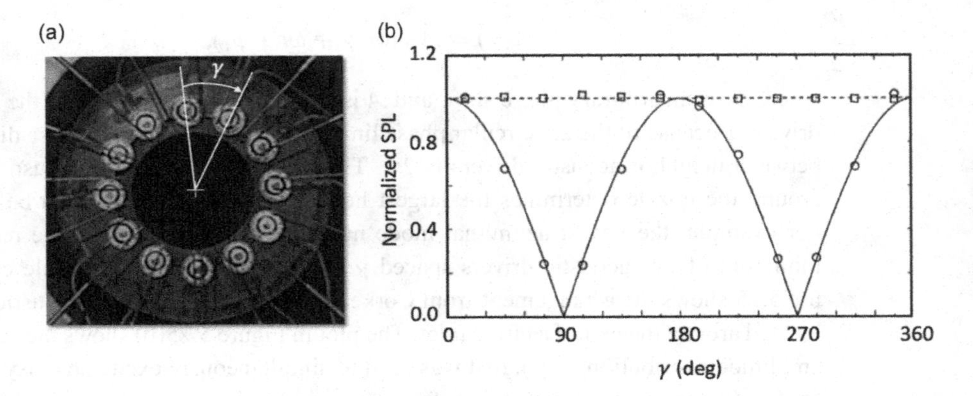

Figure 5.25 Azimuthal array of acoustic speakers around jet exit (a) and (b) azimuthal amplitude distribution, $A(\gamma)$, to simultaneously excite an axisymmetric mode, $m = 0$ at one frequency, and a pair of helical modes with $m = \pm 1$ at a different frequency. Taken from Corke and Kusek (1993).

where $b' = bA_{A_0}/C_r\alpha_{i_A}$. From their experiments, Corke and Kusek (1993) found that $b' = 3.76 \times 10^{-2}$ or $b/C_r\alpha_{i_A} = 19.41$, which represents a dimensionless coupling coefficient of the parametric resonance.

As a result of subharmonic resonance in this case, the helical mode grows at an enhanced, *double-exponential* rate and the shear layer rapidly thickens, entraining and mixing the jet core and surrounding fluids. However, similar to the passive elliptic nozzle jets presented in Section 5.2, the degree to which the shear layer thickens depends on the azimuthal location.

An example of the azimuthal variation of the momentum thickness of the jet shear layer resulting from the *resonant growth* of subharmonic $m = \pm 1$ helical modes is shown in Figure 5.26. This corresponds to excitation of the most amplified axisymmetric mode with azimuthal wave number and frequency, $(f, 0)$, and a pair of helical modes with azimuthal wave number and frequency, $(f/2, \pm 1)$, that was designed to lead to the resonant growth of the subharmonic helical mode.

The resonant growth of the helical mode will result in a number of nonlinear interactions, one of which is a modification of the mean flow. For helical mode pairs with azimuthal wave number $\pm m$ and streamwise wave number, $\alpha_r = c_r/2\pi f$, the difference interaction, $(\alpha_r, \pm m) - (\alpha_r, \pm m) = (0, \pm 2m)$, will produce an azimuthal variation of the mean flow with azimuthal wave number of $\pm 2m$. Physically, this will appear as a $\cos 2\gamma$ variation of the shear layer thickness.

In the case of Figure 5.26, the simultaneous excitation of the $(f, 0)$ and $(f/2, \pm 1)$ modes leading to the resonant growth of the pair of the $m = \pm 1$ helical modes should produce an azimuthal variation of the mean flow with azimuthal wave number of $m = \pm 2$. This would appear as two maxima and minima in the azimuthal variation of momentum thickness, which is observed in Figure 5.26.

Figure 5.27 compares the development of the azimuthally summed momentum thickness for three jet shear layer excitation cases consisting of the baseline (no excitation), excitation of the axisymmetric mode, $(f, 0)$, alone, and the combination of the

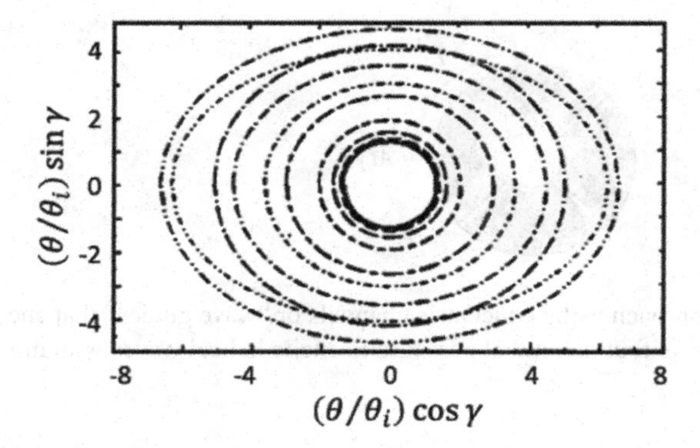

Figure 5.26 Streamwise development of the azimuthal variation of the momentum thickness with jet shear layer excitation of a fundamental axisymmetric mode, $(f, 0)$, and a pair of helical modes at the subharmonic frequency, $(f/2, \pm 1)$. The different x-positions are denoted by different line types. These cover the streamwise distances $0.118 \leq x/D \leq 0.315$ in which $x/D = 0.118$ and $x/D = 0.315$ correspond to the inner-most and outer-most curves from the center, respectively. Taken from Corke and Kusek (1993)

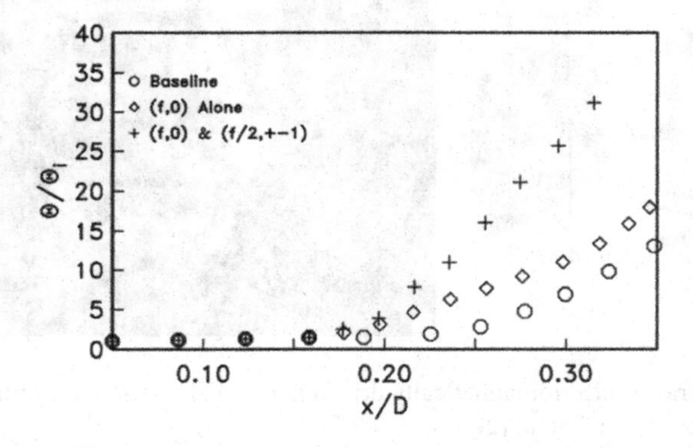

Figure 5.27 Streamwise development of the azimuthally summed momentum thickness for different cases of jet shear layer excitation. Data from Corke and Kusek (1993).

fundamental axisymmetric and pair of subharmonic $m = \pm 1$ helical modes that produced the azimuthal variation shown in Figure 5.26. This indicates the substantially faster streamwise growth in the shear layer thickness and subsequently jet spreading, which results from the resonant growth of the helical modes. This provides an example of utilizing the instabilities, both linear and nonlinear, to produce an *order one effect with order epsilon energy input.*

An alternative to placing acoustic drivers around the exterior of the jet nozzle is to utilize concentric channels or "wave guides" that encircle the nozzle. An example is illustrated in Figure 5.28 in which the channels are segmented into four 90° arcs around the nozzle. A method similar to this was used by Kibens (1980) to excite axisymmetric instability waves. Another approach used by Suzuki et al. (2004) is an

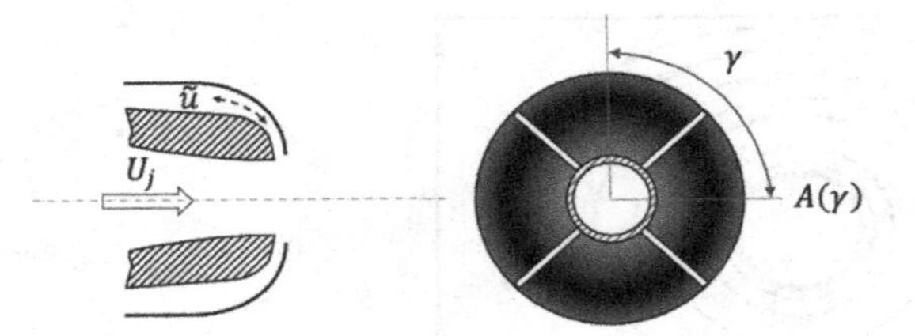

Figure 5.28 Jet forcing approach using concentric channels or "wave guides" that encircle the nozzle. The example shows four azimuthal passages to excite helical waves with mode numbers up to ± 2.

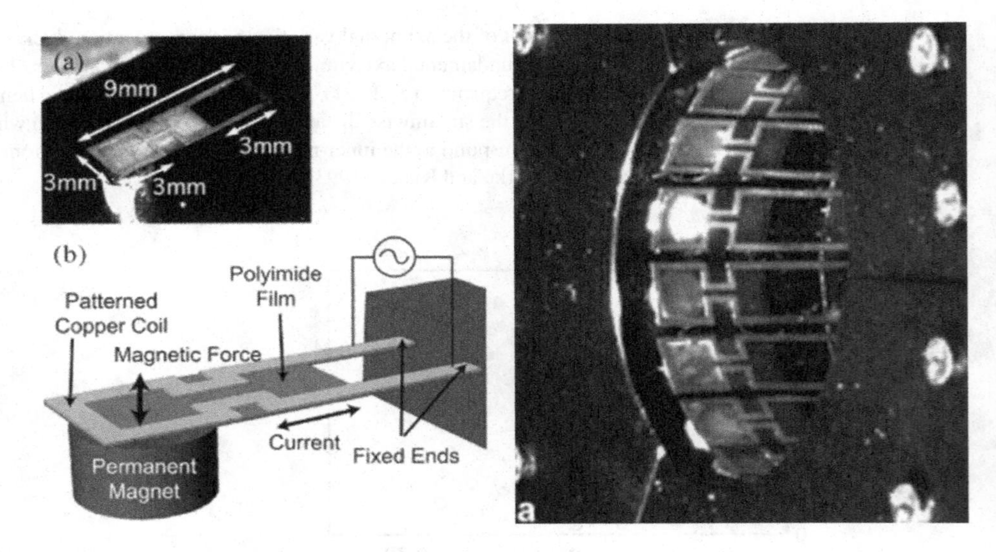

Figure 5.29 Azimuthal array of electromagnetically driven flap actuators used in exciting jet helical modes. Taken from Suzuki et al. (2004).

azimuthal array of electromagnetically activated flap actuators that line the inside surface of the jet nozzle. Figure 5.29 shows a photograph for the jet from that reference that included 18 flap actuators around the nozzle interior. The actuators deflect in the wall-normal direction to produce vortical disturbance that can be amplified by the exiting shear layer. Suzuki et al. (2004) performed experiments in water with combinations of excited axisymmetric and helical instability modes. One case involved exciting all the actuators at the same amplitude but with the first nine (upper 180° location) actuators being 180° phase shifted from the last nine (lower 180° location) actuators. The result caused the jet to bifurcate in a manner that has been extensively studied by Parekh et al. (1989) and Reynolds et al. (2003). A flow visualization photograph of a bifurcating jet from Parekh et al. (1989) is shown in Figure 5.30.

A bifurcating jet like that shown in Figure 5.30 results from simultaneously exciting an axisymmetric and $m = 1$ helical mode, where the frequency of the helical mode

Figure 5.30 Flow visualization of a bifurcating jet produced by simultaneously exciting an axisymmetric mode at frequency f_A and an $m = 1$ helical mode at frequency F_H where $f_A/f_H = 2$. Taken from Parekh et al. (1989).

is half that of the axisymmetric mode. The result is to cause successive vortex rings to be placed slightly off the jet axis, such that adjacent vortex rings are eccentric with respect to one another. As a result of a mutual induction, the eccentric vortex rings tend to tilt one another further, causing both to move further away from the jet axis. The number of bifurcated streams depends on the ratio of the axisymmetric mode frequency, f_A, to the helical mode frequency, f_H. If $f_A/f_H = 2$, the jet will bifurcate into two streams. This is the case shown in Figure 5.30. If $f_A/f_H = 3$ or 4, the jet can be made to divide into three or four separate vortex streams (Reynolds et al., 2003). Furthermore, if f_A/f_H is not an integer, no vortex ring will exactly follow another previously formed, and the jet can produce a shower of vortex rings that Reynolds et al. (2003) dubbed a "blooming jet."

As would be expected, bifurcating and blooming jets substantially increase mixing between the jet core and surrounding fluid. However, these have been primarily demonstrated at lower Reynolds numbers, $Re_D \leq 20{,}000$. At higher Reynolds numbers, the reduced boundary-layer thickness at the nozzle exit requires larger excitation amplitudes to produce these effects. Efforts to thicken the boundary layer have been employed to some success (Reynolds et al., 2003). The most notable of these utilized a Coanda effect at the nozzle exit to radially expand the flow. A schematic of a so-called Coanda jet is shown in Figure 5.31. Reynolds et al. (2003) observed that velocity ratio, $U_b/U_j \simeq 2$, was able to fully turn the core jet flow 90° around the circular radius. This velocity ratio however will depend on the radius of curvature.

As previously discussed in Section 5.1.4, dynamic events in the shear layer such as the first vortex roll-up and vortex pairing produce pressure waves that travel upstream to the point of acoustic receptivity where they are converted into vortical fluctuations that can enhance phase locking, and in special cases lead to global feedback resonance. The $Re_D = 42{,}000$ jet with the lowest turbulence level of Drubka (1981) is a case in point (see Figure 5.9).

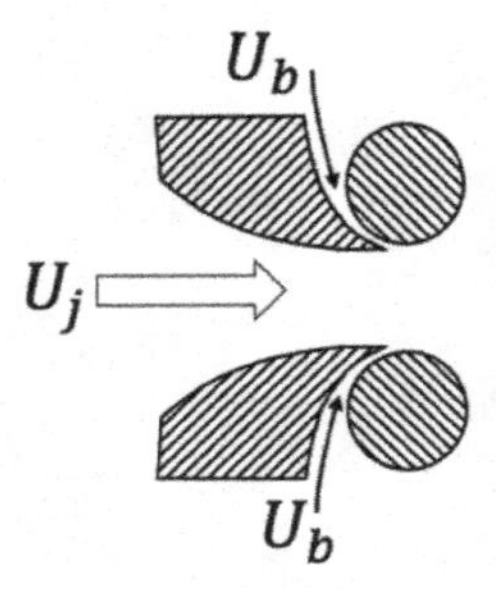

Figure 5.31 Schematic drawing of a Coanda jet concept for enhanced jet spreading.

An alternative method for promoting feedback resonance in jets is shown in Figure 5.32. This utilizes a hot-wire velocity sensor placed at a downstream location that produces voltage fluctuations proportional to velocity fluctuations in the shear layer, which are then fed back as input to acoustic drivers located at the jet exit. The experimental setup was the same as that used by Corke and Kusek (1993) as shown in Figure 5.25. The velocity fluctuations were sensed by a single hot-wire sensor (HW_{fb}) that was placed in the shear layer, near the edge of the potential core, at a streamwise location near the end of the linear growth region. The voltage fluctuations, proportional to the streamwise velocity fluctuations, were then simultaneously input to two analog variable gain amplifiers. These set the feedback gain of axisymmetric (G_A) and helical modes (G_H). The output from these gain circuits was then summed, amplified further (G), and then input to the acoustic driver array located at the exit of the jet nozzle. For the helical modes, an exit condition on the azimuthal mode number ($\pm m$) was imposed by pre-adjusting signal gains to each of the 12 acoustic drivers according to $A(\gamma) = \mathcal{A}\cos(m\gamma + \phi_2)$. A computer-controlled switch selectively opened or closed the electronic feedback loop.

The feedback arrangement is ideal for examining models for the temporal interaction of multiple instability modes of the type introduced by Stuart (1962) and discussed in Section 5.1.2. In this example, we consider two modes with amplitudes A and B, along with their respective subharmonic modes with amplitudes C and D that are coupled via amplitude evolution equations of the following form:

$$\frac{dA}{dT} = a_1A + b_1A^3 + c_1AB^2 + d_1AC \tag{5.18}$$

$$\frac{dB}{dT} = a_2B + b_2B^3 + c_2BA^2 + d_2BD \tag{5.19}$$

$$\frac{dC}{dT} = a_3C + b_3C^3 + c_3CD^2 + d_3CA \tag{5.20}$$

$$\frac{dD}{dT} = a_4D + b_4D^3 + c_4DC^2 + d_4DB. \tag{5.21}$$

In this formulation, the first term represents the linear instability growth, the second term represents the cubic self-interaction, the third term represents a higher-order

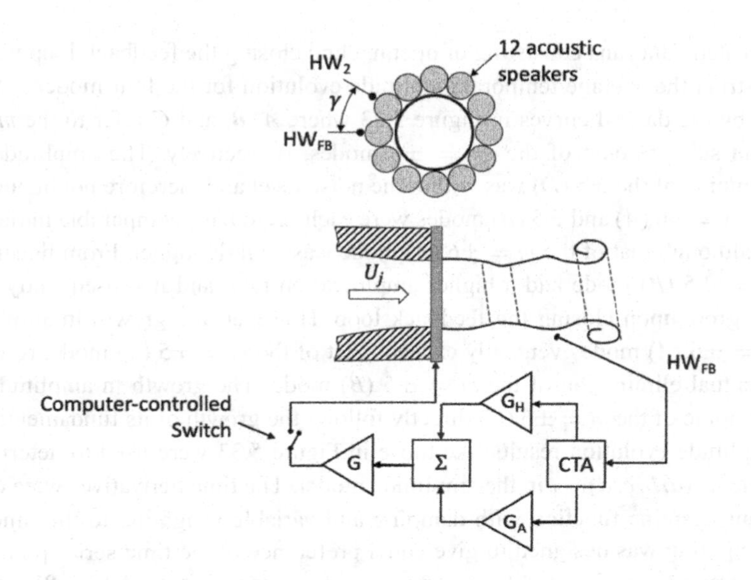

Figure 5.32 Jet closed-loop feedback control schematic. Taken from Corke et al. (1994).

interaction with the other primary modes, and the last term represents a parametric interaction between the primary modes and their respective subharmonic modes.

In its simplest configuration, the azimuthal amplitude distribution to the acoustic driver array can be adjusted to excite only axisymmetric ($m = 0$) modes. With the feedback loop closed, there exists a critical gain, $G_{A_{\text{crit}}}$, above which disturbances grow to produce a saturated limit cycle at a single frequency. The eigenfrequency is a function of the most amplified shear layer frequency and the number of integer wavelengths of that mode from the jet exit to the feedback sensor, x_{fb}. Using this setup, Reisenthel (1988) had documented that near criticality, the system exhibited a supercritical Hopf bifurcation leading to the exponential growth-in-time of self-sustained oscillations.

The dynamical systems model considers multiple azimuthal mode (m) combinations. A separate gain setting for each azimuthal mode number, $G_H/G_{H_{\text{crit}}}$, sets an initial condition *weighting* for each of the helical modes. The actual frequency and azimuthal mode number the system "selects" in the steady state is determined by the ratio of the gains, $(G_H/G_{H_{\text{crit}}})/(G_A/G_{A_{\text{crit}}})$.

The feedback arrangement shown in Figure 5.32 can be used to examine any number of azimuthal mode combinations from $0 \leq m \leq \pm 6$. In the results to follow, the azimuthal amplitude, $A(\gamma)$, was set to favor azimuthal mode numbers of $m = \pm 5$ and $m = \pm 6$. This choice was motivated by experiments of Glauser et al. (1991) that indicated that these higher azimuthal mode numbers were most important in transferring energy to lower and higher wave numbers. Frequency spectra with closed-loop feedback resulting in steady limit cycles identified the $m = \pm 6$ and $m = \pm 5$ modes to have respective frequencies of 1480 and 1120 Hz. Their corresponding subharmonics were 740 and 560 Hz, respectively.

Repeated data runs consisting of opening and closing the feedback loop were used to construct the average temporal amplitude evolution for the four modes. These are shown by the dashed curves in Figure 5.33 where A, B, and C refer to the $m = \pm 6$, ± 5, and subharmonic of the $m = \pm 6$ modes, respectively. The amplitude of the subharmonic of the ± 5 (D) was within the noise level and therefore not included.

The $m = \pm 6$ (A) and ± 5 (B) modes were each set to have comparable initial amplitudes, although that of the $m = \pm 6$ (A) mode was slightly higher. From linear theory, the $m = \pm 5$ (B) mode had a higher amplification rate, and it subsequently was the first to grow upon closing the feedback loop. However, the growth in amplitude of the $m = \pm 6$ (A) mode eventually exceeds that of the $m = \pm 5$ (B) mode, resulting in the eventual elimination of the $m = \pm 5$ (B) mode. The growth in amplitude of the subharmonic of the $m = \pm 6$ (C) directly follows the growth of its fundamental mode.

Amplitude evolution results like those in Figure 5.33 were used to determine the coefficients $(a, b, c, d)_{1...4}$ in the amplitude model. The time derivatives were obtained by fitting a spline function with damping and variable weighting to the time series. The weighting was designed to give equal preference to the time series points in the shorter transient region and longer steady-state region. The model coefficients were determined by minimizing the global least-squares error in a fit of the time series to the model equations. Before fitting, the amplitudes were normalized by the maximum amplitude so that they are dimensionless and fall within a unit circle. The final coefficients represent an average based on a statistical ensemble. The result is shown in the following three equations (Corke et al., 1994):

$$\frac{dA}{dT} = 0.03234A - 0.02831A^3 + 0.1670AB^2 - 0.0321AC \tag{5.22}$$

$$\frac{dB}{dT} = 0.08928B - 0.26370B^3 - 0.3240BA^2 \tag{5.23}$$

$$\frac{dC}{dT} = 0.04110C - 0.50650C^3 - 0.0286CA. \tag{5.24}$$

Note that this does not include the subharmonic of the $m = \pm 5$ (D) mode since its amplitude was not measurable above the noise. The amplitude evolution derived from the model equations is shown by the solid line curves in Figure 5.33.

In some of the limit cycle realizations used to determine the amplitude model coefficients, the initial conditions were such that the $m = \pm 5$ (B) mode became dominant and suppressed the growth of the $m = \pm 6$ (A) mode. This behavior was captured in the amplitude model. This is illustrated in Figure 5.34(a) which shows the solution space for the amplitudes of the $m = \pm 6$ (A) and $m = \pm 5$ (B) modes. The different trajectories are due to the ratio of their initial amplitudes. There is a clear dividing line at which the larger amplitude of one of the modes suppresses that of the other mode. Figure 5.34(b) shows amplitude trajectories for the two modes taken from the experiment. The agreement with the amplitude model is very good. Such validation provides a predictive capability that can be applied to feedback control schemes used in jet applications for enhanced mixing or jet-generated acoustic control.

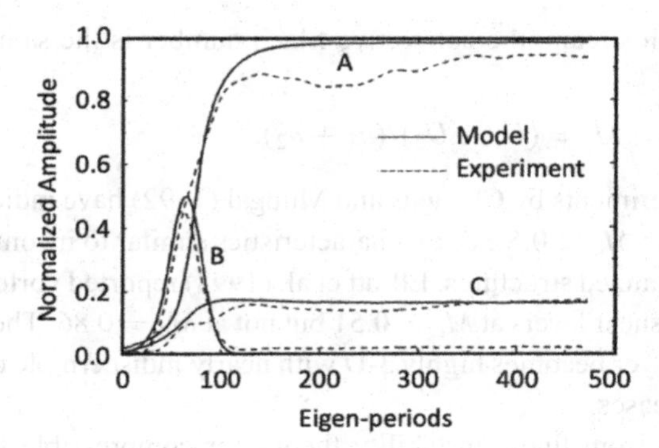

Figure 5.33 Closed-loop temporal amplitude evolution of the $m = \pm 6$ (A) and ± 5 (B) modes and the subharmonic of the $m = \pm 6$ mode (C) with the fundamental modes having comparable initial amplitudes. Taken from Corke et al. (1994).

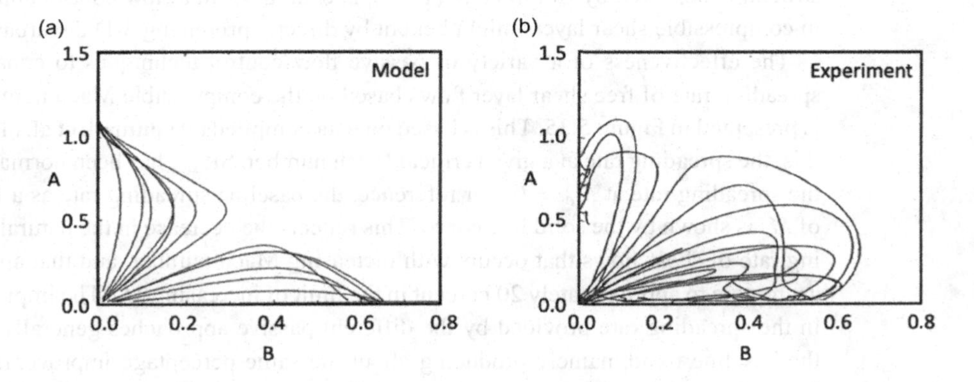

Figure 5.34 Comparison between the amplitude model and experiment of the closed-loop amplitude evolution of the $m = \pm 6$ (A) and ± 5 (B) modes for different ratios of their respective initial amplitudes. Here (a) and (b), respectively, refer to the amplitudes of the $m = \pm 6$ and ± 5 modes. Taken from Corke et al. (1994).

5.3 Supersonic Free Shear Flows

The dynamics of a supersonic shear layer depends on the compressibility effects which can be expressed by the "convective Mach number," M_c. The convective Mach number is defined as

$$M_c = (U - U_c)/a, \tag{5.25}$$

where U_c is the convection velocity of large-scale structures resulting from instabilities in the shear layer, U is the mean free-stream velocity, and a is the speed of sound based on conditions in the free-stream. In a shear layer defined by two streamwise velocities such as in Figure 5.1, and in which the static pressures and specific heat ratios are the same in each stream, the convective Mach number is the same in both streams and given as

$$M_c = (U_1 - U_2)/(a_1 + a_2). \tag{5.26}$$

Flow visualization experiments by Clemens and Mungal (1992) have indicated that supersonic shear flows with $M_c < 0.5$ exhibit characteristics similar to incompressible shear layers, with 2-D organized structures. Elliott et al. (1993) reported vortex pairing typical of incompressible shear layers at $M_c = 0.51$ but not at $M_c = 0.86$. They further reported that the mixing layer becomes highly 3-D with nearly indiscernible organized 2-D structures as M_c increases.

An explanation comes from linear instability theory for compressible shear layers. Contrary to subsonic shear layers where the most unstable linear modes are 2-D, at compressible Mach numbers, 3-D disturbances (oblique waves) have higher amplification rates (Sandham and Reynolds, 1990). This can explain the lack of 2-D structures observed by Elliott et al. (1993), and suggests that flow control approaches in compressible shear layers might benefit by directly producing 3-D disturbances.

The effectiveness of a variety of passive flow control techniques to enhance the spreading rate of free shear layer flows based on the compressible Mach number, M_c, is presented in Figure 5.35. This is based on data compiled by Gutmark et al. (1995). In this, the spreading rate at a given critical Mach number, SR_{M_c}, has been normalized by the spreading rate at $M_c = 0$. For reference, the baseline spreading rate as a function of M_c is shown by the solid line curve. This reflects the decrease in the natural spreading rate of shear flows that occurs with increasing Mach number, and that appears to asymptote to approximately 20 percent in the limit of increasing M_c. The improvement in the spreading rate provided by the different passive approaches generally follows the baseline trend, namely producing about the same percentage improvement with increasing Mach number.

The effect of increasing Mach number carries over to jet flows as well, with a stabilization of the axisymmetric mode (Michalke, 1971), and amplification of 3-D (helical) modes (Viswanathan and Morris, 1992). In contrast to the free shear-layer, with compressible jets, the normalized baseline spreading rate asymptotes to approximately 50 percent with increasing M_c (Viswanathan and Morris, 1992).

5.3.1 Passive Control

As indicated in Figure 5.35, a primary motivation for flow control of high Mach number shear flows and jets is toward enhanced mixing. As with low Mach number flow control approaches presented in previous sections, the techniques generally fall within passive and active methods. Some of the passive approaches presented for low Mach number flows in Figures 5.12 and 5.13 are applicable to high Mach number flows, particularly because of the more amplified 3-D instability.

A technique for natural excitation of the shear layer by Yu et al. (1994) involved placing a rectangular cavity adjacent to the shear layer exit plane. Figure 5.36 shows a schematic of the concept. The principle is to generate a cavity resonance with a

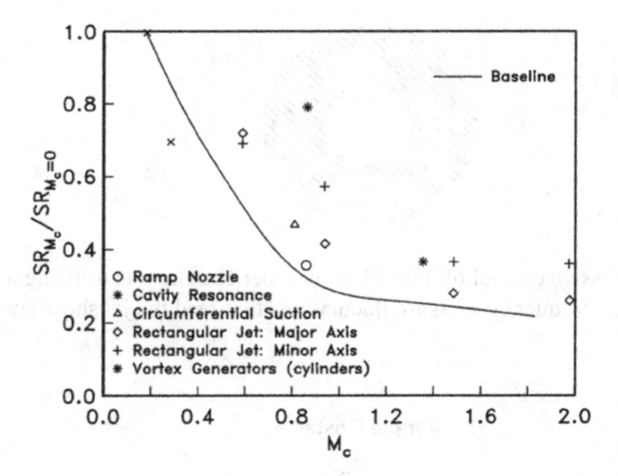

Figure 5.35 Effect of different passive flow control devices on free shear flow spreading rate (SR) as a function of the compressible Mach number, M_C. Solid line represents the baseline (no flow control) spreading rate. Data from Gutmark et al. (1995, figure 1).

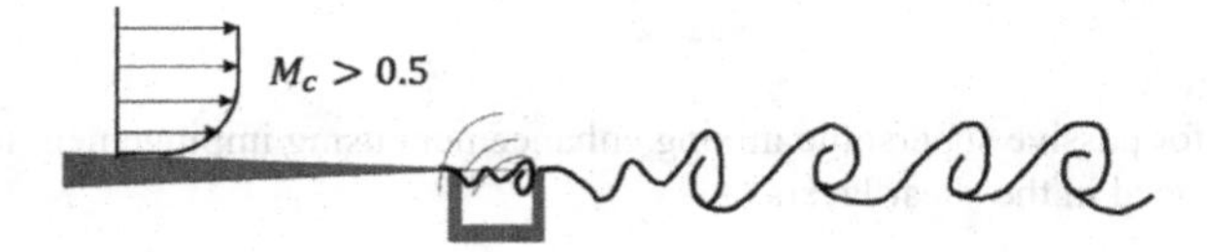

Figure 5.36 Concept for passive control of high Mach number shear layers utilizing a cavity resonance to produce high-frequency pressure fluctuations that feed the shear layer instability.

frequency that could be amplified by the shear layer. Using this approach, Yu et al. (1994) reported a 300 percent increase in the shear layer spreading rate. An embodiment of the concept for axisymmetric jets is shown in Figure 5.37. In either case of plane shear layers or jets, the cavity resonance frequency depends on the local velocity and therefore needs to be scaled for the application conditions.

Another passive method aimed at enhancing mixing in supersonic jets involves placing obstacles downstream of the exit, in the shear layers. This was originally investigated by Rice and Raman (1993) and Raman and Rice (1995). A schematic representation of this technique is shown in Figure 5.38. The underlying principle of the method is that when the vortical fluctuations that evolve from the initial shear layer instability impact the obstruction, it produces unsteady forces that radiate pressure waves. Some of the pressure waves radiate upstream in the ambient (subsonic) flow to reach the jet nozzle exit point of acoustic receptivity. As in the manner illustrated in Figure 5.32, this pressure feedback reenforces an instability frequency that is both in the amplified region and satisfies the requirement of an integer number of wavelengths between the nozzle exit and obstacle.

Raman and Rice (1995) applied the method with a variety of obstacle shapes (shown in Figure 5.38(b)) in a rectangular supersonic jet. They documented that the introduction of the obstacles produced an "intense tone" and a resulting large

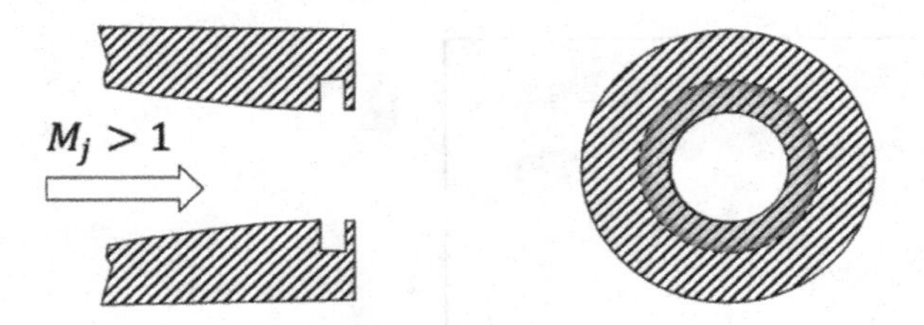

Figure 5.37 Concept for passive control of high Mach number circular jets utilizing a cavity resonance to produce high-frequency pressure fluctuations that feed the jet shear layer instability.

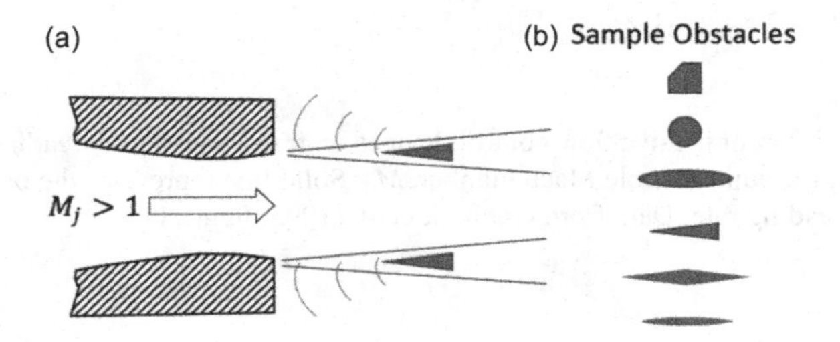

Figure 5.38 Concept for passive supersonic mixing enhancement using impingement tones (a) from obstacles (b) placed in the shear layers.

amplitude "flapping," with a subsequent enhanced spreading of the jet flow. Flow visualization images in Figure 5.41 from Rice and Raman (1993) compare a Mach 1.396 jet flow (a) without and (b) with the obstacles present in the jet shear layer at $x/H_{\text{throat}} = 8.2$ (b) that reveals the resulting intense jet flapping. Figure 5.40 compares the ratio of the total mass flow at different axial locations, $\dot{m}(x)$, and the original mass flow exiting from the jet nozzle, $\dot{m}(0)$. At any axial location, the mass flow increased due to entrainment of the surrounding air. Figure 5.40 reveals that the entrainment was substantially increased by the flapping motion.

From the practical aspect of a thrusting jet, Raman and Rice (1995) introduced a "mixing benefit parameter," MBP, given as

$$\text{MBP} = \frac{(\dot{m}(x)/\dot{m}(0))_{\text{withobstacle}} - (\dot{m}(x)/\dot{m}(0))_{\text{withoutobstacle}}}{(\dot{m}(x)/\dot{m}(0))_{\text{withoutobstacle}}}, \tag{5.27}$$

where $\dot{m}(x)/\dot{m}(0)$ is the ratio of the mass flux at any axial location compared to the mass flux exiting the nozzle. Not surprisingly, the largest mixing benefit came from inserting shapes with the largest leading-edge bluntness, for example, the top two shapes in Figure 5.38(b). However, when normalizing by the percent thrust loss, the optimum shapes were the more aerodynamic profiles with sharp leading edges, such as the triangle and diamond profiles.

The "intermediate-origin" nozzles shown in Figure 5.17 and discussed in Section 5.2 for passive flow control in jets at lower Mach number flows also provide

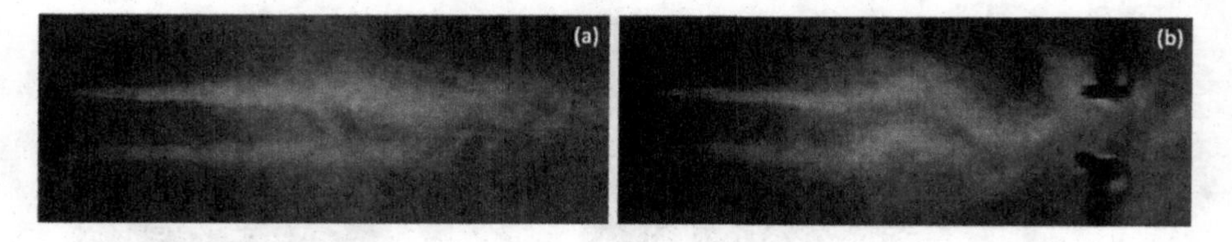

Figure 5.39 Flow visualization images of a Mach 1.396 rectangular jet flow without (a) and with the obstacles present in the jet shear layer at $x/H_{\text{throat}} = 8.2$ (b). Taken from Rice and Raman (1993).

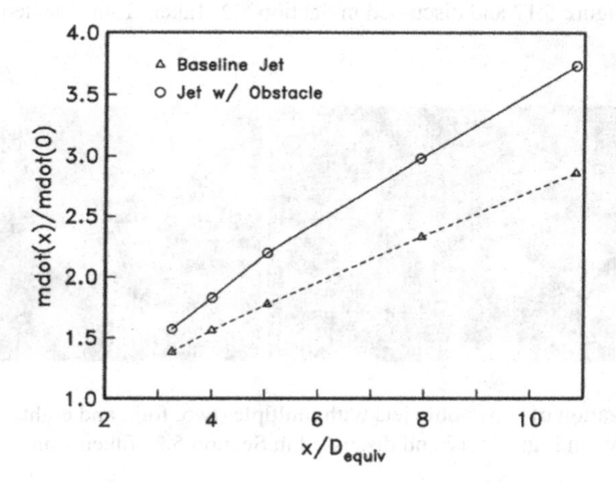

Figure 5.40 Mass flow ratios in the axial direction that correspond to the jet condition that produced the images in Figure 5.39. Data from Rice and Raman (1993).

an approach at supersonic Mach numbers. Examples of this are shown in Figures 5.41 and 5.42 which show Schlieren flow visualization images from Wlezien and Kibens (1988).

Figure 5.41 shows Schlieren images for two inclined exit nozzles. As was previously illustrated in Figure 5.18(a), the inclination angles are defined by a rotation of the exit plane that displaces the top- and bottom-dead-center positions by a fraction of the nozzle exit diameter, *D*. Figure 5.41(a) and (b) respectively corresponds to nozzles shown in Figure 5.17 having inclined exits D/2, and D. In each case, the Schlieren visualization reveals a diamond shock cell pattern that is a characteristic of under-expanded supersonic jets. The inclined exit nozzles are observed to cause a deflection and divergence of the jet that increases with increasing exit angle. The divergence was accompanied by a rearrangement of the shock cells that broke from the characteristic symmetric pattern.

Figure 5.42 shows flow visualization images with nozzles having periodic cutouts to form multiple tabs at the exit. Examples of this were illustrated in Figure 5.17. The images in Figure 5.42 correspond to two-, four-, and eight-tab configurations. As opposed to the inclined exit nozzles, the azimuthal symmetry of the tabs prevents any mean deflections of the jet. These tabbed nozzles primarily increase the jet divergence.

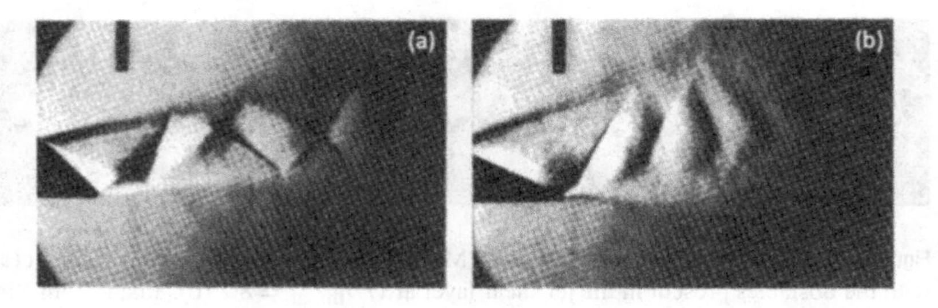

Figure 5.41 Flow visualization of supersonic jets with inclined exit nozzles D/2 (a) and D (b) that were illustrated in Figure 5.17 and discussed in Section 5.2. Taken from Wlezien and Kibens (1988).

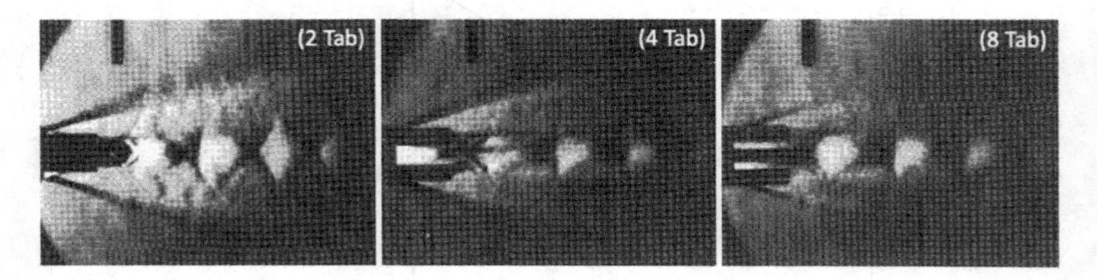

Figure 5.42 Flow visualization of supersonic jets with multiple (two, four, and eight) tab nozzles of a design shown in Figure 5.17 and discussed in Section 5.2. Taken from Wlezien and Kibens (1988).

The two-tab nozzle generates a divergent flow that is similar to that of the inclined exit nozzles, although the internal shock pattern is observed to be significantly different. The jet divergence was observed to decrease as the number of tabs increased, reflecting the smaller open space (gap) between the tabs. Wlezien and Kibens (1988) had further observed that the shock pattern for the four- and eight-tab nozzles was similar to that of a jet without tabs.

5.3.2 Active Control

As with subsonic flows, active control of supersonic shear layer and jets utilizes most of the same methods illustrated in Figures 5.21–5.29 and discussed in Section 5.2. With regard to acoustic excitation, Tam (1978) found that a narrow beam of sound aimed at an angle of 50°–80° was most effective in exciting instability waves in compressible shear layers. However, in situ methods of excitation are likely a more practical method. As an example, Kibens and Glezer (1992) utilized piezoelectric actuators to excite flapping modes in a jet for Mach numbers from 0.7 to 1.2. The piezoelectric actuators were operated at their resonant frequency to maximize their amplitude. They found that jet flapping could be produced when the ratio of the excitation frequency, f_{ex}, to the shear layer instability frequency, f_i, was less than 1. They also found that for $f_{ex}/f_i > 1$, the jet orientation could be controlled.

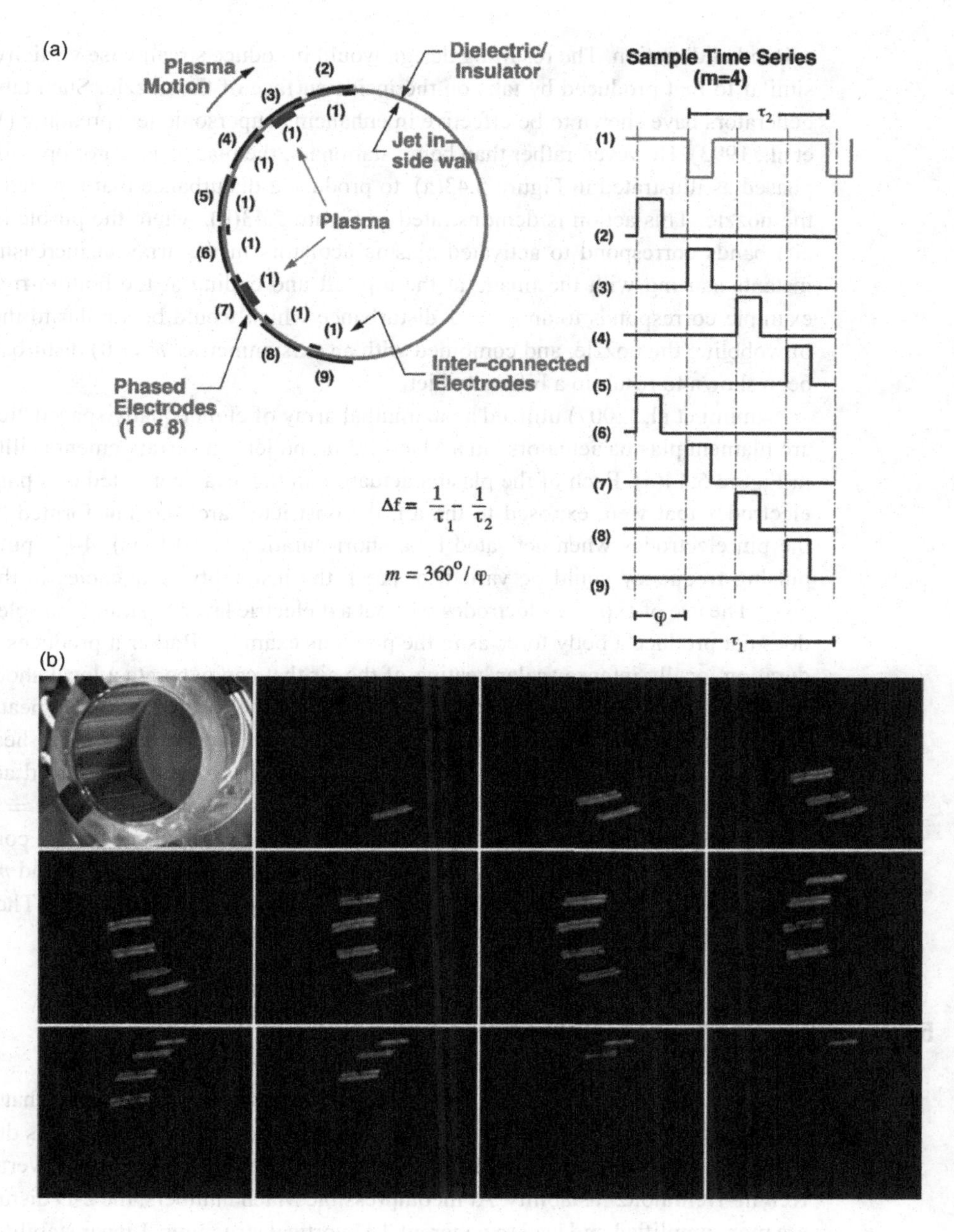

Figure 5.43 Schematic of azimuthal phased array of 16 single dielectric barrier discharge plasma actuators (a) and sequential images illustrating phased plasma activation motion to excite an $m = 1$ mode (b).

Plasma actuators have also been used to excite instabilities in supersonic jets. Figure 5.43 illustrates a concept involving an azimuthal array of 16 single DBD plasma actuators. The plasma actuators are elongated in the streamwise direction and have an electrode geometry that would produce a body force vector field that is directed in the azimuthal direction. The resulting design would introduce streamwise vorticity that is similar to that produced by tabs on the inside surface of the nozzle. Such tab vortex generators have shown to be effective in enhancing supersonic jet spreading (Wishart et al., 1993). However, rather than being stationary, the plasma actuator operation was phased as illustrated in Figure 5.43(a), to produce a disturbance that traveled around the nozzle. This action is demonstrated in Figure 5.43(b), where the purple (ionized air) bands correspond to activated plasma actuators in the array at increasing time instants starting with the image at the top-left and ending at the bottom-right. The example corresponds to an $m = 1$ disturbance which would be similar to the effect of wobbling the nozzle, and combined with an axisymmetric ($m = 0$) disturbance has been shown to result in a bifurcated jet.

Samimi et al. (2007) utilized an azimuthal array of eight equally spaced “localized arc filament plasma actuators” in a Mach 1.3 round jet. The arrangement is illustrated in Figure 5.44(a). Each of the plasma actuators in the array consisted of a pair of pin electrodes that were exposed to the air. A constricted arc filament formed between the pin electrodes when activated by a short-duration ($\simeq$ 0.1 μs), 4-kV pulse. The pulsing frequency could be varied to match the instability frequencies in the shear layer. The use of exposed electrodes without a dielectric layer separating the electrodes does not produce a body force as in the previous example. Rather it produces a short-duration locally intense scalar heating of the air that can generate a local shock wave that expands from the source. The combination of the localized intense heating and accompanying shock/pressure wave is the mechanism for exciting the jet shear layer. In the arrangement of Samimi et al. (2007), the eight azimuthally spaced actuators were operated with phase shifts to excite azimuthal mode numbers from $m = 0$ to the maximum with eight actuators of $m = 4$. Flow visualization images for conditions that were designed to separately excite an axisymmetric ($m = 0$) mode and $m = \pm 1$ pairs of helical modes are shown in Figure 5.44(b) and (c), respectively. The ability for the helical modes to enhance the jet spreading rate is clearly evident.

5.4 Summary: Free Shear Layer Control

The basis of flow control in free shear layers is the instability mechanisms that underly the growth of disturbances as well as the sensitivity to the environment. As discussed in this chapter, the initial growth of disturbances in free shear layers is governed by a Kelvin–Helmholtz instability. At incompressible Mach numbers, the 2-D disturbances are more amplified and lead to coherent 2-D vortical structures. Linear stability theory predicts well the most unstable frequencies and phase speeds. The 2-D structures that evolve in a convective manner go through a process of one or more vortex pairings. In each case, the passage frequency decreases by a factor of two. Vortex pairing is an energetic event that enhances mixing. The process of pairing also generates acoustic pressure waves that are felt back at the shear layer virtual origin, which generally corresponds to the trailing edge of the physical boundary separating the two flow streams. The long wavelength acoustic waves are transformed into shorter wavelength vortical waves through the acoustic receptivity mechanism. This is the key mechanism for

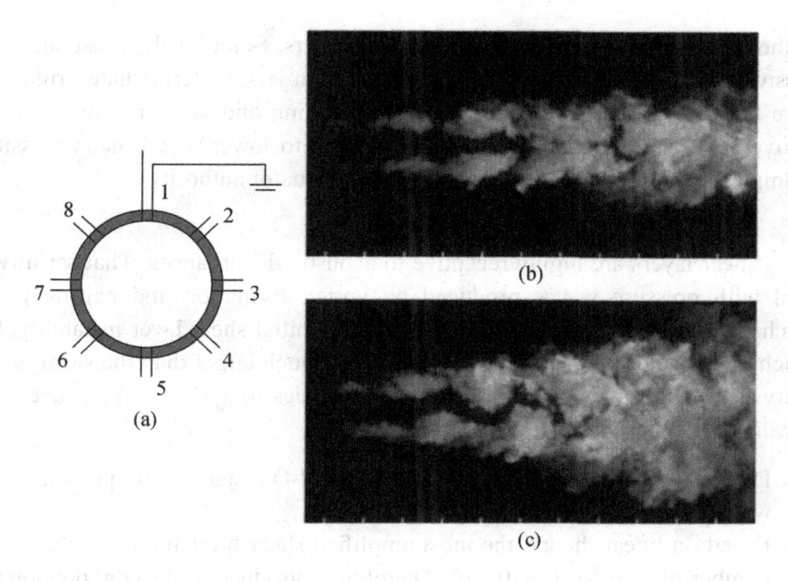

Figure 5.44 Azimuthal array of eight "localized arc filament plasma actuators" (LAFPA) (a) used to excite (b) axisymmetric ($m = 0$) and (c) $m = \pm 1$ pairs of helical modes in a Mach 1.3 round jet. Taken from Samimi et al. (2007).

controlling free shear layers with acoustic drivers. The natural acoustic feedback that occurs can reenforce the initial amplitude of instabilities and lead to a global instability in some instances.

Jets embody all the physics of free shear layers but add a second instability mechanism with a frequency that scales with the jet diameter. This so-called column mode can couple with the generally higher-frequency shear layer through a succession of shear layer vortex pairings that end with the column mode frequency. This can occur naturally or be artificially driven. Such coupling can greatly enhance jet spreading and mixing.

Although the 2-D mode is most amplified at incompressible Mach numbers, 3-D modes in the form of oblique waves are also amplified. If 3-D disturbance levels are large enough, they can compensate for the lower amplification rates of the 3-D modes, making them dominant. This can also be promoted by introducing a 3-D distortion of the mean flow using arrays of vortex generators or ramps. A more elegant method to promote the growth of the oblique wave modes is by simultaneously introducing a 2-D mode that can resonantly interact with the 3-D mode causing it to grow in amplitude at a faster than linear rate. In jets, the oblique waves take the form of a helix. These can form singly, turning in either clockwise or counterclockwise directions, or in pairs with both directions of turning. The controlled interaction between the axisymmetric and helical modes in jets can have dramatic effects including bifurcating streams to "blooming" with greatly enhanced spreading.

At compressible Mach numbers, the most amplified instability modes in free shear layer are 3-D, or more specifically oblique waves. Therefore, flow control in this regime can directly focus on introducing 3-D disturbances. The approaches are similar to those used with incompressible Mach numbers. Some of the most successful are passive modifications such as ramps and tabs. In jets, "intermediate-origin" nozzles have been very successful in enhancing spreading and vectoring the jet core flow. Active approaches are more limited compared to lower-speed incompressible flow examples, generally because of limits in flow actuator authority.

Problems

5.1 Shear layers are highly receptive to acoustic disturbances. That sensitivity coupled with pressure waves produced by vortex formation and pairing provides a mechanism for a downstream influence of the initial shear layer instability. In a low Mach number flow, the acoustic wavelength is much larger than the shear layer instability wavelength, and therefore Eq. (5.9) reduces to $x_j/\lambda_j = N$, where x_j are the streamwise locations of the Nth vortex roll-ups.

1. Derive an expression for x_j in terms of the 2-D instability frequency, $f_{i,0}$, and the wave phase speed, c_r.
2. Based on linear theory, the most amplified shear layer frequency has a Strouhal number of $\mathrm{St} = f\theta/U = 0.032$. Therefore, introduce St into the previous expression for x_j.
3. In the jet shear layer of Corke and Kusek (1993) with $U = 20.3$ m/s, the initial shear layer momentum thickness was $\theta_i = 0.1$ mm, and $c_r = 0.5U$. Based on these conditions, list the streamwise locations of the initial shear layer roll-up and first and second vortex pairings based on the feedback model.
4. How would these change if the shear layer were excited at half the most amplified frequency? What effect would this have on the streamwise spreading of the shear layer?

5.2 Jets introduce another instability that scales with the jet velocity and diameter with the most typical Strouhal number being $\mathrm{St}_c = fD/U = 0.4$. Although there is a large separation between the Strouhal number of the jet shear layer, $\mathrm{St}_{i,0}$ and that of the jet "column mode," it is possible for the two to interact. The condition for this to occur is that $\mathrm{St}_n = f_{i,0}D/U_j = 2^N \mathrm{St}_c$, or namely that the initial shear layer goes through a succession of vortex pairings to reach the column mode frequency. Note that both of the instability frequencies are normalized by the jet diameter, D.

1. Show that the condition for resonant coupling between the initial shear layer and jet column mode can be written in terms of the jet Reynolds number, as $\mathrm{Re}_D = UD/\nu = C(2^N \mathrm{St}_c)^2$ where C is a constant.
2. In the experiment of Drubka et al. (1989), $D = 5.08$ cm. For their "1L case" where the initial shear layer thickness $\theta_i = 0.20$ mm, resonance was observed at $\mathrm{Re}_D = 42{,}000$. Based on $\mathrm{St}_D = 0.42$, verify that $N = 3$ satisfies the resonance criteria and gives a value of $C = 3720$.
3. While keeping the same $\mathrm{Re}_D = 42{,}000$, Drubka et al. (1989) found that increasing the disturbance levels in the jet eliminated the resonance mechanism. In that case with the higher disturbance level, the initial shear layer thickness was reduced by 3.6 percent to $\theta_i = 0.193$ mm. Can you explain the elimination of the resonance

on the basis of the change in the shear layer Strouhal number? Show all of your steps in drawing a conclusion.

5.3 The closed-loop feedback setup that was shown in Figure 5.32 was used to study the temporal growth and interaction of multiple jet shear layer instability modes. A specific example involved the excitation of two helical modes with azimuthal mode numbers of $m = \pm 5$ and ± 6. Their temporal growth and interaction was modeled via amplitude evolution equations that were listed in Eq. (5.23) and Eq. (5.24). For this,

1. Determine formulation for the saturation amplitudes, $d|\ |/dt = 0$, for the $m = \pm 6$ (A) and $m = \pm 5$ (B).
2. How does the saturation value of A vary if the amplitude ratio $A/B = 1.2$? What if $A/B = 0.8$?
3. How does the saturation value of C vary if the amplitude ratio $A/B = 1.2$ or $A/B = 0.8$?

5.4 Figure 5.38 presents a concept for enhancing the mixing of a supersonic jet that involves placing obstacles within the shear layer.

1. Could the same approach work to enhance the spreading of a subsonic shear layer? If so, what is the mechanism?
2. What would be the optimum streamwise location to place an obstacle in order to *enhance* the growth of the initial fundamental shear layer instability?
3. Based on Section 5.1.3, what would be the best location for the obstacle in order to *suppress* the growth of the initial fundamental shear layer frequency?

5.5 The underlying mechanism of the angled and step-back jet nozzles of Wlezien and Kibens (1986) is to produce an azimuthal variation in the shear layer initial momentum thickness, θ_i.

1. If a stepped nozzle like that shown in Figure 5.17(G) results in a 10 percent difference in the momentum thickness, what will be the difference in the shear layer instability frequencies on either side of the jet exit?
2. What will be the difference in the shear layer instability wavelengths on either side of the jet?
3. What effect would this have on the jet spreading or vector angle?

6 Two-Dimensional Laminar Boundary Layers

The problem of transition from a laminar to turbulent state in viscous boundary layers is of great practical interest. For example, on aircraft, the low skin-friction coefficient of laminar boundary layers is very attractive to those sizing the propulsion system or determining fuel requirements. In contrast, the low mixing of fluid properties such as chemical species, heat, or momentum that occurs in laminar boundary layers may be intolerable to designs where flows are on the verge of separation (See Chapter 4). These contrasting requirements make it important to both predict and control the laminar or turbulent state of a boundary layer.

The objective of this chapter is to provide the basic ideas and results of boundary-layer stability in order to understand transition mechanisms, transition control, and transition prediction of boundary layers. The focus in this chapter is on boundary layers in which the mean flow is two-dimensional (2-D) (no cross-flow). The following chapter will focus on three-dimensional (3-D) boundary layers that include a mean cross-flow component. In both cases, the current state-of-the-art of boundary-layer stability is reviewed and by using recent results, a number of unique transition mechanisms that can play a role in the breakdown to turbulence are illustrated. Ultimately, the control of the stability and turbulent transition characteristics of boundary layers requires such a thorough understanding of these mechanisms.

In fluids, turbulent motion is usually observed rather than laminar motion because the Reynolds number range of laminar motion is generally limited. The transition from laminar to turbulent flow occurs because of an incipient instability of the basic flow field. This instability intimately depends on subtle, and sometimes obscure, details of the flow. Figure 6.1 provides an early (1985) "road map" of paths to turbulence for laminar 2-D boundary layers (Morkovin, 1985). Although this has evolved over the years with our added understanding, much of it is still relevant and sets a framework for the control of turbulence onset.

As illustrated in the path to turbulence "road map" in Figure 6.1, boundary-layer transition generally begins as a result of disturbances such as sound or vorticity that emanate from the free-stream and enter the boundary layer as steady and/or unsteady fluctuations of the basic state. The initial growth of these disturbances is described by linear stability theory (i.e., linearized, unsteady, and Navier–Stokes (N-S)). The growth of these disturbances is weak, occurring over a viscous length scale, and can be modulated by pressure gradients, surface mass transfer, temperature gradients, etc.

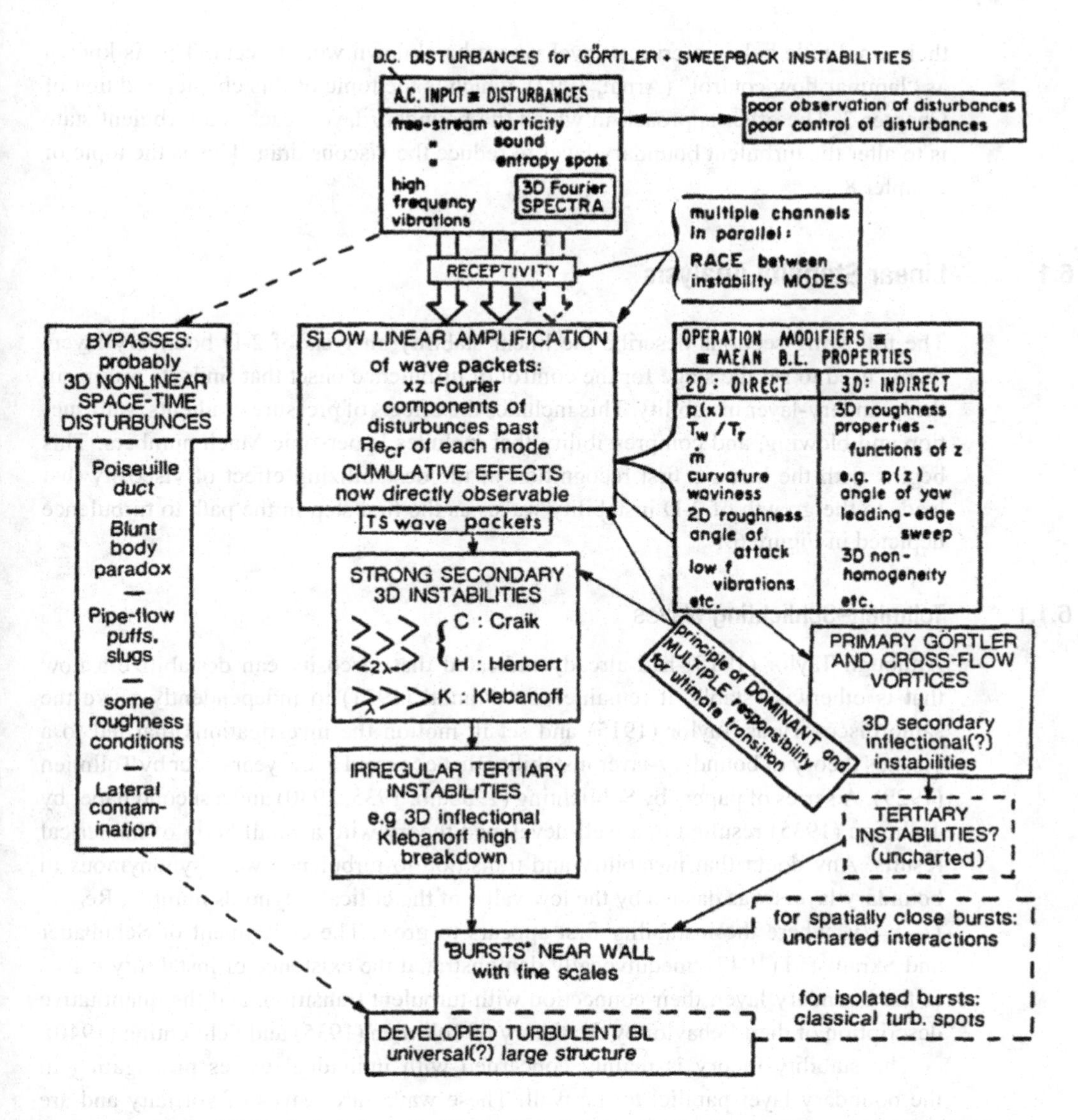

Figure 6.1 Early "road map" of paths to turbulence for laminar 2-D boundary layers. Taken from Morkovin (1985).

If the disturbances grow to larger amplitudes (beyond linear theory assumptions), it can spawn 3-D nonlinear interactions that generate mean flow distortions leading to secondary instabilities. At this last stage, the disturbance growth is very rapid with turbulent breakdown occurring in short succession, over a convective length scale.

Since the linear stability behavior can be calculated directly, transition prediction schemes are usually based on linear theory. The control of turbulent skin friction in transitional flows has two general approaches. The first is to prevent the boundary layer from becoming turbulent by limiting the growth of linear disturbances, keeping their amplitude below a critical level where breakdown would occur. This is known as "laminar flow control" (Arnal, 1992), which is the topic of this chapter and that of Chapter 7. The other approach in which the boundary layer reaches a turbulent state is to alter the turbulent boundary layer to reduce the viscous drag. This is the topic of Chapter 8.

6.1 Linear Stability Analysis

The following sections describe the linear stability analysis of 2-D boundary layers that is used to set the stage for the control of turbulence onset that finds its origins in the boundary-layer instability. This includes the effects of pressure gradients, wall suction and blowing, and compressibility that includes hypersonic Mach numbers. This begins with the historic first recognition of the destabilizing effect of viscosity that leads to the growth of 2-D instability waves as the first step in the path to turbulence depicted in Figure 6.1.

6.1.1 Tollmien–Schlichting Waves

Although Taylor (1915) had already indicated that viscosity can destabilize a flow that is otherwise stable, it remained for Prandtl (1921) to independently make the same discovery as Taylor (1915) and set in motion the investigations that led to a viscous theory of boundary-layer instability that occurred a few years later by Tollmien (1929). A series of papers by Schlichting (1933a,b, 1935, 1940) and a second paper by Tollmien (1935) resulted in a well-developed theory with a small body of numerical results. Any doubt that instability and transition to turbulence were synonymous in boundary layers was dashed by the low value of the critical Reynolds number, $\mathrm{Re}_{\mathrm{cr}} = U_\infty x_{\mathrm{cr}}/\nu$, where the instability first appears to grow. The experiment of Schubauer and Skramstad (1947) unequivocally demonstrated the existence of instability waves in the boundary layer, their connection with turbulent transition, and the quantitative description of their behavior by the theory of Tollmien (1935) and Schlichting (1940).

The stability theory is mainly concerned with individual waves propagating in the boundary layer parallel to the wall. These waves are waves of vorticity and are commonly referred to as Tollmien–Schlichting waves, or T-S waves, or simply as instability waves. The amplitudes of the waves, which vary through the boundary layer and die off exponentially in the free-stream, are small enough so that a linear theory may be used. The frequency of a wave is ω and the wave number is $k = 2\pi/\lambda$, where λ is the wavelength. The wave may be 2-D, with the lines of constant phase normal to the free-stream direction (and parallel to the wall), or it may be oblique, in which case the wave number is a vector k at an angle ψ to the free-stream direction with a streamwise (x) component α and a spanwise (z) component β. The phase velocity c is always less than the free-stream velocity U_∞, so that at some point in the boundary layer, the mean velocity is equal to c. This point is called the "critical point," or "critical layer," and it plays a central role in the mathematical theory, as well as in T-S wave control. The wave amplitude usually has a maximum near the critical layer.

The stability analysis in this chapter involves 3-D disturbances in an incompressible parallel boundary-layer flow without curvature. The basic state consists of a 1-D flow where

$$U = U(y), \quad V = 0, \text{ and } W = W(y), \tag{6.1}$$

where U is the streamwise velocity component, W is the spanwise velocity component, and y is the coordinate normal to the surface.

The parallel flow assumption is a first approximation to the basic state that is reasonable at large Reynolds numbers, and in boundary layers with favorable pressure gradients. Although the nonparallel effects are noticeable in Blasius boundary layers, it is primarily evident at higher dimensionless frequencies that are not amplified under the parallel flow assumption, and decreases the minimum critical Reynolds number by about 100 (Saric and Nayfeh, 1975). From the viewpoint of instability as a precursor of transition, a more important range of frequencies is that located in the center of the unstable region where growth rates are large. In this latter region, there is little difference between the results of parallel and nonparallel treatments. Therefore, the parallel-flow calculations are adequate for considerations relevant to transition.

The stability equations are obtained by superposing small disturbances onto the basic state, namely

$$u^*/U_\infty = U + u'(x,y,z,t) \tag{6.2}$$

$$v^*/U_\infty = v'(x,y,z,t) \tag{6.3}$$

$$w^*/U_\infty = W + w'(x,y,z,t) \tag{6.4}$$

$$p^*/(\rho U_\infty^2) = P + p'(x,y,z,t), \tag{6.5}$$

where u^*, v^*, w^*, and p^* satisfy the dimensional N-S equations, u', v', w', and p' are the respective velocity and pressure disturbances to the basic flow quantities, U, W, and P. These forms of the disturbances are substituted into the dimensionless form of the unsteady N-S equations to obtain the “disturbance equations.” Because the basic flow (U, W, and P) identically satisfies the N-S equations, its terms drop out in the final form of the disturbance equations. Finally, the equations are *linearized* by neglecting products of disturbance terms to yield the following linearized disturbance equations:

$$u'_x + v'_y + w'_z = 0 \tag{6.6}$$

$$u'_t + Uu'_x + Wu'_z + U_y v' + p'_x - \nabla^2 u'/R = 0 \tag{6.7}$$

$$v'_t + Uv'_x + Wv'_z + p'_y - \nabla^2 v'/R = 0 \tag{6.8}$$

$$w'_t + Uw'_x + Ww'_z + W_y v' + p_z - \nabla^2 w'/R = 0. \tag{6.9}$$

The subscripts in the disturbance equations denote partial differentiation, and $R = U_\infty L/\nu$ where L is the length scale used in the non-dimensionalization of the N-S equations.

The disturbances are assumed to take a normal-mode form

$$q'(x,y,z,t) = q(y)\exp[i(\alpha x + \beta z - \omega t)] + \text{c.c.}, \tag{6.10}$$

where q' represents any of the disturbance quantities, α is the streamwise wave number, β is the spanwise wave number, ω is the frequency, and c.c. stands for complex conjugate. The wave numbers are complex where $\alpha = \alpha_r + i\alpha_i$ and likewise $\beta = \beta_r + i\beta_i$, and $i = \sqrt{-1}$. As a result, $q(y)$ is complex, but q' is real.

In the absence of the parallel flow assumption, α, β, and R depend on the streamwise location, x^*. In this regard, a commonly chosen reference length that accounts for the boundary layer growth is $L = \delta_r = (\nu x^*/U_\infty)^{1/2}$.

Substituting the normal mode form into the linearized disturbance equations results in the following equations:

$$i\alpha u + i\beta w + Dv = 0 \tag{6.11}$$

$$i(\alpha U + \beta W - \omega)u + vDU + i\alpha p - (D^2 - k^2)u/R = 0 \tag{6.12}$$

$$i(\alpha U + \beta W - \omega)v + Dp - (D^2 - k^2)v/R = 0 \tag{6.13}$$

$$i(\alpha U + \beta W - \omega)w + vDW + i\beta p - (D^2 - k^2)w/R = 0. \tag{6.14}$$

Here, u, v, and w are the disturbance quantities, where the $'$-notation was dropped, $k^2 \equiv \alpha^2 + \beta^2$ is the 2-D wave vector, and $D = d/dy$.

These equations can be combined into a single fourth-order equation famously known as the *Orr–Sommerfeld* equation, which is given in the following:

$$D^4\phi - 2k^2D^2\phi + k^4\phi - iR[(\alpha U + \beta W - \omega)(D^2\phi - k^2\phi) - \alpha(D^2U)\phi - \beta(D^2W)\phi] = 0, \tag{6.15}$$

where $v = \phi$ and the boundary conditions are

$$\phi(y = 0) = D\phi(y = 0) = 0, \quad \phi(y \to \infty) \to 0. \tag{6.16}$$

Equation (6.15) is linear and homogeneous and therefore forms an eigenvalue problem. For boundary layers, there is a finite discrete set of eigenvalues and a continuous spectrum. The eigenfunctions are called *modes* and are superposed to construct an arbitrary disturbance profile. The least stable mode is called the *first mode*.

Squire's transformation applied to Eq. (6.15) shows that 2-D disturbances are more amplified than 3-D disturbances and that the minimum critical Reynolds number for the growth of disturbances is given directly by 2-D analysis. It is important to point out that this only applies to incompressible Mach numbers. In addition, 3-D disturbances do play a major role in the turbulent transition process.

Therefore, considering only 2-D disturbances, where $W = 0$, $\beta = 0$, and $k = \alpha$, Eq. (6.15) takes the following form:

$$(D^2 - \alpha^2)^2\phi - i\alpha R[(U - \omega/\alpha)(D(2 - \alpha^2) - D^2U]\phi = 0. \tag{6.17}$$

For this, the normal mode form is a 2-D wave

$$q'(x, y, z, t) = q(y)\exp(-\alpha_i x)\exp[i(\alpha_r - \omega t)] + \text{c.c.} \tag{6.18}$$

Here, $\alpha = \alpha_r + i\alpha_i$ and ω is real. The spatial growth rate is therefore $-\alpha_i$, so that disturbances are amplified when $-\alpha_i < 0$, damped when $-\alpha_i > 0$, and neutrally growing when $-\alpha_i = 0$.

The eigenvalue problem is expressed as $\alpha = f(\omega, R)$. The phase speed is given as $c = \omega/\alpha_r$.

Rather than α being complex, the spatial growth can be transformed into a temporal growth where α is real, and ω is complex. In this case, $c = \omega/\alpha = c_r + ic_i$, and the normal mode form is

$$q'(x,y,t) = q(y)\exp[i\alpha(x - ct)] + \text{c.c.} \tag{6.19}$$

Disturbances are amplified when $c_i > 0$, damped when $c_i < 0$, and neutrally growing when $c_i = 0$. The eigenvalue problem is then $c = f(\alpha, R)$, and the phase speed is c_r. This temporal formulation has advantages because the eigenvalue, c, appears linearly in the equation.

The laminar boundary layer is a dispersive medium for the propagation of instability waves. That is, different frequencies propagate with different phase velocities so that individual harmonic components in a group of waves will displace from each other over time. The group velocity can be considered to be a property of the individual waves. If an observer is moving at the group velocity of a normal mode, the wave in the moving frame of reference will appear to grow in time (temporal growth). In contrast, in the stationary frame of reference, the wave appears to grow in space. Therefore, this yields the relation between temporal and spatial growths, namely

$$d/dt = c_r d/dx_g, \tag{6.20}$$

where c_r is the temporal phase speed, and x_g is the coordinate in the phase speed direction. Therefore, for the temporal amplification rate, ω_i, the spatial amplification rate in the direction parallel to c_r is

$$-(\alpha_i)_g = \omega_i/c_r. \tag{6.21}$$

This temporal to spatial amplification conversion is attributed to Gaster (1962), and is known as the *Gaster transformation*. It is important to note that the Gaster transformation is only valid for small amplification rates where within the approximation, the frequency and wave number of the spatial wave are the same as the temporal wave.

In boundary layers, the critical parameter in determining instability is the Reynolds number, which is usually defined as

$$R = U_\infty \delta_r/\nu. \tag{6.22}$$

The length scale, δ_r, is the boundary-layer thickness, which, for a Blasius boundary layer, is $\delta_r = (\nu x/U_\infty)^{1/2}$. Incorporating this into Eq. (6.22) provides a useful streamwise (x) dependence on the critical parameter, namely

$$R = (R_x)^{1/2}. \tag{6.23}$$

The instability wave frequency is generally represented in dimensionless form as

$$F = \omega/R = 2\pi f \nu/U_\infty^2, \tag{6.24}$$

where f is the frequency with units of cycles/second (Hertz). F is constant for a given flow condition.

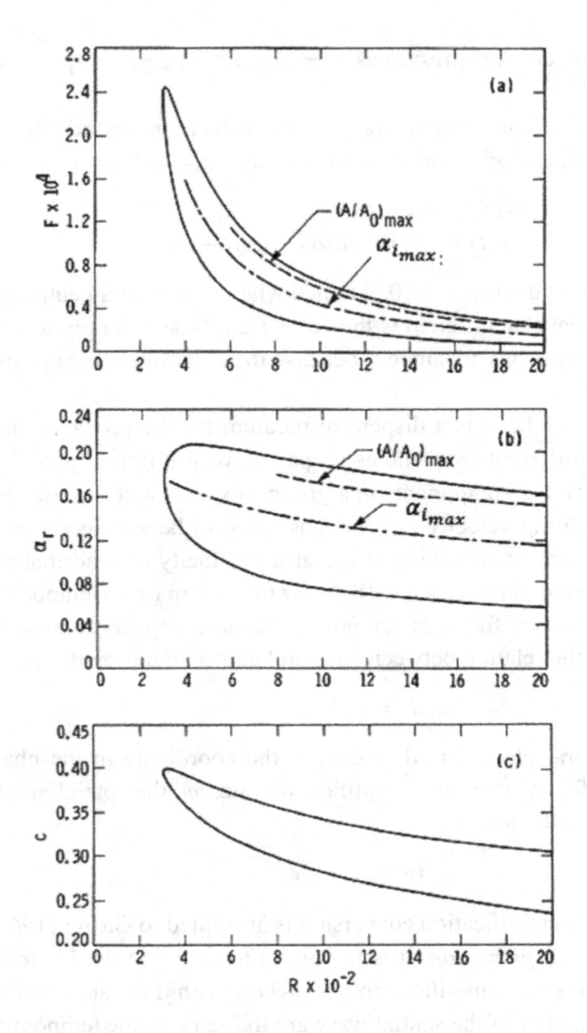

Figure 6.2 Neutral stability curves for Blasius boundary layer. (a) Dimensionless frequency, (b) wave numbers, and (c) phase velocity. Taken from Mack (1984).

Because of its simplicity, and that the basic state (mean profile) is easily computed, the Blasius boundary layer is a good reference flow for transition studies. Figure 6.2 shows neutral ($\alpha_i = 0$) stability curves for a Blasius boundary layer based on *parallel flow assumption*. In each of the plots, the abscissa is Reynolds number, R, given by Eq. (6.23). In Figure 6.2(a), the ordinate is the dimensionless frequency, F, defined in Eq. (6.24). Disturbances at Reynolds numbers or dimensionless frequencies, F, that occur outside of the neutral growth curve are damped. Those that occur inside of the neutral growth curve are amplified. Included in Figure 6.2(a) are two additional curves that give the frequencies of the maximum amplification rate, $-\alpha_i$, and the maximum amplitude ratio, A/A_0, where A_0 is the amplitude at the lower branch neutral point at the corresponding frequency. Both maxima are with respect to frequency at

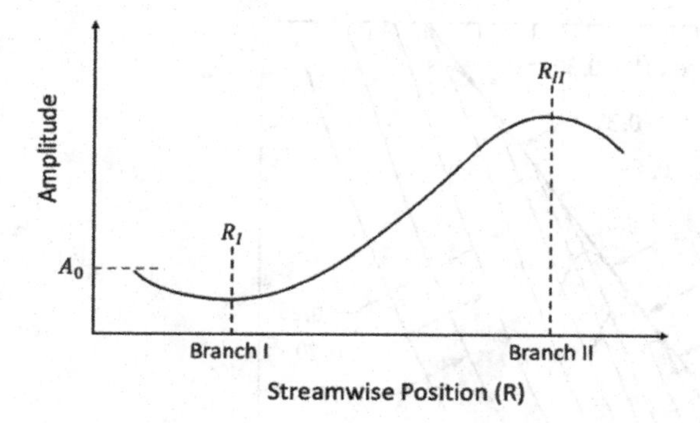

Figure 6.3 Illustration showing the streamwise growth in amplitude of a disturbances at a constant frequency for a Blasius boundary layer based on linear stability theory.

constant Reynolds number. The corresponding wave numbers, α_r, are plotted in Figure 6.2(b). The T-S instability wavelength is $\lambda = 2\pi/\alpha_r$. For the range of amplified wave numbers in Figure 6.2(b) at $R = 1000$, the T-S wavelengths range from approximately 20 to 50 boundary-layer thicknesses (Mack, 1984). Figure 6.2(c) shows the phase velocities of amplified T-S waves. These are observed to range between 0.28 and $0.33U_\infty$.

In the classic experiments by Schubauer and Skramstad (1947), a 2-D vibrating ribbon was used to introduce disturbances at a fixed frequency near the leading edge of a flat plate over which a Blasius boundary layer developed. The amplitude of the disturbance would then be measured at a fixed height from the surface at different downstream distances from where the disturbance was introduced. Based on Figure 6.2(a), a disturbance starting in the range of amplified dimensionless frequencies, and with an initial amplitude of A_0, will decay exponentially until it reaches the x-Reynolds number where it crosses the lower branch (Branch I) of the neutral growth curve. Having moved into the range of Reynolds numbers beyond the neutral curve Branch I, the disturbance will grow exponentially until it reaches the upper branch (Branch II) of the neutral growth curve, where it will then decay exponentially. This streamwise evolution of a disturbance at a fixed frequency is illustrated in Figure 6.3. By varying the disturbance frequency and noting those x-Reynolds numbers where the disturbance growth first amplified and then decayed, Schubauer and Skramstad (1947) were able to determine the neutral curve for the Blasius boundary layer and validate the linear stability theory of Tollmien (1935) and Schlichting (1940).

Figure 6.4 shows the amplitude ratio, A/A_0, as a function of the x-Reynolds number, R, for a range of fixed frequencies. The amplitude is presented on a log scale to account for the exponential growth of the disturbances based on linear theory. Based on the neutral curve that was shown in Figure 6.2, the higher frequencies are amplified first, but their streamwise extent (distance between R_{I} and R_{II}) of growth is shorter, so that they do not reach as high amplitude as the lower frequencies, which have a longer extent of growth. Assuming that the disturbances contain a broad range of frequencies, one would expect that their respective growth in amplitude would form an

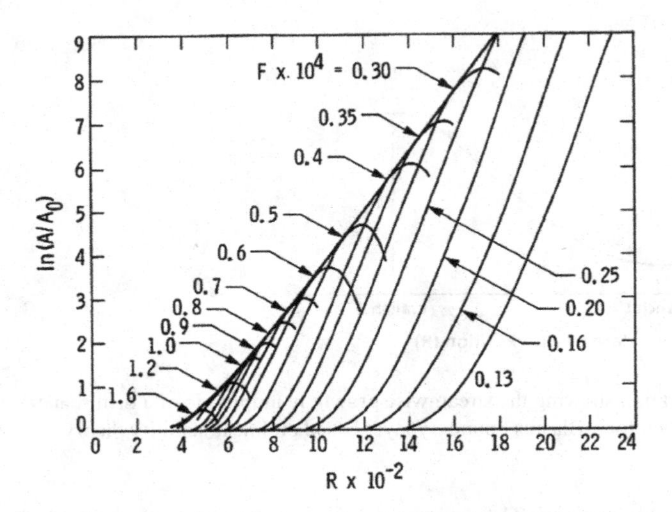

Figure 6.4 Streamwise growth in amplitude of disturbances at a constant frequency for a Blasius boundary layer based on linear stability theory with parallel flow assumption. Taken from Mack (1984).

envelope that gives the maximum amplitude ratio possible at any x-Reynolds number. That envelope is shown in the figure by the thicker curve that follows the peaks of the amplitude growth curves at different frequencies.

Regarding the prediction of transition to turbulence, a conjecture is that it occurs when the disturbance amplitude reaches a *critical* level. With regard to linear stability theory, it is important to note that it can only predict a growth in amplitude relative to an initial amplitude, namely A/A_0. Therefore, the absolute amplitude is not known without knowing A_0.

Based on the spatial growth of disturbances given by Eq. (6.18), the growth in the amplitude between an initial location, x_0, and an arbitrary streamwise location, x, is

$$A(x)/A_0 - A(x_0)/A_0 = \exp\left[\int_{x_0}^{x} -\alpha_i x dx\right], \tag{6.25}$$

where $\alpha_i = f(F, R)$. Changing the independent variable from x to R where we note that $R^2 = U_\infty x/\nu$, one obtains

$$A(R)/A_0 - A(R_0)/A_0 = \exp\left[\int_{R_0}^{R} -2\alpha_i R dR\right]. \tag{6.26}$$

Taking the natural log of both sides, one obtains the *amplification factor*, $N = \ln(A/A_0)$, where

$$N = \ln(A/A_0) = \int_{R_0}^{R} -2\alpha_i R dR \tag{6.27}$$

and R_0 is the Reynolds number where the disturbance at dimensionless frequency, F, first becomes unstable (Branch I of the neutral curve).

If the boundary layer is not self-similar, integration with respect to the physical distance, x, may be more appropriate. Mack (1984) suggests

$$N = \ln(A/A_0) = \int_{x_0}^{x} -(R_{\infty L} U_e \alpha_i / R) dx, \tag{6.28}$$

where $R_{\infty L} = U_\infty L/\nu$, U_e is the local velocity at the edge of the boundary layer, and $R = (U_e x/\nu)^{1/2}$ is the local x-Reynolds number.

Having a definition for N, the task is to determine how that correlates with the Reynolds number at transition, R_{trans}. This has led to the so-called e^N method of Smith and Van Ingen (Arnal, 1984, 1992; Mack, 1984). The e^N method provides a simple turbulence transition prediction tool. The question is the value of N. Typical values range from 7 to 10. The fact that it has a chance of success as a prediction tool is that the large amplitudes generated at the end of linear growth result in a mean flow distortion that often includes an inflectional mean profile that is inviscidly unstable, and rapidly break down into turbulence in a short distance.

However, using a given"N-factor" as a reference, *transition control* seeks to change the conditions in order to keep N below a value that is expected to result in turbulent transition. If the laminar flow is maintained, and the disturbances remain linear, the N-factor can be a reasonable metric of merit for transition control.

An important indicator of T-S instability in an experimental flow field is the wall-normal fluctuation amplitude and phase distribution. In the normal mode formulation expressed, for example, in Eq. (6.18), this is $q(y)$, which in the eigenvalue expression corresponds to the eigenfunction. The first eigenfunctions were obtained by Schlichting (1935). These were found to agree well with measurements of Schubauer and Skramstad (1947), which was important in validating linear stability theory. Examples of the eigenfunction amplitude and phase at three Reynolds numbers are presented in Figure 6.5. The amplitudes shown in Figure 6.5(b) have not been normalized, but it is customary to normalize them by the free-stream velocity, U_∞. The wall-normal coordinate has been normalized by the local boundary layer thickness, δ. The minimum amplitude that occurs in the outer part of the boundary layer corresponds to a 180° phase change that is documented for the three Reynolds numbers in Figure 6.5(b).

For the conditions used to produce Figure 6.5, the critical layer is at approximately $y/\delta = 0.15$. Recall that the location of the critical layer corresponds to the distance from the wall where the mean velocity, $U(y)$, equals the T-S wave velocity such as that given in Figure 6.2(c). The height of the critical layer does not vary significantly with Reynolds number. However, as illustrated in Figure 6.5(a), the location of the amplitude peak is a strong function of the Reynolds number.

Effect of Pressure Gradient

The effect of pressure gradients on boundary-layer stability is most conveniently represented by means of the Falkner–Skan–Hartree family of self-similar boundary layers, where the Hartree parameter, β_h, is the pressure gradient parameter. The values of β_h range from -0.19883774 at which the boundary layer separates to $\beta_h = 0$ which corresponds to a Blasius profile and to $\beta_h = 1$ which corresponds to a stagnation

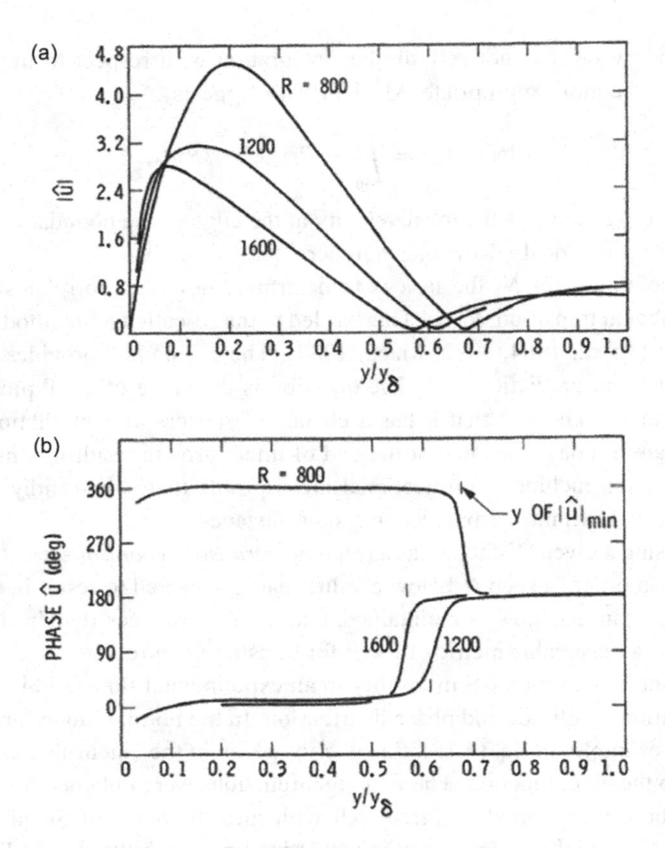

Figure 6.5 Wall-normal profiles of T-S eigenfunction, u' (a), and phase (b) for $F = 0.30 \times 10^{-4}$ at $R = 800$, 1,200, and 1,600. Taken from Mack (1984).

point profile. Extensive numerical calculations for Falkner–Skan profiles have been carried out by Obremski et al. (1969). The effect of the pressure gradient on the N-factor envelope curve is shown in Figure 6.6. The $\beta = 0$ corresponds to the condition that was shown in Figure 6.4. It is clear based on the slope, $d\log(A/A_0)/dR$, that a favorable pressure gradient, $\beta > 0$, stabilizes the boundary layer. Conversely, an adverse pressure gradient, $\beta < 0$, destabilizes the boundary layer. The strong instability for adverse pressure gradients is caused by an inflection point in the mean velocity profile that moves away from the wall as β_h becomes more negative. This is further discussed from the point of view of boundary layer separation in Chapter 4.

Effect of Suction and Blowing

It is well-documented that wall suction stabilizes boundary layers and, conversely, wall blowing destabilizes boundary layers. Wall suction is even effective in stabilizing boundary layers exhibiting highly unstable inflectional mean velocity profiles. The underlying stability mechanism in each is similar to the effect of pressure gradients. Wall suction provides "fuller" mean velocity profiles with stronger wall shear that

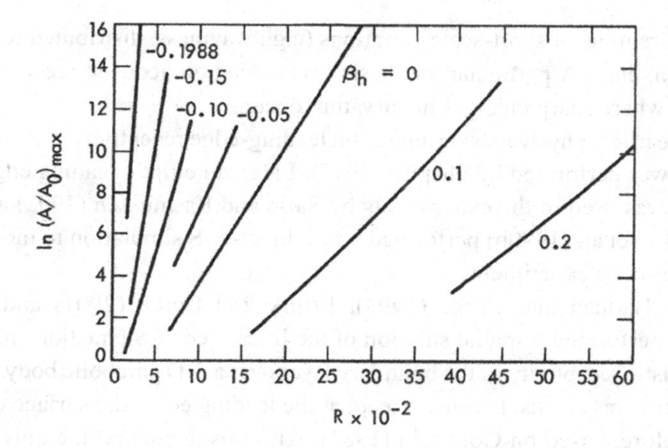

Figure 6.6 N-factor envelope curves for the Falkner–Skan family of boundary layers. Taken from Mack (1984).

is similar to the effect of a favorable pressure gradient. Wall blowing produces an inflectional mean velocity profile that is similar to that produced by an adverse pressure gradient. A history of the suction-type laminar flow control with emphasis on flight research is presented by Braslow (1999). Further examples of this are presented in Section 6.2 on laminar flow control.

Effect of Acoustic Disturbances

The boundary-layer transition process generally begins through the receptivity of the laminar boundary layer to external perturbations. The receptivity process generally comes about through nonparallel mean flow effects, which may arise either in the leading-edge region or in a localized region farther downstream in the boundary layer (Goldstein and Hultgren, 1989). Receptivity plays a crucial role in any transition prediction tool that utilizes an amplitude-based (N-factor) approach since it determines the initial amplitude of the instability waves, while the flow instability governs the growth of the excited waves. Therefore, in order to control laminar flows efficiently, one needs to have a better understanding of the receptivity as well as the instability characteristics. In particular, the receptivity coefficient, defined as the ratio of the amplitude of instability wave to that of the external disturbance, is quite important.

Goldstein (1983, 1985) developed one of the earliest asymptotic analysis for leading-edge receptivity on a semi-infinite zero-thickness plate. This work defined two general classes of receptivity regions: (1) the leading-edge regions where the boundary layer is thin and growing rapidly and (2) regions further downstream where the boundary layer is forced to make a rapid adjustment. Class (2) can be further subdivided into localized and non-localized receptivity. Localized receptivity results from an interaction between the free-stream disturbances and steady localized disturbance generated by a surface inhomogeneity (e.g., humps, gaps, or suction/blowing slots). Non-localized receptivity stems from an interaction in boundary-layer flow with unsteady free-stream disturbances and the steady disturbance created over surfaces with extended regions of short-scale variations (e.g., waviness, distributed roughness, uneven suction, etc.). A particular emphasis was placed on acoustic receptivity sites being regions where sharp changes in curvature occur.

One of the earliest physical experiments on leading-edge receptivity was for a sharp flat plate that was performed by Shapiro (1977). Later, an elliptic leading edge joined to a flat plate was used in the experiments by Saric and Rasmussen (1992) and Saric et al. (1994). Lin et al. (1990b) performed a full direct-N-S simulation to model those elliptic leading-edge experiments.

Studies by Haddad and Corke (1998), Erturk and Corke (2001) and Haddad et al. (2005b) performed a spatial solution of the linearized N-S equations in order to study the acoustic receptivity of the boundary layer over a 2-D parabolic body. The parabolic geometry was chosen because except at the leading edge, the surface curvature is zero. Therefore, based on Goldstein (1985), with this geometry, the only acoustic receptivity site is the leading edge. Later, Wanderley and Corke (2001) performed simulations to study the boundary-layer receptivity to free-stream sound on elliptic leading edges of flat plates that related to the experiments of Saric and Rasmussen (1992) and Saric et al. (1994).

In the most general case, Haddad et al. (2005b) examined the effects of leading-edge radius and streamwise pressure gradient on the acoustic receptivity. The pressure gradient was controlled by imposing an angle of attack with respect to the free-stream on the parabolic body. Figure 6.7 shows a general schematic diagram of the parabolic body at an angle of attack, α, in a uniform stream, and acoustic disturbance at an incident angle, α_s, with respect to body centerline.

In general, the receptivity coefficient at any streamwise location is defined as the ratio of the maximum T-S amplitude to that of the free-stream disturbance, namely $K_x \equiv |u_{\text{TS}}|/|u_\infty|$. An example showing the effect of leading-edge nose radius and angle of attack on the leading-edge receptivity coefficient, K_{LE}, is shown in Figure 6.8.

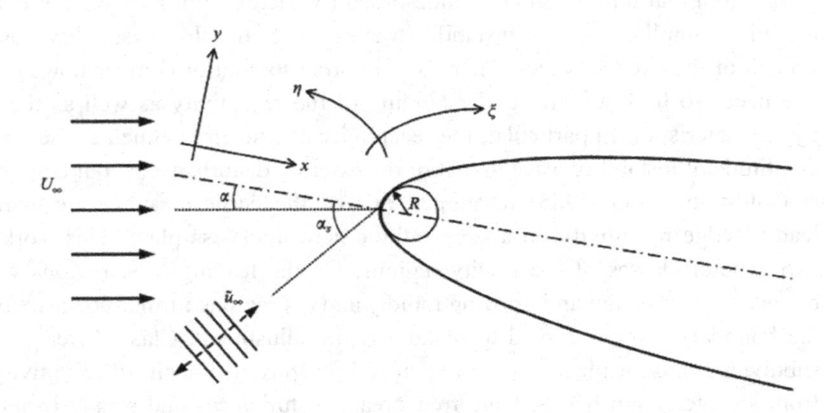

Figure 6.7 General schematic diagram of parabolic body at an angle of attack in a uniform stream, and acoustic disturbance at an incident angle with respect to body centerline. Taken from Haddad et al. (2005b).

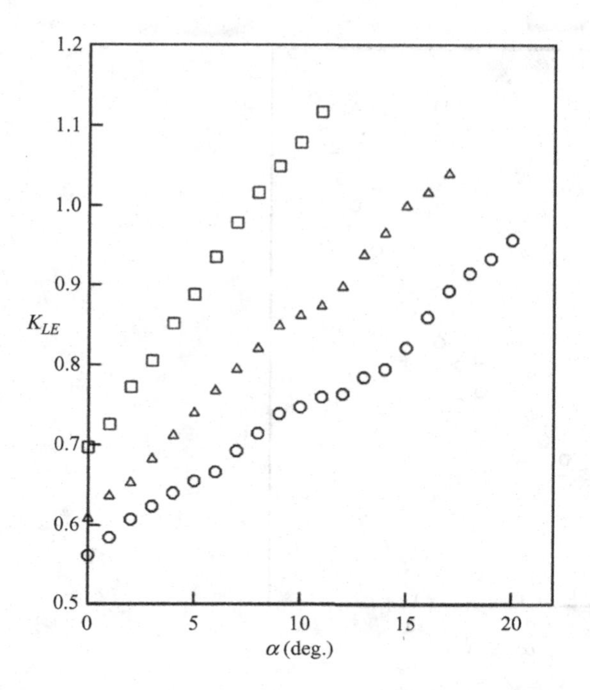

Figure 6.8 Leading-edge receptivity coefficients for three leading-edge radii, Re = 10 (squares), 50 (triangles), and 100 (circles) as a function of the parabolic body angle of attack prior to flow separation. Taken from Haddad et al. (2005b).

For this, the nose radii are equivalent to the nose radius Reynolds number, Re $= R_{\mathrm{LE}} = U_\infty/\nu$. In the results in this figure, the ranges of angles of attack were below which the boundary layer would separate at the leading edge. The effect of flow separation is addressed in a subsequent figure.

There are two important conclusions to be drawn from Figure 6.8. The first is that the leading-edge receptivity increased with *decreasing* nose radius. Thus, adding leading-edge bluntness can reduce the impact of acoustic disturbances on boundary-layer transition. The second conclusion that can be drawn is that the leading-edge receptivity increases with increasing adverse pressure gradient, which, in this case, was imposed by increasing the body angle of attack. The effect of adverse pressure gradient is magnified with decreasing nose radius. Thus, leading-edge shapes that provide a favorable pressure gradient will also reduce the impact of acoustic disturbances on boundary-layer transition.

Because of the parabolic geometry used in these simulations, the only site of acoustic receptivity is at the leading edge. However, as discussed in Chapter 4, "Separated Flows," a boundary-layer separation bubble will provide a receptivity site that is similar to the leading edge. For the parabolic body at larger angles of attack, a separation bubble would form slightly downstream of the leading edge. As a result, the receptivity of the boundary layer due to the separation bubble receptivity focuses on the maximum T-S wave amplitude that occurs downstream, specifically at the Branch II location.

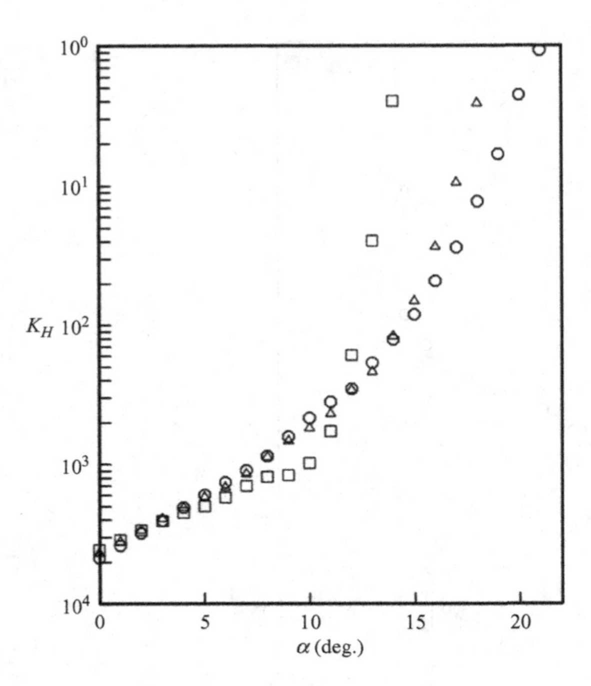

Figure 6.9 Branch II receptivity coefficients as a function of the parabolic body angle of incidence for three leading-edge radii, Re = 10 (squares), 50 (triangles), and 100 (circles). Taken from Haddad et al. (2005b).

For this, we define a Branch II receptivity coefficient, $K_{\text{II}} \equiv |u_{\text{TS}}|_{\text{II}}/|u_\infty|$. From a practical point, the Branch II receptivity coefficient is the most convenient quantity to relate to experiments because the larger amplitudes make it easier and more accurate to measure. K_{II} can be related to the leading-edge receptivity by determining the equivalent Branch I amplitude based on linear-theory growth, and extending from that point to the leading edge through an appropriate theory or N-S calculations for the leading-edge flow.

The K_{II} receptivity coefficients as a function of the body angle of attack for the three nose radii (Reynolds numbers) are shown in Figure 6.9. Considering first the largest nose radius, Re $= 100$, one observes a smooth increase in K_{II} with increasing angle of attack. The angle of attack where the flow separated near the leading edge depended on the nose radius, with the angle of attack at separation decreasing with decreasing nose radius (Haddad et al., 2005). With the largest nose radius, Re $= 100$, the boundary layer first separated at $\alpha = 21°$, which is the last angle in the dataset.

The values for K_{II} for the Re $= 50$ leading edge lie on top of those of the Re $= 100$ leading edge up to $\alpha = 15°$. Those for the Re $= 50$ leading edge lie on top of those for the other two nose radii up until $\alpha = 6°$. The significance of $\alpha = 6°$ and $15°$ is that those are the angles of attack at which a separation bubble forms near the leading edge for the Re $= 10$ and 50 leading edges. The result of the formation of a separation bubble is to dramatically increase the amplitude of the T-S waves as measured at Branch II. In the case of the smallest leading-edge radius, this increases the T-S amplitude at Branch II by almost two orders of magnitude compared to the largest leading-edge radius. In a real application, those Branch II amplitude levels would have certainly resulted in turbulent transition.

6.1.2 Compressible Stability Analysis

The state-of-the-art of compressible boundary-layer instability remains to be that of Mack (1984). The linear stability analysis described therein highlights three major differences compared to the previously described instability of incompressible boundary layers.

In a compressible flow, the basic state profile of a boundary layer on an adiabatic wall always has an inflection point somewhere. The extension of the Rayleigh inflection point criterion to compressible boundary layers as a sufficient condition for the existence of an instability is

$$D(\rho DU)) = 0, \quad \text{at} \quad y = y_s > y_0, \tag{6.29}$$

where y_s is the "generalized inflection point," y_0 is the point where $U = 1 - 1/M_e$, and M_e is the edge Mach number. With increasing Mach number, y_s moves away from the wall, providing more weight to the effect of the inviscid instability on transition. At approximately Mach 5, the inviscid instability becomes dominant over the viscous instability, which consists of oblique T-S waves. As a result, for stability analysis in compressible boundary layers, inviscid theory is singularly more useful.

Following Mack (1984), the inviscid equations can be written in a simplified form by introducing the Mach number

$$\overline{M} = (\alpha U + \beta W - \omega) M_e / (k^2 T)^{1/2}, \tag{6.30}$$

where $k^2 = (\alpha^2 + \beta^2)$. For a temporal neutral wave, $\overline{M}$ is real and equal to the local Mach number of the mean flow in the direction of the wave number vector k relative to the phase velocity, ω_r/k. In all other cases, $\overline{M}$ is complex.

With some manipulation(Mack, 1984), the inviscid form of the normal-mode disturbance equations reduces to a single second-order equation for v

$$D^2\Psi + D\{\ln[\overline{M}^2/(1 - \overline{M}^2)]\}D\Psi - k^2(1 - \overline{M}^2)\Psi = 0, \tag{6.31}$$

where

$$\Psi \equiv v/(\alpha U + \beta W - \omega). \tag{6.32}$$

For insight into the solution of Eq. (6.31), Mack (1984) considered the case when α^2 was large, in which case the first derivative term can be neglected such that

$$D\Psi - k^2(1 - \overline{M}^2)\Psi = 0. \tag{6.33}$$

When $\overline{M}^2 < 1$, the solutions to Eq. (6.33) are elliptic. More importantly, when $\overline{M}^2 > 1$, Eq. (6.33) becomes a wave equation that supports an infinite sequence of discrete wave numbers, $\alpha = f(\beta, \omega, R, \overline{M})$, that satisfy the boundary conditions. The

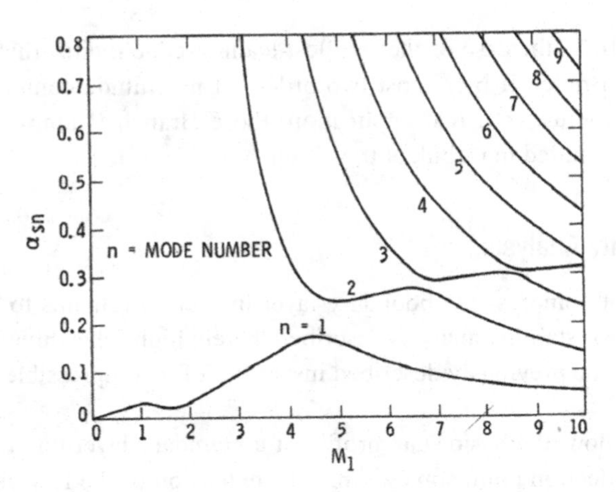

Figure 6.10 Multiple wave numbers of 2-D inflectional neutral curves $(c - c_s)$ for an insulated wall. Taken from Mack (1984).

"sonic point" in the basic state profile where $\overline{M} = 1$ is defined as $y = y_a$, which is also referred to as the "turning point." For $y < y_a$, the solutions are neutrally growing oscillations. For $y > y_a$, the oscillatory modes grow exponentially. The wave numbers of the first mode (Lees and Reshotko, 1962) occur at $M_e = 2.2$ $(y_a = 0)$.

The sequences of discrete higher modes have wave number values of $2\alpha_{sn}/\pi = 1, 3, 5, 7, \ldots$, where the s-subscript denotes a neutral subsonic solution, and the n-subscript refers to the multiple mode numbers. Physically, the disturbances travel at a wave speed, c_r (normalized with U_e), that is subsonic relative to the edge velocity, U_e, but supersonic relative to the region near the wall $(y < y_a)$. In contrast to the most amplified T-S (first mode) waves at compressible Mach numbers that are oblique, the higher modes are 2-D waves. They are not T-S waves by character or behavior. They represent sound waves that reflect inviscidly between the solid wall and the relative sonic line in the boundary layer. Because of their distinctive nature and the important contribution to their understanding that came from Mack (1984), the higher modes are referred to as "*Mack modes*."

The range of wave numbers, α_{sn}, for Mack modes 1–9 are shown in Figure 6.10. With increasing Mach number, M_e, the relative sonic point, y_a, moves out into the boundary layer, and α_{sn} varies in inverse proportion to y_a. No higher modes were found for $M_e < 2.2$ (Mack, 1984).

Figure 6.11 presents the maximum temporal amplification rates and the corresponding wave frequencies for 2-D waves of the first four Mack modes. This indicates that below $M_e = 2.2$, the boundary layer is stable to inviscid 2-D waves, and above $M_e = 2.2$, the second mode has the largest temporal amplification rate. The second mode amplification rates can be significant. At $M_e = 5$, the amplitude growth of the second mode over that of the first mode is approximately double what is possible in a Blasius boundary layer at the Reynolds number of the maximum amplification rate, and about 25 percent of the maximum growth in a Falkner–Skan boundary layer at separation (Mack, 1984).

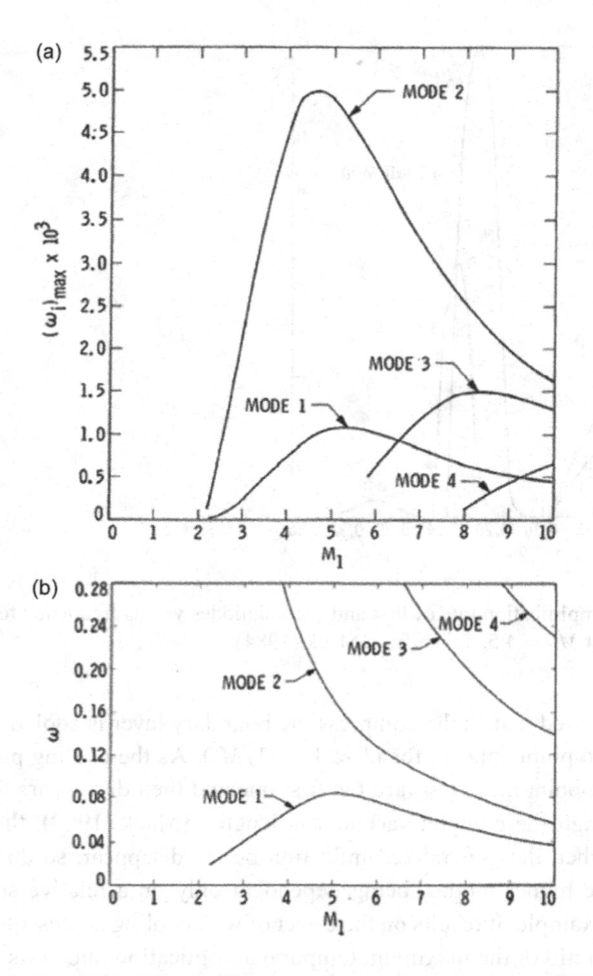

Figure 6.11 Effect of Mach number on the temporal amplification rate (a) and frequency (b) of 2-D waves of the first four Mack modes. Taken from Mack (1984).

Figure 6.12 presents temporal amplification rates of the first and second modes as a function of wave angles at $M_e = 4.5$. This illustrates that 3-D first-mode waves are more unstable than the corresponding 2-D waves. The most amplified first-mode wave occurs at a wave angle of approximately 60°. The amplification rate of the 60° wave is approximately twice that of the 2-D wave, making the former most likely to be observed in supersonic boundary layers. In contrast, 3-D second-mode waves are less amplified than the corresponding 2-D second-mode waves. This is the case for all of the higher modes.

With regard to the second mode, there is a strong tuning with the boundary-layer thickness, so that the frequency of the most amplified disturbance may be predicted from this flow parameter. In particular, the fluctuation wavelength is approximately twice the boundary-layer thickness. This implies that if the boundary-layer thickness is changed, for example, by cooling, a corresponding, predictable change in frequency should be observed.

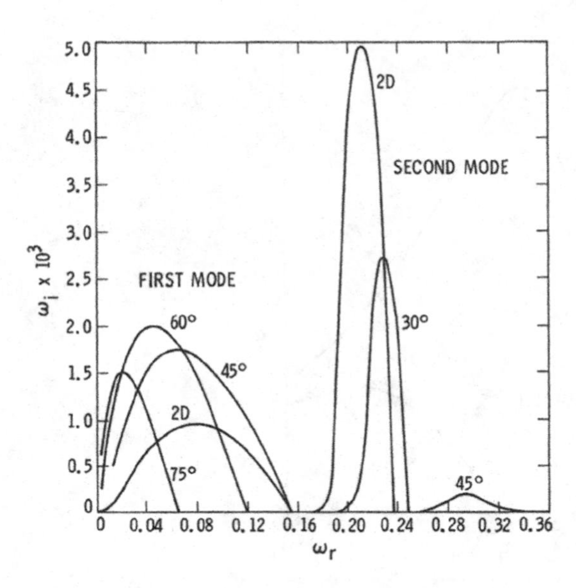

Figure 6.12 Temporal amplification rate of first and second modes versus frequency for different wave angles at $M_e = 4.5$. Taken from Mack (1984).

Mack (1984) observed that as the compressible boundary layer is cooled, a second generalized inflection point appears for $U < 1 - (1/M_e)$. As the cooling progresses, the second inflection point moves toward the first one and then disappears for highly cooled walls. Although the complete account is lengthy (Mack, 1969), the general conclusion is that when the generalized inflection points disappear, so do the first-mode waves, but the higher modes, being dependent only on a relative supersonic region, remain. An example of results on the effect of wall cooling is presented in Figure 6.13. There, the ratio of the maximum temporal amplification rate to its uncooled value is plotted versus the ratio of the wall temperature, T_w, the recovery temperature, T_r, at $M_e = 3.0$, 4.5, and 5.8 for the 3-D first mode, and at $M_e = 5.8$ for the 2-D second mode. In each case, the wave angle of the respective modes is the one that is most unstable. Figure 6.13 illustrates that the first mode can be completely stabilized by wall cooling. However, with wall cooling, the second mode is destabilized. In that instance, if the amplification rate is normalized by the boundary layer thickness, the increase in ω_i is just about compensated by the reduction in y_0 so that $\omega_i y_\delta$ is virtually unchanged by the cooling (Mack, 1984).

The previous figures on compressible boundary-layer stability have focused on temporal amplification. From the point of view of the maximum amplification rate, Figure 6.14 shows the differences between temporal and spatial theories. As a first point, the spatial theory and the temporal theories along with the Gaster transformation give almost identical values of the amplitude ratio and therefore either are relevant to transition prediction (Mack, 1984). Figure 6.14 documents the effect of Mach number on the maximum amplification rate of the first two Mack modes. For the first Mack mode, it also indicates the most amplified wave angle. As previously

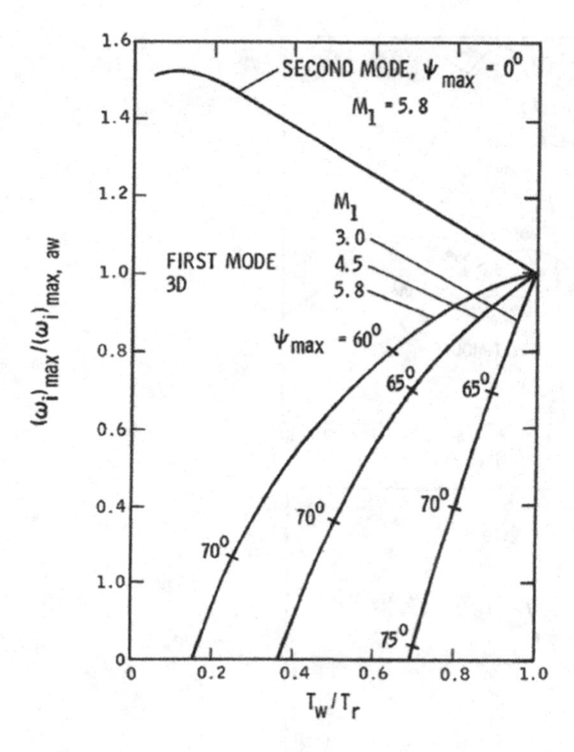

Figure 6.13 Temporal amplification rate of first and second modes versus frequency for different wave angles at $M_e = 4.5$. Taken from Mack (1984).

pointed out, the most amplified second mode is 2-D. For the first mode, for $M_e > 1$, the most amplified wave is oblique. At incompressible Mach numbers, as noted earlier, the most amplified T-S mode is 2-D. Thus, as Mach number increases, the wave angle is observed to transition to larger angles, reaching a maximum at about Mach 3 and then decreasing slowly at higher Mach numbers to approximately $\psi = 50°$ at Mach 7. Finally, the most important aspect of the results in Figure 6.14 is the weakening of the viscous instability (first mode) followed by the increasing dominance of the inviscid instability, peaking at Mach 6. Ultimately, the thickening of the boundary layer at higher Mach numbers causes a proportional reduction in the amplification rate of the second mode (Mack, 1984).

To summarize the effect of compressibility on boundary-layer instability:

1. it is formulated on generalized inflection point criterion that gives rise to inviscid disturbances;
2. the primary (mode 1) viscous instability consists of oblique waves;
3. the existence of higher inviscid modes (second, third, ...) that consist of 2-D waves can be more unstable than the primary instability at $M_e > 4.5$;
4. wall cooling suppresses the growth of the primary viscous instability but enhances the growth of the higher inviscid instability.

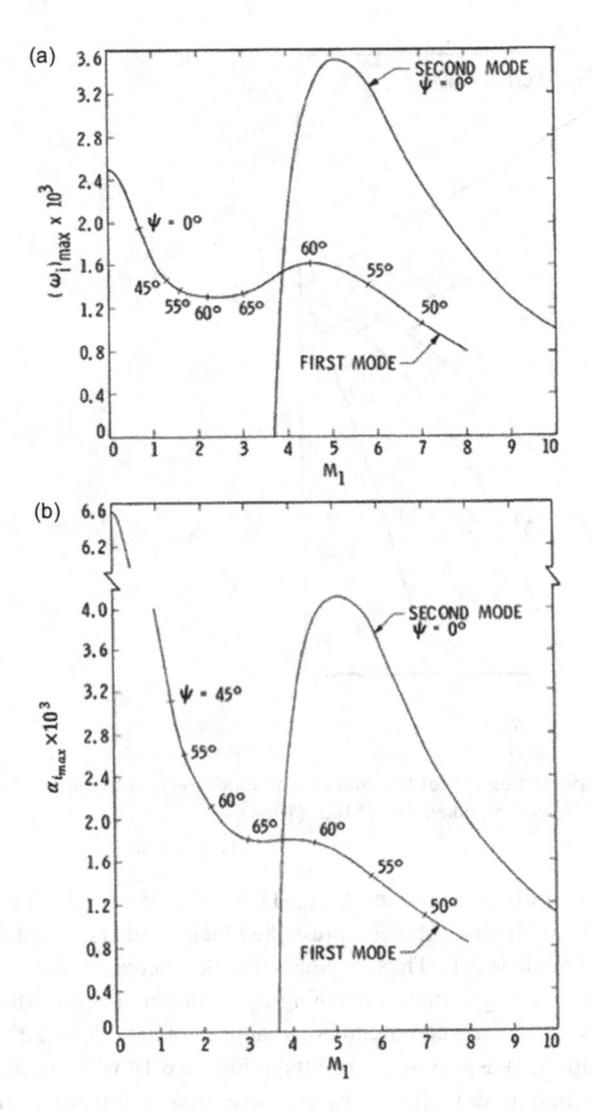

Figure 6.14 Effect of Mach number on the maximum temporal amplification rate (a) and maximum spatial amplification rate (b) for first and second Mach modes with an insulated wall. Taken from Mack (1984).

6.1.3 Secondary Instabilities and Transition to Turbulence

Experiments on 2-D incompressible boundary-layer transition performed in low-disturbance wind tunnels have established three basic regimes of breakdown from laminar to turbulent flow. The first regime consists of plane T-S waves that are invariant in the spanwise direction and propagate with the flow. As previously discussed, frequencies and growth rates of these waves are readily predicted from linear stability theory. In the second regime, spanwise-periodic 3-D deformations of the nominally

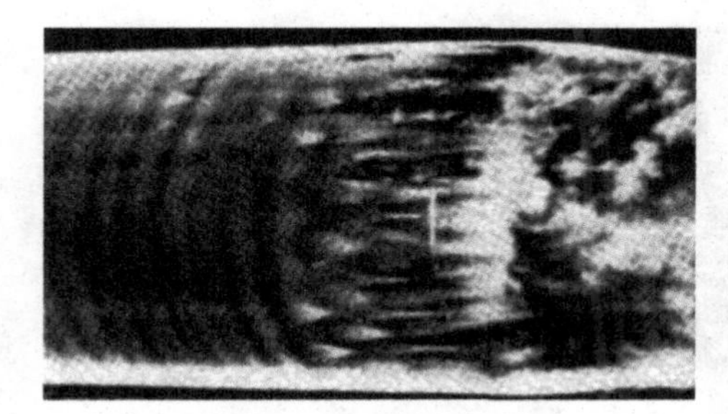

Figure 6.15 Smoke visualization documenting instability stages in boundary layer over an ogive body of revolution leading to the formation of staggered peak–valley structure. $\mathrm{Re}_L = 1.03 \times 10^6$. Taken from Kegelman et al. (1983).

2-D TS waves appear. Finally in the third regime, rapid streamwise stretching and secondary instabilities of the lifted 3-D structures lead to the generation of smaller scale random motions and the final stages of breakdown. A flow visualization image from Kegelman et al. (1983), that illustrates this process is shown in Figure 6.15. This corresponds to a boundary layer on an ogive-shaped body of revolution at a Reynolds number based on the length of the body of 1.03×10^6.

The most commonly cited early experiments that focused on the development of three-dimensionality in unstable boundary layers are due to Klebanoff and Tidstrom (1959) and Klebanoff et al. (1962). In their pioneering investigations, they established the existence of downstream-growing spanwise variations in the mean and fluctuating velocity components and their fundamental role in the consistent features of the developing instability. In addition, they made attempts in one of the first applications of transition control in boundary layers, to locally fix the significant 3-D features of turbulent breakdown by creating artificial disturbances of spanwise periodic nature. Under reasonably large 2-D input intensities (order of 1 percent u/U_∞), they documented the warping of initially plane T-S waves into peak–valley pairs that were aligned in the flow direction. This yielded a particularly catastrophic breakdown characterized by the sudden appearance of "spikes" in the time traces of streamwise velocity fluctuations, and the growth of high-frequency oscillations and turbulent "spots." As a result of the detailed description of this process given by Klebanoff et al. (1962), it has come to be called "K-type" breakdown (Herbert and Morkovin, 1980). Figure 6.16 shows a smoke-wire visualization image that documents K-type structures that evolve from the plane T-S waves.

The initial experimental results of Klebanoff and Tidstrom (1959) and later of Hama (1960) sparked early theoretical attempts to model the progressive growth of three-dimensionality in boundary layers. Benney and Lin (1960) and later Benney (1964), considered the second-order nonlinear interaction of 2-D T-S waves with 3-D waves of a given spanwise periodicity. They found that such interactions promoted the growth of spanwise periodic longitudinal vortical patterns that were qualitatively similar to the K-type structure. Although the Benney–Lin model pointed to a mechanism for the observed growth of three-dimensionality in unstable boundary

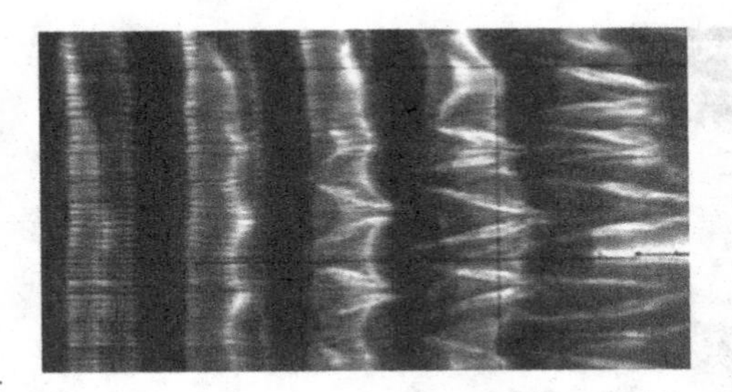

Figure 6.16 Smoke visualization documenting K-type aligned peak–valley structure in boundary layer over a flat plate. Flow direction is from left to right. Taken from Herbert (1988) and attributed to Saric (private communication).

layers, it furnished no estimate of the preferred spanwise periodicity observed in the experiments.

As a result of this and other more serious shortcomings, Craik (1971) proposed a model that would favor the *selective growth* of 3-D disturbances. This involved resonant interactions among a suitable triad of T-S waves. Such interactions were expected to be quite strong owing to a phase-coupled energy interchange among wave components with the potential transfer of energy from the primary shear flow to the disturbance in the region of the critical layer. The importance of resonant subharmonic wave interactions in free shear layers had been pointed out by Kelly (1968), and was found to be an important mechanism governing the initial growth of instabilities and feedback in jets (Corke et al., 1991). Further discussion on this aspect is covered in Chapter 5, "Free Shear Layers and Jets."

The Craik model considered the disturbance field to be the result of a plane T-S wave and two oblique T-S waves propagating at equal and opposite angles to the flow direction in the form of a wave *triad*. Resonance occurs when the phase velocities of the plane and oblique wave components are matched. This generally involves the most amplified 2-D T-S wave with wave number α and pairs of oblique waves with a streamwise wave number $\alpha/2$.

The Craik model provides a mechanism for selective amplification of a pair of oblique waves even in situations where such waves may be damped according to linear theory. Since it involves the subharmonic, it is often referred to as a subharmonic interaction and is often implied when describing Craik or "C-type" subharmonic transition. An important implication of this mechanism is that because it involves *linear* stability modes, their initial amplitudes can be asymptotically small. This has ramifications on transition control methods that seek to phase cancel the plane T-S wave for transition control that will be discussed later in this chapter.

The first experimental observance of a T-S subharmonic in a transitioning boundary layer was made by Kachanov et al. (1977). The subharmonic was determined from spectra of streamwise velocity fluctuations. Further investigations by Kachanov and Levchenko (1984) provided detailed phase and amplitude distributions that confirmed the existence of a synchronized phase locking between a 2-D T-S wave and a pair of oblique waves with equal and opposite angles to the flow. They believed this to be evidence of the Craik modes even though the angles of the subharmonic oblique waves differed substantially from the theoretical value satisfying the Orr–Sommerfeld equation. A possible explanation was later provided by Herbert (1983).

Using a vibrating ribbon to excite plane T-S waves in a 2-D boundary layer, Saric and Thomas (1984) observed that the T-S mode initial amplitude results in peak–valley structures with different spanwise wavelengths. As an explanation, Herbert (1983) examined the stability to 3-D disturbances of a basic state consisting of a Blasius flow and plane T-S waves. The unstable 3-D disturbances were found to be Squire modes, not Orr-Summerfeld modes as in the case of the Craik mechanism. By this mechanism, the phase velocity of the 3-D modes is independent of the spanwise wave number, $\beta = 2\pi/\lambda_z$, and is close to that of the fundamental T-S wave. As a result, near-resonant conditions could exist independent of β, as opposed to the C-type modes that only exist for unique values of spanwise wave numbers. This mechanism is referred to as "H-type" transition.

In order to further explore the different mechanisms for the growth of 3-D disturbances, Corke and Mangano (1989) performed experiments that introduced simultaneous phase-coupled 2-D and 3-D waves in a 2-D boundary layer. The simultaneous 2-D and 3-D wave generation was accomplished using a spanwise array of line heaters suspended just above the wall at the approximate height of the critical layer in the laminar boundary layer. These were operated to produce, through local heating, time-periodic spanwise-phase-varying velocity perturbations. A schematic of the experimental setup is shown in Figure 6.17(a). Figure 6.18 indicates the location of the wave excitation (heating wires array), smoke-wire visualization location, measurement locations, and excited frequencies relative to the T-S spatial neutral growth curve.

An example of the effect of the plane T-S initial amplitude on the resonant growth of the oblique subharmonic mode pairs is shown in Figure 6.19. The abscissa is a measure of the excitation amplitude of the plane T-S mode. The ordinate is the respective streamwise velocity fluctuations of the fundamental plane T-S wave and that of the subharmonic oblique waves. The fluctuation amplitudes were measured at a fixed downstream distance from the disturbance input. The initial amplitude of the subharmonic waves was kept fixed. As the initial amplitude of the plane T-S mode was increased, the response was a linear increase in the amplitude of the associated fundamental frequency velocity fluctuations. The linear response is consistent with linear stability. In contrast, the linear increase in the amplitude of the plane T-S wave resulted in an exponential increase in the amplitude of the subharmonic oblique waves.

The most relevant model for subharmonic resonance in boundary layers comes from Goldstein and Lee (1992). Others include the nonlinear model of Mankbadi (1990), Mankbadi (1991) and Mankbadi et al. (1993) which were derived for a Blasius layer, and a weakly nonlinear (linear-quadratic) model by Zelman and Maslennikova (1993). Each of these has a parametric resonance leading to the growth of a subharmonic 3-D mode. The Goldstein and Lee (1992) and Mankbadi et al. (1993) models also include a quartic dependence of the fundamental mode amplitude on the amplitude of the subharmonic.

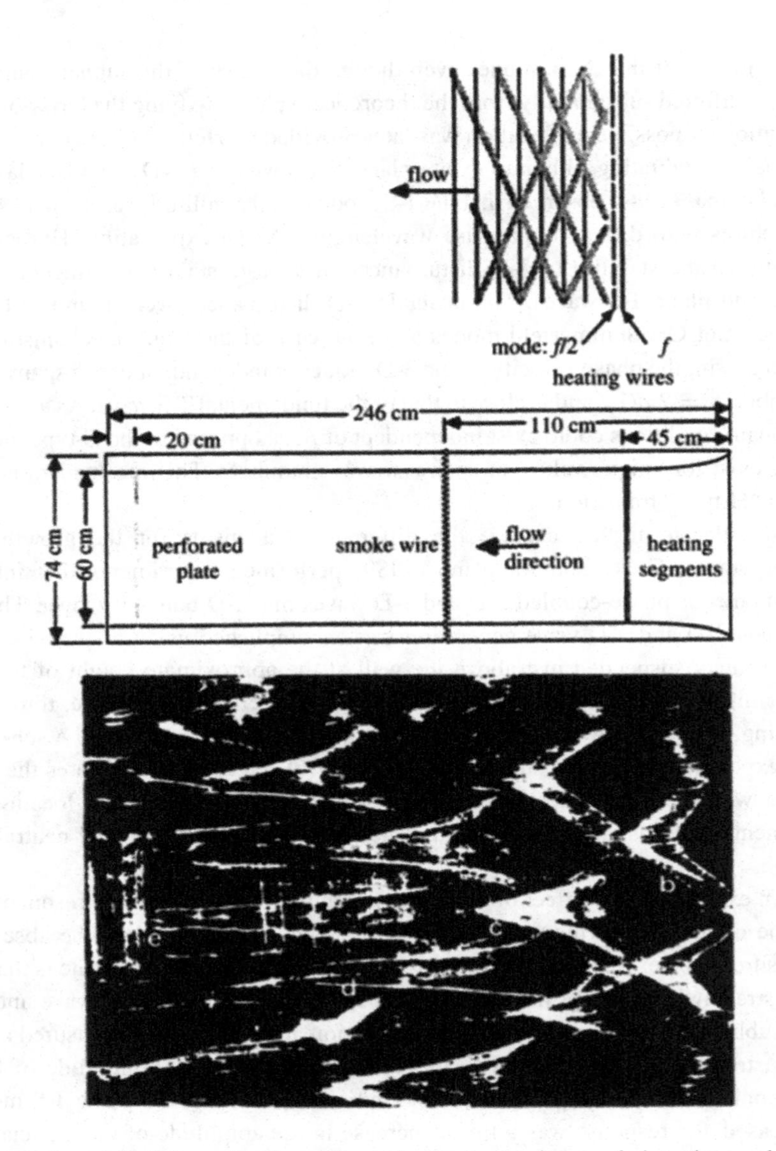

Figure 6.17 Schematic of experimental setup to simultaneously introduce a plane T-S wave and pairs of oblique waves with equal-opposite wave angles at different amplitudes, frequency, and initial phase. Taken from Corke and Mangano (1989).

A convenient starting point is the form of the evolution equations given by Mankbadi (1991). These are in complex form,

$$\frac{dA}{dR} = \sigma^4(1 + R_i^2/R^2)[(K_0 + iK_i)A + iM_1(\mathcal{R}^{4.5}/\lambda^{1.5})ABB^* - iM_2(\mathcal{R}^6/\lambda^2)B^*B^3]$$
$$\frac{dB}{dR} = \sigma^4(1 + R_i^2/R^2)[K_{0b}B + 0.3\pi(\mathcal{R}^3/\lambda)iB^*A - 0.4iM(\mathcal{R}^{4.5}/\lambda^{1.5})B^2B^*], \tag{6.34}$$

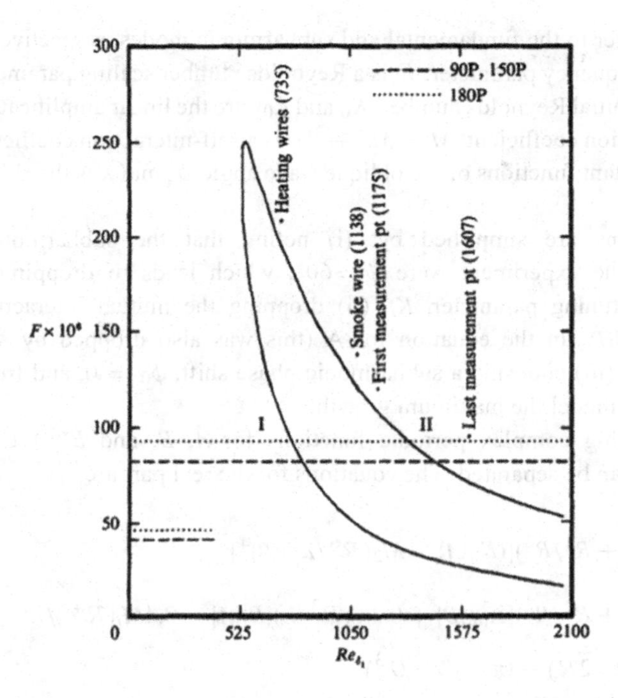

Figure 6.18 Location of the wave excitation (heating wires array), smoke-wire visualization location, measurement locations, and excited frequencies relative to the T-S spatial neutral growth curve used in the experimental setup shown in Figure 6.17. Taken from Corke and Mangano (1989).

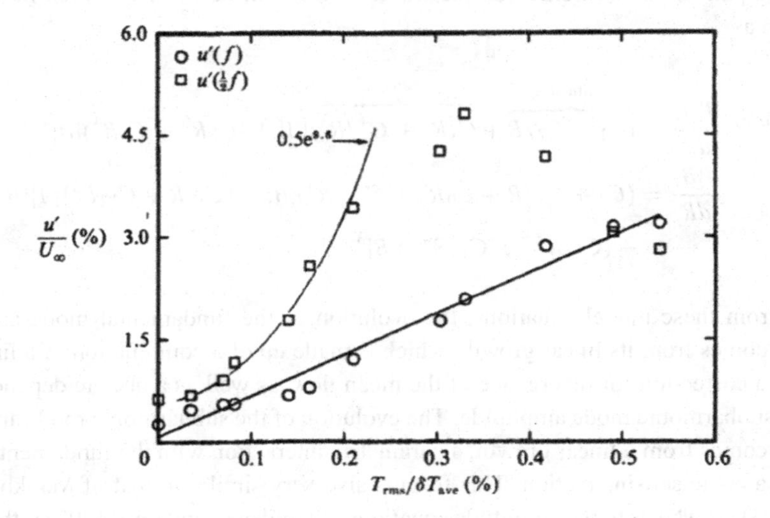

Figure 6.19 Response of the amplitude of oblique T-S wave pairs at frequency $f/2$ at a fixed downstream distance to the initial amplitude of a plane T-S wave at frequency f for conditions of Craik-type triad resonance. Taken from Corke and Mangano (1989).

where A and B refer to the fundamental and subharmonic modes, respectively; σ is a constant small-frequency parameter; $\mathcal{R}$ is a Reynolds number scaling parameter equal to $\sigma^{10}R$; R_i is an initial Reynolds number; K_0 and K_{0b} are the linear amplification rates; M_2 is a back-reaction coefficient; $M = Mr + iM_i$ is a self-interaction coefficient, both of which are constant functions of the oblique wave angle ψ; and λ is the skin friction coefficient.

These equations are simplified by: (i) noting that the subharmonic mode wave angles in the experiment were $\psi = 60°$, which leads to dropping the linear frequency detuning parameter, K_i, (ii) dropping the mutual interaction term, $iM_1(\mathcal{R}^{4.5}/\lambda^{1.5})ABB^*$, in the equation for A (this was also dropped by Mankbadi et al. (1993), and (iii) choosing a subharmonic phase shift, $\phi_0 = 0$, and frequencies $\omega_1 = 2\omega_2$, which model the maximum growth.

After substituting complex periodic functions for A, B, and B^*, the real and imaginary parts can be separated. The equations for the real part are

$$\begin{aligned}
\frac{d|A|}{dR} &= \sigma^4(1 + R_i^2/R^2)[(K_0|A| - M_2(\mathcal{R}^6/\lambda^2)|B|^4] \\
\frac{d|B|}{dR} &= \sigma^4(1 + R_i^2/R^2)[K_{0b}|B| + 0.3\pi(\mathcal{R}^3/\lambda)|B||A| - 0.4M_i(\mathcal{R}^{4.5}/\lambda^{1.5})|B|^3] \\
K_0 &= \lambda^2/(2\sqrt{2}\mathcal{R}) - (\pi/8)(\mathcal{R}^2/\lambda^3) \\
K_{0b} &= [\lambda^2(\cos\psi)^{1/2}/\mathcal{R} - (\pi/8)(\mathcal{R}^2/\lambda^3\cos\psi)]/(\cos\psi + 1/\cos\psi).
\end{aligned} \tag{6.35}$$

These equations represent the amplitude envelope as a function of Reynolds number, $R = \sqrt{U_e x/\nu}$, for the two modes, A and B, representing a fundamental plane wave and a pair of subharmonic oblique waves. These can be further written in a general form as

$$\begin{aligned}
\frac{d|A|}{dR} &= (\overbrace{C_1 + C_2/R}^{\text{linear}} + \overbrace{C_3R^2 + C^4/R^3}^{\text{mean div.}})|A| + (C_5R^6 + C_6R^4)|B|^4 \\
\frac{d|B|}{dR} &= (C_7 + C_8/R + C_9R^2 + C_{10}/R^3)|B| + (C_{11}R + C_{12}R^3)|A||B| \\
&\quad - (C_{13}R^{2.5} + C_{14}R^{4.5})|B|^3.
\end{aligned} \tag{6.36}$$

From these model equations, the evolution of the fundamental mode amplitude, $|A|$, comes from its linear growth, which is made up of a combination of a linear part and a correction for divergence of the mean flow, as well as a quartic dependence on the subharmonic mode amplitude. The evolution of the subharmonic mode amplitude, $|B|$, comes from a linear growth, a parametric interaction with the fundamental mode, and a cubic self-interaction. This form is also very similar to that of Mankbadi et al. (1993), as well as to the amplitude equations of Goldstein and Lee (1992) if the history effects are neglected.

Experiments were performed by Corke and Gruber (1996) to investigate the ability of the coupled amplitude model to predict the streamwise amplitude development of a fundamental plane wave and pairs of subharmonic oblique wave pairs in Falkner–Skan–Hartree boundary layers with $\beta_H = -0.06$ and -0.09. The results are shown in Figure 6.20. In these, the experimental data is shown by the symbols. The solid and long-dashed curves correspond to a fit of a cubic spline to the experimental data. The dotted curves correspond to the coupled amplitude model. Figures 6.20(a) and (b) are the results for the two adverse pressure gradients, $\beta_H = -0.06$ and -0.09, respectively. Figure 6.20(c) corresponds to $\beta_H = -0.09$ where the mean flow divergence term was neglected. This obviously results in a poor representation of the experimental results.

Overall, the model equations of this form worked well and confirmed that they contain the proper physics for the sensitive coupled interaction present in 3-D mode subharmonic resonance. They indicate the importance of the mean flow divergence in the linear growth of the modes, and the quartic dependence of the fundamental mode amplitude on the subharmonic amplitude when the two modes become fully coupled. Understanding these effects is important in devising methods to either *suppress or accelerate* turbulent transition in 2-D boundary layers.

The particularly "catastrophic" breakdown observed by Klebanoff et al. (1962) that consisted of a sudden appearance of large, high-frequency oscillations and turbulent "spots" is likely associated with their use of large levels (1 percent of U_∞) of plane wave forcing and the imposed spanwise mean flow distortion. In contrast, a C-type path to transition can begin with asymptotically small amplitude levels. In that case, the transition process may be different. For example, under conditions designed to promote a Craik-type resonance starting with linear amplitudes that were an order of magnitude lower (0.07 percent of U_∞), Corke and Mangano (1989) observed a gradual spectral broadening at low, intermediate, and higher frequencies in a process of "mode detuning."

In this, the initially exact fundamental subharmonic interaction becomes gradually detuned, leading to the generation of frequency sidebands. These sideband modes derive energy from the primary modes and through sum and difference interactions produce new modes. Each successive generation of modes interacts with the others until the spectrum becomes relatively broadband. A bellwether of this process is the appearance of a low-frequency component corresponding to the lowest difference mode. Such an increase in the energy in velocity fluctuations at low frequencies had been observed in the experiments of Kachanov and Levchenko (1984). Figure 6.21 presents a scenario for reaching broadband velocity fluctuations associated with the transition to turbulence, starting from a subharmonic 3-D mode resonance of the Craik or Herbert types. In this case, the pairs of oblique modes have a frequency and streamwise wave number that are exactly one-half those of the plane TS mode. As a result, the other traveling modes produced by sum and difference interactions are limited to intervals of the lowest difference frequency, $f/2$, namely, f, $3f/2$, $2f$, and $5f/2$.

Figure 6.22 illustrates the requirements for a resonant triad between a plane wave and a pair of oblique waves with equal but opposite wave angles. These are represented in the frequency-spanwise wave number, $[f, \beta]$, plane. For a *tuned resonance* (a), only *one pair* of oblique waves at $f/2$ is necessary to form a triad with a plane wave at

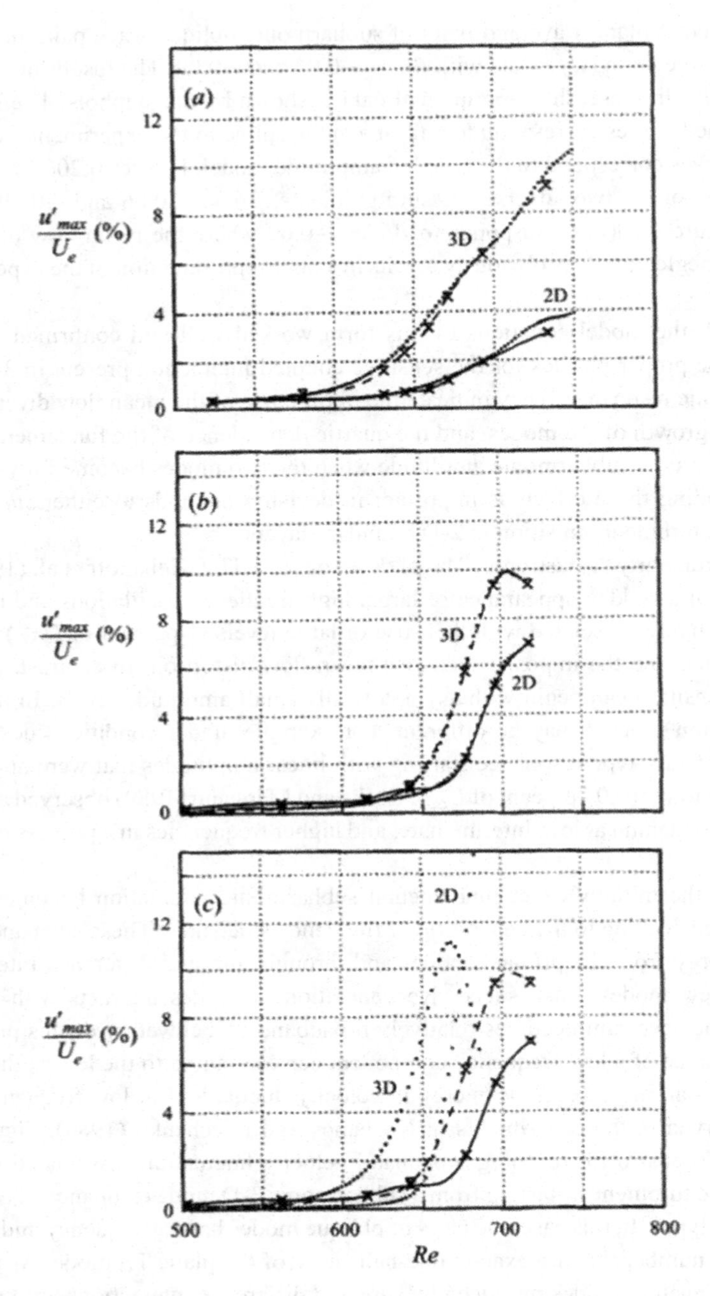

Figure 6.20 Streamwise development of maximum amplitude obtained from amplitude model in Eq. (6.34) and Eq. (6.35). Symbols are original data. Dashed and solid curves are from a fit of a cubic spline to the data. The heavy-dotted curve is from integration of the model equations. (a) $\beta_H = -0.06$. (b) and (c) are $\beta_H = -0.09$ with (b) and without (c) correction for mean flow divergence in linear growth. Taken from Corke and Gruber (1996).

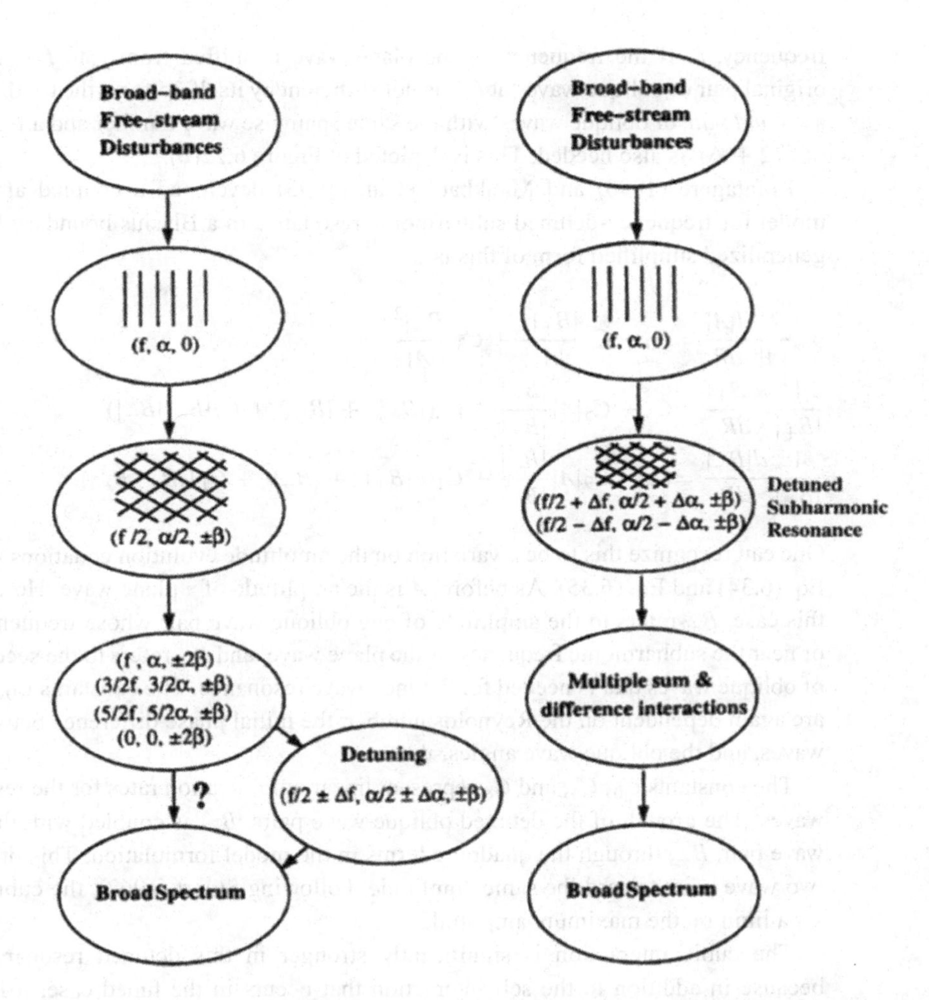

Figure 6.21 Schematic of two paths for the generation of a broadband spectrum of 3-D modes that emanate from low initial amplitude K-type or H-type mechanisms. Taken from Corke (1995).

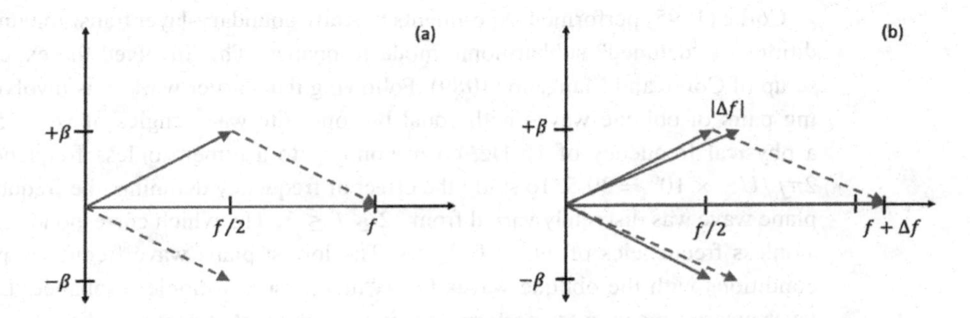

Figure 6.22 Schematic representation illustrating frequency-spanwise wave number requirements for tuned (a) and detuned (b) triad resonance. Taken from Corke (1995).

frequency, f. If the frequency of the plane wave is shifted from f to $f + \Delta f$, the original pair of oblique waves at $f/2$ is not sufficient by itself, to form the triad. Rather, a *second pair* of oblique waves with the same spanwise wave number and a frequency of $f/2 + \Delta f$ is also needed. This is depicted in Figure 6.22(b).

Koratagere (1990) and Mankbadi et al. (1993) developed a coupled amplitude model for frequency-detuned subharmonic resonance in a Blasius boundary layer. A generalized simplified form of this is

$$\begin{aligned}\frac{1}{|A|}\frac{d|A|}{dR} &= C_0 + C_1\frac{|B_+|^2}{|A|} + C_3\frac{|B_-|^2}{|A|}\\ \frac{1}{|B_+|}\frac{d|B_+|}{dR} &= C_4 - C_5|A|\frac{|B_-|}{|B_+|} + C_6(|B_+|^2 + |B_-|^2 + C_7|B_+||B_-|)\\ \frac{1}{|B_-|}\frac{d|B_-|}{dR} &= C_8 - C_9|A|\frac{|B_+|}{|B_-|} + C_10(|B_+|^2 + |B_-|^2 + C_11|B_+||B_-|).\end{aligned} \tag{6.37}$$

One can recognize this to be a variation on the amplitude evolution equations given in Eq. (6.34) and Eq. (6.35). As before, A is the amplitude of a plane wave. However, in this case, B_+ refers to the amplitude of one oblique wave pair whose frequency is at or near the subharmonic frequency of the plane wave, and B_- refers to the second pair of oblique waves that is needed for detuned wave resonance. The constants $C_0, \ldots, C_1$ are again dependent on the Reynolds number, the initial phase difference between the waves, and the oblique wave angles, ψ.

The constants C_0, C_4, and C_8 represent linear amplification rates for the respective waves. The growth of the detuned oblique wave pairs, B_-, is coupled with the tuned wave pair, B_+, through the quadratic terms in the model formulation. This drives the two wave pairs toward the same amplitude. Following Stuart (1962), the cubic terms set a limit on the maximum amplitude.

The cubic interaction is significantly stronger in this detuned resonance case because in addition to the self-interaction that occurs in the tuned case, it involves a cross-interaction between the two pairs of oblique waves. In addition, contrary to the "tuned" subharmonic resonance, the parametric growth of B_+ will depend on its initial amplitude through the coupling with B_-.

Corke (1995) performed experiments to study boundary-layer transition under conditions of "detuned" subharmonic mode resonance. This involved the experimental setup of Corke and Mangano (1989). Following that earlier work, this involved exciting pairs of oblique waves with equal but opposite wave angles of $\psi = 59°$ with a physical frequency of 16 Hz, corresponding to a dimensionless frequency, $F = 2\pi f/U_\infty^2 \times 10^6 = 39.5$. To study the effect of frequency detuning, the frequency of a plane wave was discretely varied from $32 \leq f \leq 36$ Hz, which corresponds to dimensionless frequencies of $79 \leq F \leq 88$. The lowest plane wave frequency produced conditions with the oblique waves for a "tuned" subharmonic resonance. Boundary layer measurements were performed using hot-wires that captured the time-averaged and time-resolved streamwise velocity component. Measurements were taken at the spanwise, z, location corresponding to an intersection of the opposite-angled oblique

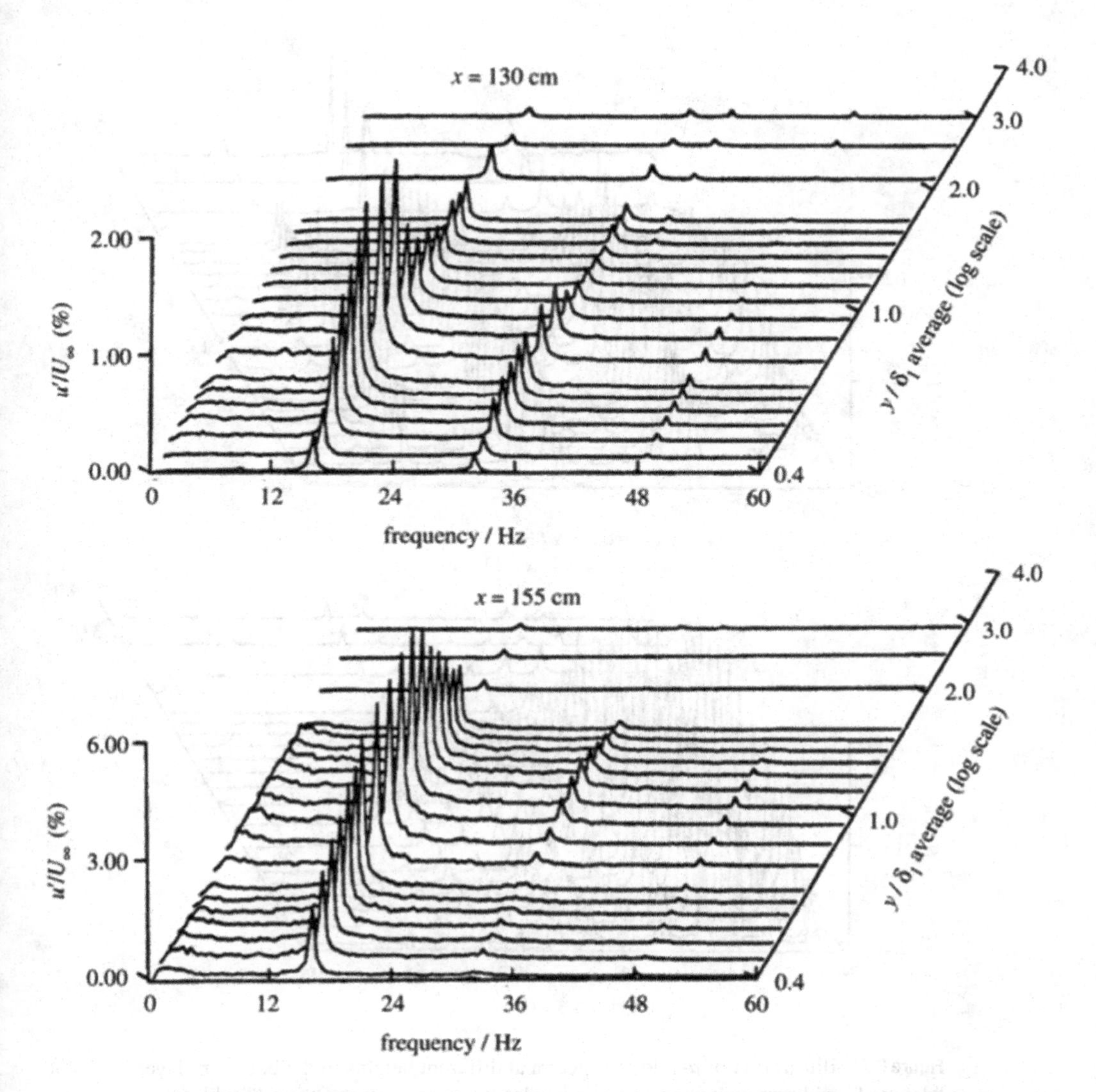

Figure 6.23 Illustration of u-velocity spectra at different heights in the boundary layer for 16:32 "tuned" subharmonic resonance at two x locations. Taken from Corke (1995).

mode pairs, where the amplitude of the initial 3-D mode is a local maximum. Otherwise, measurement points were taken as discrete locations in the wall normal, y, at different streamwise, x, positions.

An example of spectra of streamwise velocity fluctuations for a "tuned" subharmonic resonance condition is shown for two streamwise locations in Figure 6.23. The initial modes are at frequencies of 16 and 32 Hz, which correspond to the oblique wave pairs, $(1/2f_{2D}, \pm\beta_{3D})$, and the plane wave, $(f_{2D}, 0)$. These initial modes eventually result in the development of other interacted modes that appear in the spectra. These correspond to sum and difference interactions, $(3/2f_{2D}, \pm\beta_{3D})$, $(5/2f_{2D}, \pm\beta_{3D})$,

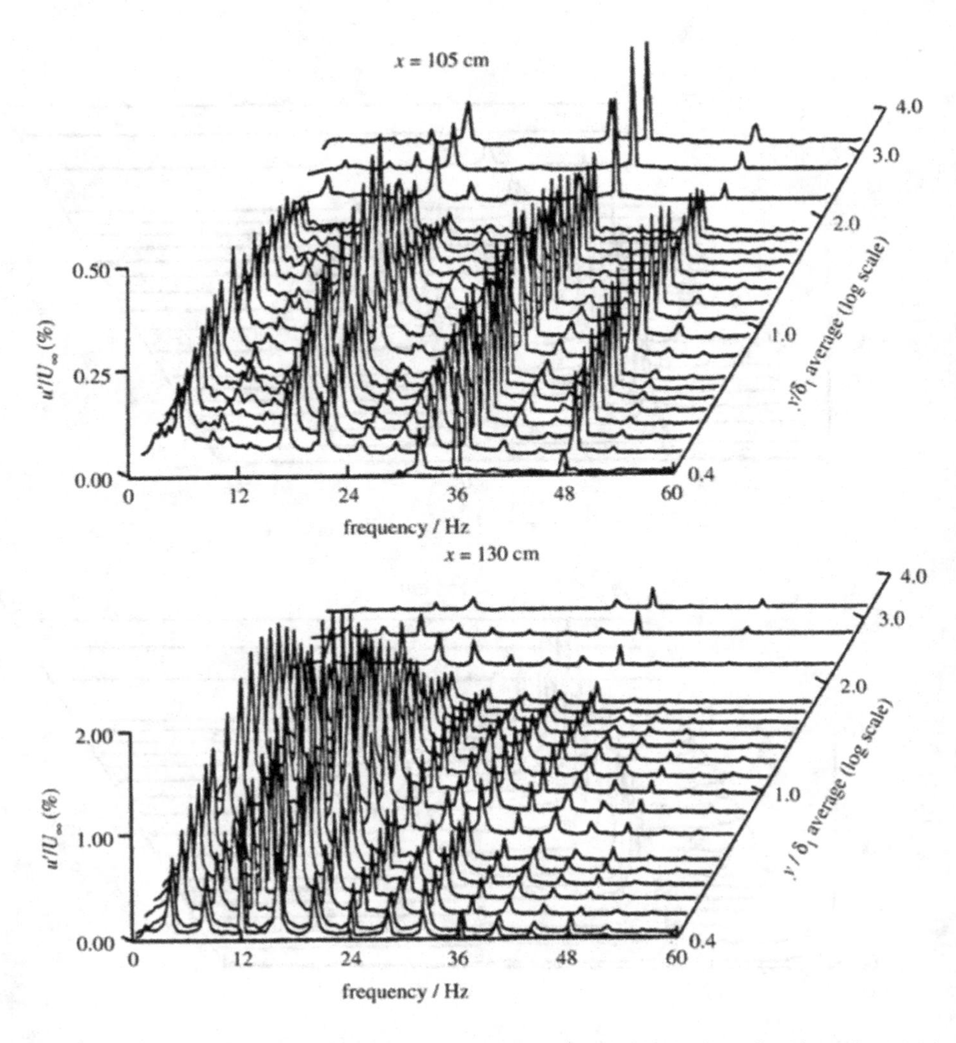

Figure 6.24 Illustration of u-velocity spectra at different heights in the boundary layer for 16:36 "detuned" subharmonic resonance at two x locations contrast the "tuned" subharmonic spectra in Figure 6.23. Taken from Corke (1995).

$(f_{2D}, \pm 2\beta_{3D})$, and $(0, \pm 2\beta_{3D})$, all of which are 3-D modes, with the last of these being a mean flow distortion. The mean flow distortion has been documented to occur when the harmonic mode reaches an amplitude, $u'/U_\infty \simeq 5\%$.

An example of spectra for a "detuned" subharmonic resonance condition is shown in Figure 6.24. In this case, the oblique wave pairs are again at a frequency of 16 Hz, but the plane wave frequency has been shifted by 4 Hz to a frequency of 36 Hz. In this detuned case, additional spectral peaks at 4, 20, 42, and 48 Hz are observed at the upstream location. As previously shown in Figure 6.22, a detuned subharmonic resonance requires a *second* pair of oblique waves. In this example, the second pair

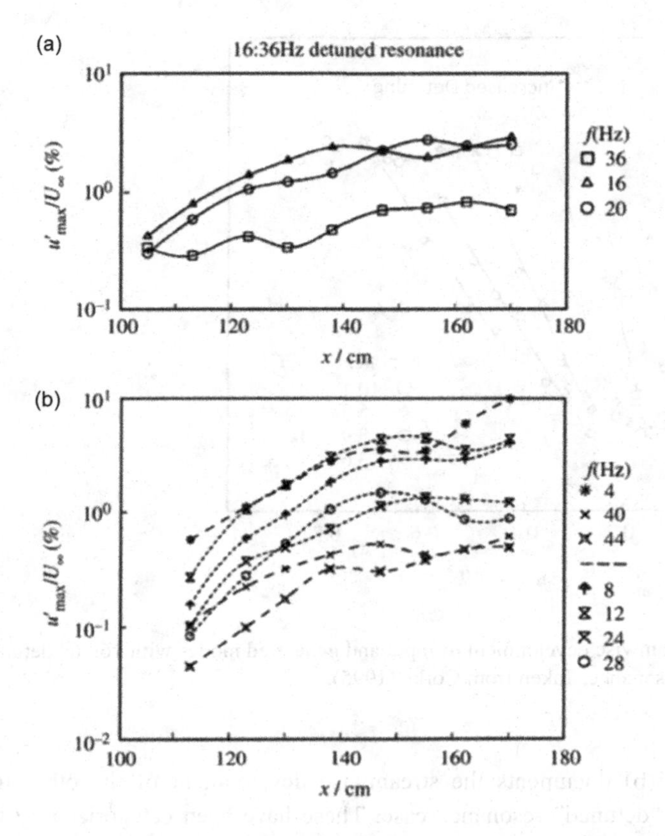

Figure 6.25 Streamwise development of input (a) and (b) generated modes with 16:36 "detuned" subharmonic resonance. Taken from Corke (1995).

of oblique waves occurs at 20 Hz. They are generated by the difference interaction, $(f_{2D}, 0) - (f_{3D}, \pm\beta) = ((f_{2D} - f_{3D}), \pm\beta)$, where $f_{2D} = 36$ Hz and $f_{3D} = 16$ Hz. A difference interaction between the two pairs of oblique modes generates the lowest difference mode at 4 Hz namely $((f_{2D}, 0) - (f_{3D}, \pm\beta)) - (f_{3D}, \pm\beta) = ((f_{2D} - 2f_{3D}), \pm 2\beta)$. The other peaks in the spectrum at 32 and 48 Hz are harmonics of the input (16 Hz) oblique wave pairs.

Based on the coupled amplitude model for the detuned subharmonic resonance, it is expected that the two pairs of oblique modes (16 and 20 Hz) will grow at the same rate. This is documented in Figure 6.25(a). This shows the streamwise development of the maximum amplitude in the plane wave (36 Hz) and two pairs of oblique waves. The primary oblique wave pairs at 16 Hz were directly input so that they had a higher initial amplitude than the 20-Hz waves. However, because of the exchange in energy between the two pairs of oblique waves predicted by the coupled amplitude model, the second pair of oblique waves at 20 Hz eventually matches the amplitude of the 16-Hz waves.

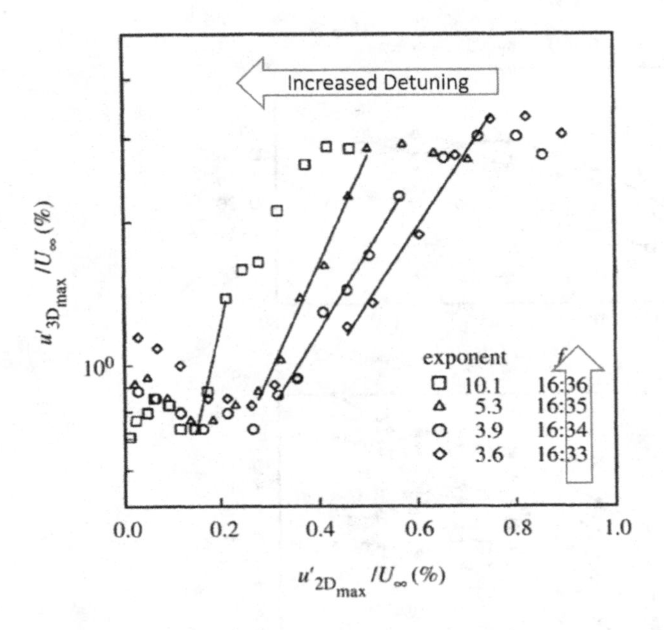

Figure 6.26 Streamwise development of input and generated modes with 16:36 "detuned" subharmonic resonance. Taken from Corke (1995).

Figure 6.25(b) documents the streamwise development of the other interacted modes in the "detuned" resonance case. These have been categorized by the order of the interaction. For example, the first group is made up of those modes that arise from a sum or difference interaction with the triad modes. These are 4 and 40 Hz. The next group comes from interactions that involve the previous group. These are at 12, 24, and 28 Hz. It is expected that the amplification rate will increase with the group order. The results (Corke, 1995) verified this to be the case, which then provides a mechanism for a rapid growth of energy at a broadband of frequencies that can occur through a detuned subharmonic resonance. This offers a *low initial disturbance* path to transition that is possibly more realistic than that first proposed by Klebanoff et al. (1962). This has important implications to transition control, particularly in delaying transition.

Corke (1995) examined a range of detuned subharmonic oblique mode interactions to examine the effect this had on the minimum 2-D mode amplitude needed to promote subharmonic resonance, the growth rate of the detuned subharmonic oblique mode, and the level of amplitude growth saturation that might be indicative of turbulence onset. The results are shown in Figure 6.26. In this, the abscissa is the controlled initial amplitude of the 2-D mode. The ordinate is the amplitude of the detuned 3-D oblique mode measured at a fixed downstream location that was within the linear growth region. The frequency of the 3-D oblique modes was fixed at 16 Hz. The detuned subharmonic interaction was performed by varying the 2-D mode frequency from the tuned frequency (32 Hz) to higher frequencies of 33–36 Hz, producing detuned subharmonics of $\Delta f = 1$–4 Hz. The arrows in the figure indicate the direction of increasing subharmonic frequency detuning. The results clearly show that the increased amount of subharmonic frequency detuning

1. *decreased* the minimum 2-D mode amplitude needed to promote the resonant growth of the detuned subharmonic 3-D modes;
2. *increased* the growth of the detuned subharmonic 3-D mode at that fixed downstream location;
3. *decreased* the saturation amplitude of the detuned subharmonic 3-D mode at that location, suggesting earlier turbulence onset.

These results indicate that except under well-controlled laboratory conditions, the most likely subharmonic transition will be "detuned" from the onset. Thus, the prospects for 2-D boundary-layer transition control based on 2-D mode phase canceling are questionable. Further, approaches of phase canceling 3-D modes will require including a broad range of detuned subharmonic frequencies, putting that approach in question as well.

6.2 Turbulent Transition Control

Interest in boundary layer transition control has spanned more than 90 years, beginning in the 1930s. This can seek to either delay turbulence onset or accelerate it. Passive techniques to delay or accelerate transition generally involve modifying the mean flow in order to respectively inhibit or accelerate the growth of disturbance waves. Active techniques to delay or accelerate transition are ones that seek to interact with and either suppress or amplify the growth of disturbance waves.

6.2.1 Passive Transition Control

Examples of passive approaches that can apply to transition *delay* or *acceleration* include pressure gradient modification, mean surface temperature control, and localized steady surface suction (Saric, 1992; Braslow, 1999). One of the simplest passive methods of controlling boundary-layer transition is through pressure gradient control. For example, as was presented in Figure 6.6, the stability N-factors of T-S waves strongly depend on the streamwise pressure gradient, where an adverse pressure gradient will lower transition N-factors, and a favorable pressure gradient will have the opposite effect. The latter has been exploited to reduce drag on airfoil sections by increasing the nose radius, or more generally the airfoil section shape. An example of the latter approach is the national laminar flow airfoil developed by the NACA in the 1930s and first used on the North American P-51 Mustang aircraft.

Passive Transition Delay

Wall suction or blowing are two other approaches for boundary-layer transition control. These might be considered an active approach since it requires energy to move

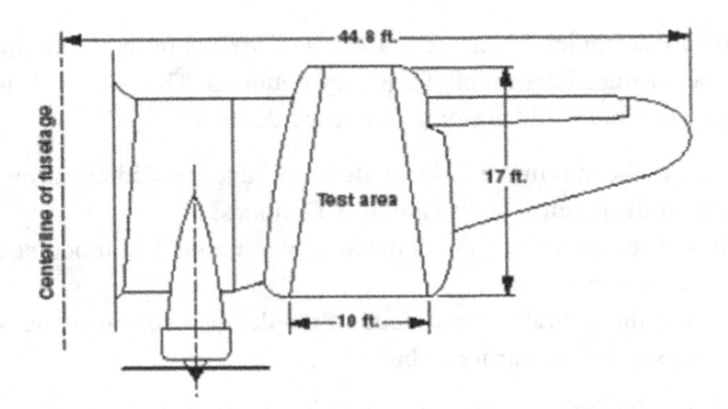

Figure 6.27 Schematic and photograph showing wing glove on B-18 aircraft for the first validation of transition control using wall suction. NASA Photo L-25336 from Braslow (1999).

flow through a porous surface. However, because it modifies the mean flow, in this chapter, it is considered to be a passive approach. As was presented in Figure 6.2, there exists a minimum critical Reynolds number for the growth of instability (T-S) waves. Although Figure 6.2 puts this in terms of Re_x, it can easily be represented through bulk properties of the boundary layer, for example, local displacement and momentum thicknesses, δ^* and θ. Wall *blowing* will act to thicken the boundary layer, locally increasing the momentum thickness Reynolds numbers to be above critical values needed to amplify disturbances. In contrast, wall *suction* will thin the boundary layer, reducing the momentum thickness Reynolds number to below critical values and causing linear disturbances to be damped.

The earliest known experimental work on the use of wall suction for aircraft was done in the late 1930s and the 1940s. The first flight experiments were performed in 1941 on a "glove" that covered a portion of a wing on a B-18 aircraft. A schematic and photograph of the aircraft with laminar flow control (LFC) glove is shown in Figure 6.27. The glove had 17 suction slots located between the 20 percent and 60 percent chord positions of the wing. These flight experiments validated the concept. A number of other flight experiments on various aircraft were also performed. The Boeing Company performed flight research in 1985 on the wing of a 757 aircraft to determine the possible effects of the acoustic environment on the LFC performance. The concern was acoustic receptivity to sound generated by the engines. Other practical concerns that have occurred over the years included clogging of suction holes due to insects, ice, rain, and paint. Although wall suction has been validated to reduce viscous drag during cruise, it has never been applied to a commercial fleet, citing the practical concerns as reasons.

Passive Transition Acceleration

There are a wide range of passive boundary-layer "tripping" techniques used to accelerate the streamwise development to a fully turbulent regime. These include distributed roughness particles, dimpled surfaces, and 2-D protrusions. A critical parameter for any form of roughness is the roughness Reynolds number, $Re_k = u^* k_{crit}/\nu$, where k_{crit} is the maximum roughness height that *does not affect* the turbulent transition location. The velocity used in the roughness Reynolds number is the friction velocity, $u^* = (\tau_{0k}/\rho)^{1/2}$, where τ_{0k} is the wall shear stress at the location of the roughness.

For the case of a 2-D cylindrical roughness element, the minimum Reynolds number to have an effect on turbulent transition (Tani et al., 1940; Fage and Preston, 1941) is $15 \leq Re_k \leq 20$. In the case of sharp protrusions, the minimum Re_k could be lower.

Dryden (1953) developed a relation between the streamwise transition location, x_{tr}, and the height of a roughness element, k, and its streamwise position, x_k. The correlating factor was the transition Reynolds number, $Re_{tr} = U\delta^*_{tr}/\nu$ where δ^*_{tr} is the boundary-layer displacement thickness at the transition location. Dryden (1953) found that

$$\frac{U\delta^*_{tr}}{\nu} = 3.0\frac{k}{\delta^*}\frac{x_k}{k}, \tag{6.38}$$

where x_k is the streamwise location of the roughness element. Kramer (1961a) found that a 2-D cylindrical roughness at an arbitrary position is fully effective in tripping the flow to turbulence if $Uk/\nu \geq 900$.

Even in the case of a fully effective boundary-layer trip, there is a streamwise distance between where the flow becomes fully turbulent, x_{tr}, and the location of the trip, x_k. Based on Kramer (1961a), the dimensionless distance is given by the relation

$$\frac{U(x_{tr} - x_k)}{\nu} = 2 \times 10^4. \tag{6.39}$$

The criteria for turbulent trips consisting of distributed "sand paper" roughness is not as clear as that for 2-D roughness. For example, in the definition of the roughness height, k, it can refer to the mean height of the particles, the peak height of particles, or the root mean square of the roughness height distribution. Assuming a periodic variation (waviness) of the roughness distribution, the three statistics are related. Another approach is to consider each roughness particle to be a 2-D element, each generating an effect that somehow scales like the previous 2-D trips. Perhaps a more physical view is to consider each roughness particle as an unsteady disturbance source with broad frequency content. These disturbances will be amplified according to the stability of the basic flow. For the Falkner–Scan family of boundary layers, this is expected to be 2-D T-S waves. Therefore, if the distributed roughness does not modify the mean flow, its effect is to increase the initial amplitude of disturbances that feed the T-S waves,

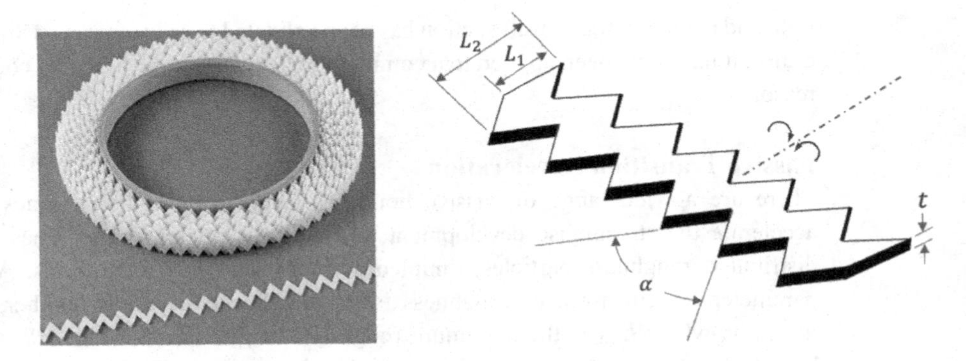

Figure 6.28 Photograph and schematic drawing of a "zig-zag" strip used as a boundary layer turbulent trip.

causing them to reach nonlinear amplitudes in a shorter streamwise distance, x_{tr}. This is largely to depend on the type, height, and distribution of the roughness particles. In seeking some criteria, one can consider an old standard from Feindt (1957) where the roughness has no effect on turbulent transition when

$$\frac{Uk}{\nu} < 120. \tag{6.40}$$

Finally, a method of turbulent transition enhancement that has been a common feature on wings of sail planes to maintain attached flow is a "zig-zag" tape. A photograph and schematic drawing of the tape is shown in Figure 6.28. This is a variation on the classical 2-D trip with the addition of a shape that will promote counter-rotating streamwise vortices that will cause a rapid thickening of the boundary layer and subsequently turbulent transition. The tape is somewhat more convenient to applying discrete vortex generators as turbulent trips, but the impact is likely the same. For reference, details on streamwise vortex generators used for flow separation control are discussed in Chapter 4. For the zig-zag strips, the parameters include its thickness, t, streamwise width, L_2, and zig-zag angle, α. All of these can vary, but most typically $\alpha = 60°$, and t ranges from 1 to 2 mm, although larger thicknesses can be obtained by stacking strips. Given the finite thickness of the strips, they can have a tripping effect that is similar to a 2-D cylindrical roughness. However, the streamwise vortices that emanate from the zig-zag pattern are likely to be the overriding factor on the location of the transition onset.

6.2.2 Active Transition Control

Active transition control requires some form of actuator input power and control. This can be used to modify the basic flow or dynamically intervene with the development of the instability that leads to boundary-layer transition. The following sections present a number of active approaches for transition control of 2-D boundary layers.

Active Transition Delay

As previously presented in Section 6.1, boundary-layer instability waves grow from infinitesimal and generally uncontrollable disturbances. At Mach numbers below Mach 5, the waves are due to a T-S instability. At subsonic Mach numbers, these are plane (2-D) waves. At supersonic Mach numbers, these are oblique waves.

In a low noise environment, laminar–turbulent transition arises from the growth of instability waves. In this instance, in the linear instability regime, active control can be used to suppress the development of the instability waves by dynamically interfering with their growth. The most common approach to delaying turbulent transition of T-S waves has been to actively excite an independent wave that has an amplitude that matches the naturally developing wave but is 180° out of phase, in order to linearly cancel the naturally developing wave. It obviously can only work as long as the amplitude of the natural wave is in the linear range ($u'/U_\infty < 0.01$). One advantage of this technique is that effective control can be achieved with only a small expenditure of energy. A challenge in this approach is that the amplitude and phase difference have to be maintained perfectly. The practical efficacy of this approach to laminar flow control was examined by Thomas (1983).

There have been a number of extensive experimental studies on the control of plane T-S instabilities in laminar boundary layers using this wave superposition principle. Milling (1981) was one of the first to conduct a T-S wave cancelation experiment on a flat-plate boundary layer in a low-turbulence water channel. He used two vibrating wires, one located upstream and the other located some distance downstream. Both wires were designed to vibrate in the direction perpendicular to the wall. This is generally accomplished by placing a magnet in the surface of the boundary-layer plate and applying an AC current through the wire that will cause it to vibrate at the AC frequency. Because the ends of the wire are generally fixed, the spanwise amplitude of the wire motion is a maximum at the midpoint, so the excited T-S wave amplitude is not exactly uniform in span. An example of such a setup is shown in Figure 6.29. The upstream vibrating wire was used to generate a well-controlled upstream disturbance with known amplitude and phase angle. The same approach using a vibrating *ribbon* was used in the experiments by Schubauer and Skramstad (1947) to validate T-S linear stability theory. In Milling (1981), the downstream wire was used to provide a disturbance that could linearly cancel the generated T-S wave at the downstream wire

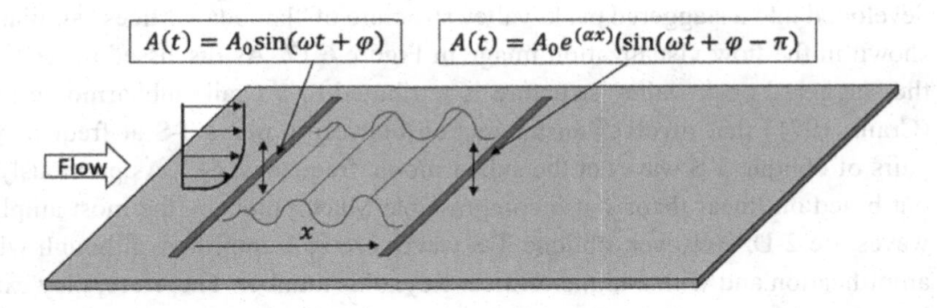

Figure 6.29 Schematic drawing of a typical plane T-S wave phase cancelation setup.

location. This had to be adjusted to have the exact amplitude of the approaching wave, which would have grown in amplitude based on its x-development, and have a π phase shift to the approaching wave to produce a linear phase cancelation. Based on linear theory, the angular phase of the approaching wave would be constant over the streamwise distance between the two vibrating wires so that the downstream wire motion phase angle could simply be π phase shifted from the upstream wire motion. Using this approach, Milling (1981) observed a significant reduction in the T-S wave amplitude past the downstream vibrating wire. It is possibly fair to say that this was more of a validation of linear stability theory than a viable active flow control approach. This latter point will be further examined.

Based on a similar approach, Liepmann et al. (1982) and Liepmann and Nosenchuck (1982) utilized periodically heated surface strips to introduce T-S waves in the boundary layer on a flat plate in a water tunnel. This wave-generation technique is easier in water than in air because of the very good thermal coupling that can be achieved with water and the strong temperature dependency of its viscosity. Because the heating strips were fixed on the surface of the wall, the disturbance amplitude was uniform in the spanwise direction, which is an improvement over the vibrating wires or ribbons. The latter paper (Liepmann and Nosenchuck, 1982) particularly focused on phase cancelation of the upstream excited T-S waves. The results were similar to those of Milling (1981), with a significant reduction of the upstream excited T-S wave amplitude.

The practical limitations of 2-D T-S wave cancelation for transition control were brought out in experiments by Thomas (1983). As in the previously cited experiments, he utilized vibrating ribbons over a flat plate in a wind tunnel, with the upstream ribbon introducing a controlled T-S wave and the downstream ribbon operated to linearly cancel the approaching T-S wave amplitude. He utilized the smoke-wire technique (Corke et al., 1977) to visualize the developing T-S waves. The smoke wire was placed above the wall surface at the height of the T-S critical layer. A similar approach was used in obtaining the flow visualization image shown in Figure 6.17. As in similar experiments (Milling, 1981; Liepmann and Nosenchuck, 1982), a reduction in the amplitude of the T-S wave was observed through the motion of the downstream ribbon. However, the smoke-wire flow visualization revealed that with the reduction in the plane T-S wave, 3-D waves emerged.

The 3-D waves observed by Thomas (1983) in his phase-canceling experiments developed into a staggered peak–valley structure of "lambda vortices" similar to those shown in the flow visualization image in Figure 6.17. As discussed in Section 6.1.3, the staggered peak–valley structure is attributed to a Craik subharmonic resonance (Craik, 1971) that involves an interaction between a plane T-S at frequency, f, and pairs of oblique T-S waves at the subharmonic frequency, $f/2$. As previously pointed out based on linear theory, at incompressible Mach numbers, the most amplified T-S waves are 2-D. However, oblique T-S waves are still amplified, although with lower amplification and with a higher critical Reynolds number. Therefore, they exist in the boundary layer, but their effect is initially masked by the 2-D wave development.

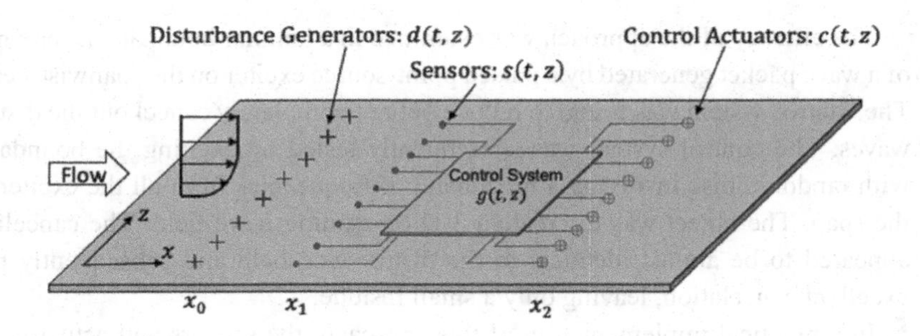

Figure 6.30 Schematic drawing of a 3-D disturbance control setup from Li and Gaster (2006).

The Craik mechanism can occur at asymptotically small amplitudes of the interacting waves. In this simple approach of plane T-S wave cancelation, any amplitude of the plane T-S wave that remains is sufficient to interact with the less amplified oblique T-S waves and lead to enhanced (double exponential) growth of the oblique waves. The observations of Thomas (1983) confirmed this and placed a damper on any practical use of plane-T-S wave cancelation for 2-D boundary-layer transition control.

Acknowledging the limitations of transition control that focuses on only the plane waves, Li and Gaster (2006) investigated the control of 3-D disturbances in incompressible boundary layers. For this, disturbances were artificially introduced into the boundary layer so as to create simulated natural disturbance excitations. This consisted of a spanwise array of buried speakers that were intended to mimic the receptivity process of the naturally excited disturbances that were to be canceled. As in the previously cited use of upstream disturbance generators, this provided deterministic control of the imposed wave field. The difference in the Li and Gaster (2006) experiment was that the disturbances could be 3-D. A schematic of the setup, which is generally applicable to 3-D disturbance generation and control, is shown in Figure 6.30. The disturbances were introduced by a spanwise array of speakers. These operated to produce periodic wall-normal jets. In the case of Corke and Mangano (1989), the spanwise array of disturbance generators was heating wire segments suspended at the height of the critical layer.

The sensors were flush-mounted pressure transducers, although they could be hot-wires or Pitot probes, with the requirement that they have a sufficient frequency bandwidth to resolve the instability frequencies. For Li and Gaster (2006), the spanwise array of control actuators were similar to the disturbance actuators.

In order to account for the 3-D characteristics of the disturbances and growing instability waves, Li and Gaster (2006) constructed a multi-input multi-output control system. For this, the sensor signal, $s(t,z)$, at each spanwise position was convolved in both space and time with the transfer function, $g(t,z)$, of the control system to provide the control signal, $d(t,z)$, driving the downstream actuator array. The canceling signals, $c(t,z)$, at the downstream actuator array was obtained by convolving the driving signal, $d(t,z)$, with the full impulse response of the boundary layer.

The efficacy of the approach was examined in a laminar boundary-layer simulation of a wave packet generated by a pulsed point-source exciter on the spanwise centerline. The control system was found to reduce but not completely cancel out the disturbance waves. The control system was subsequently tested by exciting the boundary layer with random noise involving a broadband of frequencies from all the exciters across the span. The object was to create a 3-D on-coming wave field. The canceling field appeared to be almost identical to the disturbance field and subsequently provided excellent cancelation, leaving only a small residue.

In a practical implementation of this approach, the sensors and actuators need to be strategically positioned across the span. In particular, the sensors need to be able to resolve the 3-D characteristics of any growing disturbances. Following the Nyquist criterion, their spanwise spacing needs to be smaller than half the minimum spanwise wavelength of the original disturbances. The control actuator array needs to be similarly spaced to be able to act on the smallest wavelength of the 3-D disturbances.

Li and Gaster (2006) found that the full transfer function, $g(t, z)$, required spanwise cross-linking of all the sensors and actuators, so that each actuator would need to use the information from the neighboring sensors as well as the one directly upstream. The cross-linking hugely increased the complexity of the control and therefore proved to be impractical.

In an approach to minimize the computational overhead of a feedback boundary-layer transition control system, Fan et al. (1995) introduced the use of a neural network controller. The concept was initially investigated through boundary-layer simulations in which the generated instability waves provided training for the neural network. The approach was classic "feed-forward" control in which the neural network took input from an upstream sensor and based on its training, generated a response signal to a downstream control actuator to cancel the disturbance without any feedback as to the response of the control. A schematic of the feed-forward arrangement is shown in Figure 6.31. In the simulations, the feed-forward control was successful in canceling the incoming waves, with a residual of only a few percent of the original wave amplitude.

The approach was subsequently investigated in wind tunnel tests in the boundary layer over a flat plate (Fan, 1995). This used a single disturbance generator, sensors, and control actuator that were all located along the spanwise centerline of the boundary-layer plate. The sensors were microphones located flush with the surface of the plate. The sensors and control actuator were all located downstream of the disturbance generator. One sensor was located at a streamwise location between the disturbance generator and the control actuator. The other sensor was located downstream of the control actuator. In the feed-forward control arrangement, the downstream sensor simply documented the response of the control. The control actuator consisted of a wall element that deflected in the wall-normal direction. The disturbance generator was operated to produce a continuous train of wave packets, similar to the evaluation approach of Li and Gaster (2006).

In an approach similar to the simulation training (Fan et al., 1995), the upstream sensor was used to train the neural network controller. Following training, the controller was able to cancel the incoming T-S wave packets. However, the wind tunnel tests

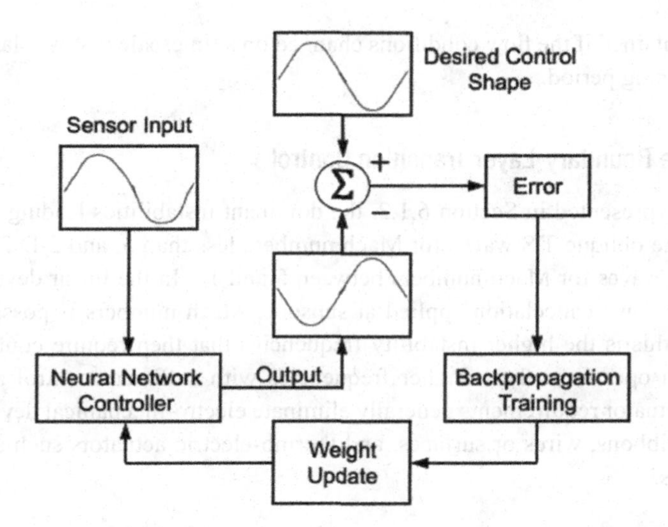

Figure 6.31 Schematic showing example of feed-forward neural network control arrangement used for T-S wave cancelation.

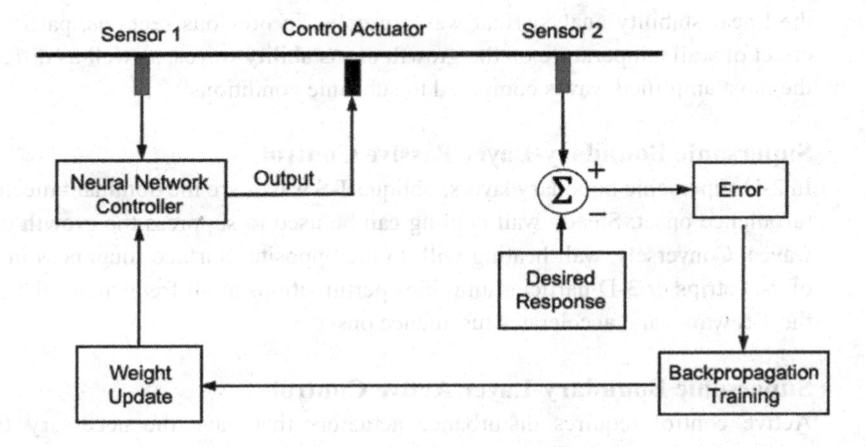

Figure 6.32 Schematic showing example feedback neural network control arrangement used for T-S wave cancelation.

indicated that backpropagation (feedback control) was needed to overcome noise and reduce the necessary amount of training of the neural network. In this case, the sensor located downstream of the control actuator was used to assess the degree of wave cancelation. A schematic of this feedback control arrangement is shown in Figure 6.32. In this case, the training data consisted of real-time measurements of action/response pairs. This approach also enabled the continuous training by updating the controller with the most recent history of the action/response pairs. With this approach, Fan (1995) observed that following an initial training period of about 100 T-S cycles, almost complete T-S wave cancelation occurred. Furthermore, the low residual level could be maintained if the flow conditions changed on a timescale that was larger than the initial training period.

6.2.3 Compressible Boundary-Layer Transition Control

As previously presented in Section 6.1.2, the dominant instabilities leading to turbulence onset are oblique T-S waves for Mach numbers less than 5, and 2-D 2nd mode second-mode waves for Mach numbers between 5 and 10. In the linear development region, active wave cancelation applied at subsonic Mach numbers is possible. The challenge in this is the higher instability frequencies that then require control actuators that can operate at these higher frequencies with sufficient control authority. These flow actuator requirements generally eliminate electro-mechanical devices such as vibrating ribbons, wires or surfaces, and thermo-electric actuators such as heated strips or wires.

Supersonic Boundary Layers

This section introduces passive and active flow control methods for 2-D boundary-layer instability control at supersonic Mach numbers. The approaches are based on the linear stability analysis that was presented in previous sections, particularly the effect of wall temperature on the growth of instability waves, as well as differences in the most amplified waves compared to subsonic conditions.

Supersonic Boundary-Layer Passive Control

In 2-D supersonic boundary layers, oblique T-S waves are the dominant mechanism of turbulence onset. Steady wall cooling can be used to suppress the growth of the T-S waves. Conversely, wall heating will do the opposite. Surface roughness in the form of 2-D strips or 3-D particles amplifies perturbations at all frequencies that feed into the T-S waves and accelerates turbulence onset.

Supersonic Boundary-Layer Active Control

Active control requires disturbance actuators that have the necessary frequency response to match the T-S instability frequencies. One of the earliest viable actuators that met the criteria for high-speed instability control was developed by Kosinov et al. (1990). A schematic of the actuator design is shown in Figure 6.33. It consisted of an electrode that was located in a cavity below the surface of a metal boundary-layer plate. The electrode was powered by an AC voltage source. The reference for the voltage source was earth ground. The metal boundary-layer plate was also connected to earth ground. A sufficiently large AC voltage caused a plasma discharge to form between the electrode tip and the inside surface of the cavity. This produced an unsteady pressure in the cavity that communicated to the flow-side surface of the plate through a small hole.

Kosinov et al. (1990) used this actuator to excite T-S waves over a flat plate in air at Mach 2. Only a single actuator was used, which meant that the disturbance would excite the most amplified band of oblique waves. The excitation frequencies

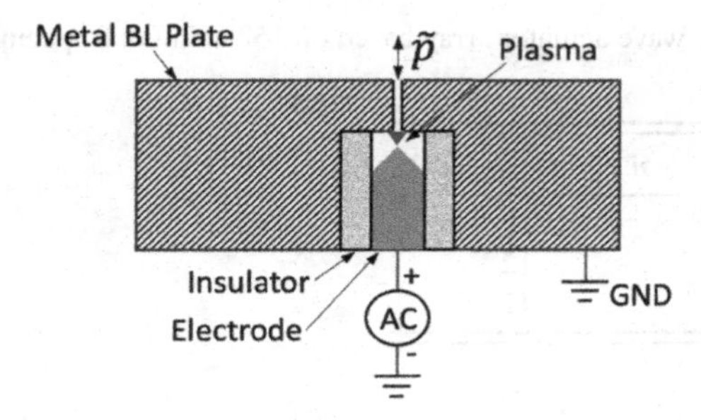

Figure 6.33 Schematic drawing of disturbance actuator used in supersonic boundary-layer experiments of Kosinov et al. (1990).

were 10, 20, and 30 kHz. The amplitude of the disturbances was controlled by the AC voltage amplitude. Their main objective was to verify linear theory predictions for supersonic flows. They found that for $(\mathrm{Re}_x)^{1/2} \leq 1,000$, agreement with linear theory occurred with excitation amplitudes less than 15 times the background (free-stream) disturbance levels. The measurements consisted of ensemble-averaged velocity measurements that were conditioned on the phase of the input AC disturbances. The measurements were performed with a hot-wire anemometer. Measurements taken in a spanwise array documented the growth of pairs of oblique waves with equal but opposite wave angles that were excited by the point disturbance. The wave angles were found to be between 41° and 70°, which was within the band of most amplified wave angles of oblique T-S waves predicted by linear theory. Although aimed at validating linear stability theory, the experiment provided a means for exciting instability waves in boundary layers at compressible Mach numbers.

The disturbance excitation at a point by Kosinov et al. (1990) produced a full spectrum of T-S waves with different wave angles. Ultimately, the most amplified wave emerged and would lead to turbulence onset. As an extension of the Kosinov et al. (1990) disturbance actuation concept, Corke et al. (2002) utilized an array of plasma disturbance generators to excite specific wave angles of oblique T-S wave. This was applied to the boundary layer over a 7° half-angle right circular cone at zero angle of attack at Mach 3.5. A schematic drawing of the azimuthal disturbance actuator array is shown in Figure 6.34. A photograph of the cone model indicating the location of the disturbance actuator array is shown in Figure 6.35.

The design of the actuator array was based on linear theory predictions for the boundary layer over the cone at Mach 3.5. These indicated a most amplified range of dimensionless frequencies $10 \leq F \times 10^6 \leq 30$, with a wave angle of $\psi = 60°$. An important aspect of the actuator array design was the spacing (azimuthal in this case) of the actuator elements in order to produce the necessary oblique wave angles. The choice in this case was an azimuthal spacing of 15°. Based on the most amplified range of T-S instability frequencies, Table 6.1 lists the corresponding wave angles that could be produced by the azimuthal array. Note that the maximum angle is based on a maximum 2π phase shift between neighbor actuators.

Table 6.1 Conditions for oblique wave actuator array based on 15° azimuthal spacing between individual disturbance actuators.

$F \times 10^6$	f (kHz)	ψ (°)	$n = 2\pi r_{\text{actuator}}/\lambda_\theta$
11.9	50	66	12
14.2	60	62	12
16.7	70	58	12

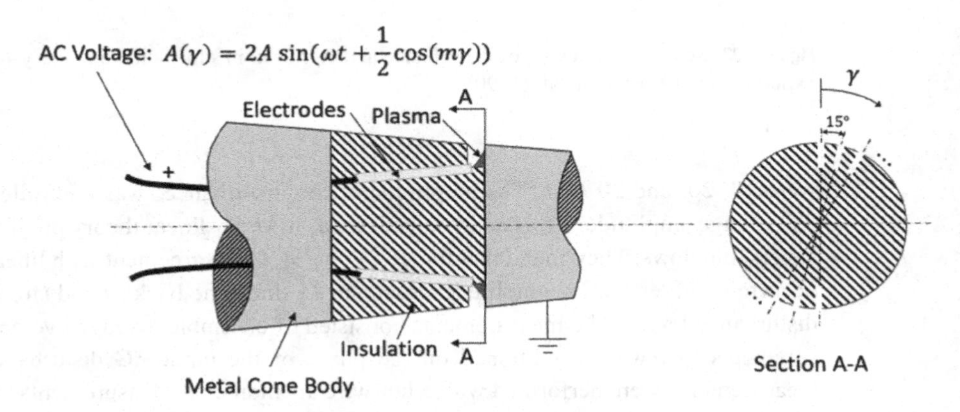

Figure 6.34 Schematic drawing of azimuthal disturbance actuator used in supersonic boundary-layer experiments of Corke et al. (2002).

Figure 6.35 Photograph of 7° half-angle right circular cone showing the location of the azimuthal disturbance actuator. Taken from Corke et al. (2002).

Following Kosinov et al. (1990), the actuator design was based on a glow discharge that was capable of producing disturbances at the necessary frequencies. The actuator was made up of an array of individual anodes that were individually supplied with an AC voltage source of controllable amplitude. The frequency of the AC source corresponded to the instability excitation frequency.

Linear stability does not favor the sign (+ or −) of the oblique wave angle. Each is equally amplified and it is conceivable that both will exist simultaneously. Evidence of this follows observations in subsonic boundary layers where the subharmonic resonance involves pairs of oblique waves at equal and opposite wave angles producing the staggered delta vortex patterns that were shown in Figure 6.17. Therefore, the actuator array was operated to produce oblique wave pairs at equal but opposite angles, $\pm\psi$.

To excite an oblique instability wave with a positive wave angle and frequency, $\omega = 2\pi f$, the periodic voltage time series supplied to a disturbance actuator located at an azimuthal angle, γ, would be

$$A(\gamma) = \mathcal{A}\sin(n\gamma + \omega t + \phi_1), \tag{6.41}$$

where ϕ_1 is an arbitrary phase shift, and $\mathcal{A}$ is the maximum amplitude to the actuator array. Because of the 2π circular phase limit, the largest azimuthal phase difference between neighbor disturbance actuators is 2π. Similarly, the voltage time series supplied to a disturbance actuator at an azimuthal angle, γ, that is needed to produce an oblique wave with a negative wave angle is

$$A(\gamma) = \mathcal{A}\sin(-n\gamma + \omega t + \phi_2). \tag{6.42}$$

To simultaneously produce oblique wave pairs with equal and opposite wave angles, the two periodic functions are added to obtain an azimuthal amplitude distribution given as

$$A(\gamma) = 2\mathcal{A}\sin\left(\omega t + \frac{1}{2}(\phi_1 + \phi_2)\cos(n\gamma + \frac{1}{2}(\phi_1 - \phi_2)\right). \tag{6.43}$$

Assuming an arbitrary phase difference, $\phi_1 = \phi_2 = 0$, pairs of oblique waves with equal and opposite wave angles at frequency ω can be produced by supplying a disturbance actuator located at an azimuthal angle, γ, with an amplitude $A(\gamma) = 2\mathcal{A}\cos(n\gamma)$. This represents a stationary cosine amplitude pattern around the actuator array with a peak amplitude of $2\mathcal{A}$, which, based on Kosinov et al. (1990), needs to be less than approximately 12 times the background disturbances to satisfy linear theory.

In the case of Corke et al. (2002), the actuator array was located on the cone near the location of Branch I of the linear stability neutral growth curve. If placed upstream of Branch I, the introduced disturbances would be exponentially damped. If it was located too far downstream of Branch I, the introduced disturbances would be competing with already growing disturbances, and therefore require larger forcing amplitudes that possibly would exceed linear stability requirements.

The actuator design and experiment of Corke et al. (2002) verified the ability to excite pairs of oblique T-S waves of different angles and frequencies in accordance with linear theory predictions. In principle, this actuator array design could be applied to T-S wave cancelation control following methods applied to subsonic boundary layers that were described in Section 6.2.2. However, this is still to be demonstrated.

Hypersonic Boundary Layers

As discussed in Section 6.1.2, at hypersonic Mach numbers ($M_\infty > 5$), turbulence onset of 2-D boundary layers is dominated by the second mode. In contrast to the most amplified T-S (first mode) waves at compressible Mach numbers that are oblique, the higher modes (second, third, etc.) are 2-D waves. They represent sound waves that reflect inviscidly between the solid wall and the relative sonic line in the boundary layer.

Hypersonic Boundary-Layer Passive Control

Most of the hypersonic 2-D boundary-layer turbulence onset control has focused on passive approaches. This is undoubtedly the result of the limitations of active disturbance actuators in generating the much higher instability frequencies (order of 500 kHz) of the second-mode waves. Examples of passive approaches that *suppress* the growth of the second-mode instability include leading-edge bluntness, steady wall heating, acoustic absorbing wall liners, and strategically placed surface roughness or wall waviness.

Fedorov and Malmuth (2001) were the first to propose that the transition in hypersonic flow can be delayed by suppressing the second-mode disturbances using a suitably designed passive ultrasonically absorptive porous coating. This passive, ultrasonically absorptive coating consists of a thin porous layer that is flush-mounted onto the wall solid surface. Utilizing an analytical model of flow within thin blind pores with approximately 10–20 pores per second-mode wavelength, and an approximately 25 percent porosity, Fedorov and Malmuth (2001) predicted a factor-of-two reduction in the second-mode growth rate. The first experimental verification of this passive transition control concept was performed by Rasheed et al. (2002).

The acoustic absorbing theory presumes that the incident waves are plane monochromatic ultrasonic acoustic waves with wavelengths that are much larger than the size of a single pore or groove. Figure 6.36 provides a general illustration of a notional porous surface.

From a design aspect, the acoustic admittance is given as

$$A = \frac{n}{Z_0} \tanh md, \tag{6.44}$$

where the pores are assumed to be equally spaced blind cylindrical pores with a depth, d, a radius, r, and a porosity, n. The characteristic impedance, Z_0, and the propagation constant, m, are

$$Z_0 = -\frac{\sqrt{\hat{\rho}/\hat{C}}}{M\sqrt{T_w}} \tag{6.45}$$

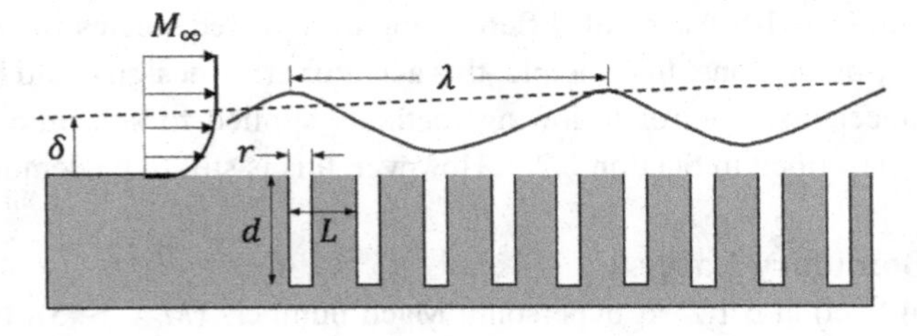

Figure 6.36 Representation of equally spaced blind cylindrical porous surface used for second-mode suppression.

and

$$m = \frac{i\omega M\sqrt{\hat{\rho}\hat{C}}}{\sqrt{T_w}}, \tag{6.46}$$

where $\hat{\rho}$ is the dimensionless complex dynamic density and $\hat{C}$ is the dynamic compressibility, both given as

$$\hat{\rho} = \frac{1}{1 - F_1(\Gamma)} \tag{6.47}$$

and

$$\hat{C} = 1 + (\gamma - 1)F_2(\hat{\Gamma}). \tag{6.48}$$

The function F and the macroscopic parameter Γ depend on the pore shape, namely

$$F(\Gamma) = \frac{2J_1(\Gamma)}{\Gamma J_0(\Gamma)}, \tag{6.49}$$

where J_0 and J_1 are Bessel functions of the arguments Γ and $\hat{\Gamma}$, where

$$\Gamma = r\sqrt{\frac{i\omega\rho_0}{\mu}} \tag{6.50}$$

and

$$\hat{\Gamma} = \Gamma\sqrt{\mathrm{Pr}}, \tag{6.51}$$

where ω is the dimensionless angular frequency and Pr is the Prandtl number. M is taken to be the boundary-layer edge Mach number.

A demonstration of the effect of a porous surface on the suppression of second-mode growth was performed in experiments by Wartemann et al. (2011) and Wartemann et al. (2012). This involved a 7° half-angle right-circular cone with exchangeable nose tips with radii of 2.5 and 5 mm. One-third of the surface of the cone was covered with a fiber-reinforced ceramic (carbon–carbon) material that has a natural random porosity. For this material, the median hydraulic pore diameter was 13 μm. The porosity was approximately 15 percent of the median core diameter. The thickness of the carbon–carbon ceramic coating was 5 mm. It was inset in the cone wall so that it was flushed with the surrounding wall surface. A schematic drawing of the cone model is shown in Figure 6.37. The cone was placed in a Mach 7.5 wind tunnel. The experiments documented a 70 percent reduction in the second-mode amplitude and a complete suppression of the turbulence onset over the porous surface. For the same conditions, the boundary layer over the smooth surface was fully turbulent at the downstream end of the cone.

A recent numerical study by Duan et al. (2013) indicated that 2-D roughness elements placed downstream of the instability "synchronization point" in a hypersonic boundary layer would cause the second-mode instability to be damped. To understand this, we refer to Federov (1997) who defined the regions of first- and second-mode development in a hypersonic flat plate boundary layer. The so-called first-mode was distinguished by a phase speed $c_r \rightarrow 1 + 1/M_e$ at $x \rightarrow 0$. This has also been referred to

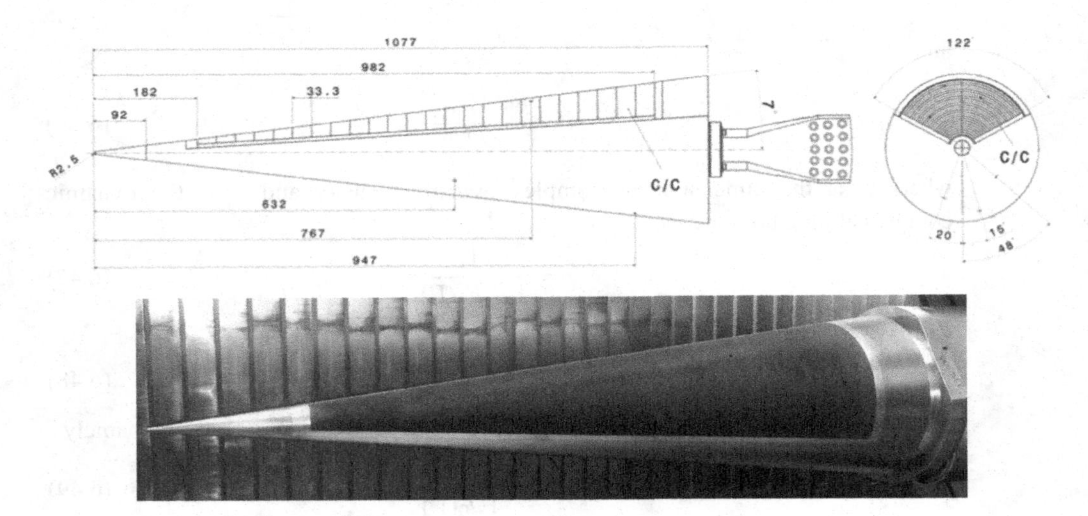

Figure 6.37 Schematic drawing and photograph of cone model with a partial covering of a porous ceramic (carbon–carbon) material intended to suppress second-mode instability growth. Taken from Wartemann et al. (2014).

as the "fast mode" or mode "F." However, the second-mode was distinguished by the phase speed $c_r \to 1 - 1/M_e$ at $x \to 0$. Near the leading edge, synchronization of the first- and second-mode waves with acoustic waves with phase speeds, $c_r = 1 \pm 1/M_e$, can result in a strong excitation of instability and early turbulence onset. This is what has motivated the development of "quiet" hypersonic wind tunnels. Federov (1997) refers to the region near the leading edge where a synchronization between the instability waves and the acoustic waves can occur as synchronization "Region 1." Further downstream from the leading edge, Federov (1997) defines a synchronization "Region 2" where synchronization between the first mode, and vorticity and entropy waves with $C_r = 1$ can occur. He points out that in this region, the free-stream vorticity and entropy disturbances may effectively generate the first mode and conversely the first mode may induce vorticity and entropy disturbances. Federov (1997) points out that near the Region 2 synchronism point, the first mode penetrates outside the boundary layer, which is a feature that can be exploited to identify Region 2 by measuring the correlation between hot-wire signals inside and outside the boundary layer.

Further downstream, Federov (1997) defined a synchronization"Region 3" in which the phase speeds of the first and second modes are very close to each other. In this region, the first and second modes are synchronized near the branch point of the discrete spectrum. Federov (1997) points out that in this region, the boundary-layer disturbance field has abnormal behavior, which could be detected by hot-wire measurements. The "synchronization point" referred to by Duan et al. (2013) is this synchronization Region 3 defined by Federov (1997). The premise is 2-D roughness placed closely downstream of this region can suppress the growth of the second mode.

Fong (2017) performed an extensive numerical study on the effect and design of 2-D roughness placed at the synchronization Region 3 to suppress second-mode growth. This led to the following 2-D roughness design approach for the second-mode transition control.

1. Determine the most dominant (generally most unstable) second-mode frequency responsible for turbulence onset. This can be determined using the e^N method for linear stability calculations.
2. Determine the synchronization Region 3 point for that second-mode frequency. This will correspond to the location where the phase speeds of the first (Mode F) and second (Mode S) modes are the same.
3. Determine the boundary-layer thickness at that synchronization Region 3 point.
4. Design the 2-D roughness element. The height of the roughness element should be from 40 to 60 percent of the local boundary-layer thickness to avoid "tripping" the boundary layer. The streamwise width of the roughness element should be on the order of two boundary-layer thicknesses.
5. Place a *single* 2-D roughness element at the location of the synchronization Region 3 to damp the most unstable second mode *or* place an *array* of 2-D roughness elements downstream of the synchronization Region 3 to damp disturbances that have lower frequencies than those at the synchronization Region 3.

A validation experiment of the effect of 2-D roughness strips on second-mode stability control was performed in the Sandia Laboratory Hypersonic Wind Tunnel at Mach 5 and 8. A description of the experiment is given by Casper et al. (2016). The test article consisted of a 7° half-angle cone that was installed with a series of roughness strips. A photograph of the test article is shown in Figure 6.38. This shows a series of roughness strips. The experiment also included the use of a single roughness strip. In this case, the strip was placed 7 cm downstream of the Region 3 synchronization point. Based on the local boundary-layer thickness, the height and width of the roughness strip were $h/\delta = 0.5$ and $w/\delta = 2.0$, respectively. When multiple strips were used, the heights and widths of the successive strips were scaled to be the same

Figure 6.38 Photograph of Sandia Laboratory cone model with axisymmetric strips designed to study their effect on 2nd Mode instability growth. Taken from Haley et al. (2018).

proportion of the boundary-layer thickness as it developed downstream. The streamwise spacing between the strips corresponded to 10 widths of the nearest upstream strip.

The experiment at Mach 8 indicated a delay in the transition onset that was tied to a reduction in the growth of the dominant second-mode wave produced by the roughness strips. Unfortunately, the experiment at Mach 5 indicated that the roughness strips did not delay turbulence onset. In that case, the roughness strips were believed to have modified the basic flow (nonlinear effect) and effectively "tripped" the boundary layer into a fully turbulent state. This might suggest that the criteria for the strip height may need to be reevaluated. From a practical aspect, before this can be a viable approach for flight vehicles, the sensitivity of the strip(s) placement to changing flight conditions needs to be evaluated. The underlying physics is without question. The practicality of the approach still needs to be examined.

A related approach to the 2-D roughness involves locating a 2-D cavity slightly downstream of the Region 3 stabilization point. This has been investigated in numerical simulations by Hao and Wen (2021). The results indicated that the cavity location played an important role in the development of Mode S (second mode), stabilizing it when placed downstream of the synchronization point, and destabilizing it when placed upstream of that point. The cavity depth also played a role. The stabilization of the second mode increased with increasing cavity depth as long as the depth was less than the local boundary-layer thickness. Depths larger than the boundary-layer thickness decreased its effectiveness. In the simulations, the length-to-depth ratio of the cavity was 6.67. The sensitivity to this ratio was not examined. Overall, the use of the cavity for second-mode stabilization is an analog to the 2-D strips and governed by the same physics.

Finally, with regard to passive control of second-mode transition, studies (Stetson and Rushton, 1967; Stainbeck, 1969; Stetson et al., 1985) have also shown that increased nosetip bluntness has a stabilizing effect which is consistent with the observations that the onset of transition is displaced downstream as the nose bluntness is increased. However, while the boundary-layer flow continues to become more stable with increasing nose bluntness, experiments indicate that the downstream movement in transition actually slows down and eventually reverses as the nose bluntness exceeds a certain critical range of values. The observed reversal in transition onset at large values of nose bluntness is contrary to the predictions of linear stability theory and, therefore, must be explained using a different paradigm.

The effect of transition Reynolds number, Re_T, on cone nose radius Reynolds number, Re_r based on Stetson (1983) is shown in Figure 6.39. This illustrates the approximately exponential (linear on the log-scale) increase in the transition Reynolds number as the cone nose radius on the cone is increased. This, however, abruptly stops at $\mathrm{Re}_r \simeq 10^7$, where the transition Reynolds number rapidly reverses. The unexpected transition delay reversal has been referred as the "bluntness paradox" which is mentioned in under "bypasses" in the boundary-layer transition flowchart that is shown in Figure 6.1. There is some speculation that the cause of the transition reversal

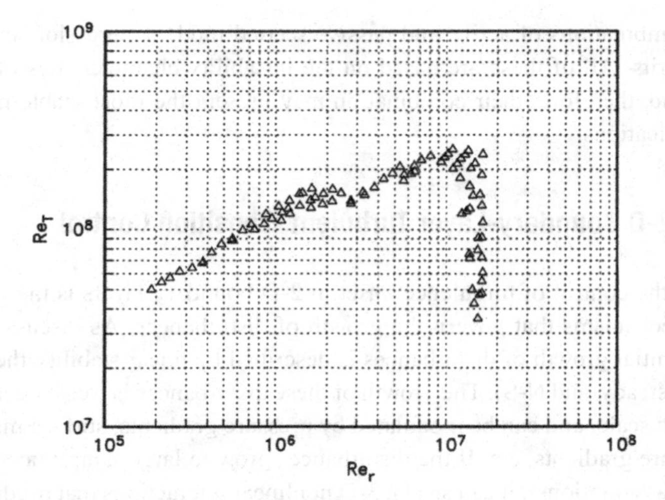

Figure 6.39 Effect of transition Reynolds number, Re_T, on cone nose radius Reynolds number, Re_r. Data from Stetson (1983).

results from added sensitivity to surface roughness on the cone nose. Nonetheless, prior to the transition reversal, nose bluntness offers a passive means for controlling second-mode transitions.

Hypersonic Boundary-Layer Active Control

The previous examples fall under the category of *passive* second-mode transition control. Examples of validated *active* control approaches include wall heating and CO_2 injection.

It has long been known through studies by Lees (1947) and Mack (1985) of hypersonic flows that wall cooling stabilizes the first Mode and destabilizes the second mode. Thus, *wall heating* presents a method to delay turbulence onset resulting from second mode growth. This is a common practice to suppress second mode instability growth in the boundary layers of nozzle expansion section of "quiet" hypersonic wind tunnels.

It is known that CO_2 gas has acoustic damping properties. This suggests that injection of CO_2 gas into the boundary layer could suppress the growth of the second mode in a similar manner to acoustic absorbing liners. Evidence of this was reported by Jewell et al. (2013). This involved a 5° half-angle right circular cone in a hypervelocity flow in a gas mixture consisting of 50 percent each of air and CO_2 by mass. Compared to pure air, the gas mixture with CO_2 was found to increase the location of turbulence onset by 30 percent. Although this involved a fully mixed gas stream, it is conceivable that CO_2 injected into the boundary layer could be as effective, and certainly more efficient. Possibly injection into the most sensitive location with regard to second-mode growth, namely the Region 3 synchronization point, could be the most efficient approach and minimize the necessary CO_2 mass flow rate.

Finally, combinations of active and passive second-mode suppression approaches may have merits. All of these are based on the instability characteristics of the second mode, and, therefore, their combination may present the most viable means for practical applications.

6.3 Summary: 2-D Boundary-Layer Turbulent Transition Control

The basis of the control of turbulence onset in 2-D boundary layers is the control of instability mechanisms that govern the growth of disturbances. As discussed in this chapter, the initial growth of disturbances is described by linear stability theory (i.e., linearized, unsteady, and N-S). The growth of these disturbances is weak, occurs over a viscous length scale, and can be modulated by pressure gradients, surface mass transfer, temperature gradients, etc. If the disturbances grow to larger amplitudes (beyond linear theory assumptions), it can spawn 3-D nonlinear interactions that produce mean flow distortions leading to secondary instabilities. At this last stage, the disturbance growth is very rapid with turbulent breakdown in short succession (occurring over a convective length scale).

In subsonic 2-D boundary layers, turbulence onset begins with the exponential growth of 2-D T-S waves. Although the most amplified T-S waves are 2-D, pairs of 3-D waves with equal and opposite spanwise wave numbers are also amplified. As pointed out by Craik (1971), the 2-D and 3-D T-S waves can interact through a subharmonic resonance even when the amplitude of the 2-D wave is asymptotically small. As a result, methods of active transition control based on 2-D T-S wave cancelation are impractical since any residual levels of the 2-D wave are sufficient to interact with the 3-D waves to quickly form 3-D lambda vortex structures, local mean flow distortion, secondary inviscid instabilities, and turbulence onset. Given this, the most practical means of transition control in subsonic 2-D boundary layers remains to be passive approaches: surface roughness and adverse pressure gradient to accelerate transition, and favorable pressure gradient and wall suction to delay transition.

In 2-D supersonic boundary layers ($M \leq 5$), the most amplified T-S waves are 3-D. These consist of oblique waves with equal but opposite spanwise wave numbers. Depending on the Mach number, the wave angles range from 45° to 60°. The T-S wave initial amplitude is sensitive to acoustic disturbances through the receptivity mechanism at leading edges. The sensitivity increases as the leading-edge radius decreases. Steady wall cooling can be used to suppress the growth of the T-S waves. Wall heating will do the opposite. Surface roughness in any form (2-D strips or 3-D particles) will accelerate turbulence onset.

Active disturbance generators that have sufficient frequency response for use in supersonic boundary layers have generally utilized electric discharges (arcs and plasma). Because of the 3-D nature of the waves, any analog to the wave cancelation approach used in subsonic boundary layers must introduce controlled 3-D disturbances that excite controlled oblique waves. This is unlikely to be successful or practical, to delay transition. It, however, can be used to accelerate turbulence onset.

In 2-D hypersonic boundary layers ($M > 5$), in addition to the 3-D T-S waves, there exist more amplified 2-D acoustic instability modes. With the 3-D T-S waves being labeled as the first mode, the 2-D acoustic instability modes are labeled as the second, third, etc. modes. Of the higher modes, the second mode is the first amplified. At these hypersonic Mach numbers, the amplification rate of the second mode is significantly larger than that of the first mode. As a result, the approach to delaying turbulence onset focuses on the control of second-mode growth.

Most of the second-mode control has focused on passive approaches. This undoubtedly results from the frequency limitations of active disturbance actuators. Examples of passive approaches that suppress the growth of the second-mode instability include leading-edge bluntness, steady wall heating, acoustic absorbing wall liners, and strategically placed (synchronization Region 3) surface roughness or wall waviness. All have been demonstrated in a laboratory environment to produce some reduction in second-mode growth. This and other approaches remain to be a very active area of research. Unfortunately, the sensitivity of these approaches to changing conditions undermines their practicality to a flight vehicle.

As will be discussed in Chapter 7, "3-D Laminar-Turbulent Transition," the 2-D second mode may not be the dominant mechanism for turbulence onset on lift-generating hypersonic vehicles that are dominated by a mean cross-flow and cross-flow instability. As a result, from a practical viewpoint, second-mode transition control may be redundant.

Problems

6.1 The "road map" of paths to turbulence shown in Figure 6.1 is a snapshot of the best understanding of *incompressible 2-D boundary-layer* transition from 40 years ago.

1. Based on our current understanding, draw an updated and *simplified* flowchart.
2. Of the four "bypasses" listed in the road map, can any now be explained? Can any new ones be added?
3. List the types of disturbance sources that can play a role in turbulence transition? List these in the order of least to most impact on transition.

6.2 Although for convenience, most flow actuators used to control T-S are located on the wall, it is suggested in this text book that the optimum location is at the height of the *critical layer*.

1. What are the characteristics of the critical layer?
2. What makes the critical layer the best choice in which to introduce controlled disturbances?
3. If one were to move a sensor in the flow direction at the T-S phase velocity, what would be the relative flow direction just above and below the critical layer?
4. Why does a sheet of smoke tracers introduced into a boundary layer form into light and dark bands like those shown in Figure 6.15, marking the 2-D T-S waves?

6.3 Given the linear neutral stability curve for an incompressible 2-D boundary layers shown in Figure 6.2, design an experiment to demonstrate linear phase canceling of 2-D T-S waves.

1. Draw on the plot of F versus R the exact points where the initial and phase-canceling 2-D disturbances would be introduced. Explain your choices.
2. In an open-loop control, based on linear theory, determine the amplitude, frequency, and phase of the downstream 2-D disturbance generator to perfectly phase cancel the initial 2-D T-S wave.
3. Does this appear to be a viable method of delaying turbulence transition in 2-D boundary layers? See, for example, Saric and Thomas (1984).

6.4 All of the experiments used to validate linear theory of 2-D boundary layers have been performed in wind tunnels that include both vortical and acoustic background disturbances. As in the experiments of Schubauer and Skramstad (1947), the boundary layer was developed on a suspended flat plate with a leading edge.

1. In consideration of the background disturbances in the wind tunnel, describe the considerations you would make in the shape of the boundary-layer plate leading edge.
2. In a "natural" transition experiment, describe how the background disturbances would affect the most amplified frequency, amplification rates, and N-factor for turbulence transition.
3. How can the effect of background disturbances be overcome in an experiment with "controlled" disturbance input?
4. How can one prove that the "controlled" disturbance amplitudes satisfy linear theory assumptions?

6.5 In designing a "quiet" *hypersonic* wind tunnel, it is necessary that the boundary layer in the diverging part of the converging–diverging nozzle remain laminar. In this process, it is necessary to delay turbulent transition due to T-S and second-mode instabilities.

1. In this application, describe passive and active methods that can be used to suppress the growth of T-S waves.
2. Similarly in this application, describe passive and active methods that can be used to suppress the growth of second-mode waves.
3. Do any of the methods for one adversely affect the other? If so, is there a suitable compromise?
4. Does surface roughness have an impact on achieving a "quiet" tunnel? If so, how?

6.6 A passive approach to second-mode transition control involves the use of surface acoustic liners. In this application, parameters in the design of the acoustic liners include the number and spacing of pores (i.e., surface porosity), and their depth. These are to be "tuned" to the characteristics of the second mode, e.g., frequency and wavelength.

1. List the steps in the design of an acoustic liner to suppress the growth of second-mode waves in a boundary layer at Mach 6.
2. List details of the design: number of pores per second-mode wavelength and depth of the pores. Assume the pores have a circular cross section.
3. The second-mode frequency scales with the boundary-layer thickness. What effect would this have on the acoustic liner design?
4. The T-S mode transition is sensitive to surface roughness. Should the surface roughness of the surface liner be a factor at Mach 6?

7 Three-Dimensional Laminar Boundary Layers

A fully three-dimensional (3-D) boundary-layer flow exhibits quite different behavior from that of a corresponding two-dimensional (2-D) flow. Of particular interest to turbulent transition are the stability characteristics of 3-D boundary layers, which can produce instability growth rates that are larger than the usual Tollmien–Schlichting (T-S) waves, or at hypersonic Mach numbers, second-mode waves that were discussed in Chapter 6. Examples of 3-D flows of practical interest include swept wings, rotating cones, cylinders and spheres, corners, inlets, and rotating disks. Each of these flows exhibits instability behavior that is general to 3-D boundary layers, with a consistent characteristic of the presence of streamwise vorticity that produces a strong spanwise modulation of the basic state that can give rise to secondary instabilities.

7.1 Swept Wing Flows

The flow over a swept wing is a typical example of a 3-D boundary layer. This type of 3-D flow is susceptible to four types of instabilities that lead to turbulence transition. These are a leading-edge instability and contamination, a streamwise instability, a centrifugal instability, and a cross-flow instability. Leading-edge instability and contamination occurs along the leading-edge attachment line and is associated with an instability of the flow attachment line. The streamwise instability is associated with the chordwise component of flow over the wing. Transition in this case is similar to that in 2-D flows, where a T-S instability is the dominant mechanism. It usually occurs in zero or mild positive pressure-gradient regions on the wing. Centrifugal instabilities occur in boundary layers over a concave surface. The dominant mechanism for turbulence transition is through the Görtler instability. Finally, the cross-flow instability occurs in strong pressure-gradient regions on swept wings. In other than moderate wing sweep angles, cross-flow or T-S instability mechanisms will dominate over the Görtler instability.

On a swept wing, the combination of pressure gradient and wing sweep causes the inviscid-flow (boundary layer edge) streamlines to deflect inboard. This is illustrated in Figure 7.1. This effect is more pronounced in the low momentum fluid in the boundary layer near the wall, where the pressure gradient produces a cross-flow velocity component that is perpendicular to the local inviscid flow velocity component. This results in

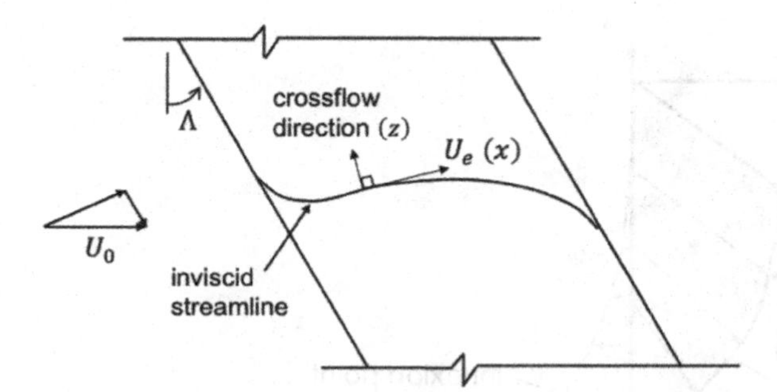

Figure 7.1 Illustration of an inviscid streamline that results through a combination of pressure gradient and wing sweep.

the characteristic cross-flow boundary-layer mean velocity profile shown in Figure 7.2. In this case, U_e is the local edge velocity. The local cross-flow velocity component is in the z direction, perpendicular to the local U_e. The combination results in the mean velocity profile shown by the shaded region.

The cross-flow profile has a maximum velocity somewhere in the middle of the boundary layer, going to zero on the body surface and at the boundary-layer edge. Its requirement for zero cross-flow at the wall (no slip) and at the edge of the boundary layer results in an inflection point somewhere in the profile. An inflectional mean velocity profile is a characteristic of free shear layers (see Chapter 5), and inviscidly unstable to infinitesimal disturbances. The cross-flow instability results in both traveling and stationary waves. The stationary waves grow to form a regular pattern of stationary corotating cross-flow vortices with axes aligned in the inviscid flow direction. An example of this for a swept wing flow that was obtained by Saric and Yeates (1985) using naphthalene surface flow visualization is shown in Figure 7.3. The stationary cross-flow vortices are corotating. The stationary pattern of dark and light bands in the image in Figure 7.3 results from the removal (dark bands) of the naphthalene surface layer directly under the corotating vortices.

Reibert et al. (1996) investigated the nonlinear saturation of stationary waves on the swept wing. In contrast to the natural transition case that is shown in Figure 7.3, a full-span array of micron-sized discrete roughness elements (DREs) was used to impose a uniform spanwise periodicity of the stationary cross-flow mode. The result is shown in Figure 7.4 which presents constant-level contours of the normalized streamwise velocity, u/U_o, measured in a 2-D cross section of the cross-flow vortices. The mean cross-flow direction is from left to right. This reveals the corotation direction of the vortices.

A larger body of experimental work on the nature of the cross-flow instability exists for swept wings. These include Muller and Bippes (1988), Bippes (1991), Dagenhart et al. (1989), Kohama et al. (1991), Radeztsky et al. (1993) and Arnal and Juillen (1987). The experiment of Muller and Bippes (1988) is somewhat singular because it was one of the few to document both stationary and traveling cross-flow modes. In that case, they did not observe the generation of a high-frequency secondary

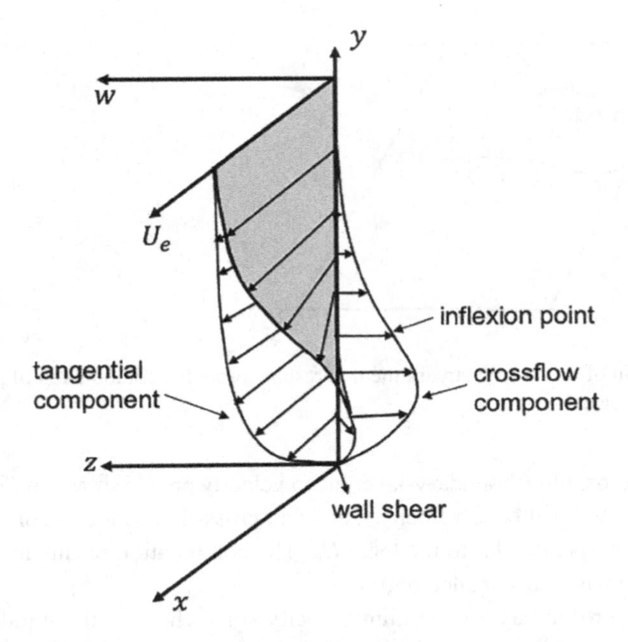

Figure 7.2 Wall-normal profiles of velocity components in a boundary layer with cross-flow.

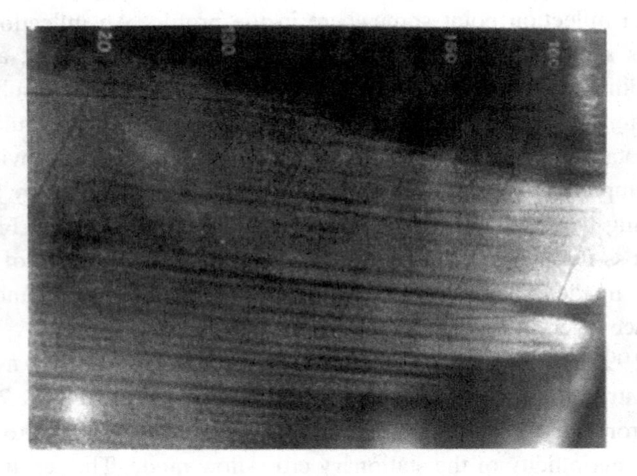

Figure 7.3 Photograph of surface visualization over a swept wing that reveals stationary cross–flow vortices. Taken from Saric and Yeates (1985).

instability mode leading to transition, which has been generally attributed to the breakdown of cross-flow vortices on disks (Kohama, 1984, 1987) and on a swept cylinder (Poll, 1985). However, in the stationary-mode-dominated swept wing experiment of Kohama et al. (1991), a high-frequency secondary instability was observed just prior to transition.

The apparent conflict in the two experiments can be reconciled by the noted differences in the amplitude level of the traveling modes. Muller and Bippes (1988) studied the effect of the free-stream turbulence level on the amplitude of the stationary and

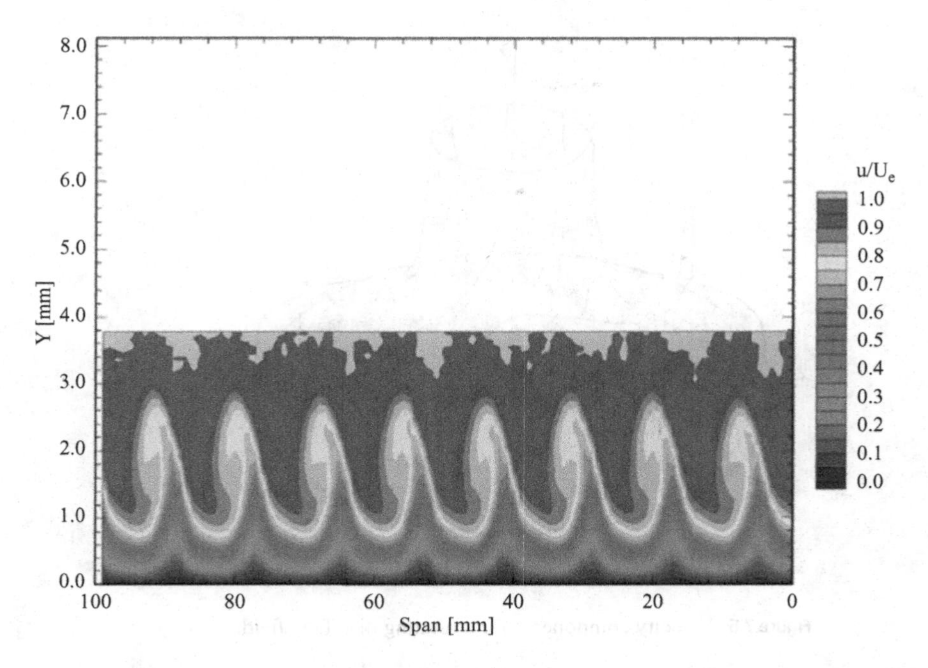

Figure 7.4 Contours of normalized streamwise velocity, u/U_O, in a 2-D cross section of the stationary cross-flow vortices over a swept wing. The highly uniform spanwise periodicity was achieved by placing a uniform pattern of discrete roughness near the leading edge of the wing surface. Taken from Reibert et al. (1996).

traveling modes. They found that at low turbulence intensity levels ($u'/U_\infty = 0.05$ percent), those were comparable to the Dagenhart et al. (1989), Kohama et al. (1991), and Radeztsky et al. (1993) experiments, and turbulence onset was dominated by stationary cross-flow modes. However, at higher turbulence intensity levels from 0.15 percent to 0.30 percent, the traveling cross-flow modes became more dominant. In these cases, although the growth of the stationary cross-flow mode was less dominant, transition occurred earlier. From this, Muller and Bippes (1988) concluded that the traveling cross-flow modes were more important to transition. These results, on which the stationary and traveling cross-flow modes exist at comparable levels, open the possibility of a nonlinear interaction between them. Lekoudis (1979) was one of the first to consider the possibility of a triad-type interaction between traveling cross-flow waves in the 3-D boundary layer on a swept wing. Later in rotating disk experiments, Corke and Knasiak (1998) and Corke et al. (2007b) confirmed a nonlinear resonance between traveling and stationary cross-flow modes.

7.2 Rotating Disk Flows

The cross-flow instability of the boundary layer over a rotating disk is very similar to that over the swept wing and has been frequently used as a canonical 3-D flow that exemplifies the cross-flow instability. Figure 7.5 shows the velocity components for

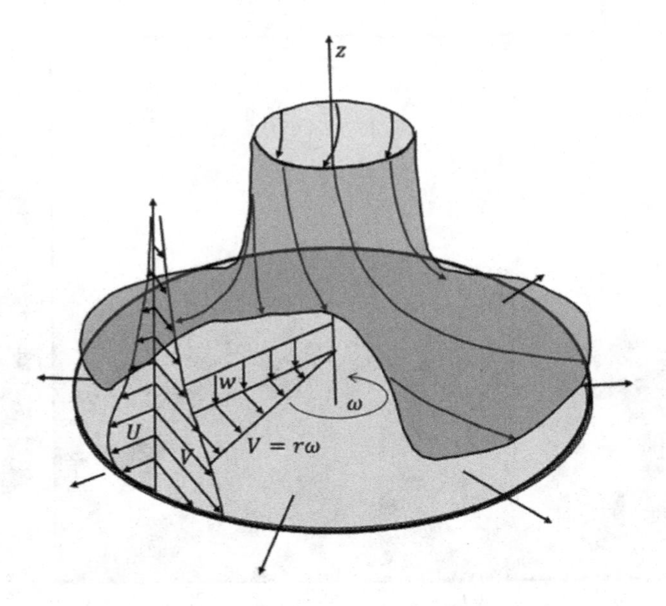

Figure 7.5 Velocity components for a rotating disk flow field.

the flow over the rotating disk. This is very similar to the mean velocity profile of the boundary layer over a swept wing that is shown in Figure 7.1. The key difference is that the disk experiences the effect of Coriolis forces due to the disk rotation. A key advantage to the rotating disk is that the mean (basic) flow over the disk has an exact solution, making instability analysis more accurate. An advantage for experiments is that except for very close to the rotation axis, the boundary-layer thickness is constant in the radial direction. Thus, for a fixed rotation speed, ω, the radial location, r, is equivalent to the Reynolds number, $R = r\sqrt{\omega/\nu}$.

The first analysis of the laminar boundary layer on a rotating disk in a quiescent fluid was performed by Karman (1921). The introduction of a similarity variable, $\eta = \sqrt{\omega/\nu}z$, led to a set of ordinary differential equations governing the basic flow, namely

$$2F + H' = 0 \tag{7.1}$$

$$F^2 - G^2 + F'H = F'' \tag{7.2}$$

$$2FG + G'H = G'' \tag{7.3}$$

$$P' + HH' - H'' = 0, \tag{7.4}$$

where $F(\eta) = u_r/(r\omega)$, $G(\eta) = u_\theta/(r\omega)$, $H(\eta) = w/\sqrt{\nu\omega}$, and $P(\eta) = (p - p_0)/(\rho\nu\omega)$.

The boundary conditions for $\eta > 0$ are

$$F = 0,\ G = 1,\ H = 0,\ P = 0 \text{ for } \eta = 0 \tag{7.5}$$

$$F = 0,\ G = 0 \text{ for } \eta \to \infty. \tag{7.6}$$

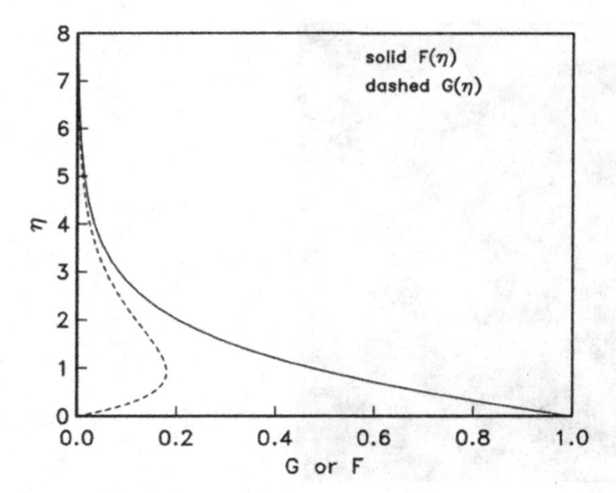

Figure 7.6 Similarity solution profiles for rotating disk boundary layer flow.

The coupled ordinary differential equations can easily be solved numerically, for example, using a Runge–Kutta–Gill integration approach. The similarity wall-normal profiles for $G(\eta)$ and $F(\eta)$, which are respectively the normalized azimuthal and radial velocity components, u_θ and u_r, are shown in Figure 7.6.

With the rotating disk, there are two types of cross-flow instabilities: Type I and Type II. The Type I cross-flow instability is analogous to that which occurs in the boundary layer on a swept wing. With the rotating disk flow, the Type I cross-flow instability appears as outward-spiraling waves. These are evident in the classic flow visualization image from Kobayashi et al. (1980) that is shown in Figure 7.7. These waves were first detected experimentally in the hot-wire measurements of Smith (1946). The theoretical analysis accompanying experimental results followed later in the classic paper by Gregory et al. (1955). In that, they surmised a number of key elements about this instability such as the role that small surface roughness could have in fixing the locations of stationary modes, and the expected existence of traveling modes that could not be detected in their surface flow visualization.

The Type I instability has been shown (Wilkinson and Malik, 1985) to have a critical Reynolds number, $R_{c_I} = 285$. In addition, Malik et al. (1981) have shown that the number of Type I instability waves around the disk (mode number) to increase linearly with radial location, or $n = \beta R$, where n is the azimuthal mode number (an integer) and β is the azimuthal wave number. At the critical Reynolds number, the most amplified azimuthal wave number is $\beta = 0.0698$, and the resulting spiral wave angle $\psi = \tan^{-1}(\beta/\alpha) = 11.2°$. This is depicted in Figure 7.8.

Even though linear theory predicts the traveling modes to be more amplified, a majority of the experiments have focused on stationary cross-flow modes. The reasons for this are twofold. The first is that many of the studies have been based on flow visualization (Gregory et al., 1955; Federov et al., 1976; Kobayashi et al., 1980; Kohama, 1984) that tends to emphasize stationary features. The second is the sensitivity of this instability to surface roughness which can bias the larger-amplitude modes

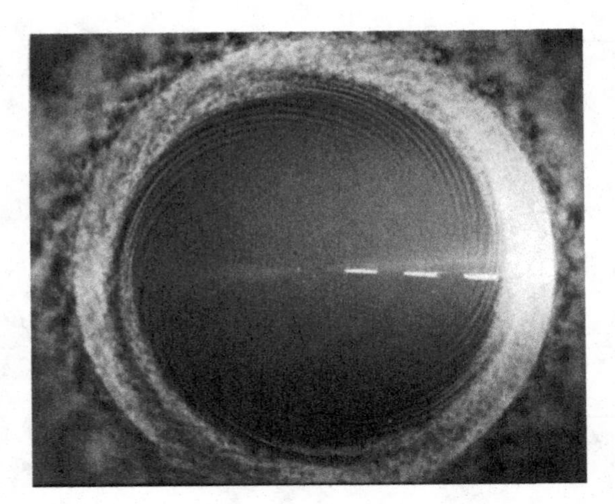

Figure 7.7 Smoke visualization photograph of outward spiraling stationary cross-flow vortices leading to turbulent transition in the boundary layer over a rotating disk. Taken from Kobayashi et al. (1980).

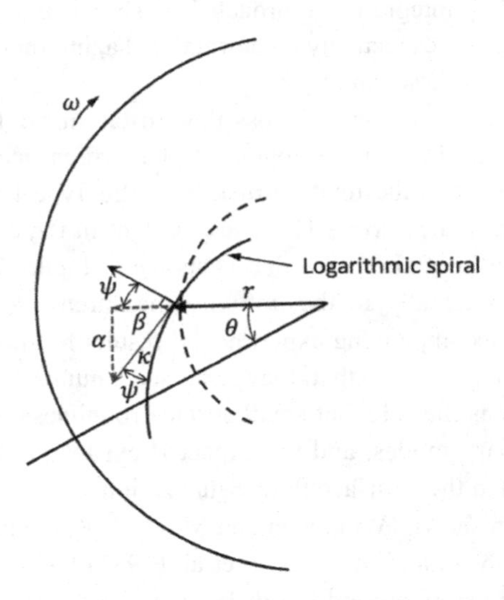

Figure 7.8 Schematic defining Type I instability log-spiral angle, ψ, and wave numbers, α and β.

toward the stationary type. For example, Wilkinson and Malik (1985) observed that stationary modes originated from small randomly placed dust particles on the surface of a clean disk. This prompted them to study the effect of a single isolated surface roughness. From this they observed the growth of wave packets that rapidly spread around the disk to eventually fill the entire circumference.

Although the traveling Type I waves on a swept wing can have both positive and negative phase speeds, on the rotating disk, the traveling waves only have a negative phase speed, meaning that they are traveling at a speed that is less than the speed of the disk surface. Faller (1991) estimated the traveling Type I waves to be traveling approximately 5 percent slower than the disk speed. By definition, the stationary Type I waves are traveling at the surface speed of the disk.

The Type II cross-flow instability is related to the Coriolis/centrifugal forces that act on the flow over the disk surface. This also results in spiral waves, although the angle of the waves, ψ, is the opposite to that of the Type I waves. Faller and Kaylor (1966) showed that the Type II instability was easily excited by external disturbances. Based on Faller (1991), the Type II instability critical Reynolds number is $R_{c_{II}} = 69$. Therefore, it is expected to appear inboard of the Type I instability on the disk. However, the amplification rate of the Type II instability is much lower than that of the Type I instability (Faller, 1991) so that unless there are large disturbances that would increase the initial amplitude of the Type II instability, the Type I instability is the dominant mechanism for turbulence onset.

Corke and Knasiak (1998) performed experiments on a rotating disk flow that used patterns of roughness dots on the surface in order to excite specific azimuthal wave numbers of stationary cross-flow modes. The fluid was air at standard atmospheric pressure. The roughness dots consisted of drops of ink that were applied by an ink drafting pen that was attached to motorized computer-controlled traversing mechanism. Figure 7.9 shows a sample photograph of the dots on the disk surface. These were located at the radial location of the critical Reynolds number, $R_{c_I} = 285$, and have a spacing and arrangement that corresponds to the most amplified mode number and spiral angle of $n = 19$ and $\psi = 11.2°$, respectively, based on linear stability theory.

The height of the roughness dots was small enough to not alter the mean flow that determines the linear stability characteristics. This is documented in Figure 7.10, which shows the normalized mean azimuthal velocity, $u_\theta/(r\omega)$, at different radial locations relative to the critical radius, r_c, that corresponds to $R_{c_I} = 285$. The ordinate is the nondimensional distance from the disk surface, $z^* = z\sqrt{\omega/\nu}$. The solid curve through the data corresponds to the similarity profile, $F(\eta)$, shown in Figure 7.6. As shown in Figure 7.10, the measured mean profile agrees well with the Karman (1921) similarity profile at inboard radii, $r/r_c \leq 1.501$. The deviation of the mean flow from the similarity profile at further outboard radii is an indication of the effect of nonlinear amplitudes reached by the spatially developing cross-flow modes.

A velocity sensor such as a hot-wire sensor, placed in the boundary layer over the rotating disk will detect the passage of the spiral cross-flow instability waves (Figure 7.7) as they convect past the sensor. These will consist of stationary waves traveling at the speed of the disk and traveling waves with a wave speed that is lower than the disk speed. The two wave types can be separated from the velocity time series by ensemble averaging the time series that are triggered on each disk rotation. The traveling cross-flow waves that are not correlated with the disk rotation will be averaged out of a large number of ensembles, leaving only the time series that corresponds to stationary cross-flow waves. This is the process used by Corke and Knasiak (1998) to obtain the azimuthal velocity time series shown in Figure 7.11. The time

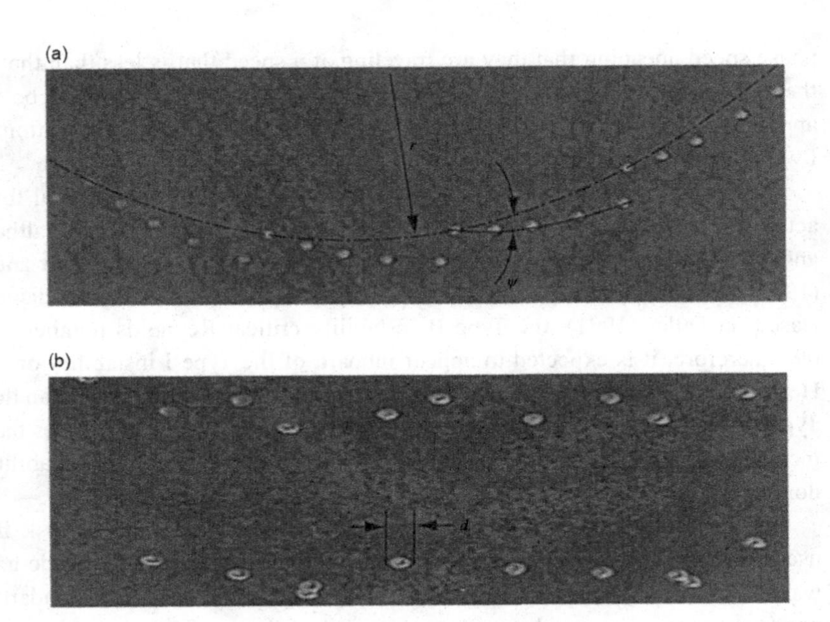

Figure 7.9 Example of ink dots applied to surface of a rotating disk to excite specific azimuthal mode numbers of stationary Type I cross-flow waves. The dot diameter was $d = 1.6$ mm, and the dot height was $h = 0.06$ mm, giving a dimensionless height of $h\sqrt{\omega/\nu} = 0.16$. Taken from Corke and Knasiak (1998).

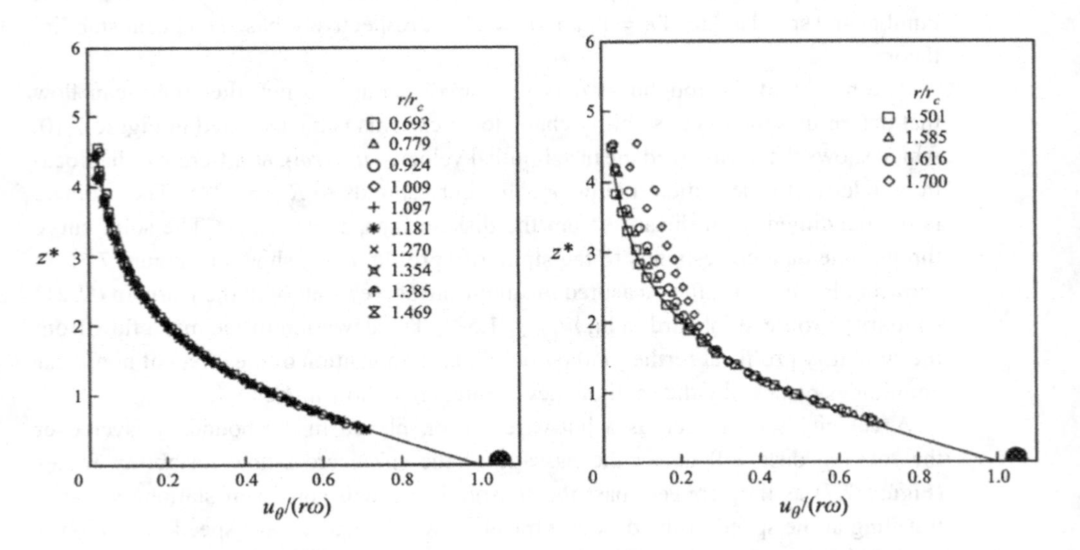

Figure 7.10 Mean azimuthal velocity profiles in similarity variables at different radial locations on a rotating disk having discrete roughness dots to excite a specific azimuthal mode number of Type I cross-flow modes. Taken from Corke and Knasiak (1998).

series correspond to different radial locations, with each location indicated where the horizontal dashed lines intersect the r/r_c values listed on the ordinate. The abscissa

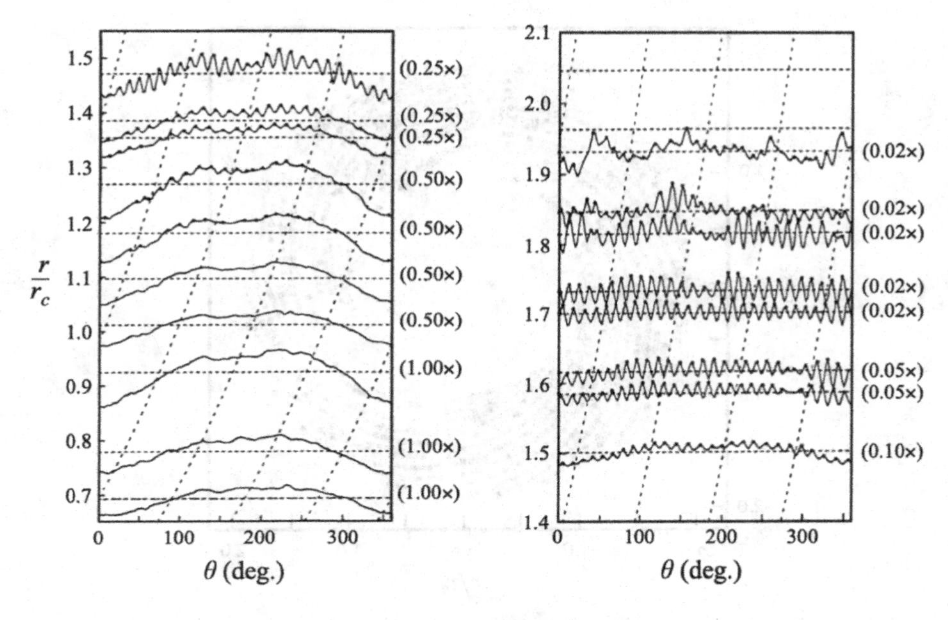

Figure 7.11 Azimuthal velocity time series measured with a hot-wire located at $z^* = 1$ above the disk surface that corresponds to stationary Type I cross-flow waves over one disk rotation at different radial locations. Taken from Corke and Knasiak (1998).

corresponds to one disk rotation. The amplitudes of the time series have been scaled, with scaling factors shown on the right ordinate of each plot. This accounts for the radial growth in amplitude of the stationary cross-flow waves.

These time series correspond to a "smooth" disk surface without the discrete roughness that was shown in Figure 7.9. The velocity time series document quasi-periodic fluctuations that correspond to the passage of stationary cross-flow waves convecting past the stationary sensor. Filled symbols have been applied to mark the peaks in the waves. The curved dashed lines through the dots correspond to the $\psi = 11.2°$ spiral predicted by linear theory.

The locations of the peaks (filled symbols) in the time series in Figure 7.11 are shown in the disk circular coordinate in Figure 7.12. These again are indicated by the filled symbols. The disk rotation is clockwise. The open symbols in this representation correspond to $\psi = 11.2°$ spiral waves that originate from each of the filled symbols. These appear to connect the space between the filled symbols fairly well, supporting the linear theory predictions for wave angle for the Type I cross-flow instability.

The sensitivity of the Type I cross-flow instability to minute roughness makes experiments on "smooth" disks difficult. As an example, Figure 7.13(a) plots the Type I instability wave number as a function of the radial position on a smooth clean disk (Corke and Knasiak, 1998). Malik et al. (1981) had shown that the number of Type I instability waves around the disk (mode number) should increase linearly with radial location, namely $n = \beta R$. Based on linear theory, the most amplified wave occurs for $\beta = 0.0698$. In Figure 7.13(a), although the wave number does vary linearly with

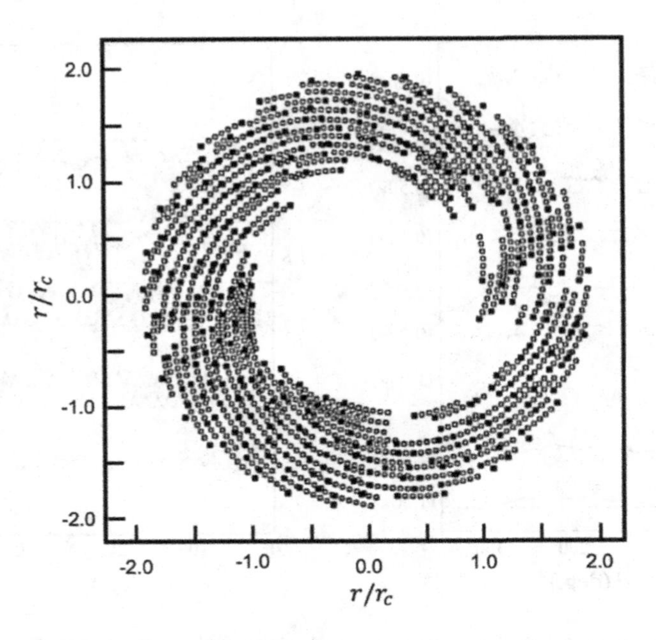

Figure 7.12 Time series peak locations from Figure 7.11 shown by filled symbols in disk circular coordinate. Open symbols correspond to spiral waves that originate from the filled symbols. The disk rotation is clockwise. Taken from Corke and Knasiak (1998).

radius and thereby Reynolds number, for that instance with a "smooth clean" disk, the waves that emerge have a slightly higher $\beta = 0.0796$. This is not an error in linear theory but rather exemplifies how under the best conditions, minute roughness (dust?) can excite a less amplified wave. In contrast, in Figure 7.13(b), when discrete roughness dots with the most amplified azimuthal mode number of 19 were applied at the critical radius, r_c, the waves that emerged had the expected azimuthal wave number of $\beta = 0.0696$ predicted by linear theory. This sensitivity of stationary cross-flow modes to discrete roughness is exploited in cross-flow transition control that was pioneered by Reibert (1996) and Saric et al. (1998a) that will be discussed later in this chapter.

As an additional check on linear theory predictions for the rotating disk boundary-layer cross-flow instability, the cross-correlation between two simultaneous hot-wire sensors was performed by Matlis (1997) and Corke et al. (2007b) to determine the 2-D wave number vector, κ, at each frequency. This also involved placing discrete roughness dots with an azimuthal mode number of 19 on the disk surface at the critical radius, r_c. The wave number vector is defined in Figure 7.8. Dispersion curves for β are presented in Figure 7.14(a). The stationary dispersion curves are used as a check on the accuracy of the wave number measurements. Because the rotation ensemble-averaged flow is a priori stationary, its phase speed is exactly that of the disk. The results verified the expected dispersion relation, $n = \beta R$, for all mode numbers, n.

The dispersion curves were also used to determine the radial wave number, α. These were used along with those for β to determine the spiral angle ψ of the instability wave angle, where $\psi = \tan^{-1}(\beta/\alpha)$. These are presented in the right column of

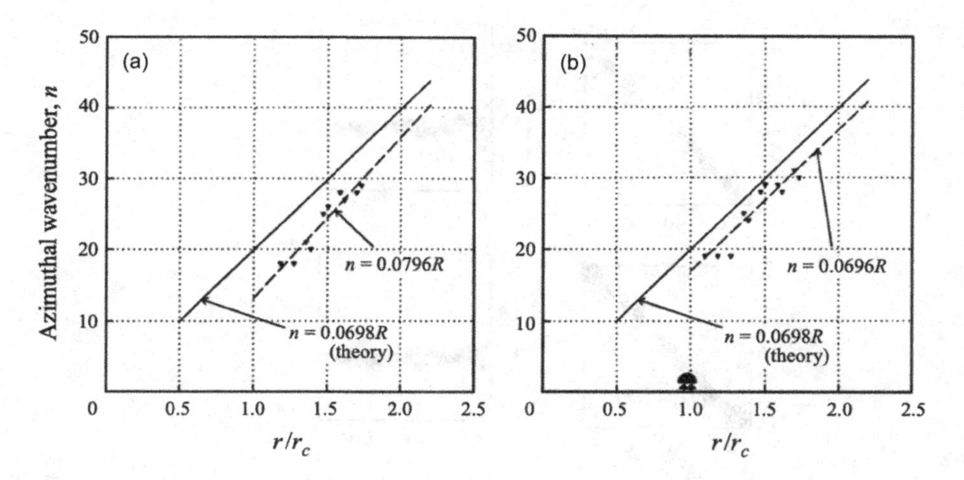

Figure 7.13 Radial development of the azimuthal wave number for the "smooth" disk surface (a) and with discrete roughness having an azimuthal mode number of $n = 19$ located at the critical Reynolds number radius, r_C (b). Taken from Corke and Knasiak (1998).

Figure 7.14(b). This verifies that all of the stationary modes in this case follow the theoretically preferred angle of 11.2°, shown by the solid line through the data points.

7.2.1 Type I Cross-Flow Turbulence Transition

The simultaneous existence of stationary and traveling cross-flow modes of comparable amplitudes opens the possibility of a nonlinear interaction between them that can affect turbulent transition. Lekoudis (1979) was one of the first to consider the possibility of a triad-type resonant interaction between traveling cross-flow waves in the 3-D boundary layer on a swept wing. For the same flow, Malik et al. (1994) used a nonlinear model to study the interaction between traveling and stationary cross-flow modes. They observed that the modes could interact and that the final stage depended on their relative initial amplitudes. For example, when the initial amplitudes of the traveling and the stationary modes were the same, the traveling modes, owing to their larger growth rates, would dominate most of the flow development. With this in mind, Corke and Knasiak (1998) documented a triple phase locking between traveling and stationary cross-flow modes through cross-bicoherence (CBC) analysis of velocity time series that provided the first experimental evidence of a triad resonance mechanism in this flow.

As a follow-up to Corke and Knasiak (1998), Matlis (1997) and Corke et al. (2007b) sought further evidence of triad resonance between the stationary and traveling cross-flow modes, and the impact this had on turbulence transition. For this, they also introduced a controlled disturbance by placing discrete roughness with $n = 19$ at the r_C radius. The azimuthal velocity fluctuations were measured with a hot-wire sensor, and following the method of Corke and Knasiak (1998), the velocity time series were decomposed into stationary and traveling components. Spectra of the stationary

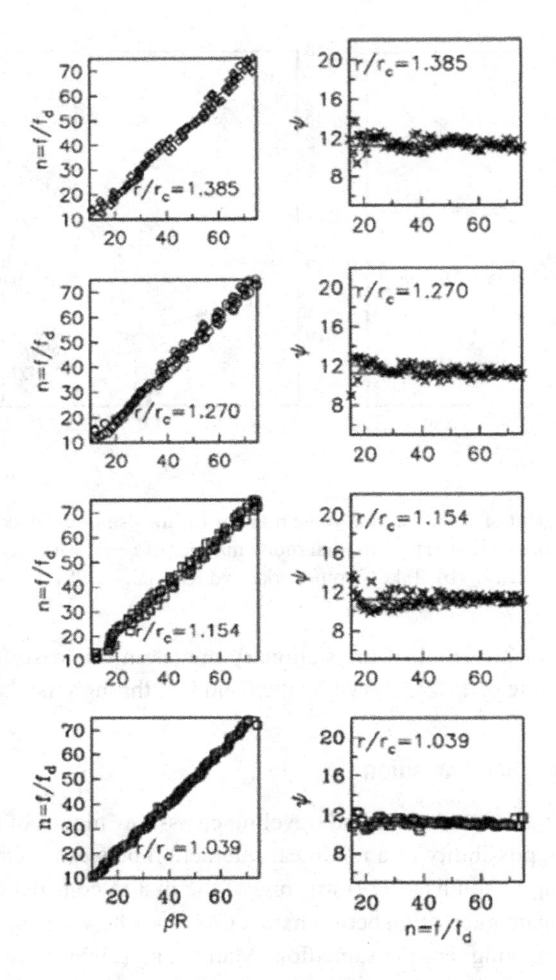

Figure 7.14 Stationary wave dispersion curves (a) for azimuthal wave number, β, and (b) spiral wave angles, ψ, at different radial locations with discrete roughness having an azimuthal wave number of $n = 19$ located at the critical Reynolds number radius, r_C. Taken from Matlis (1997) and Corke et al. (2007b).

and traveling azimuthal velocity fluctuations measured at different radial locations are shown in Figure 7.15. Figure 7.15(a) corresponds to the stationary modes, with the abscissa indicating the mode number, n. The first radial location is just outboard of the discrete roughness. At this location, the blue arrow denotes $n = 19$ that is excited by the discrete roughness. At further outboard locations, the blue arrows mark the most amplified mode numbers based on linear theory, $n = \beta R$. These radial locations were specifically selected because they are the onset of nonlinear cross-flow mode development. The right column shows spectra of the traveling cross-flow modes at the corresponding radial locations. The abscissa for these spectra is frequency.

At the further outboard radii, the spectra for the stationary mode indicates the emergence of a low-frequency stationary mode. The red arrow corresponds to $n = 4$. This

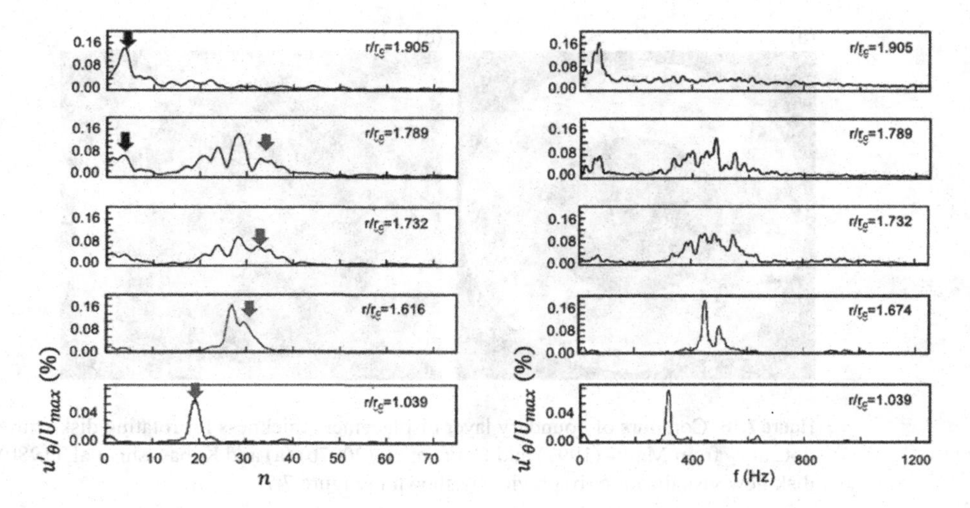

Figure 7.15 Spectra of velocity fluctuations at different radial positions for stationary (a) and traveling (b) crossflow waves with discrete roughness having an azimuthal mode number of $n = 19$ located at the critical Reynolds number radius, r_C. Blue arrows mark linear most amplified values. Red arrow denotes $n = 4$. Taken from Matlis (1997) and Corke et al. (2007b).

stationary mode was found to be visible in the azimuthal variation of the boundary-layer displacement thickness, $\delta^*(r,\theta)/\overline{\delta^*(r)}$, in the cylindrical coordinate of the disk (Matlis, 1997; Corke et al., 2007b). The displacement thickness values are shown as constant-level grayscale contours. The darker shaded regions correspond to where the displacement thickness is less than the mean. The lighter shaded regions are where it is greater than the mean. An $n = 4$ variation, labeled 1–4, is clearly visible as the four alternating light/dark regions around the disk at the outer radii. With this insight, the flow visualization image from Kobayashi et al. (1980) that is shown in Figure 7.7 is shown for reference. There in this classic image, it is possible to also identify an $n = 4$ (labeled) stationary pattern in the smoke visualization.

The origin of the $n = 4$ mode can be traced through a nonlinear interaction between the traveling and stationary cross-flow modes. Matlis (1997) and Corke et al. (2007b) documented this through a CBC analysis that examined the triple phase locking between the traveling modes and the stationary modes. Specifically, it was found to originate from an interaction between traveling modes at 517 and 449 Hz whose difference was 67 Hz which, when divided by the disk rotation frequency (16.667 Hz), gave the stationary mode number, $n = 4$. Dispersion measurements using two hot-wire sensors performed by Matlis (1997) and Corke et al. (2007b) confirmed that in addition to the triple frequency phase locking, the difference in the traveling mode *wave numbers* matched the wave number of the stationary mode, confirming resonance.

The possible resonance interaction between the traveling and stationary cross-flow modes equally applies to the swept wing and therefore presents a powerful mechanism to accelerate turbulence onset if desired. In contrast, if the intent is to delay turbulence onset, conditions that support this resonance mechanism need to be avoided.

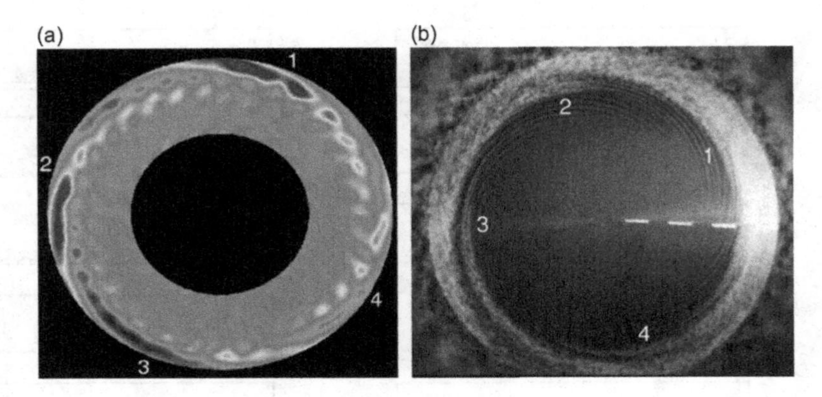

Figure 7.16 Contours of boundary layer displacement thickness for rotating disk with $n = 19$ dots at r_C from Matlis (1997) and Corke et al. (2007b) (a) and Kobayashi et al. (1980) rotating disk flow visualization (b) previously shown in Figure 7.7.

7.2.2 Type I Absolute Instability

The previous section described a path to turbulence onset for the Type I cross-flow instability that is convective, namely with disturbances growing in space along a path that follows the mean flow direction. In contrast to that, Lingwood (1995) indicated that the rotating disk flow is absolutely unstable, in which case disturbances could grow in both space and time, with the latter resulting in "explosive" growth and virtually immediate turbulence onset. As observational evidence of this, Lingwood suggested that values of the transition Reynolds number taken from experimental studies, particularly including that of Kobayashi et al. (1980) that is primarily based on flow visualization, varied by less than 3 percent of the absolute instability critical Reynolds number, $R_{C_A} = 513$ (later corrected to be 507 by Lingwood (1997)). The coincidence of the transition Reynolds number was, however, not supported by the results from Wilkinson and Malik (1985) who found in careful low-disturbance experiments using hot-wire evidence to document transition at $543 \leq R_{\text{tr}} \leq 556$, placing the average about 8 percent higher than the Lingwood (1997) R_{C_A}.

The Lingwood (1995) prediction came from a linear stability analysis of the rotating disk boundary layer that considered the growth of stationary and traveling Type I waves. The viscous analysis assumed a locally parallel flow, streamline curvature, and Coriolis effects. It indicated that a critical Reynolds number (or critical radius for a given rotation speed) existed at which disturbances grew temporally, leading to an unbounded linear response and presumably immediate turbulence onset. The analytic prediction of the absolute instability critical Reynolds number has subsequently been verified by Pier (2003) and Davies and Carpenter (2003), and is not in dispute. The *issue is its role in transition* to turbulence of the disk flow.

Lingwood (1996) performed an experimental study designed to capture the temporal growth associated with the absolute instability. This involved introducing unsteady disturbances into the boundary layer and following their development in space and time. The unsteady disturbances were a short duration air pulse that emanated from a hole in the disk surface. The pulse occurred once every disk rotation. The location of the pulse was just outboard of the minimum critical radius for Type I cross-flow modes. Lingwood followed the evolution of the azimuthal velocity fluctuations with a hot-wire sensor placed at different radial and azimuthal distances from the air pulse. Ensemble averages of the time series, correlated with the azimuthal position of the air pulse, revealed wave packets. When the leading and trailing edges of the wave packets were presented in terms of their Reynolds number (radius) and time (azimuthal position with respect to the disk rotation speed), they revealed a tendency for an accelerated advancing of the trailing edge. Unfortunately, Lingwood's measurements stopped short of the critical radius for the predicted absolute instability. Nonetheless, Lingwood reported that the well-defined structure of the wave packet disintegrated when the critical Reynolds number was approached. The assumption was that this was due to the absolute instability, which caused the turbulence onset.

A different picture emerged following numerical simulations by Davies and Carpenter (2003). This involved solving the fully linearized Navier–Stokes equations for conditions of the rotating disk flow. In their simulation, they introduced impulse-like disturbances that led to the growth of wave packets. At lower Reynolds numbers, the results were found to reproduce the behavior observed by Lingwood (1996). In particular, there was close agreement to the space–time development of the leading and trailing edges of the wave packets found in the experiment. However, in the absolutely unstable region, the strong temporal growth and upstream propagation was not sustained for long times, and the convective instabilities eventually dominated. Thus, they concluded that the absolute instability of the rotating disk boundary layer does not produce a linear amplified global mode but seemed to be more associated with a *transient temporal growth*. As such, it could not explain or solely account for the turbulence transition locations observed in experiments.

An answer to this was suggested by Pier (2003) who examined the secondary instability of finite-amplitude waves of the rotating disk flow. The analysis revealed that the primary saturated waves initiated at the critical radius of the absolute instability are already absolutely unstable with respect to secondary perturbations. In this scenario, the primary nonlinear waves are a prerequisite for the development of the secondary instability that leads to transition to turbulence. The primary waves in this case are traveling with respect to the disk frame and have an azimuthal wave number (n) of 68 ($\omega = 50$).

Othman and Corke (2006) performed a series of experiments to add further clarity to the role of the absolute instability on turbulence transition of the Type I cross-flow instability. The experiment made use of the same rotating disk facility used by Corke and Knasiak (1998), Matlis (1997) and Corke et al. (2007b). This allowed a direct comparison of the basic flow and natural cross-flow instability development that had been documented in those previous experiments. Temporal disturbances were introduced using a pulsed micro air jet that was suspended above the disk from the traversing mechanism. A photograph and schematic of the air jet is shown in Figure 7.17.

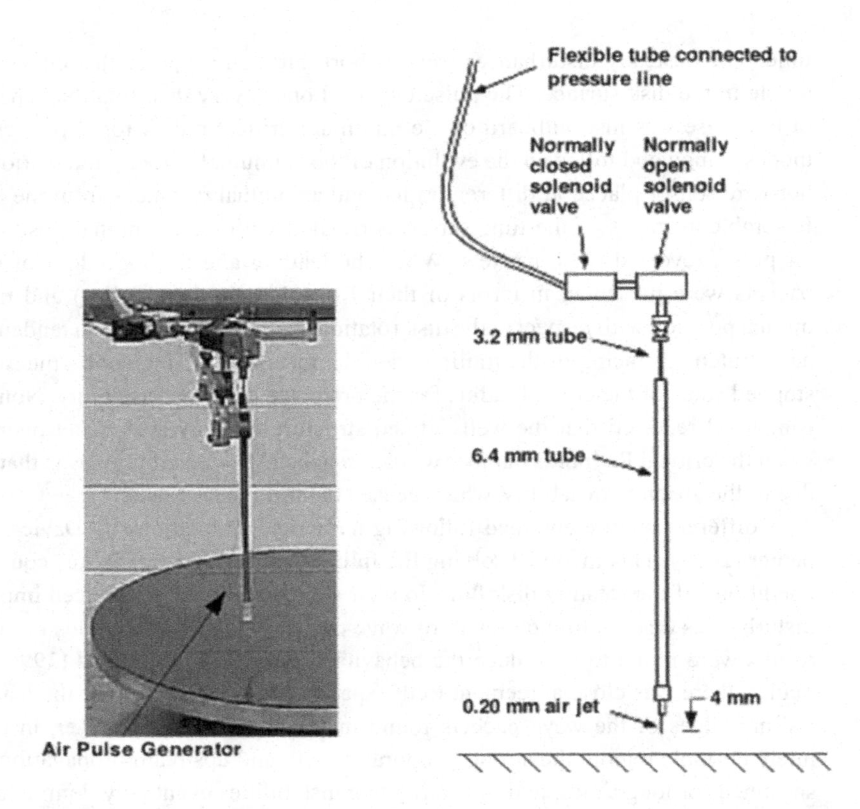

Figure 7.17 Photograph and schematic of air pulse generator used by Othman (2005) to introduce temporal disturbances in rotating disk boundary layer.

The air exited from a hypodermic tube with a 0.203-mm inside diameter. Two computer-controlled electronic solenoid valves, one normally closed and the other normally open, were used to generate repeatable, short-duration air pulses. The full duration of the pulse corresponded to 20 percent of the disk rotation. The exit of the hypodermic tube was located 4 mm above the disk surface, which was approximately twice the boundary-layer thickness. The radial location of the pulsed air jet corresponded to $R = 311$, which was slightly outboard of the Type I instability critical Reynolds number, R_{c_I}. This was the same radial location as the surface-mounted air pulse generator used by Lingwood (1996). Two different pulse amplitudes were used. The lower amplitude air pulse was designed to produce disturbance wave packets with linear characteristics. This was verified by Othman and Corke (2006) in which the wall-normal amplitude distributions of the disturbance wave packets produced by the air pulses were found to match the linear theory eigenfunction.

The velocity measurements consisted of digitally sampling contiguous voltage time series points proportional to the azimuthal velocity, U_θ. The location of the hot-wire relative to the position of the air pulse generator varied. This is represented in the schematic shown in Figure 7.18. Although the radial location of the pulse generator remained fixed at $R = 311$, its azimuthal position was varied to produce different azimuthal spacings, θ, between the pulse generator and the hot-wire sensor.

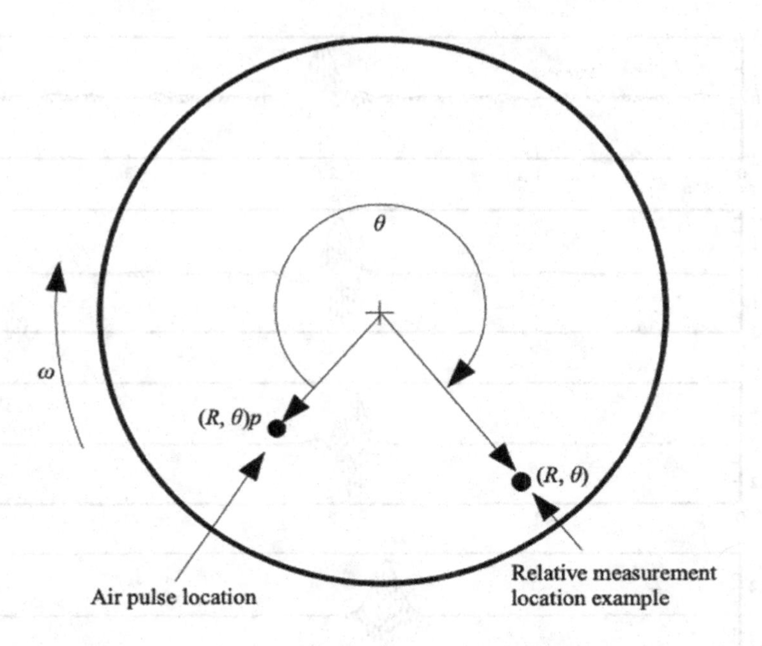

Figure 7.18 Schematic defining the relative positions of the air pulse generator and hot-wire sensor used in documenting the space–time development of disturbance wave packets. Taken from Othman and Corke (2006).

The velocity time series acquisition was coordinated with the pulse initiation. The approach was to start the acquisition of the hot-wire sensor voltage a short time before triggering the air pulse. The voltage signal used to initiate the air pulse was also simultaneously sampled with the hot-wire time series so that it could be used as a time reference for ensemble averaging. Typically, 1,000 records were used to form the ensemble average. There was no correlation between the time-series acquisition and the disk rotational position. Therefore, the ensemble average yielded only features that were related to the disturbance pulse, and traveling with respect to the disk rotation frame of reference.

The evolution of disturbance wave packets was documented by the ensemble-averaged velocity measurements taken at different radial and azimuthal distances from the air pulse disturbance generator. A sample of these is shown in Figure 7.19, which are at $R = 514$, which is just supercritical for absolute instability growth based on Lingwood (1996). The different plots correspond to increasing azimuthal angles from the disturbance generator. The time taken for the wave packets to reach the velocity sensor increases with the azimuthal spacing. The time axis has been normalized by the disk rotation time, T. For time increasing from left to right, the leading edges of the wave packets are on the left-hand side, and the trailing edges are on the right. Note that at this radius for the sensor angle locations, the wave packet has traveled from one to two times around the disk.

Of particular interest with the absolute instability mechanism is the spreading of the wave packets in time. Othman and Corke (2006) determined this by analyzing the evolution of wave packets at all of the locations where leading and trailing edges

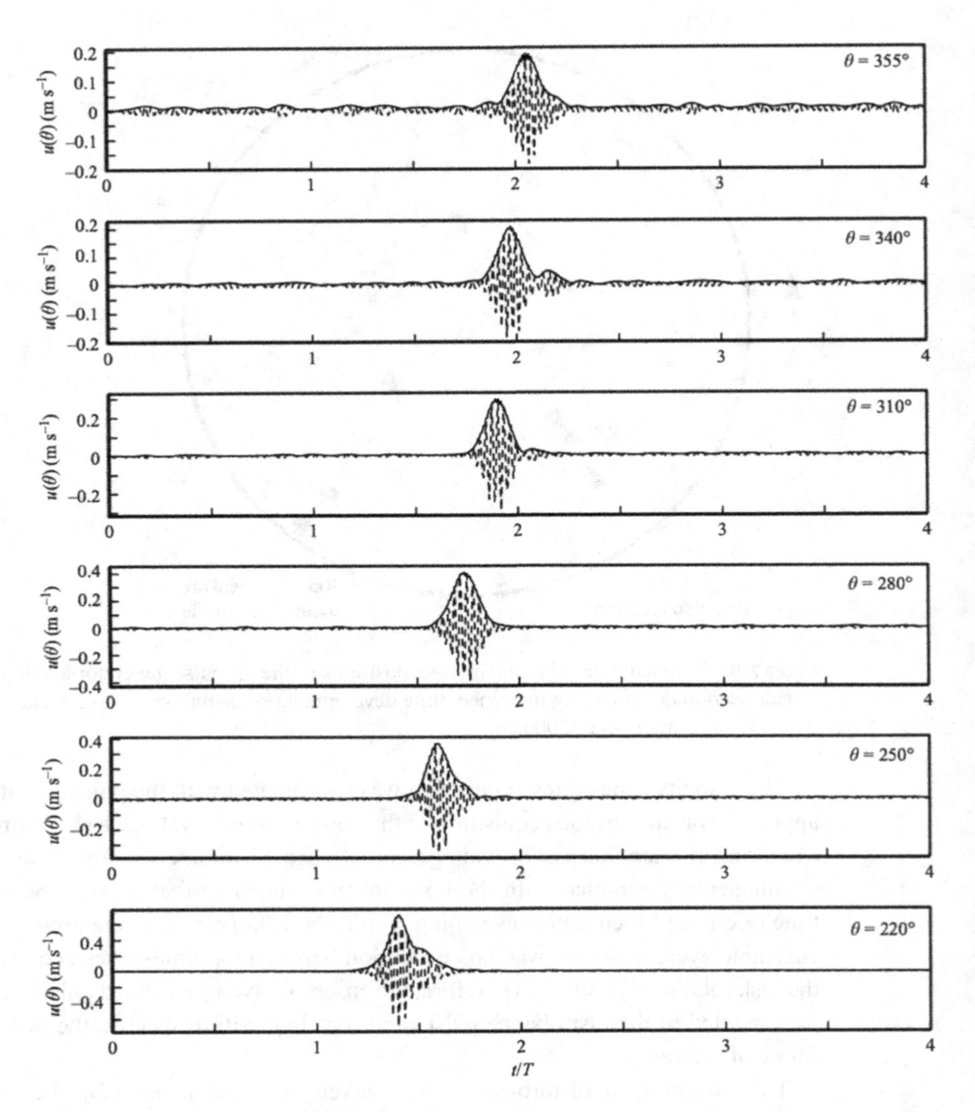

Figure 7.19 Ensemble-averaged azimuthal velocity time series disturbance wave packets measured at different azimuthal angles (θ) at $R = 514$. Dashed curve is ensemble-averaged velocity. Solid curve is amplitude envelope produced by a specially designed digital Hilbert filter. Taken from Othman and Corke (2006).

were identifiable. The processes for determining the edges of the wave packet were (1) defining the amplitude envelope calculated using a specially designed Hilbert transform, (2) locating the peak in the envelope amplitude, and then (3) moving forward and backward in time to find the point where the amplitude envelope reached the disturbance background level. The result of determining the wave packet leading and trailing edges is presented in Figure 7.20. The square symbols mark all of the leading edges of the wave packets at every position. The triangle symbols mark all of

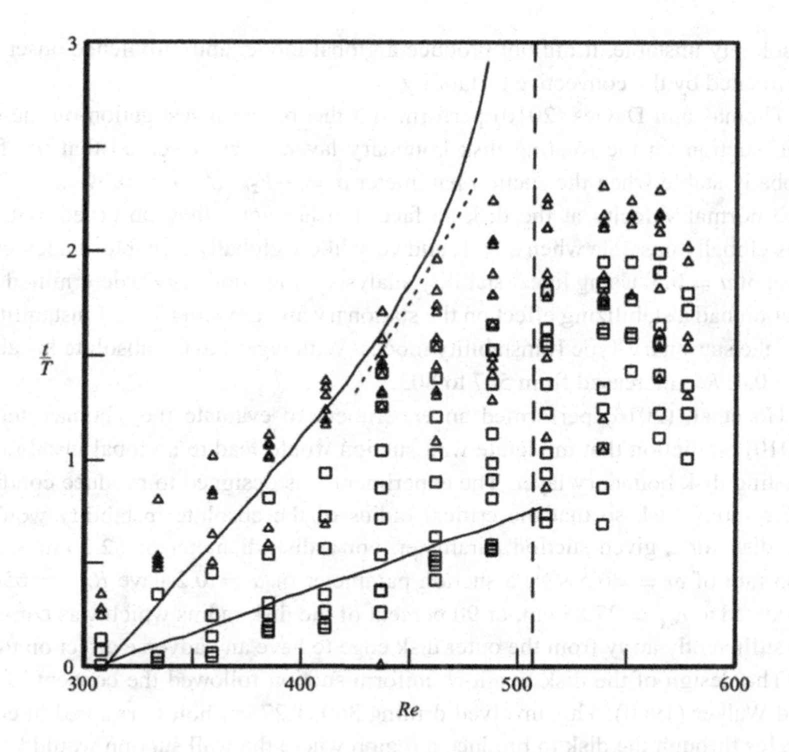

Figure 7.20 Space–time map of leading-edge (squares) and trailing-edge (triangles) locations of disturbance wave packets generated by air pulse. Taken from Othman and Corke (2006). Note: Solid curves are the trend from Lingwood (1996). The dashed line is from the linear simulation of Davies and Carpenter (2003).

the trailing edges. The distances between the leading and trailing-edge pairs represent individual wave packets. Singly they might appear as fingers that are elongated in the R-direction and slightly inclined in the time-direction.

As the wave packets develop radially, they merge together. Therefore, what is ultimately important is the upper bound in time of the wave packet trailing edges, and the lower bound in time of the wave packet leading edges. For reference, the solid curves mark the bounds drawn by Lingwood (1996). The dotted curve corresponds to the space–time development of the trailing edge of an $n = 67$ disturbance for the linear simulation of Davies and Carpenter (2003). The vertical dashed line marks the critical Reynolds number for the absolute instability from Lingwood (1997), $R_{c_A} = 507$.

The absolute instability analysis of Lingwood (1995) and the accompanying experiment of Lingwood (1996) predicted an acceleration of the temporal spreading of the wave packet trailing edge. In the experiment (Lingwood, 1996), this was largely extrapolated since the largest measurement Reynolds number was approximately 480. In contrast, Figure 7.20 documents a deceleration of the trailing-edge temporal spreading, with it asymptotically approaching a constant as predicted by Davies and Carpenter (2003). The conclusion was that although the higher frequencies are absolutely unstable, it did not produce a global mode, and turbulence onset was still dominated by the convective instability.

Thomas and Davies (2010) performed a theoretical investigation on the effect of wall suction on the rotating disk boundary layer. They observed that the flow was globally stable when the suction parameter $a = -V_{z0}/\sqrt{\nu\omega} \leq 0$, where $-V_{z_0}$ is the wall-normal velocity at the disk surface. Furthermore, they observed that the flow was globally unstable when $a = 1$, and very likely globally unstable at a lesser suction level of $a = 0.5$. Using linear stability analysis, Lingwood (1997) determined that wall suction had a stabilizing effect on the stationary and traveling Type I instability modes and the stationary Type II instability modes. With regard to the absolute instability, for $a = 0.4$, R_{c_A} increased from 507 to 803.

Ho et al. (2016) performed an experiment to evaluate the Thomas and Davies (2010) prediction that moderate wall suction would lead to a global instability of the rotating disk boundary layer. The experiment was designed to produce conditions on the rotating disk so that the critical radius of the absolute instability would be on the disk for a given suction parameter. For a disk diameter of 62.2 cm and a rotation rate of $\omega = 86.5$ s^{-1}, a suction parameter of $a = 0.2$ gave $R_{c_A} = 650$, which occurred at $r_{c_A} = 27.85$ cm, or 90 percent of the disk radius which was considered to be sufficiently away from the outer disk edge to have any adverse effect on the flow.

The design of the disk to allow uniform suction followed the concept of Gregory and Walker (1960). This involved drilling 360, 1.27 cm holes arranged in concentric circles through the disk to produce a region where the wall suction would be applied. This is shown in the photograph in Figure 7.21(a). The locations of the holes start at the radial location $r = 14.29$ cm and continue to $r = 29.21$ cm. At the design disk rotation rate, $\omega = 86.5$ s^{-1}, these radial locations corresponded to $R = 317$ and $R = 696.5$. This placed the suction portion of the disk inboard of R_{c_A} with zero suction and slightly outboard of R_{c_A} with a suction parameter of $a = 0.2$.

The holes through the disk were covered with a sheet of pressed wire mesh that added a specific amount of pressure drop to help uniformize the suction on the surface of the disk. This is shown in the photograph in Figure 7.21(b). This wire mesh surface was not smooth enough for the flow to remain laminar before reaching the absolute instability critical radius with a suction parameter of $a = 0.2$. Therefore, a smooth, porous covering consisting of a fine weave silk cloth was stretched over the disk. A special tensioning device as shown in Figure 7.21(c) was used to stretch the cloth over the disk. The cloth was held tight by a clamping ring that was attached to the outer edge of the disk. The final disk assembly with the stretched cloth is shown in Figure 7.21(d). A variable speed vacuum pump drew air through the porous surface of the disk.

Ho et al. (2016) utilized the same air pulse generator developed by Othman and Corke (2006) to introduce temporal disturbances. Also following Othman and Corke (2006), a hot-wire oriented to be primarily sensitive to U_θ was located at different (r, θ) locations to capture the developing of the disturbance wave packets. The determination of the wave packet leading and trailing edges involved match-filtering of the time series. The matched filter provided a measure of the temporal correlation between the

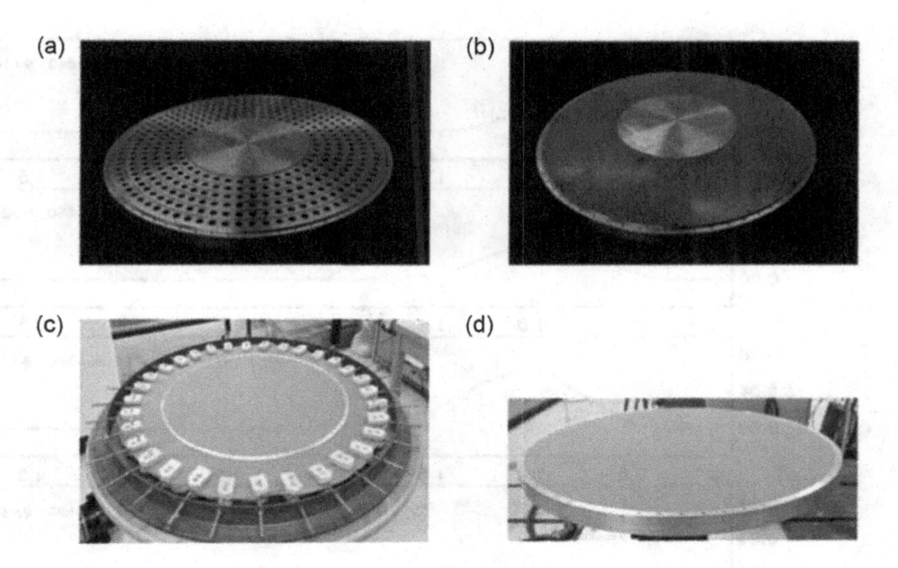

Figure 7.21 Construction steps of disk designed for suction showing array of holes (a), wire mesh (b), cloth stretching device (c), and final disk (d). Taken from Ho (2014).

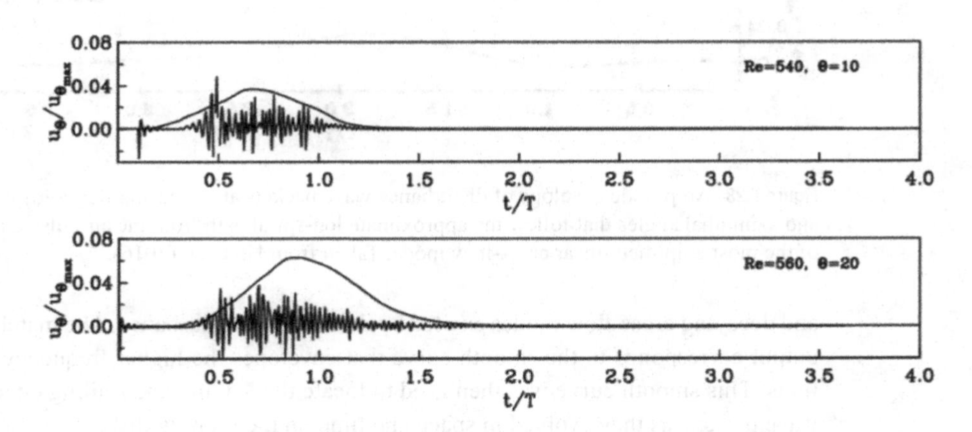

Figure 7.22 Ensemble-averaged time series and corresponding matched filter output based on azimuthal velocity fluctuations of disturbance wave packets measured at two radii and azimuthal positions on the rotating disk. Taken from Ho et al. (2016).

time series that were used to identify the bounds of the wave packets. The correlations were performed in the frequency domain using a digital fast Fourier transform. The time series each corresponded to four disk rotations. The matched filter outputs were averaged over approximately 1,000 air pulse events. No other filtering of the time series was performed. Examples of the output from the matched filtering (Ho et al., 2016) are shown in Figure 7.22.

The ensemble averaged time series is made up of higher frequency fluctuations that are representative of the azimuthal velocity fluctuations associated with the growing

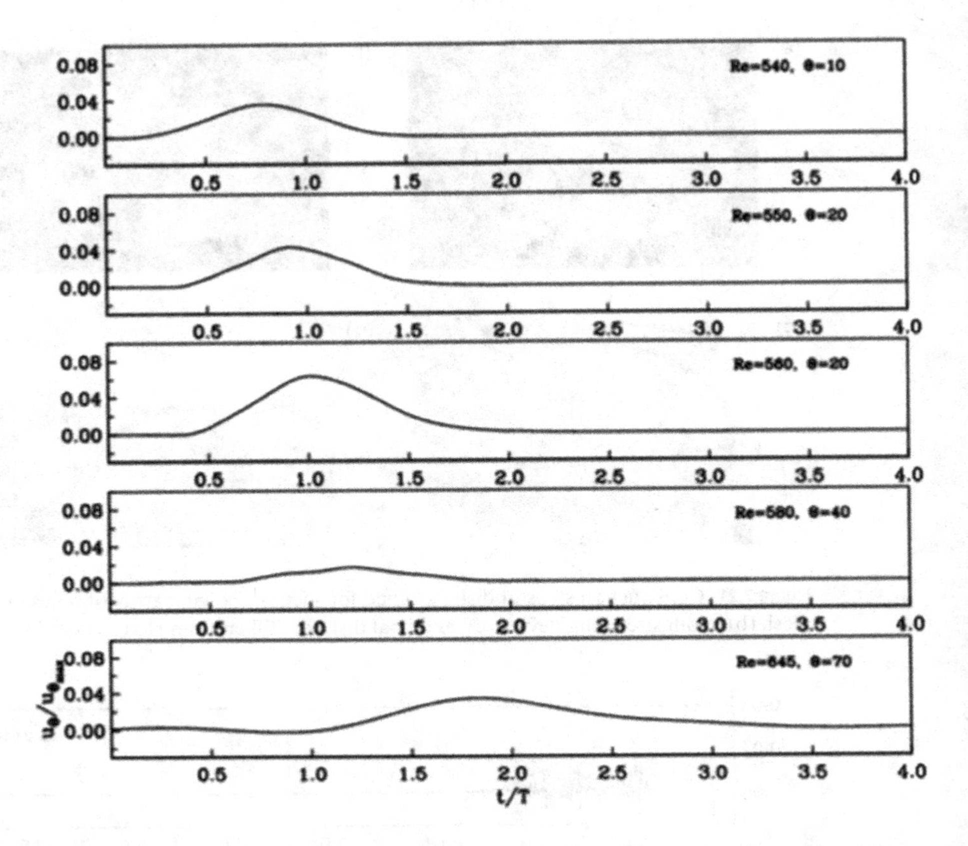

Figure 7.23 Amplitude envelope of disturbance wave packets at increasing Reynolds numbers and azimuthal angles that follow the approximate log-spiral path from the air pulse generator of the most amplified linear cross-flow mode. Taken from Ho et al. (2016).

and decaying cross-flow modes excited by the air pulse disturbance. The matched filter output corresponds to the smooth curve that envelopes the higher frequency fluctuations. This smooth curve was then used to locate the leading and trailing edges of the wave packets as they evolved in space and time on the rotating disk.

An example of the radial development of the disturbance wave packet envelope along a log-spiral trajectory based on the most amplified linear mode is shown in Figure 7.23. The abscissa in the figure is time normalized by a disk rotation, T. This shows a progression where, in addition to the spatial transport, there is a temporal spreading of the leading and trailing edges of the wave packet. At the largest radius, the wave packet is observed to have spread to encompass approximately 2.5 rotations of the disk.

Ho et al. (2016) compiled the space–time development of wave packet leading and trailing edges for a large set of locations on the rotating disk. The results are presented in Figure 7.24. The open and closed symbols in the figure respectively correspond to the leading and trailing edges of the wave packets. The upward arrow symbols signify data points where the location of the trailing edge was beyond the four-rotation time

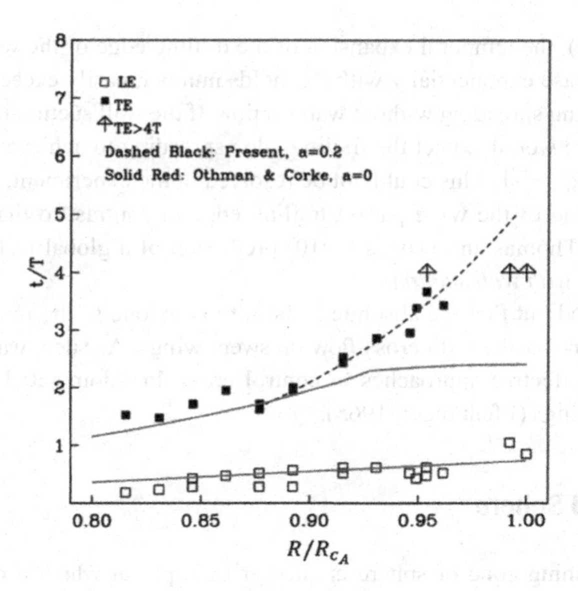

Figure 7.24 Space–time map of leading- and trailing-edge locations of disturbance wave packets for rotating disk with $\omega = 86.5\ \mathrm{s}^{-1}$ and $a = 0.2$. Solid red curve is space–time map from Othman and Corke (2006) with $a = 0$. Upward-arrow data points signify that trailing-edge location exceeded the four-disk-rotation time series limit. Dashed curve is an exponential fit to the trailing-edge data from $0.88 \le R/R_{c_A} \le 0.96$ that was extrapolated to $R_{c_A} = 1.0$. Taken from Ho et al. (2016).

series limit ($t/T > 4$). Also included are the results from Othman and Corke (2006) that were shown in Figure 7.20 for the same conditions but without wall suction, namely $a = 0$. Those results are shown by the solid red curves. In order to compare to the previous wave packet measurements without disk suction, the Reynolds number was normalized by the absolute instability critical Reynolds number, R_{c_A}, where without suction, $R_{c_A} = 507$, and with a suction parameter of $a = 0.2$, $R_{c_A} = 650$. On the ordinate, time is normalized by the disk rotation period, T.

The results show a good agreement of the leading-edge development (open square symbols) between the two experiments with $a = 0$ and $a = 0.2$. For the trailing-edge development (closed square symbols), there is good agreement between the two experiments up to $R/R_{c_A} = 0.876$ ($R = 570$). Outboard of that location, the case with wall suction shows continual spreading of the wave packet trailing edge that was not observed without wall suction. The dashed curve is an exponential fit of the trailing-edge data from $0.876 \le R/R_{c_A} \le 0.962$ that was then extrapolated to $R_{c_A} = 1.0$. The three cases denoted by the upward arrow symbols signify that the trailing-edge locations exceeded the four-disk-rotation limit of the contiguous time series.

As evident by the red curve for the wave packet trailing-edge development without suction from Othman and Corke (2006), the temporal spreading of the wave packets had asymptoted even as R_{c_A} was approached, which, as previously mentioned, agreed with the simulation observations of Davies and Carpenter (2003). In contrast, with wall suction ($a = 0.2$), the temporal expansion of the trailing edge of the wave packets continued to increase exponentially with Reynolds number, easily exceeding by a factor of two the trailing spreading without wall suction. If the wall suction resulted in a global instability, we would expect the trailing-edge spreading to achieve a vertical temporal slope at $R/R_{c_A} = 1$. This could not be resolved in the experiment. However, the enhanced spreading of the wave packet trailing edge in contrast to that without suction supports the Thomas and Davies (2010) prediction of a global instability of the rotating disk flow *with wall suction*.

It should be pointed out that the absolute instability is unique to the rotating disk with suction. It does not occur with cross-flow on swept wings. As such, wall suction presents one of the effective approaches to control cross-flow-dominated turbulent transition on swept wings (Pfenninger, 1965).

7.3 Rotating Cone and Sphere

The flow over a spinning cone or sphere is another example in which a cross-flow instability is the dominant mechanism of turbulence onset. Early studies on rotating cones in still air were performed by Kreith et al. (1962) and Tien and Campbell (1963). Their focus was on measuring the transition Reynolds number, but they offered no evidence on the mechanism of transition. An early work by Salzberg and Kezios (1965) investigated the transition Reynolds number on spinning cones in a free-stream flowfield. As a follow-up experiment, Okamoto et al. (1976) found that turbulent transition depended on the rotational parameter, $\omega r/U_o$, where ω is the rotation speed, r is the local radial distance from the axis of symmetry, and U_0 is the free-stream velocity. Later, Kobayashi (1981) suggested a slightly revised rotational parameter of $\omega r_b/U_e$, where r_b is the radial distance from the axis of symmetry at the base of the cone, and U_e is the local flow speed at the edge of the boundary layer.

The details of the mechanism for transition on rotating cones first came through flow visualization experiments by Mueller et al. (1981), Kegelman et al. (1983), Kobayashi and Izumi (1983b), Kobayashi and Kohama (1985), and Kohama (1985). An example visualization from Kobayashi and Izumi (1983b) is shown in Figure 7.25. Similar spirals were observed by Kohama and Kobayashi (1983a) for rotating spheres. An example is shown in Figure 7.26. Although the streaks that appear for the spinning cone in a uniform flow appear to be similar to stationary cross-flow vortices, Kohama (1985) found them to be counter-rotating rather than corotating cross-flow vortices. This was similarly observed for the rotating sphere by Kobayashi and Izumi (1983a). For a cone rotating in quiescent fluid, Kohama and Kobayashi (1983b) observed spiral counter-rotating vortices for total included angles between 0° and 60° and corotating vortices for angles between 60° and 180°. The former range is dominated by a centrifugal instability, whereas the latter is dominated by a cross-flow instability.

Based on their work on the rotating cone and sphere, Kohama and Kobayashi (1983) and Kohama and Kobayashi (1983b) determined that there exist two kinds of spiral vortices on rotating bodies; one that winds in the same direction as neighboring vortices (A), and one that winds in the opposite direction (B). Type A vortices, observed on the rotating sphere, are similar to those found on a rotating disk. Type B vortices

Figure 7.25 Flow visualization for a rotating cone revealing spiral streaks. Taken from Kobayashi and Izumi (1983b).

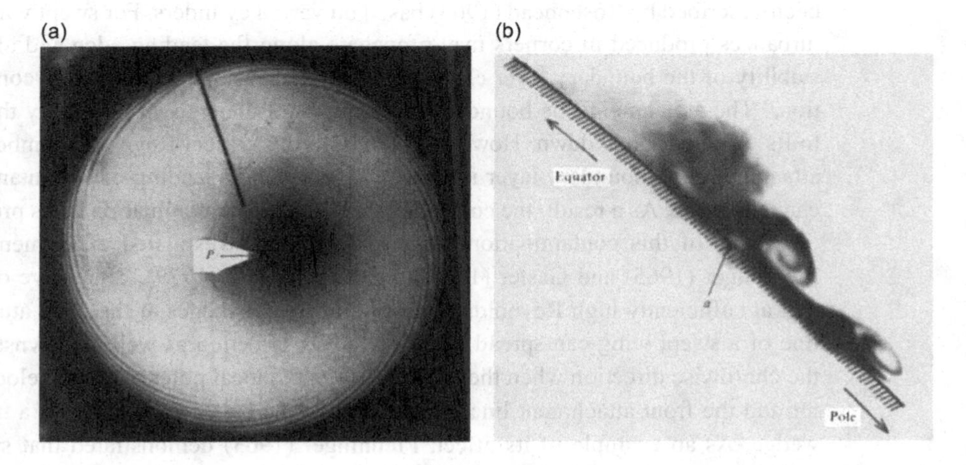

Figure 7.26 Flow visualization for a rotating sphere revealing spiral streaks (b) with a cross section (b) revealing corotating vortices. Taken from Kohama and Kobayashi (1983a).

transform to type A on a rotating cone when the total included angle and the axial-flow velocity exceed certain values. Types A and B are otherwise known as corotating and counter-rotating, respectively.

Kobayashi et al. (1987) studied the effect of free-stream turbulence on boundary-layer transition on a rotating cone in an axial flow. The free-stream turbulence levels (u'/U_∞) varied from 0.04 percent to 3.5 percent. These free-stream turbulence levels were found to have no effect on the transition Reynolds number.

7.4 Attachment Line Flow

Attachment line flow is formed at leading edges of swept wings where the flow splits over and under the wing. A schematic that illustrates this effect is shown in Figure 7.27 which is based on Poll (1979). The characteristics of this boundary-layer flow have

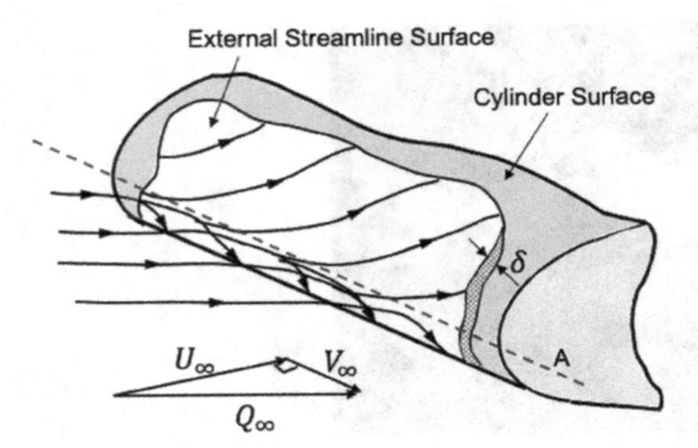

Figure 7.27 Schematic of the flow near the leading edge of a swept wing. Based on Poll (1979).

been described by Rosenhead (1963) based on yawed cylinders. For swept wings, disturbances produced in corners may propagate along the leading edge and affect the stability of the boundary layer elsewhere, the result being "leading-edge contamination." The attachment-line boundary layer can itself undergo an instability that leads to its turbulent breakdown. However, the instability critical Reynolds number of the attachment-line boundary layer is above the limit where leading-edge contamination can take place. As a result, the control of leading-edge contamination takes precedent.

Details of this contamination process were first investigated experimentally by Pfenninger (1965) and Gaster (1967), and later by Poll (1979). They have observed that at sufficiently high Reynolds numbers, turbulent wedges at the front attachment line of a swept wing can spread in the spanwise direction as well as downstream in the chordwise direction when the angle between the local potential-flow velocity vector and the front attachment line is smaller than the half-spread angle of a turbulent wedge. As an example of its effect, Pfenninger (1963) demonstrated that spanwise turbulent contamination along the 33° swept X-21 leading edge was responsible for the loss of laminar flow. He established the conditions for the existence of spanwise contamination in the presence of an initially turbulent attachment-line boundary layer, namely

$$R_{\theta_{AL}} = V_e \theta_{AL}/\nu = 90 - 100, \tag{7.7}$$

where V_e is the spanwise component of the local potential-flow velocity and θ_{AL} is the local boundary-layer momentum thickness. Considering an incompressible laminar boundary layer, the attachment-line Reynolds number can be redefined as (Sears, 1948)

$$R_{\theta_{AL}} = 0.404 \left[V_0 r \sin^2 \Gamma/(1+\epsilon)\nu \cos \Gamma \right]^{1/2}, \tag{7.8}$$

where the front part of the wing has been replaced by an equivalent ellipse of the same leading-edge radius, Γ is the sweep angle, V_0 is the undisturbed free-stream velocity, r is the leading-edge radius, and ϵ is the thickness of the equivalent ellipse.

Gaster (1967) studied the effect of placing trip wires normal to the attachment line as a means of controlling attachment-line contamination. For this, he defined a length scale, η, given as

$$\eta = [\nu/(dU_e/dx)_{x=0}]^{1/2}, \tag{7.9}$$

where U_e is the edge velocity in the streamwise direction, ν is the kinematic viscosity, and x is in the chordwise direction. Gaster found that for large trip wires, the transition Reynolds number was $R_{\text{trans}} = 104$. For smaller diameter wires, Gaster found that the transition Reynolds number scaled with the wire diameter, d, as

$$R_{\text{trans}} = 364(\eta/d)^2. \tag{7.10}$$

This is known as "Gaster's criterion." Gaster also showed that attachment-line contamination could be prevented by the use of a specially designed "Gaster Bump" at the leading edge near the wing root.

In a series of experiments on yawed cylinders, Poll (1977, 1978, 1979), investigated the Gaster d/η parameter on attachment-line transition. For $0.8 < d/\eta < 1.55$, the trip wires produced disturbances that developed along the attachment line that eventually developed turbulent onset. For $d/\eta \geq 1.55$, turbulence onset occurred directly downstream of the trip wire. For reference to the boundary-layer characteristics, for the yawed cylinder, the momentum thickness is $\theta = 0.404\eta$. Thus, the trip wire diameters can be put in terms of the boundary momentum thickness. Based on the minimum effective $d/\eta = 0.8$, the critical Reynolds number based on the edge velocity parallel to the leading edge, and the length scale, η, is $Re_{\eta_{\text{trans}}} = 250$.

With regard to attachment flow instability, it does not produce streamwise vortices of the cross-flow type. Hall et al. (1984) modeled the stagnation flow as a "swept" Hiemenz-type flow that admitted similarity solutions with a nonzero velocity normal to the wall. The analysis showed that the most unstable linear disturbance is a traveling wave of the T-S type. Hall et al. (1984) subsequently presented neutral-stability calculations for different values of suction and blowing. Their linear results were in good agreement with the experimental results of Poll (1979).

7.5 Compressible 3-D Boundary Layers

A majority of the experiments on cross-flow-dominated transition to turbulence have been performed on swept wings at low, subsonic Mach numbers. However, the underlying physics of the cross-flow instability is the same at compressible Mach numbers.

Swept wing investigations at supersonic Mach numbers include those of Saric and Reed (2002), Saric et al. (2004), Semionov et al. (2006) and Semionov and Kosinov (2007). Saric and Reed (2002) for example, investigated a cooled swept wing in a wind tunnel experiment at Mach 2.4. The wing sweep angle was large enough to produce a subsonic leading edge. Infrared thermographic images were used to indicate the onset

Figure 7.28 Surface flow visualization for a right-circular cone at a 4.2° angle of attack in a Mach 3.5 stream that reveals the characteristic pattern of stationary cross-flow vortices. Taken from Schuele et al. (2013).

of transition, which appeared as an increase in the wall temperature due to enhanced turbulent heat transfer.

A number of high Mach number cross-flow experiments have also been conducted on right-circular cones at angles of attack. The conical flow at incidence generates a streamline-normal pressure gradient that, if properly designed, can lead to a mean cross-flow that is similar to that on a swept wing. Examples of cone experiments include King (1992) at Mach 3.5, Wolf and Laub (1997) at Mach 1.6, Schuele et al. (2013) at Mach 3.5, and Corke et al. (2018) and Arndt et al. (2020) at Mach 6.0. Of these, King (1992) was the first to mention the cross-flow instability as a transition mechanism.

As an example, Figure 7.28 shows a flow visualization image from Schuele et al. (2013) for a right-circular cone at a 4.2° angle of attack in a Mach 3.5 stream that reveals the characteristic pattern of stationary cross-flow vortices. The flow visualization in this case consisted of a mixture of 15 parts 1000cSt silicon oil, 5 parts oleic acid, and 1 part titanium-dioxide powder that was applied uniformly to the surface of the cone before the tunnel run. The pattern in the surface visualization oil developed after the wind tunnel had been operating for a sufficiently long amount of time whereby the oil pattern stopped changing. This was on the order of four to five minutes. The light and dark bands result from the pumping action of the corotating stationary cross-flow vortices, with the light bands being an accumulation of the surface marker (low shear stress) and the dark bands being a depletion of the surface marker (high shear stress). Similar images were observed by King (1992).

The flow visualization image in Figure 7.28 reveals a pattern of stationary cross-flow vortices that appear in the upper azimuthal portion of the cone at an angle of attack. In general, the cross-flow instability growth rate is a maximum at the location where the cross-flow Reynolds number, $R_{CF} = V_{max}\delta/\nu$, is a maximum. V_{max} in this case is the maximum cross-flow velocity normal to the boundary-layer edge inviscid streamline. For a cone at an angle of attack, the cross-flow Reynolds number increases and then decreases from the windward side of the cone to the leeward side. The maximum cross-flow Reynolds number occurs over a range of azimuthal angles of $130° \leq \theta \leq 160°$, where $\theta = 0°$ is the windwardmost ray on the cone, and $\theta = 180°$ is the leewardmost ray.

Linear stability analysis of the boundary-layer flow over the cone for the experimental conditions of Schuele et al. (2013) was performed by Balakumar (2009). The result of the analysis is shown in Figure 7.29. The curves correspond to growth rates of cross-flow modes determined along axial rays of constant azimuthal angles of $\theta = 0°$, 45°, 90°, and 180°. The axial rays are depicted in the schematic in Figure 7.29(a). The notation in the inset of the Figure 7.29(b) plots are the dimensionless frequency, F, and the azimuthal wave number, m. The stationary mode growth curves correspond to $F = 0$ and have been highlighted with the square symbol. The x-axis is the distance along the cone axis.

As expected, from Figure 7.29, the growth rates of the traveling cross-flow modes ($F \neq 0$) are larger than those of the stationary modes. However, as previously discussed, the extreme sensitivity of the stationary cross-flow mode to surface roughness provides initial disturbance levels that favor the stationary mode. Finally, with respect to the azimuthal mode numbers for stationary modes, the initially amplified band is near a mode number, m, of approximately 50. This means that if projected around the 360° azimuth of the cone, there would be 50 stationary cross-flow vortices spaced $360°/50 = 7.2°$ apart. In actuality, because the largest amplification occurs over a range of azimuthal angles of $130° \leq \theta \leq 160°$, as evident in the flow visualization image in Figure 7.28, it is in this azimuthal region where the stationary pattern is observed. However, in that region, for $m = 50$, the expected spacing is 7.2°.

In general, turbulent transition on hypersonic lift-generating geometries is likely *dominated by a mean cross-velocity component* that is subject to a cross-flow instability. In particular, it takes only a small amount of a cross-flow for the linear amplification rate of cross-flow modes to exceed that of the second-mode instability (see Chapter 6). The dominance of cross-flow transition over other mechanisms is further accentuated by its extreme sensitivity to minute surface roughness.

Flight control within the atmosphere is an important requirement for the current generation of hypersonic vehicles. At present, this involves fins with highly swept leading edges. Like all highly swept configurations, the boundary layer over a fin develops a mean cross-flow component. A recent numerical simulation of the flow over a 70° swept fin at Mach 6 (Knutson et al., 2018) revealed a significant cross-flow velocity, with magnitudes that were three times larger than that on a 7° half-angle cone at a 6° angle of attack, where experiments by Corke et al. (2018) and Arndt et al. (2020) confirmed the cross-flow instability as the dominant mechanism of turbulence onset.

Middlebrooks (2022) performed experiments to investigate the flow field over a 7° half-angle cone that included a 70° swept fin in a Mach 6 free-stream. A special emphasis was placed on the mean cross-flow and resulting cross-flow instability that occurred in the boundary layer over the swept fin. The documentation of the flow field over the cone and fin involved infrared thermography that was used to quantify surface

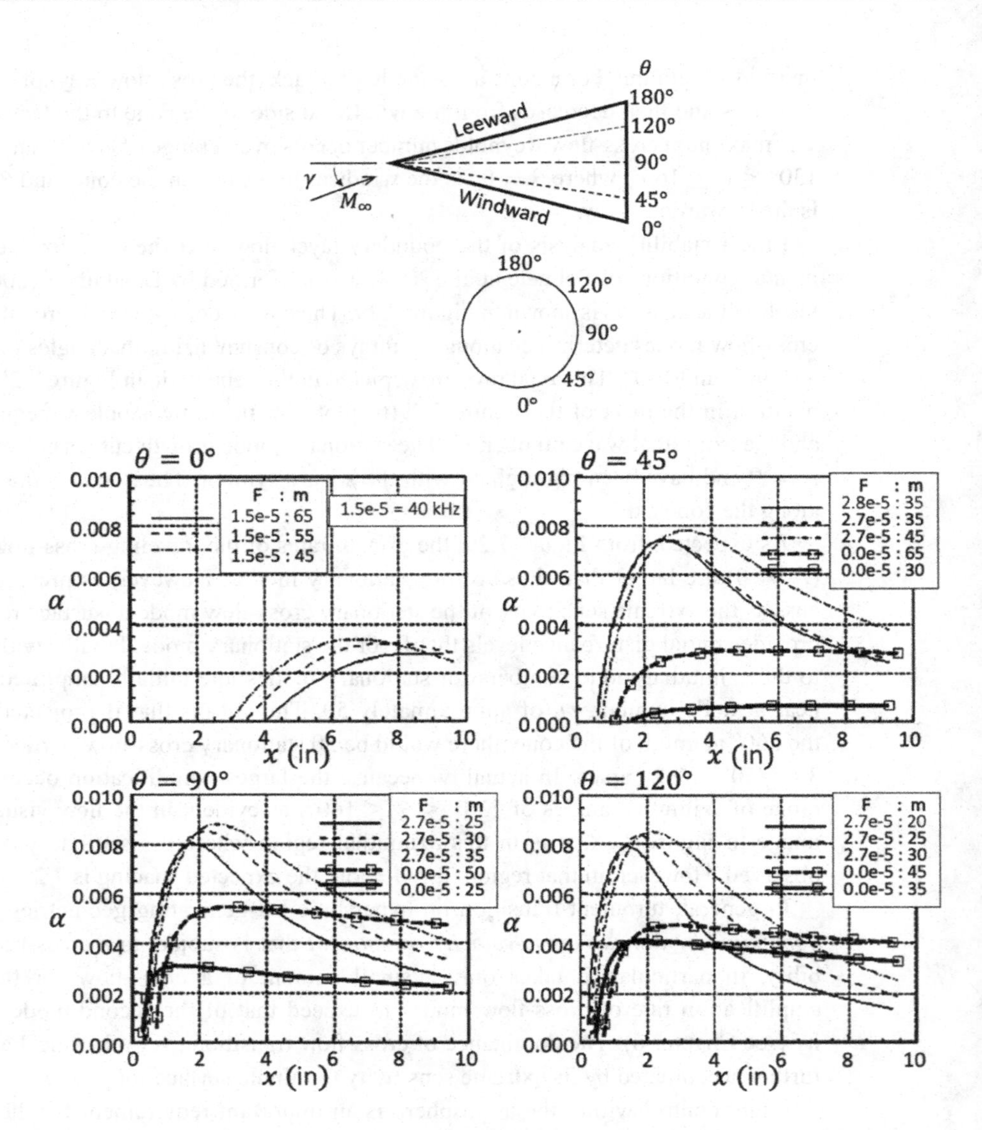

Figure 7.29 Growth rates, α, along axial rays at constant azimuthal angles for 7° half-angle right-circular cone at a 4.2° angle of attack in a Mach 3.5 stream with $Re_\infty = 9.8 \times 10^6\ m^{-1}$. Taken from Balakumar (2009).

heat flux and subsequently the turbulence transition front. The infrared thermography also revealed stationary features that could be associated with stationary cross-flow modes. For these measurements, the cone frustum and swept fin were covered by a 90-μm thick, matte black 3-M 1080 series film that was chosen because of its excellent emission properties. A photograph of the cone model with the 3-M film applied is shown in Figure 7.30. An example of an infrared image that focused on the region over the fin is shown in Figure 7.31(a). The IR thermal images were processed to obtain the corresponding images of surface heat flux. The heat flux image shown in Figure 7.31(b) corresponds to the IR image shown in Figure 7.31(a). Both images clearly reveal hot and cold streaks that are indicative of stationary cross-flow modes.

Figure 7.30 Photograph of 7° half-angle cone with 70° swept-fin covered with low emissivity film to enhance infrared imaging. Taken from Middlebrooks (2022).

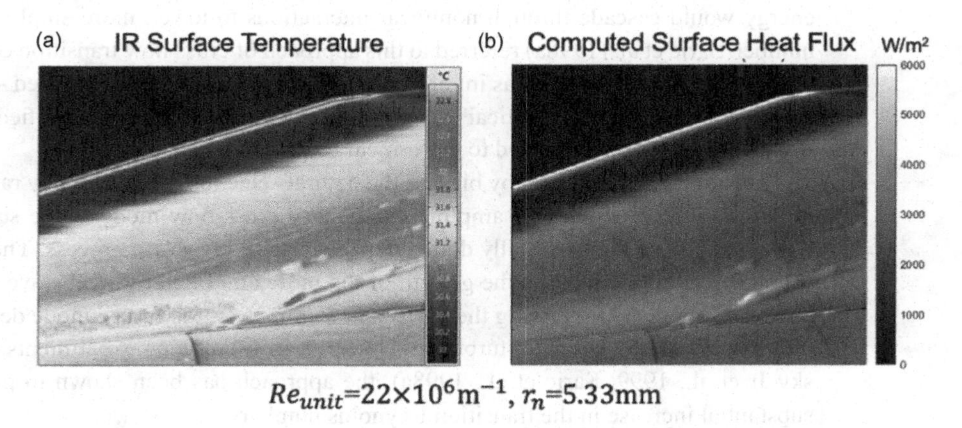

Figure 7.31 Infrared image focusing on the region of a 70° swept fin mounted on a 7° half-angle cone at Mach 6 (a) and the corresponding surface heat flux derived from the IR image (b). Taken from Middlebrooks (2022).

7.6 Cross-flow Transition Control

The following sections describe methods that are highly effective in controlling boundary-layer cross-flow instability and subsequently delaying turbulence onset. The approach is based on the extreme sensitivity of the stationary cross-flow mode to surface roughness, either passive or active. However, rather than random roughness, the design of the roughness is based on linear stability analysis that determines the most amplified wave numbers and amplification N-factors.

7.6.1 Discrete Roughness Elements (DREs)

The approach to control cross-flow-dominated transition to turbulence stems from the extreme receptivity of the instability stationary modes to surface roughness. This feature was exploited by Corke and Knasiak (1998), Matlis (1997), and Corke et al. (2007b) to excite selected wave numbers of cross-flow modes in the boundary layer over a rotating disk, which is a canonical 3-D flow that exemplifies the cross-flow instability.

Saric et al. (1998a) and Radeztsky Jr et al. (1999) exploited this property in their swept wing experiments to excite fixed spanwise wave number stationary cross-flow modes using arrays of micron-sized circular distributed roughness elements. They demonstrated that stationary cross-flow modes at the forced spanwise wave number exclusively appeared in the boundary layer. Super-harmonic wave numbers of the forced wave number were also possible to appear; however, lower wave numbers were always completely suppressed.

These observations led to the concept of exciting less amplified stationary cross-flow modes as a means of controlling transition to turbulence in these flows. The key element was that the spanwise wave number of the forced mode had to be higher than those of the naturally amplified band. This would guarantee that no disturbance energy would cascade through nonlinear interactions to lower, more amplified wave number. Saric et al. (1998a) referred to this approach of cross flow transition control as "subcritical forcing." This was in reference to the band of initially amplified spanwise wave numbers being the "critical" wave numbers. Excitation of that amplified band of wave numbers is then referred to as "critical forcing."

Transition control comes by biasing the natural selection mechanism by raising the initial amplitude of the less amplified stationary cross-flow mode at the subcritical wave number, so that it initially dominates the instability growth process. The growth of the forced mode inhibits the growth of the more amplified (critical wave number) stationary mode by modifying the basic state. Ultimately, the forced mode decays and the flow eventually becomes turbulent. However, in swept wing experiments (Radeztsky Jr et al., 1999; Saric et al., 1998a), the approach has been shown to produce a substantial increase in the transition Reynolds number.

Saric et al. (1998c) found that patterned surface roughness in the form of hemispherical "dots" with a height of 50 μm, provided an effective transition delay in low-speed swept wing experiments. Reibert (1996) observed that for a fixed (subcritical) roughness spacing of $z = 12$ mm, the total disturbance amplitude grew to a constant saturation amplitude even when the roughness height was varied from 6 to 48 μm. Although the initial disturbance amplitude increased with larger roughness, the effects downstream relaxed and yielded similar looking mode shapes. Additionally, Reibert (1996) documented that as the roughness height increased from 6 to 18 μm, in 6-μm increments obtained by stacking the roughness elements, transition for the three roughness heights occurred at the same location. One might conclude that for these three roughness heights, the disturbance they created met linear theory assumptions.

Schuele et al. (2013) investigated the use of discrete roughness to control cross-flow transition on a cone at an angle of attack at Mach 3.5. This utilized the 7° half-angle cone at a 4.2° angle of attack on which the surface flow visualization and stability analysis presented in Figures 7.28 and 7.29 were derived. The stability analysis indicated that the most amplified stationary cross-flow modes had a mode number between $m = 45$–50. In addition, the analysis indicated that the growth of cross-flow modes began at approximately 1.27 cm (0.5 in) from the cone tip. Both of these represent critical information needed to design the discrete roughness for transition control. Based on this, two discrete roughness mode numbers were selected: one with $m = 45$

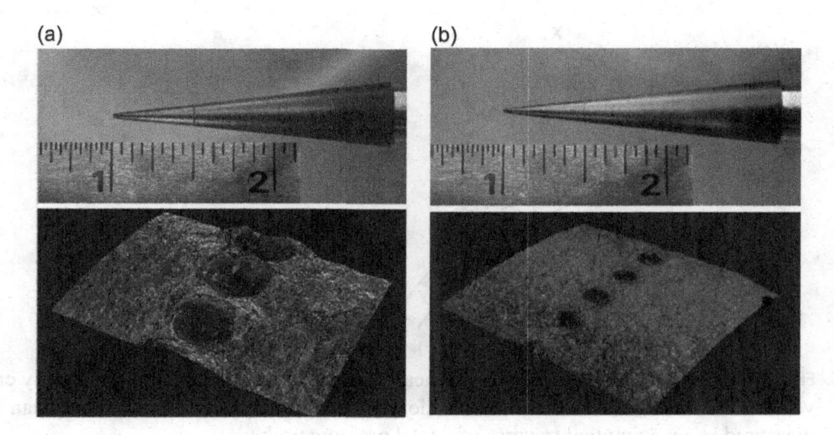

Figure 7.32 Photograph of cone tip with $m = 45$ dimples (a) and $m = 68$ dimples (b) along with their respective magnified views shown below. Taken from Schuele et al. (2013).

that represented the controlled "critical" roughness condition, and one with $m = 68$ that is 1.5 times higher than the critical azimuthal mode number (Saric and Reed, 2003) and therefore represents the "subcritical" roughness condition designed to delay turbulence onset.

The roughness was applied to two interchangeable cone nose tips as uniformly spaced indentations or "dimples." The roughness was located 1.27 cm (0.5 in) from the cone tip so that it was at the axial location where the cross-flow modes are first amplified. Placing then too far upstream would cause the disturbances to decay exponentially (linear theory) before reaching the location where the disturbances would amplify. Placing them too far downstream would cause the disturbances generated by the roughness to compete with background disturbances that would amplify further upstream. Photographs of the two cone tips with discrete roughness "dimples" are shown in Figure 7.32.

Historically, the type of patterned roughness used by others has been in the shape of spherical bumps (Radeztsky Jr et al., 1999; Saric et al., 1998a). However, because of the small circumference (9.8 mm (0.386 in)) at the axial location on the cone tip where the patterned roughness was to be placed, Schuele et al. (2013) believed that achieving consistent bumps would be more difficult than with "dimples." Saric (2008) indicated that both worked about the same.

The approach to produce the dimples was to use a pin with a conical end. The pin was plunged into the surface of the tip, with the depth determining the diameter of the hole at the surface. Control of the azimuthal spacing of the holes was done with a rotational indexing head on which the tip was mounted. A profile measurement indicated that the "dimples" on the $m = 45$ tip were slightly oval, with an axial length of 180 μm and an azimuthal width of approximately 145 μm. The ratio of the diameter to azimuthal wavelength was then 0.65, which satisfied the Saric (2008) requirement that $d/\lambda \geq 0.5$. The depth of the dimple holes was 39 μm. The diameter of the $m = 68$ dimples was approximately 80 μm. This gave a diameter to azimuthal wavelength ratio

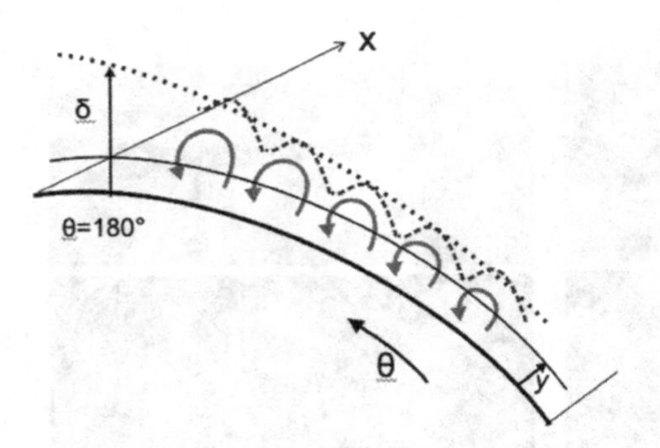

Figure 7.33 Example of boundary layer mean flow distortion produced by stationary cross-flow vortices that result in an azimuthal variation in the boundary-layer thickness that can be measured by an azimuthal traverse of a total pressure probe.

of approximately 0.6. The magnified image indicates that their size and shape were very repeatable.

The corotating stationary cross-flow vortices result in periodic bands of high and low wall shear stress that, as explained, resulted in the light and dark bands in the surface visualization that is shown in Figure 7.28. This also results in a thickening and thinning of the boundary layer that is illustrated in Figure 7.33. The thickening and thinning of the boundary layer can be detected through azimuthal surveys at a fixed height above the surface using a total-pressure Pitot probe. The periodic increase and decrease in the total pressure will reflect the thickening and thinning of the boundary layer. An example of this from Schuele et al. (2013) is shown in Figure 7.34. This corresponds to the $m = 68$ discrete roughness cone tip that is shown in Figure 7.32(b). This clearly shows the curved streamwise development of the stationary vortices that, as shown in Figure 7.35, closely follows the predicted streamline and vortex lines. The discrete roughness with a mode number $m = 68$ has an azimuthal wavelength of $360°/68 = 5.29°$. A close examination of Figure 7.34 shows that the azimuthal distance between the total pressure peaks or valleys closely matches the azimuthal spacing between the roughness elements. This observation is evidence that the boundary layer was receptive to the discrete roughness.

There are a number of ways to estimate the boundary-layer turbulence onset. A commonly used technique is based on a Preston tube (Preston, 1954). In their experiment, Schuele et al. (2013) utilized this approach in estimating the effect of the discrete roughness on delaying turbulent transition. A similar approach was used by King (1992). The basis for transition detection using a Preston tube is the occurrence of a sharp increase in the total pressure near the wall that signifies an increase in the wall shear. In this regard, it indicates the *beginning* of transition, where the jump in the wall shear stress is an indication of a mean flow distortion that signals nonlinear development.

To accomplish this, Schuele et al. (2013) measured the change in the total pressure at the cone surface along a line that followed the path of the stationary cross-flow vortices. More specifically, the measurement followed the axial location of a minimum

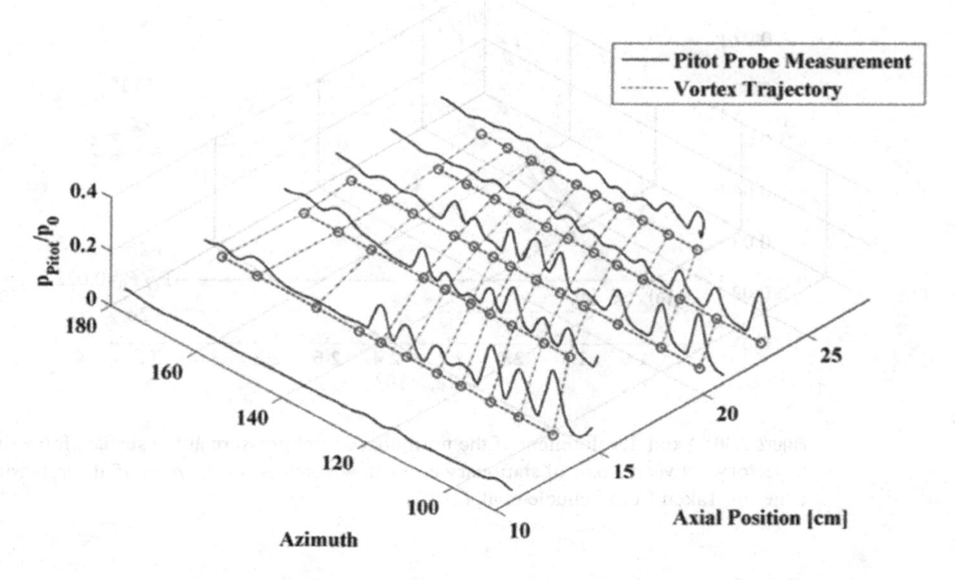

Figure 7.34 Azimuthal variation in mean pressure in the boundary layer at different axial locations for the $m = 68$ discrete roughness that is indicative of spatial development of the stationary cross-flow vortices. Taken from Schuele et al. (2013).

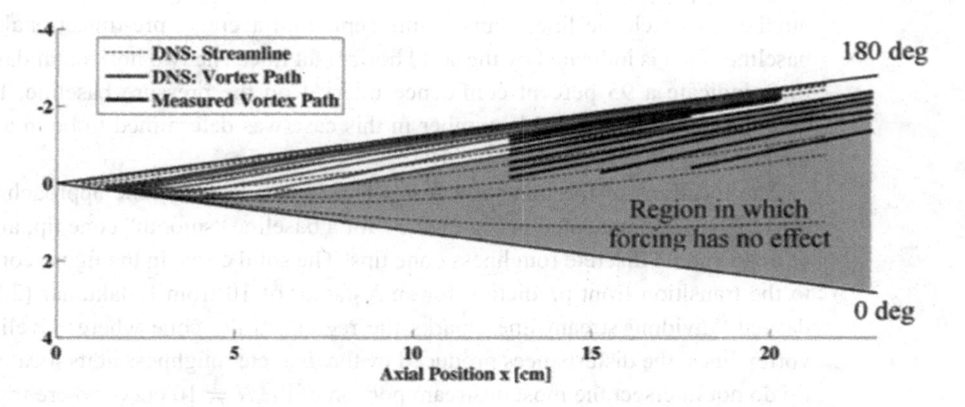

Figure 7.35 Measured stationary cross-flow vortex paths for $m = 68$ roughness from Figure 7.34, along with paths of surface streamlines of the basic flow and cross-flow modes based on DNS simulation of Balakumar (2009). Taken from Schuele et al. (2013).

pressure region between a pair of stationary cross-flow vortices that are denoted by the dashed curves in Figure 7.34. Following this approach, the axial development of the total pressure at the surface for the $m = 45$ discrete roughness tip (Schuele et al., 2013) is shown in Figure 7.36. The different data curves correspond to measurements taken at different azimuthal locations in 1° increments around the pressure minimum between a pair of cross-flow vortices. These all indicate a rapid increase in the total

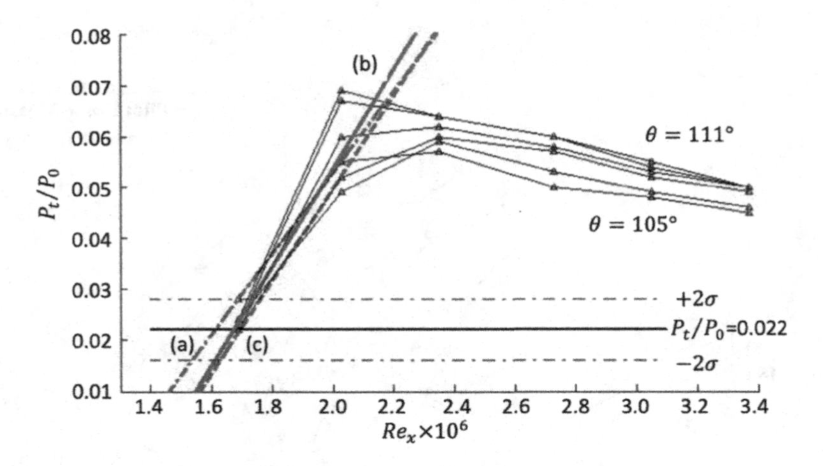

Figure 7.36 Axial development of the normalized total pressure at the surface following a trajectory between a pair of stationary cross-flow vortices for the $m = 45$ discrete roughness cone tip. Taken from Schuele et al. (2013).

pressure at the wall followed by a slow decrease that is a characteristic of a Preston tube measurement of a boundary layer undergoing turbulence onset (King, 1992). The dashed lines correspond to linear least-square-error fits to the region of the rapidly increasing pressure. The nonlinear transition onset was designated to be the Reynolds number at which the linear curves intercepted an average, pre-transitional pressure baseline. This is indicated by the solid horizontal line. The two horizontal dash-dotted lines indicate a 95 percent confidence interval on the pressure baseline. Based on this, the transition Reynolds number in this case was determined to be in a range of $1.61 \times 10^6 \leq \text{Re}_{x-\text{trans}} \leq 1.71 \times 10^6$.

The locations of the transition fronts based on a Preston tube approach (Schuele et al., 2013) are presented in Figure 7.37 for a baseline "smooth" cone tip, and for the $m = 45$ and 68 discrete roughness cone tips. The solid curve in the figure corresponds to the transition front prediction for an N-factor of 10 from Balakumar (2009). The dashed "dividing stream line" marks the region on the cone where traveling along vortex lines, the disturbances produced by the discrete roughness at its location on the tip do not intersect the most upstream portion of the $N = 10$ curve where transition is expected to first occur.

Although the stagnation pressure in the experiment was slightly lower than that in the simulation (Schuele et al., 2013), the agreement in the shape of the transition front in Figure 7.37 is reasonably good. The figure reveals that the subcritical $m = 68$ discrete roughness delayed transition by 35 percent compared to the smooth cone tip and 40 percent compared to the critical $m = 45$ discrete roughness tip. This was the first successful validation of the use of discrete roughness for cross-flow transition control at supersonic Mach numbers.

Corke et al. (2018) extended the work of Schuele et al. (2013) into the hypersonic regime to determine if cross-flow-dominated turbulent transition could be similarly

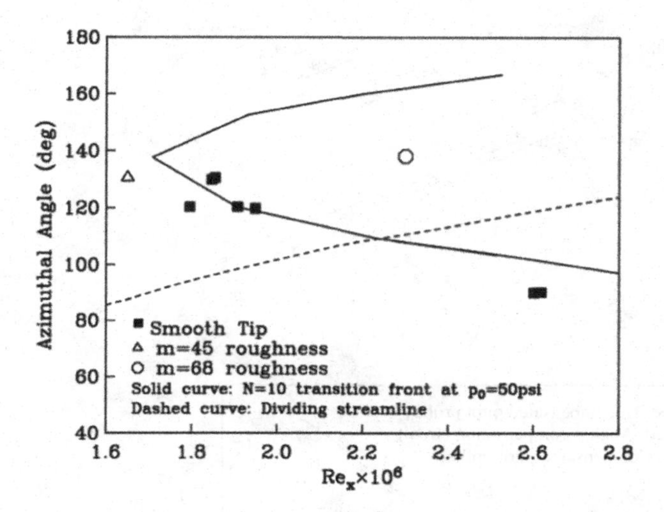

Figure 7.37 Transition locations based on a Preston tube approach for a baseline smooth cone tip, and with the $m = 45$ and 68 patterned roughness cone tips. Taken from Schuele et al. (2013).

suppressed with an appropriately patterned discrete roughness. The experiment was conducted at Mach 6. In contrast to the Schuele et al. (2013) experiment, which is conducted in a "quiet" supersonic wind tunnel designed to suppress acoustic disturbances, the Corke et al. (2018) experiments were conducted in a "conventional" Mach 6 tunnel. Thus, the experiments would examine both the effect of a higher Mach number and the receptivity of the stationary cross-flow modes to the discrete roughness under a higher free-stream disturbance condition.

The experiment used the same cone model used by Schuele et al. (2013). However, it was placed at a higher angle of attack of 6° that resulted in cross-flow stability characteristics that were identical to those at the lower Mach number. These specifically were the locations of the first neutral growth curve (Branch I at 1.27 cm from the cone tip), and the most amplified azimuthal mode numbers ($m = 45$). This allowed the use of the same roughness cone tips used by Schuele et al. (2013), allowing a direct comparison to the experiments at the lower Mach number.

The experiment used a pair of high-bandwidth total pressure probes mounted on a 3-D traversing mechanism to document the boundary layer. A schematic drawing showing assembly of dual high-bandwidth total pressure probe is shown in Figure 7.38(a). The probe voltage time series output was compensated to increase the frequency response. Figure 7.38(b) shows the frequency response of the uncompensated Pitot probes, the amplitude response of the analog frequency compensation circuit, and the final amplitude response of the frequency compensated Pitot probes. The probe frequency response was sufficient to measure pressure fluctuations associated with the traveling cross-flow modes (Corke et al., 2018).

As in the previous experiment (Schuele et al., 2013), azimuthal profiles of the time-averaged total pressure were used to document the mean flow distortion produced by

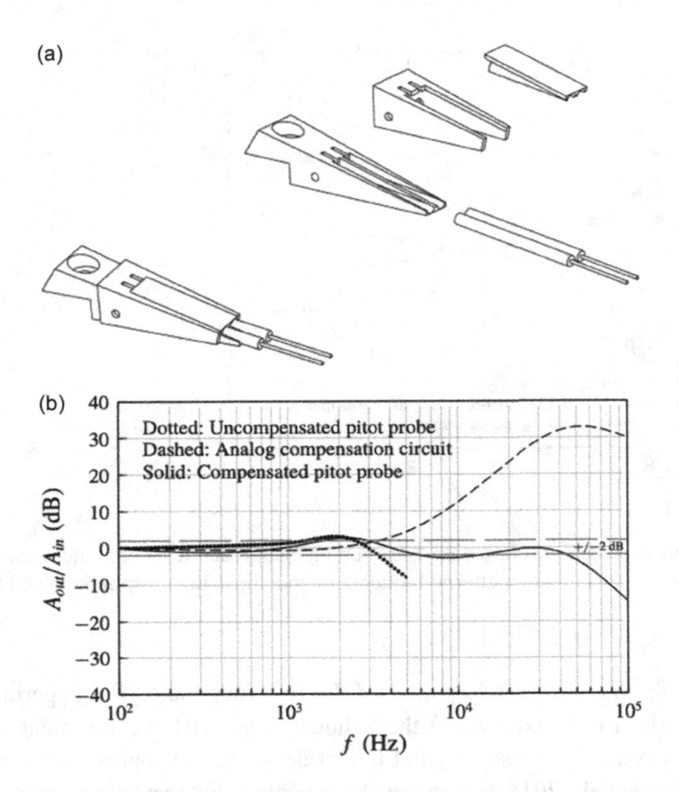

Figure 7.38 Schematic drawing showing assembly of dual high-bandwidth total pressure probe (a) and frequency response of the uncompensated Pitot probes, the analog frequency compensation circuit, and the final frequency compensated Pitot probes (b). Taken from Corke et al. (2018).

the stationary cross-flow modes. Their axial development was used to determine the transition locations for the critical and subcritical patterned roughness. Time-resolved pressure fluctuations were analyzed to document the spatial development of the traveling cross-flow modes. These provided evidence of a nonlinear interaction between the stationary and traveling cross-flow modes that had previously been documented in the boundary layer on a rotating disk (Corke and Knasiak, 1998; Matlis, 1997; Corke et al., 2007b).

An example of the azimuthal variation in the time-average total pressure at a fixed height above the cone surface with the "critical" ($m = 45$) roughness tip from Corke et al. (2018) is shown in Figure 7.39(a). A consequence of remaining at a fixed height above the cone surface was that as the boundary layer thickened in the azimuthal direction along the lee side of the cone, the sensor moved deeper in the boundary layer so that its position above the surface relative to the boundary layer edge was not constant. In order to better reveal the mean flow distortion produced by the stationary cross-flow modes, that mean trend was removed (Corke et al., 2018). The color rendering of the total pressure was then presented as the deviation (plus or minus) about the local mean value. This clearly reveals the azimuthal variation in the total pressure that is indicative of the mean flow distortion produced by the stationary cross-flow vortices. Note that negative mean-removed pressures correspond to a locally thickened portion of the boundary layer, and positive mean-removed pressures correspond to locally thinned portions. The dotted curves in the figure trace out the streamwise development of two of the azimuthally thickened portions of the boundary layer.

The azimuthal spacing between the two low pressure portions of the mean flow distortion in Figure 7.39(a) closely corresponds to 8°, which matches that of the $m = 45$ discrete roughness wave number ($360°/45 = 8°$). Further evidence of this comes from the wave number spectra of the mean flow distortion that is shown in Figure 7.39(b). This shows a dominant peak at $m = 45$. The energy in the $m = 45$ mode decays to a minimum near the axial location where the linear stability $N = 10$ factor predicts transition to occur.

The azimuthal variation in the total pressure that is indicative of the mean flow distortion produced by the "subcritical" ($m = 68$) discrete roughness is shown in Figure 7.40(a). The contrast with the "critical" ($m = 45$) roughness is immediately apparent and reflects the smaller azimuthal wave length produced by the higher azimuthal wave number discrete roughness. For reference, $m = 68$ corresponds to an azimuthal wavelength of 5.3°. The wave number spectra of this total pressure distribution is shown in Figure 7.40(b). This shows an initially dominant peak at $m = 68$. Further downstream, the peak appears to shift to higher wave numbers of the order of 100. This downstream development is consistent with Saric et al. (1998a) where "subcritical" roughness with azimuthal wave numbers of 1.5 times that of the "critical" roughness value would not generate subharmonic mode numbers. Similarly, Schuele et al. (2013) observed only higher azimuthal wave numbers generated by the "subcritical" roughness in their experiment at Mach 3.5.

The dominant peak in the wave number spectra with the subcritical roughness shown in Figure 7.40(b) extends well downstream of the linear stability $N = 10$ line, indicating that the transition Reynolds number had increased. As further evidence, the method of Schuele et al. (2013) was used to estimate transition location. This involved following the streamwise trajectory of a minimum in the total pressure distributions, shown by the dotted curves in Figures 7.39(a) and 7.40(a). The result is shown in Figure 7.41.

Following Schuele et al. (2013), the transition location in Figure 7.41 was estimated by two approaches. The first, shown in blue, followed a Preston tube approach in which a linear curve was fit to the rise in the total pressure near the wall that would accompany nonlinear mean flow distortion, signifying the beginning of transition. The streamwise, x, location where the linear curve intersected the reference level was designated to be the transition location. Based on this approach, the "subcritical" roughness resulted in a 25 percent increase in the transition location.

The second approach was based on the loss of a coherent pattern (spanwise periodicity) of the mean flow distortion. This would be evident as a saturation and abrupt decay in the axial development of the total pressure that would signify turbulence onset. Based on this approach, shown in red in Figure 7.41, the "subcritical" roughness pattern resulted in a 15 percent increase in the location of turbulence onset.

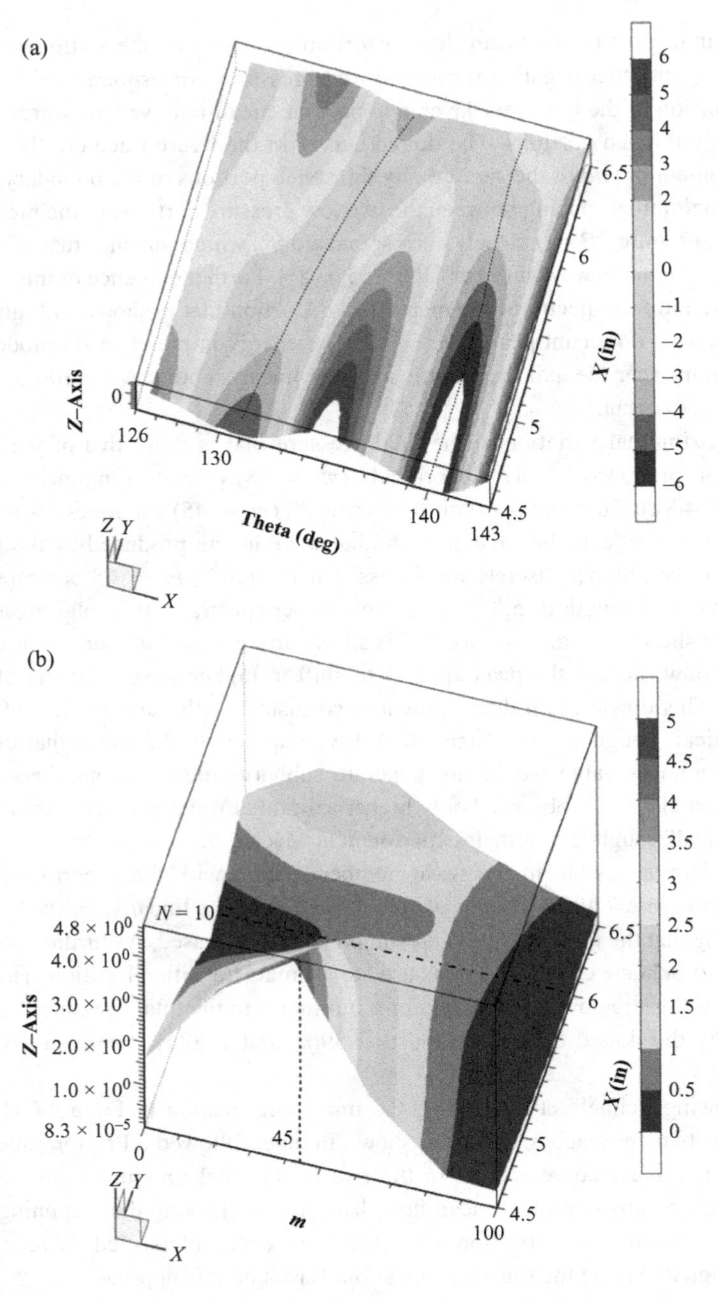

Figure 7.39 Azimuthal total pressure distributions at a constant height above the surface (a) and corresponding wave number spectra (b) for the case with the critical ($m = 45$) roughness cone tip. Taken from Corke et al. (2018).

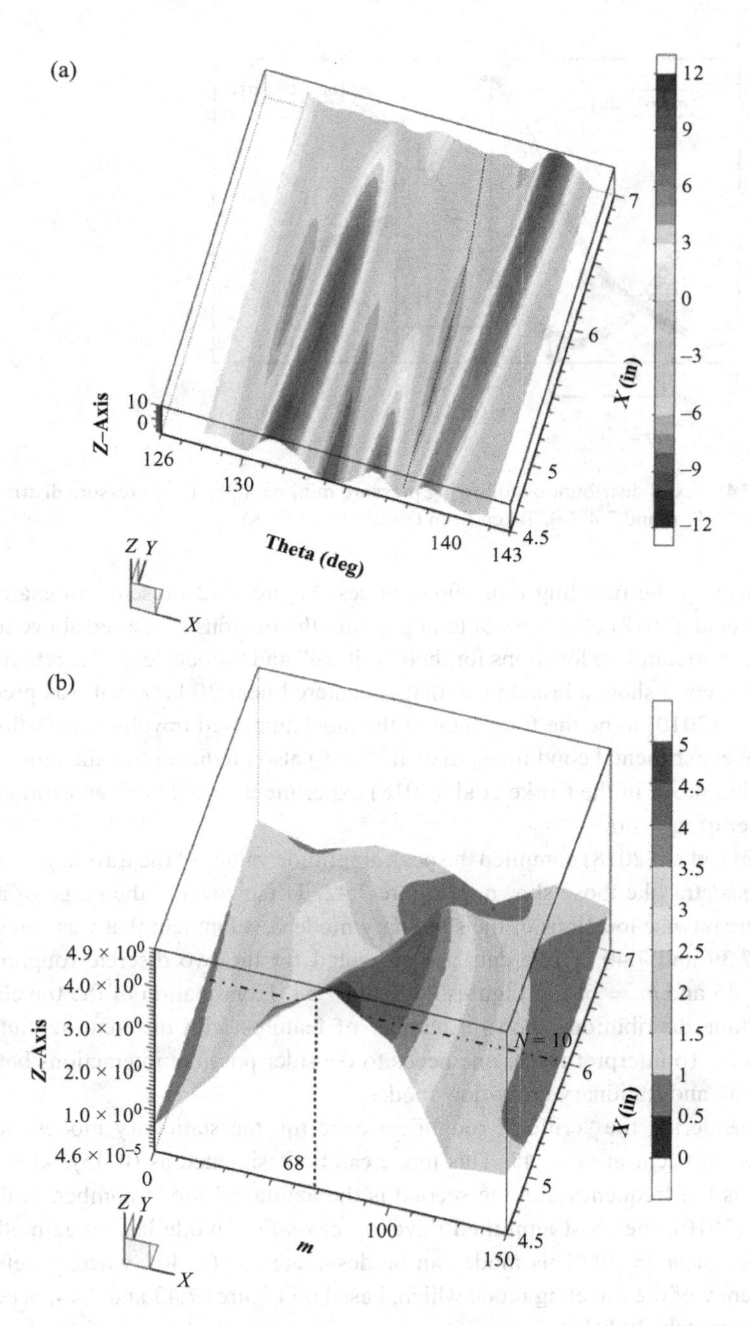

Figure 7.40 Azimuthal total pressure distributions at a constant height above the surface (a) and corresponding wave number spectra (b) for the case with the subcritical ($m = 68$) roughness cone tip. Taken from Corke et al. (2018).

As previously mentioned, Corke et al. (2018) had designed analog frequency compensation for the Pitot probes that gave them a frequency response that was capable

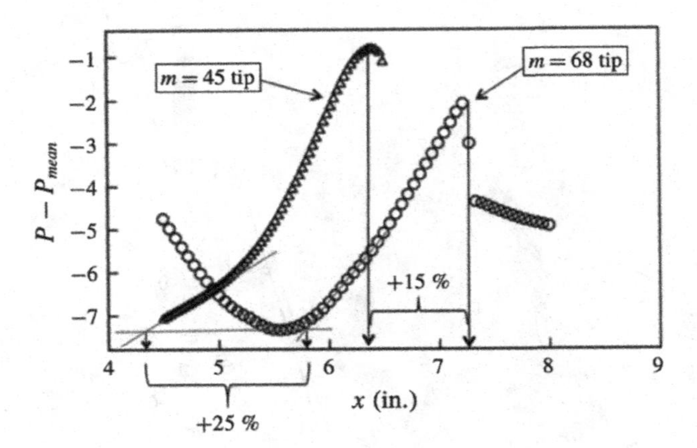

Figure 7.41 Axial distributions following pressure minima in the total pressure distributions in Figures 7.39(a) and 7.40(a). Taken from Corke et al. (2018).

of detecting the traveling cross-flow modes. Figure 7.42 presents an example from Corke et al. (2018) of spectra of total pressure fluctuations measured above the wall at different streamwise locations for their "critical" and "subcritical" discrete roughness. These spectra show a broad peak that is centered near 30 kHz that was predicted by Li et al. (2010) to be the frequency of the most amplified traveling cross-flow modes for the experimental conditions. Li et al. (2010) also predicted that the most-amplified traveling mode in the Corke et al. (2018) experiment would have an azimuthal mode number of $m = 40$.

Corke et al. (2018) compiled the peak amplitude values of the traveling disturbances from spectra like those shown in Figure 7.42. These covered the range of azimuthal and streamwise locations of the stationary mode development that was shown in Figures 7.39 and 7.40. These data are presented for the two discrete roughness cases ($m = 45$ and $m = 68$) in Figures 7.43 and 7.44. Examination of the traveling mode amplitude distributions shows a number of features with different azimuthal wave numbers. To interpret these, one needs to consider possible interactions between the traveling and stationary cross-flow modes.

Considering the "critical" roughness cone tip, the stationary modes were documented to occur at $m = 45$. This mode can be designated as $(0, 45)$, where the first index is the frequency and the second is the azimuthal mode number. Following Li et al. (2010), the most amplified traveling cross-flow mode has an azimuthal mode number of $m = 40$. This mode can be designated as $(f_t, 40)$, where f_t refers to the frequency of the traveling mode which, based on Figures 7.43 and 7.44, is centered at approximately 30 kHz.

The first interaction between these two modes will be through a sum and difference, namely

$$(0, 45) \pm (f_t, 40) = (f_t, 85) \text{ and } (f_t, 5). \tag{7.11}$$

Both $(f_t, 85)$ and $(f_t, 5)$ are traveling modes with the same frequency as the primary traveling mode. Therefore, the amplitudes of spectral peaks in Figure 7.43 should be

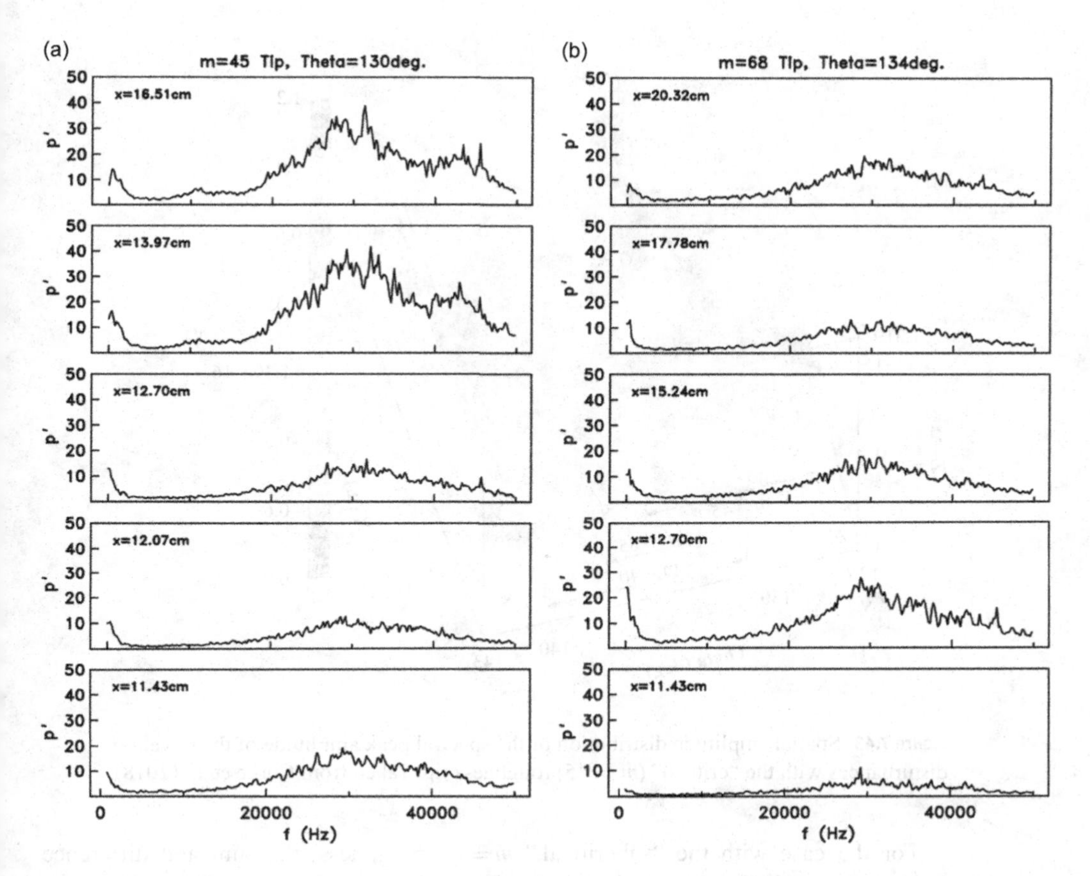

Figure 7.42 Spectra of traveling disturbances measured at different axial locations for a constant azimuthal angle for the critical ($m = 45$) roughness (a) and subcritical ($m = 68$) roughness (b). Taken from Corke et al. (2018).

representative of not only the primary traveling mode but also the sum and difference modes that occur at the same frequency.

On the basis of its azimuthal mode number, the result of the summing interaction, $(f_t, 85)$, would have an azimuthal wave angle of approximately 4° (360°/85). Similarly, the difference interaction, $(f_t, 5)$, would have an azimuthal wave angle of 72°. The 72° azimuthal wavelength was larger than the azimuthal extent of the measurements and therefore could not be detected. However, an azimuthal amplitude variation with a 4° azimuthal wavelength associated with the summing interaction, $(f_t, 85)$, is distinguishable in Figure 7.43. The wavelength is denoted on the figure as $m = 85$.

Also observable is a spatial amplitude distribution with an azimuthal mode number that matches that of the discrete roughness, $m = 45$. This is not realizable through a quadratic interaction to produce $(f_t, 45)$. Therefore, it is most likely due to a *linear* superposition that leads to a modulation of the traveling mode amplitude as it rides over the stationary mean flow distortion. Such behavior has in fact been identified through direct Navier–Stokes simulations of subsonic 3-D boundary layers undergoing cross-flow instability by Dorr et al. (2017).

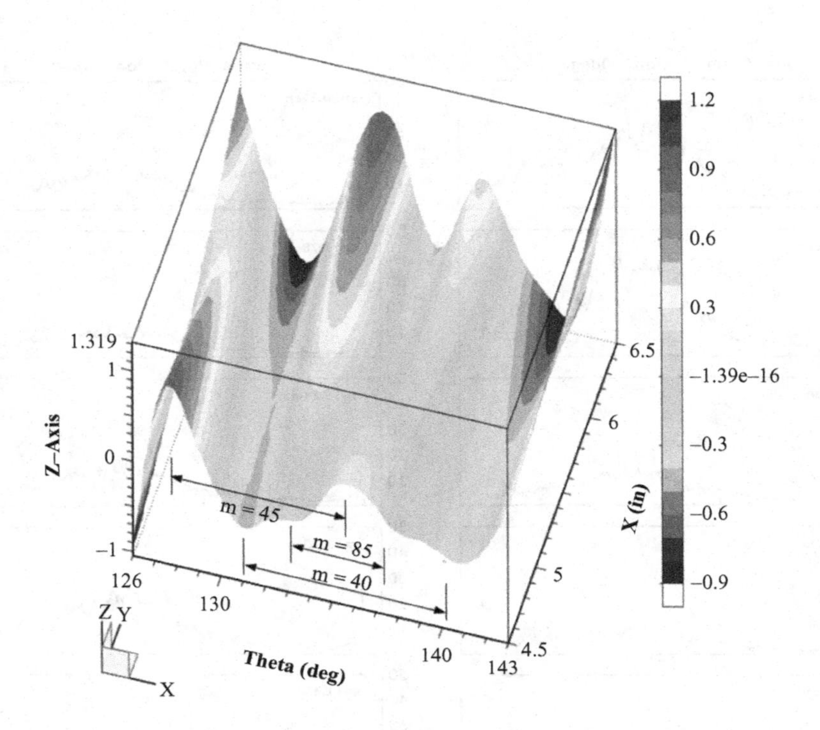

Figure 7.43 Spatial amplitude distribution of the spectral peak amplitude of the traveling disturbances with the "critical" ($m = 45$) roughness tip. Taken from Corke et al. (2018).

For the case with the "subcritical," $m = 68$ roughness, the sum and difference interaction would be

$$(0, 68) \pm (f_t, 40) = (f_t, 108) \, \text{and} \, (f_t, 28). \tag{7.12}$$

Here, $(f_t, 108)$ would have an azimuthal wave angle of approximately 3°, and $(f_t, 28)$ an azimuthal wave angle of approximately 13°. Both are traveling modes with the same frequency as the primary traveling mode. The wave angles of both of these interacted modes are evident in the spatial amplitude distribution shown in Figure 7.44. In addition, as with the previous roughness case, a spatial amplitude distribution with an azimuthal mode number that matches that of the discrete roughness, $m = 68$, is also observable. This again is most likely due to a linear superposition that produces a modulation of the traveling mode amplitude as it rides over the stationary mean flow distortion produced by the $m = 68$ roughness.

Corke et al. (2018) further quantified these sum and difference interactions by documenting the triple phase locking between the primary stationary and traveling cross-flow modes, and their sum and difference modes. This utilized the CBC statistic previously used in documenting nonlinear interaction between stationary and traveling cross-flow modes in the boundary layer over a rotating disk (Matlis, 1997; Corke et al., 2007b). The CBC indicated a significant triple phase locking between the stationary cross-flow mode generated by the "critical" $m = 45$ roughness, and the

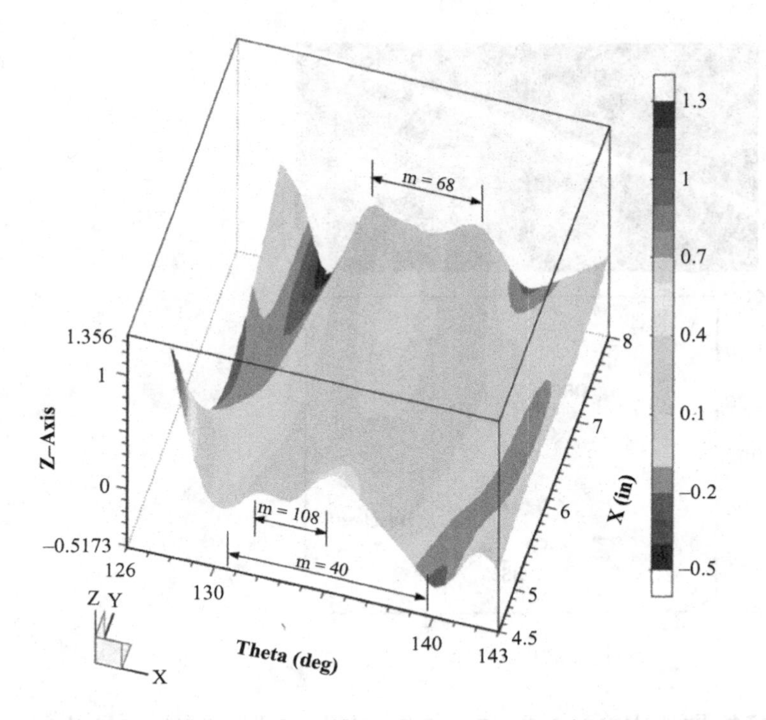

Figure 7.44 Spatial amplitude distribution of the spectral peak amplitude of the traveling disturbances with the "subcritical" ($m = 68$) roughness tip. Taken from Corke et al. (2018).

traveling cross-flow mode with an azimuthal wave number of $m = 40$ to produce a traveling cross-flow mode with an azimuthal wave number of $m = 85$. This presented a strong indication that the $m = 85$ variation in the spatial amplitude distribution in Figure 7.43 was the result of a nonlinear interaction between the stationary and traveling cross-flow modes.

Similarly, the CBC for the "subcritical" roughness case revealed a significant triple phase locking between the stationary cross-flow mode generated by the $m = 68$ roughness, and the traveling cross-flow mode with an azimuthal mode number $m = 40$ to produce a traveling cross-flow mode with an azimuthal mode number of $m = 108$. This confirmed that in this case, the $m = 108$ spatial amplitude distribution observed in Figure 7.44 was the result of a nonlinear interaction between the stationary and traveling cross-flow modes.

In another example of the use of passive discrete roughness for cross-flow transition control, Middlebrooks (2022) applied the method to the 70° swept fin that was mounted on the 7° half-angle cone used by Schuele et al. (2013), Corke et al. (2018) and Arndt et al. (2020). The size and wavelength of the discrete roughness was based on the analysis of surface heat flux images like the one shown in Figure 7.31. Based on the analysis, and following the criterion first presented by Saric et al. (1998a), the wavelength of the "subcritical" roughness was 50 percent less than that of the "critical" wavelength, resulting in a "subcritical" roughness wavelength of 13.3 mm.

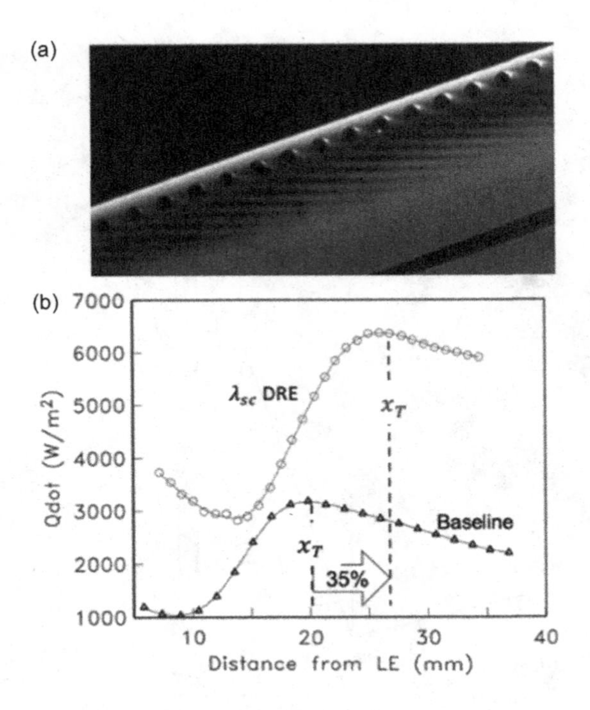

Figure 7.45 Example of cross-flow transition control on a 70° swept fin at Mach 6 using subcritical discrete roughness. (a) Surface heat flux. (b) Heat flux between two stationary cross-flow vortices that are used to determine the location of the transition onset. Taken from Middlebrooks (2022).

An example of the results is shown in Figure 7.45. This shows an image of the surface heat flux with the "subcritical" roughness located just downstream of the leading edge of the swept fin. In this case, the discrete roughness consisted of an array of circular surface depressions or dimples (similar to the approach of Schuele et al. (2013)). These were formed by laser cutting circular holes in the black film used to enhance the infrared imaging. The depth of the dimples corresponded to the thickness of the film, which was 90 μm. The diameter of the holes was half of the center-to-center spacing of the holes, giving a diameter to wavelength ratio of 0.5 that was recommended by Saric et al. (1998a).

The surface heat flux image shown in Figure 7.45(a) clearly reveals well-ordered light and dark streaks associated with the stationary cross-flow vortices. Each of the streaks are observed to originate from a respective roughness element, thereby matching the wavelength of the discrete roughness. Figure 7.45(b) shows the surface heat flux obtained by following a cross-flow vortex trajectory in the baseline (no DREs) and subcritical DRE cases. These show a pattern of increasing heat flux that marks the onset of turbulent transition, and amplitude saturation that signifies the onset of turbulence, x_T. Comparing the two x_T locations with and without control, the subcritical DRE array delayed transition by 35 percent. Further optimization is possible; in particular, the overall results indicate that the DRE depth should be reduced.

7.6.2 Plasma Actuator Arrays

The discrete roughness described in the previous section is an example of a *passive* transition control method. Successful use requires a priori knowledge of the most unstable and "critical," mode number of stationary cross-flow modes which is then used to design discrete roughness with a "subcritical" mode number. If the method were applied on a flight vehicle, that condition might only be relevant for a single flight condition that determines the cross-flow Reynolds number. Therefore, there is motivation for an *active* distributed roughness that has the potential to vary the azimuthal wave number.

Schuele et al. (2013) investigated active distributed roughness for cross-flow transition that consisted of an azimuthal array of plasma actuators. It involved two arrays on separate removable cone tips that were each designed to excite the same azimuthal wave numbers as their passive discrete roughness. The plasma actuators were more specifically single dielectric barrier discharge (SDBD) plasma actuators that utilized two electrodes that were separated by a dielectric layer. One electrode is exposed to the air. The other electrode is covered by the dielectric layer. The SDBD plasma actuator electrically functions like a capacitor and is therefore powered by an AC voltage source. When the AC voltage is large enough, the air near the exposed electrode weakly ionizes. The ionized air in the presence of the electric field produced by the electrode arrangement results in a body force vector field that acts on the neutrally charged air. An extensive review of SDBD plasma actuators was given by Corke et al. (2010a). This topic is also included in Chapter 2.

The use of SDBD plasma actuators to produce the effect of distributed roughness to control of stationary cross-flow modes was first examined by Corke (2001) and subsequently tested by Saric (2001). That design is illustrated in Figure 7.46(a). In the illustration, the long axis of the exposed electrodes is aligned with the mean flow direction. The plasma, shown in blue, is formed on the side edges of the exposed electrodes. The electric field lines are normal to the long electrode edge, which results in a body force vector field, $\vec{F}_b$, that is in the cross-flow direction. As illustrated, the body forces on each side of the long electrode act in opposite directions. The result is to produce counter-rotating streamwise vortices similar to a wedge-type vortex generator. Such passive vortex generators were discussed in Chapter 4 for controlling separated flows. In that chapter, a similarly designed plasma actuator was termed a "plasma streamwise vortex generator" or PSVG.

In order to verify the effect of the plasma actuator "comb" design, Schuele (2011) performed particle image velocimetry (PIV) measurements. A schematic of the PIV setup is shown in Figure 7.47. The arrangement of the light sheet and camera was designed to measure the velocity component parallel to the wall, u, and the wall-normal velocity component, v. The light sheet was located slightly in front of the tips of the "comb" electrodes. The measured $[u, v]$ velocity pairs were used to calculate the streamwise, z, vorticity component. The result is shown in Figure 7.48. This documents a spanwise periodic pattern of streamwise vorticity of alternating sign that is consistent with the illustration shown in Figure 7.46(a). This then represents the

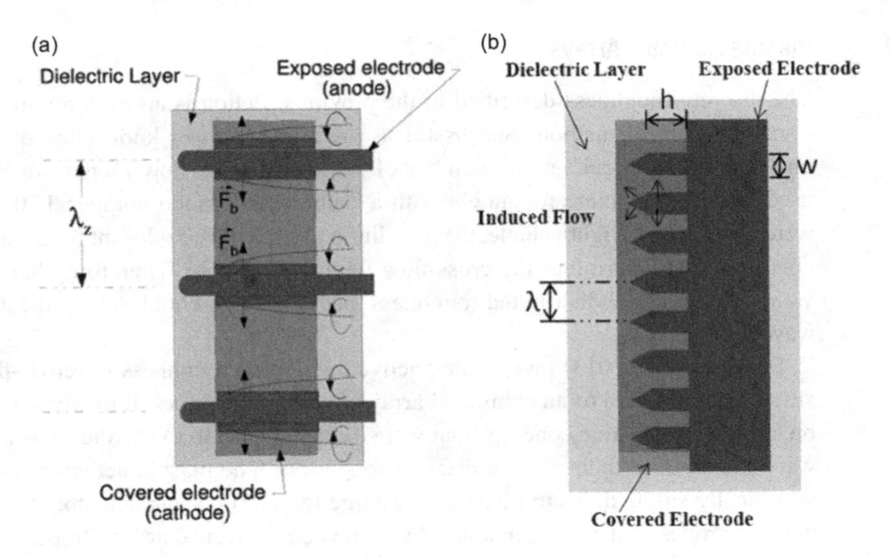

Figure 7.46 Schematic of the streamwise vorticity generator developed by Corke (2001) and tested by Saric (2001) (a), and schematic of "comb" plasma actuator array that was transferred to a cone tip for cross-flow transition control experiments by Schuele et al. (2013) (b).

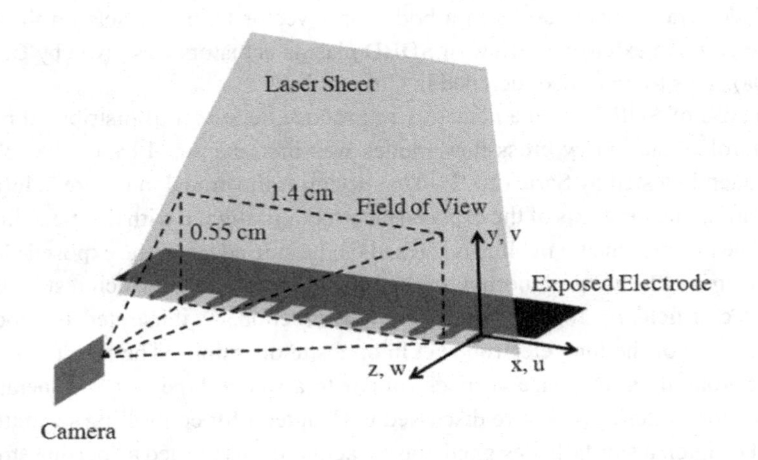

Figure 7.47 Schematic of the PIV setup used to measure velocity field induced by plasma actuator "comb" design. Taken from Schuele (2011).

initial disturbance condition produced by a plasma actuator array of this design. Choi and Kim [2018] obtained a similar picture of the vorticity field induced by a similar plasma actuator design.

Schuele et al. (2013) followed the same basic design. This is illustrated in Figure 7.46(b). However, this had a number of significant challenges. The first was that the plasma actuator electrode array could not produce any passive roughness effect when not operated. The second challenge was the physical scale of the electrodes. For the $m = 68$ case, the centerline spacing between the electrodes was 0.15 mm

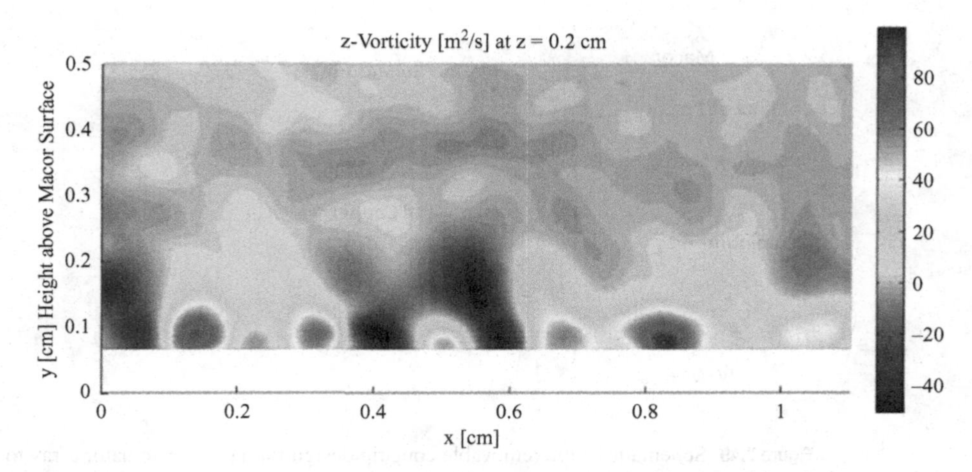

Figure 7.48 Example of streamwise-vorticity produced by plasma actuator "comb" design. Taken from Schuele (2011).

(0.0059 in). The final challenge was for the application on a right-circular cone, that the plasma actuator array was applied on a curved, sloped surface that made fabrication more difficult.

The Schuele et al. (2013) design is illustrated in Figure 7.49. The removable cone tip consisted of two materials, a machinable ceramic (Macor) toward the tip and aluminum at the base. The Macor acted as the dielectric layer for the plasma actuator. The Macor part of the cone tip was drilled out to produce a lower section with a constant wall thickness of approximately 0.38–0.51 mm (15–20 mil). This resulted in a dielectric strength of 12–15 kVpp.

Conductive silver epoxy was used to fill the top of the tapered interior of the Macor part of the tip. This acted as the common covered electrode for the exposed electrodes. The silver epoxy also made the connection to the high voltage (HV) wire lead that entered through the base of the removable cone tip. The wire was insulated by an fluorinated ethylene propylene coating. It had an outer diameter of 1.27 mm (0.05 in) and a dielectric strength of 14 kVrms.

The tapered aluminum and Macor parts of the removable cone tip were bonded together with a fast curing nonconducting epoxy. The joined parts were then final machined to the precise conical shape, smoothed with 1,800-grit sandpaper, and then lapped using a compound with an equivalent grit size of 3,200.

The exposed electrode was made by depositing copper onto the outer surface of the cone tip, on top of the Macor and aluminum parts of the cone tip. In doing so, a continuous copper layer was electrically connected to the metal base of the cone tip, which, when attached to the metal cone, was electrically grounded. The electrode "comb" pattern was then etched into the exposed copper layer using photolithography techniques.

The resulting "comb" electrode structure is shown in Figure 7.50. Figure 7.50(a) shows an overall view of the removable cone tip. The white portion of the tip is the unplated Macor. The bottom portion is the copper-plated portion of the tip. Magnified

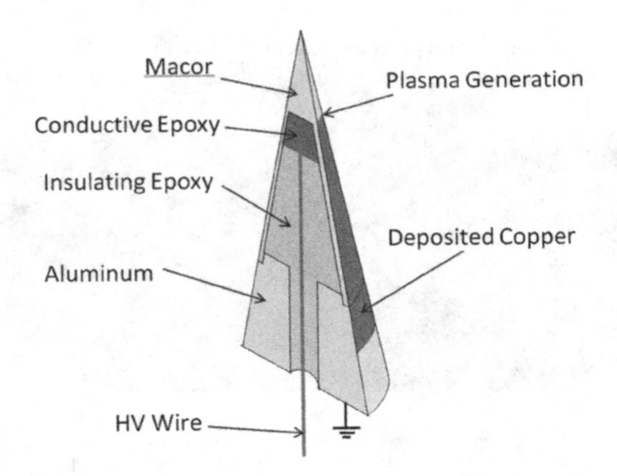

Figure 7.49 Schematic of the removable cone tip design for a plasma actuator array to excite cross-flow modes. Taken from Schuele et al. (2013).

views of the edge of the copper plating are shown in Figures 7.50(b) and (c). A most highly magnified view of the edge of the copper plating clearly reveals the "comb" electrode pattern. The dashed line in the most magnified view denotes the edge of the covered electrode. The covered electrode is under the full length of the "comb" electrodes and around the full azimuth of the tip, so that plasma will form between the electrodes.

Measurements of the copper-plated portion revealed an average plating thickness of 0.05 μm (0.197×10^{-6} in). In contrast, measurements on the unplated Macor portion of the cone tip documented the root mean square of the surface roughness to be 10 times larger than the plating thickness. Therefore, the copper electrodes were indistinguishable from the passive roughness of the Macor cone tip surface.

Photographs of the operating plasma actuators on the cone tips are shown in Figure 7.51. These were obtained by placing the cone tips in a vacuum chamber in which the pressure was lowered to 10.2 kPa (3 inHg) to simulate the static pressure in the wind tunnel at Mach 3.5 and at $P_0 = 172$ kPa. The AC input to the plasma actuator was a 5 kVp-p, 5 kHz sine wave. The ionized air (plasma) appears as blue dots around the cone tip. These coincide with the locations of the exposed electrode "comb teeth" around the cone tip. In both mode number cases, the illumination appears to be relatively uniform around the cone tip.

Schuele et al. (2013) documented the receptivity of the stationary cross-flow modes to the two plasma actuator arrays through azimuthal profiles of the total pressure similar to those that were shown in Figure 7.34. The results for the two azimuthal mode number plasma arrays are shown in Figures 7.52 and 7.53. The curves with the x-symbols correspond to the condition with the plasma roughness *not* operating. This documents any passive roughness effect of the electrode array. The curves with the filled-circle symbols correspond to when the plasma array *was operating*. The periodic maxima of the azimuthal pressure distributions at each axial location are connected to indicate the trajectories of the stationary cross-flow vortices with the plasma array OFF

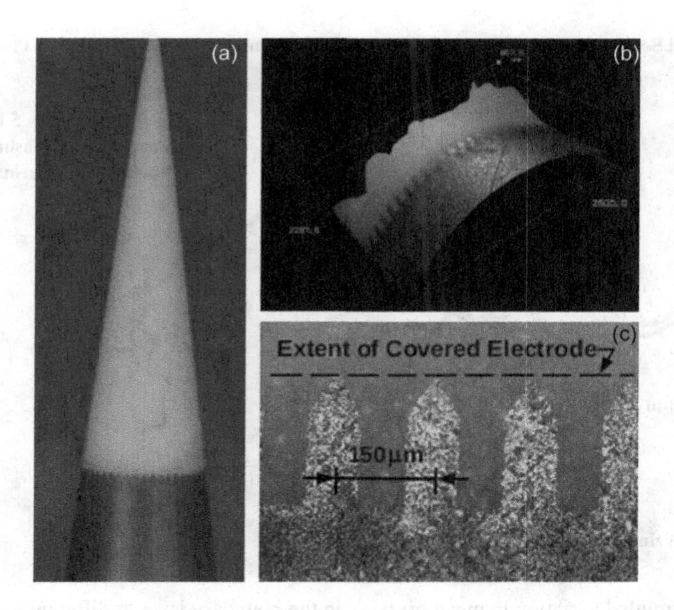

Figure 7.50 Photograph of plasma actuator cone tip (a) and magnified views (b and c) of the edge of the copper plating showing the "comb" electrode arrangement to excite $m = 68$ stationary cross-flow modes. Taken from Schuele et al. (2013).

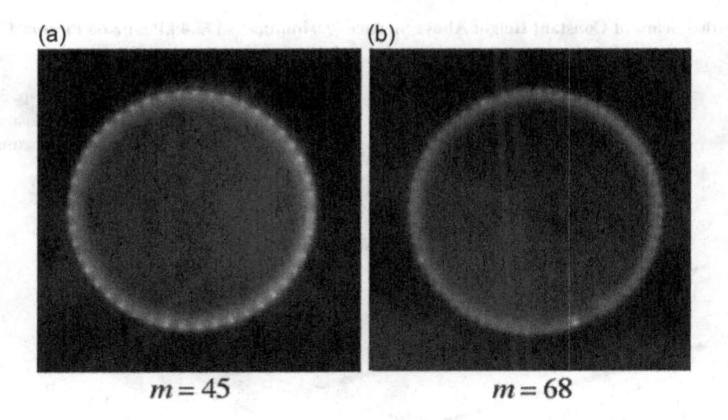

Figure 7.51 Long-time exposure of the plasma illumination pattern for the two plasma actuator cone tips designed to excite $m = 45$ (a) and $m = 68$ (b) stationary cross-flow modes. Taken from Schuele et al. (2013).

and ON conditions. The shaded region denotes the "dividing line" that was determined with the passive roughness, and below which any patterned roughness at the cone tip location is not sensed by the flow (Schuele et al., 2013).

The difference in the azimuthal variation of the total pressure indicative of the mean flow distortion produced by the stationary cross-flow vortices is clearly larger when the plasma actuator array is operating. Schuele et al. (2013) documented that the effect of the plasma array on the stationary cross-flow modes was entirely repeatable in total pressure peak–valley locations and amplitudes. With the ability to vary

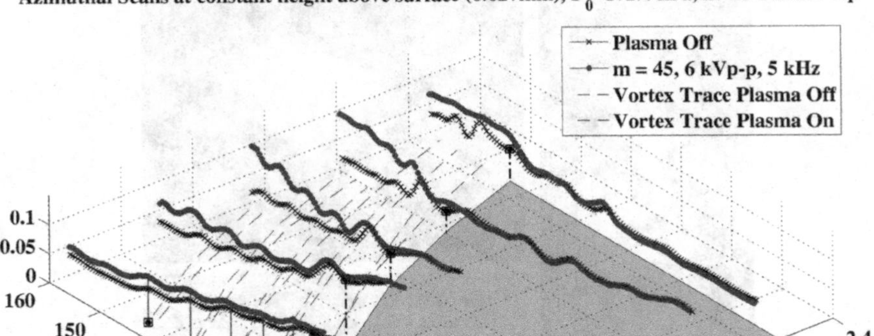

Figure 7.52 Azimuthal variation in mean pressure in the boundary layer at different axial locations for the critical wave number $m = 45$ plasma roughness operating (filled-circle symbols) and passive (x-symbols). Taken from Schuele et al. (2013).

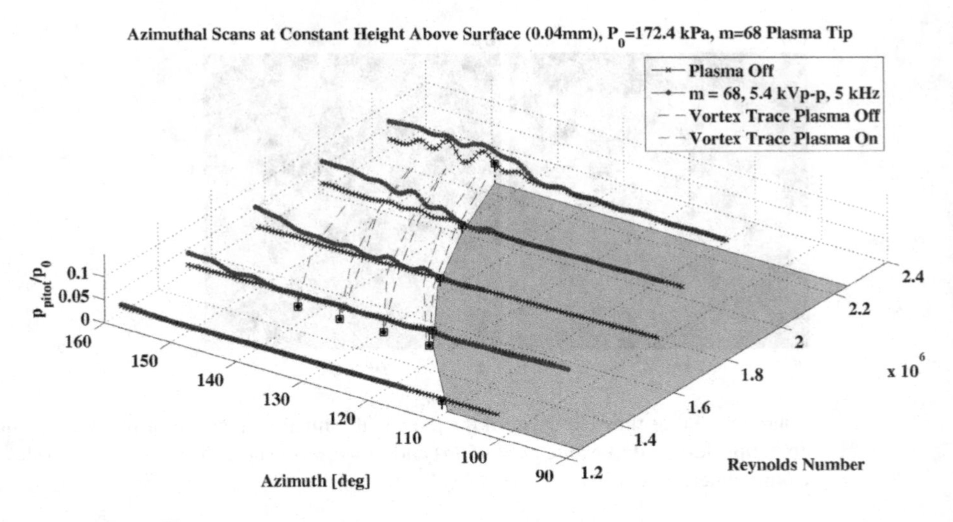

Figure 7.53 Azimuthal variation in mean pressure in the boundary layer at different axial locations for the critical wave number $m = 68$ plasma roughness operating (filled-circle symbols) and passive (x-symbols). Taken from Schuele et al. (2013).

the input voltage to the plasma array, Schuele et al. (2013) were also able to verify a linear response of the cross-flow mode at lower voltage levels, and the threshold to a nonlinear response. Similar results could not be obtained with a fixed passive roughness.

7.7 Summary

This chapter examined instabilities leading to turbulence onset in fully 3-D boundary layers that pointed to methods of transition control. This is particularly important because there are many practical applications in which 3-D flows are a dominant feature, and in which case the relevant boundary-layer instability has growth rates that are significantly larger than those exhibited in 2-D boundary layers, particularly compared to those of T-S and second-mode instabilities that were presented in Chapter 7.

The examples of 3-D flows of practical interest that were presented in this chapter include swept wings; rotating cones, cylinders, and spheres; corners; inlets; rotating disks; and right-circular cones at angles of attack or yaw angles. Each of these flows exhibits instability behavior that produces 3-D boundary layer flows with coherent streamwise vorticity that produces a strong spanwise modulation of the basic state that can give rise to secondary instabilities and turbulence onset.

The boundary layer over a swept wing is susceptible to four types of instabilities that lead to turbulence transition. These are a leading-edge instability and contamination, a streamwise instability, a centrifugal instability, and a cross-flow instability. Leading-edge instability and contamination occurs along the leading-edge attachment line and is associated with an instability of the flow attachment line. The streamwise instability is associated with the chordwise component of flow over the wing. Transition in this case is similar to that in 2-D flows, where a T-S instability is the dominant mechanism. It usually occurs in zero or mild positive pressure-gradient regions on the wing. Centrifugal instabilities occur in boundary layers over a concave surface. The dominant mechanism for turbulence transition in this case is through a Görtler instability. Finally, the cross-flow instability occurs in strong pressure-gradient regions on swept wings. In cases with swept wing angles greater than approximately 30°, cross-flow instability is the dominant mechanism of turbulence onset.

The cross-flow mean boundary layer profile has a maximum velocity somewhere in the middle of the boundary layer, going to zero on the surface and at the boundary-layer edge. Its requirement for zero cross-flow at the wall (no slip) and at the edge of the boundary layer results in an inflection point somewhere in the profile that is inviscidly unstable to infinitesimal disturbances (see Chapter 5). The cross-flow instability results in both traveling and stationary waves. Although traveling cross-flow modes are more amplified, the stationary cross-flow modes generally appear to dominate because of their significant sensitivity to minute surface roughness. The stationary waves grow to form a regular pattern of stationary corotating vortices with axes aligned in the inviscid flow direction. These have been made visible in multiple experiments using surface flow visualization techniques. They will generally appear as dark and light bands that reflect the periodic wall shear pattern produced by the corotating vortices.

The cross-flow instability of the boundary layer over a rotating disk is very similar to that over the swept wing and frequently has been used as a canonical 3-D flow that exemplifies the cross-flow instability. A key difference is that the disk experiences the effect of Coriolis forces due to the disk rotation, although the resulting instability is far less amplified than that of the cross-flow instability and therefore plays no role in turbulence onset. A key advantage of the rotating disk in studying cross-flow instability is that the mean (basic) flow over the disk has an exact solution, making instability analysis more accurate. In addition, except for very close to the rotation axis, the boundary-layer thickness is constant in the radial direction. Thus, for a fixed rotation speed, the radial location, r, is equivalent to the Reynolds number.

One feature of a rotating disk flow is the presence of a cross-flow absolute instability that is *not* a mechanism with cross-flow over a swept wing. Although theoretically proven to exist for the rotating disk, it is not a factor in turbulence onset except when sufficient wall suction is applied. In that case, the absolute instability is transformed into a global instability producing temporal growth that dominates over the otherwise convective cross-flow instability. In contrast, wall suction is a viable method for cross-flow transition control on swept wings.

The underlying physics of the cross-flow instability is the same for incompressible and compressible boundary layers, with experiments presently up to Mach 6 showing the same characteristic behavior. As such, the methods for transition control originally developed at incompressible Mach numbers work equally well at compressible Mach numbers. Examples of cross-flow transition control at supersonic and hypersonic Mach numbers that were presented in this chapter included swept wings, right-circular cones at angles of attack, and highly swept fins used for flight control. The methods demonstrated in these cases are equally applicable to supersonic and hypersonic lifting bodies where cross-flow is a factor in turbulence onset.

The approach to control cross-flow-dominated transition to turbulence stems from the extreme sensitivity of the stationary cross-flow modes to surface roughness. This was exploited by Corke and Knasiak (1998), Matlis (1997) and Corke et al. (2007b) to excite selected wave numbers of cross-flow modes in the boundary layer over a rotating disk, and by Saric et al. (1998b) and (Radeztsky Jr et al., 1999) who used micron-sized roughness to fix the spacing of cross-flow vortices in swept wing experiments.

The observations by Saric et al. (1998b) led to the transition control concept of exciting less amplified stationary cross-flow modes as a means of controlling turbulence onset in these flows. The key element was that the spanwise wave number of the forced mode had to be higher than those of the naturally amplified band. This guaranteed that no disturbance energy would cascade through nonlinear interactions to lower, more amplified wave numbers. Saric et al. (1998a) referred to this approach of cross-flow transition control as "subcritical forcing," which was in reference to most amplified being "critical" wave numbers. Thus, cross-flow transition control is based on biasing the natural selection mechanism by exciting a less amplified cross-flow mode so that it initially dominates the instability growth process.

The effectiveness of "subcritical" forcing using *passive* discrete roughness for cross-flow transition control has been demonstrated over a range of Mach numbers from incompressible to hypersonic. An *active* approach using a discrete array of plasma actuators has been demonstrated on a swept wing at an incompressible Mach number, and on a right-circular cone at an angle of attach at Mach 3.5. In that experiment, the amplitude of the "plasma roughness" was varied to verify the linear response of the stationary cross-flow instability as well as the amplitude where nonlinear effects occur. Such active control of the initial conditions (amplitude and wave number) is essential to validate linear-theory-based (N-factor) predictions of turbulence onset. Since the "critical" wave number, and thereby the appropriate "subcritical" wave number used for control, is a function of Reynolds number, in practical applications feedback control is necessary. The "plasma roughness" presented in this chapter is one example of a method in which feedback (amplitude) control has been demonstrated.

Overall, the control of cross-flow-dominated turbulence transition is now clearly established. Its success is a testament to the close collaboration that occurred between stability analysis and experiments, which is a central theme of this book.

Problems

7.1 The classic smoke visualization image of cross-flow transition on a rotating disk from Kobayashi et al. (1980) that is illustrated in Figure 7.7 shows spiraling light and dark bands that mark stationary cross-flow vortices.

1. Why do the cross-flow vortices lead to the appearance of light and dark bands in the smoke?
2. What features of the stationary cross-flow modes in the image can be used to validate linear theory for the cross-flow modes?
3. If there are discrepancies with linear theory, what could be the possible causes?
4. What could be done to improve on the experiment in making comparisons to linear theory?

7.2 Lingwood (1995) suggested that the location of turbulence onset in the visualized rotating disk flow of Kobayashi et al. (1980) provided evidence of explosive growth associated with an absolute cross-flow instability. Although the existence of an absolute instability in this case is not contested, the following questions arise.

1. What are the characteristics of an absolute instability that are different from a more common convective instability?
2. Given those differences, what type of experiment and measurements are needed to confirm its existence on the rotating disk flow?
3. If the explosive disturbance growth that results from an absolute instability is responsible for turbulence onset, how can it be suppressed? Hint: See Chapter 3, which discusses flow control of cylinder wakes that also exhibit an absolute instability.

7.3 You are asked to design a passive DREs to delay boundary layer transition on a swept wing in which turbulence onset is due to cross-flow instability.

1. Describe the experimental steps needed to determine characteristics (height, diameter, and spacing) of the DREs.
2. Describe the method used to determine their placement on the wing.
3. Describe the measurement approach to validate the boundary-layer transition delay due to the DREs.

7.4 An array of four pressure sensors are located flush with the surface of a swept wing that exhibits a strong mean cross-flow velocity component leading to a cross-flow instability. The four pressure transducers are located at four corners of a square with the purpose of determining the wavelength and phase velocity of traveling cross-flow modes. The pressure time series from each of the sensors are digitally sampled simultaneously. These time series can be processed to obtain information about the frequency content in each time series, linear coherence (measure of phase locking between frequencies across time series), and phase between frequencies across the time series.

1. Describe how the analysis of the time series from the four pressure sensors can be used to determine the wavelength and phase speed of the traveling cross-flow modes?
2. How would you decide the spacing (size of the square array) between four sensors? What is the criterion?
3. What is the minimum number of sensors needed to accomplish this?

7.5 Corke et al. (2018) were able to quantify a nonlinear interaction between stationary and traveling cross-flow modes. Using a frequency-wave-number notation, $[f, m]$, they introduced discrete roughness to excite a stationary cross-flow mode, $[0, 45]$, where the most amplified traveling cross-flow mode was $[f_t, 40]$. Through sum and difference interactions, this resulted in traveling cross-flow modes, $[0, 45] \pm [f_t, 40] = [f_t, 95]$ and $[f_t, 5]$.

1. Describe the interaction that would produce a 3-D mean flow distortion, $[0, 90]$. How would that be detected?
2. Describe the interaction that would produce $[3f, 105]$.
3. Would the excitation of conditions leading to these nonlinear interactions be useful in accelerating turbulence onset? Explain.

8 Turbulent Boundary Layers

The pursuit of viscous drag reduction in turbulent boundary layers has been a topic of interest for more than 50 years. This had been largely driven by the desire to increase aircraft performance. Flow control strategies have generally focused on either removing or altering in some way the turbulent boundary layers. In the latter approach, the focus has been on mechanisms underlying coherent motions that have been shown to contribute to sustaining turbulent energy production. This has generally involved either modifying the large vortical motions in the outer half of the boundary layer or modifying the small-scale vortical motions that occur in the lower third of the boundary layer, near the wall. Both passive and active flow control approaches that have been examined have led to varying degrees of success. These are presented in this chapter.

8.1 Background: Turbulent Boundary Layers

Research on turbulent boundary layers in the last half of the twentieth century transformed the picture given by earlier long-time averaged statistics. Once perceived as quasi-steady turbulent eddies transported by a mean shear, such flows were revealed to be highly unsteady coherent motions of strongly interacting scales. This revolution in thinking was primarily established by Townsend (1956) and Grant (1958), who, while being primarily interested in turbulent wake flows, viewed the turbulent flow field as one where the turbulent fluid is moved about by the motions of a system of large coherent eddies whose dimensions scaled with the width of the flow. The motions of these large scales carried eddies of smaller comparative size but which contained most of the turbulent energy. Townsend in particular inferred from velocity correlation measurements the presence of characteristic eddies of finite length with a structure that is inclined in the flow direction. He surmised that these large eddies in turbulent shear flows form coherent and identifiable groups, and control the overall rate of spreading by contorting the layer between the turbulent and nonturbulent fluid, and by affecting the bulk convection of turbulent energy from regions of maximum production. Later, Townsend (1970) described these structures as inclined double-roller eddies, which was consistent with the eddy pattern originally perceived by Grant (1958).

At about the same time, Kline et al. (1967a) observed through flow visualization surprisingly well-organized motions in the near-wall region of a turbulent boundary

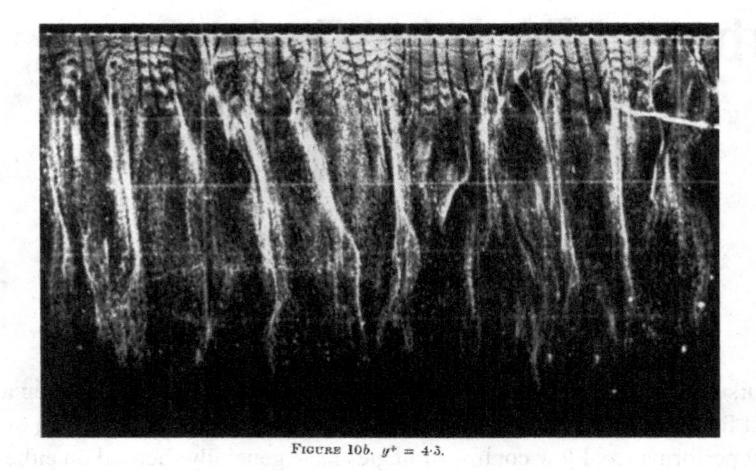

Figure 8.1 Flow visualization of wall-layer "streak" structure first discovered by Kline et al. (1967a).

layer. These motions led to the formation of low-speed wall "streaks" that eventually interacted with the outer portion of the boundary layer through a process of "gradual lift-up, sudden oscillation, bursting and ejection." This sequence became simply known as "bursting" and was felt to play an important role in the production of turbulence near the wall. Figure 8.1 shows a photograph of the wall-layer "streak" structure captured in flow visualization by Kline et al. (1967a). Mean velocity measurements in the spanwise direction cutting across the "streak" structure revealed mean flow distortion consisting of a relatively regular wall-layer thickening and thinning that would be consistent with pairs of counter-rotating pairs of streamwise vortices. The spanwise spacing was remarkably regular with an average corresponding to $\lambda^+ = \lambda u_\tau/\nu = 100$. These results are presented in Figure 8.2.

Kline et al. (1967a) carefully followed the time evolution of the "streak" structure that led to the "lift-up" and "ejection" of low momentum fluid from the wall. This is depicted as a sequence of particle streak lines in Figure 8.3(a). A remarkable result from Kline et al. (1967a) was their observation of a linear relation between the frequency of the "bursting" and the magnitude of the wall shear stress, τ_w, expressed by the wall shear velocity, u_τ. This result is shown in Figure 8.3(b). This suggests an approach that can reduce the viscous drag that could come from an approach to reduce the frequency of the "bursts."

The turbulent boundary layer is uniquely different from other turbulent flow fields in that its mean velocity profile can be divided into four main parts. The first of these is the wall region in which the flow processes are dominated by the presence of the wall. The lowest part of the wall region where $0 \leq y^+ \leq 7$ is referred to as the viscous sublayer. Here, the wall-normal distance, y, is made nondimensional using the friction velocity, $u_\tau = (\tau_w/\rho)^{1/2}$, such that $y^+ = yu_\tau/\nu$. In this region, viscosity plays a dominant role; however, turbulent fluctuations are still relatively large relative to the local mean velocity. Scaling analysis in this region yields the result that $u^+ = y^+$ in

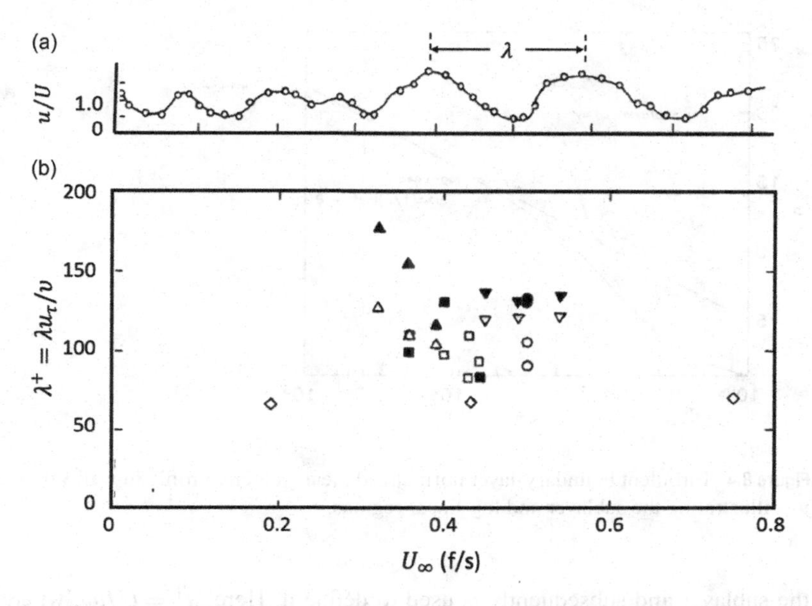

Figure 8.2 Example of spanwise mean flow distortion resulting from the wall "streak" structure (a) and "streak" spanwise wavelength scaling (b). Taken from Kline et al. (1967a).

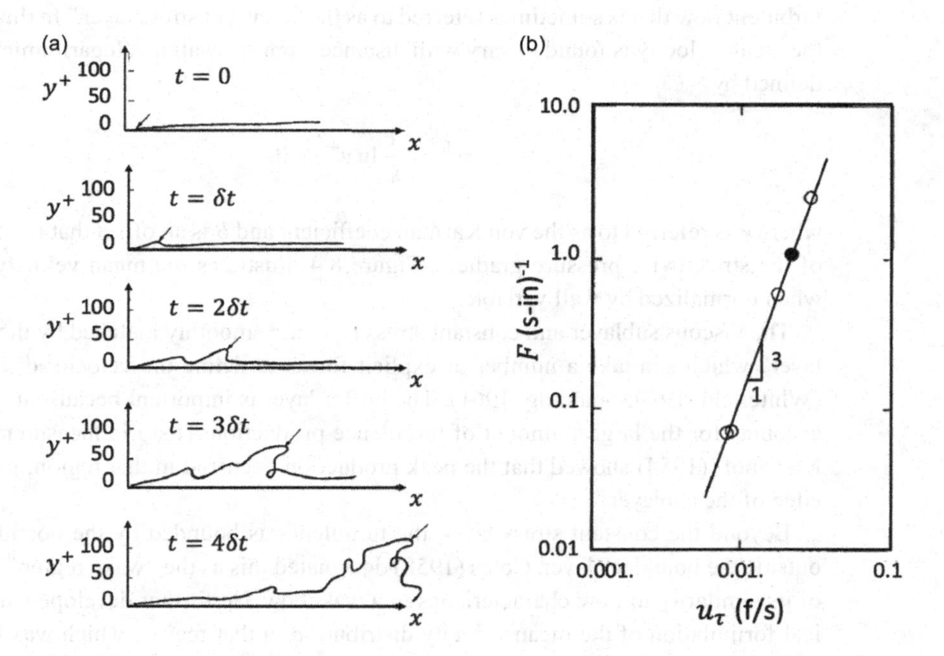

Figure 8.3 Process of "lift-up" and "ejection" of wall "streak" structure outlined by Kline et al. (1967a) (a) and their observed correlation between "streak burst" frequency and wall shear stress (b).

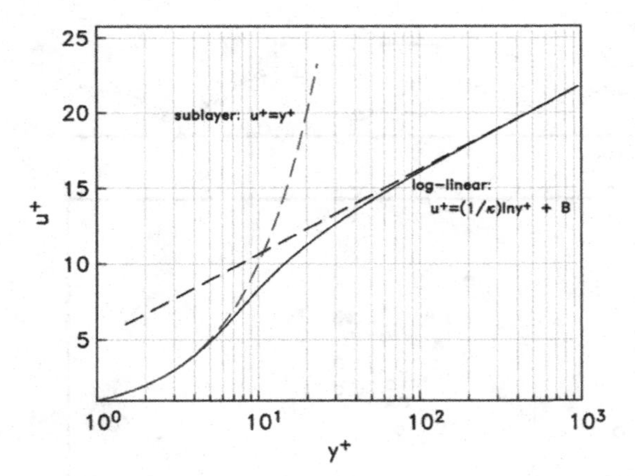

Figure 8.4 Turbulent boundary-layer normalized mean velocity profile in wall variables, u^+ and y^+, illustrating the sublayer and log-linear regions.

the sublayer and subsequently is used to define it. Here, $u^+ = U/u_\tau$, where U is the local mean velocity.

Further from the wall, $50 \leq y^+ \leq 200$, there is a region of nearly homogeneous turbulent flow that is sometimes referred to as the "constant stress layer." In this region, the mean velocity is found to vary with distance from the wall in a logarithmic manner defined by

$$u^+ = \frac{1}{\kappa} \ln y^+ + B, \tag{8.1}$$

where κ is referred to as the von Karman coefficient and B is an offset that is a function of the streamwise pressure gradient. Figure 8.4 illustrates the mean velocity profile when normalized by wall variables.

The viscous sublayer and constant stress layer are smoothly matched by the "buffer layer" which can take a number of explicit forms to define the velocity distribution (Whitefield, 1979; Spalding, 1961). The buffer layer is important because it typically accounts for the largest amount of turbulence production. Historic measurements by Klebanoff (1954) showed that the peak production occurred in that region, just at the edge of the sublayer.

Beyond the constant stress layer, the turbulence is bounded by the potential flow outside the boundary layer. Coles (1958) designated this as the "wake region" because of its similarity to flow characteristics in a wakeflow. He further developed an empirical formulation of the mean velocity distribution in that region, which was found to work well for a variety of boundary-layer flows with pressure gradients.

At the outer edge of the wake region, there exists a thin interfacial region that separates the turbulent and nonturbulent flow. This interface was termed the "superlayer" by Corrsin and Kistler (1954). It is across this region that vorticity is acquired by the potential flow and eventually incorporated into the wake flow.

Each of these regions has a characteristic length scale. In the viscous sublayer, that length is ν/u_τ. The wake region has a characteristic scale on the order of the boundary-layer thickness, δ. Kovasznay (1967) estimated the thickness of the superlayer to be $10\nu/V_p$, where V_p is the entrainment velocity.

The observation that the maximum turbulence production coincided with the region of large coherent streak activity discovered by Kline et al. (1967a) prompted other investigations of the near-wall phenomena. Corino and Brodkey (1969) visualized the near-wall motions in a turbulent pipe flow. They reported that the most important and distinguishing characteristic was the intermittent *ejection* of discrete fluid elements away from the wall. They characterized the process as a local deceleration of the fluid near the wall followed by the appearance of faster moving *sweep* of fluid that approached from the outer upstream part of the boundary layer, resulting in eruptions of the low velocity fluid away from the wall. The ejection of low-speed fluid was always found to be followed by a sweep of upstream fluid from the previous ejection event. This is generally referred to as a "sweep-ejection event." Corino and Brodkey (1969) also investigated the Reynolds number dependence of these events and found that their frequency of occurrence increased as $\text{Re}_x^{1/7}$.

Experimental discrimination of sweep-ejection events traditionally involves using an X-hotwire or a two-component laser Doppler velocimetry in order to simultaneously measure the streamwise, u, and wall-normal, v, velocity components. The velocity components are partitioned according to $[u(t) - \bar{U}] >$ or < 0, and $v(t) >$ or < 0, and sorted into a Cartesian plot of $\pm[(u(t) - \bar{U})]$ versus $\pm v(t)$. A "sweep" event is associated with quadrant 4 (Q4) where $[u(t) - \bar{U}] > 0$ and $v(t) < 0$. An "ejection" event is associated with quadrant 2 (Q2) where $[u(t) - \bar{U}] < 0$ and $v(t) > 0$. Note a positive v is defined as directed away from the wall. The method is referred to as the uv-quadrant discrimination. Lu and Willmarth (1973) were the first to show that discrete Q2 and Q4 events are the primary contributors to the Reynolds stress in a turbulent boundary layer.

Corino and Brodkey (1969) estimated that the wall sweep-ejection events accounted for 70 percent of the $\overline{uv}$ Reynolds stress in the boundary layer. This was substantiated by a visual study by Kim et al. (1971). Using a $\overline{uv}$ quadrant discrimination technique, Willmarth and Lu (1972) determined that 60 percent of the $\overline{uv}$ Reynolds stress is produced in uv-quadrant 2 (Q2) where $[u(t) - \bar{U}] < 0$ and $v(t) > 0$.

In the initial studies by Kline et al. (1967a), the "burst" frequency was associated with the wall "streak" structure and therefore thought to scale with inner-wall variables, u_τ and ν. Focusing on the large-scale motions in the outer part of the boundary layer, Rao et al. (1971) suggested the "burst" frequency scaled on outer variables, U_∞ and δ. Laufer and Narayanan (1971) suggested an outer variable scaling with a nondimensional burst period, T, of $U_\infty T/\delta \simeq 5$, which was close to the experimentally observed dimensionless period of 6 from Rao et al. (1971). Laufer and Narayanan (1971) further pointed out that the average wavelength of the turbulent–laminar interface is $2\delta/U_\infty$, which suggested an interaction between the outer boundary-layer structure and the inner wall-layer events. To seek to tie this together, Kim et al. (1971)

Figure 8.5 Smoke-wire image of a turbulent boundary layer illustrating the potential turbulence interface. Flow direction is from right to left. Taken from Corke (1981a).

suggested that the lift-up and break-up of low-speed wall streaks could be triggered by the disturbances produced by the large-scale motions.

Another "burst" detection technique that was pioneered by Blackwelder and Kaplan (1976) utilizes a single hot-wire and a discriminator based on $u'(t) > K\overline{u'}$, where $'$ indicates the root mean square (RMS), and K is a factor greater than 1. Based on this approach, they indicated that the "burst" frequency scaled with inner variables. Blackwelder and Eckelmann (1979) furthermore indicated that the low-speed streaks observed in flow visualization by Kline et al. (1967b) were the result of counter-rotating streamwise vortices. The burst ejections were then thought to be associated with wall-outward pumping action at the interface between the vortex pairs. *Interceding with this process points to an approach for turbulent boundary-layer drag control.*

In order to form a conceptualized view for "bursting" that would distinguish between inner and outer variable dependence, it is necessary to review literature on organized motions in the outer part of turbulent boundary layers. Since the early intermittency measurements in the outer part of the boundary layer by Corrsin and Kistler (1954), it was clear that the turbulent–nonturbulent interface extends well into the boundary layer. This is illustrated in a smoke-wire flow visualization image from (Corke, 1981a) that is shown in Figure 8.5. A schematic interpretation of the organized motions in the outer and wall regions is shown in Figure 8.6.

Utilizing a rake of hot-wires that spanned the whole boundary layer, Kaplan and Laufer (1969) determined that the nonturbulent intermittent regions were connected to the free-stream fluid and that the interface appeared to be highly corrugated. Utilizing conditional sampling techniques, Kovasznay et al. (1970) defined "fronts" and "backs" to the turbulent "bulges." They found that the velocity at the front of the bulges was approximately 5 percent faster than that at the back. The average velocity (convection speed) of the bulges was on the order of $0.93U_\infty$. Based on velocity correlation measurements, they defined a mean motion inside a turbulent "bulge" which was rotational with a mean vorticity component with the same sign as the mean shear. They estimated that the "lifetime" of the turbulent bulges was from 5–10δ. They further

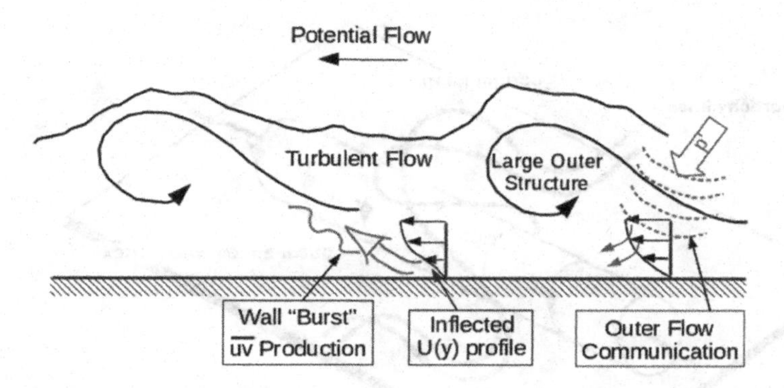

Figure 8.6 Schematic illustration of outer and wall coherent eddies and possible mechanism for interaction and turbulent production related to wall shear stress.

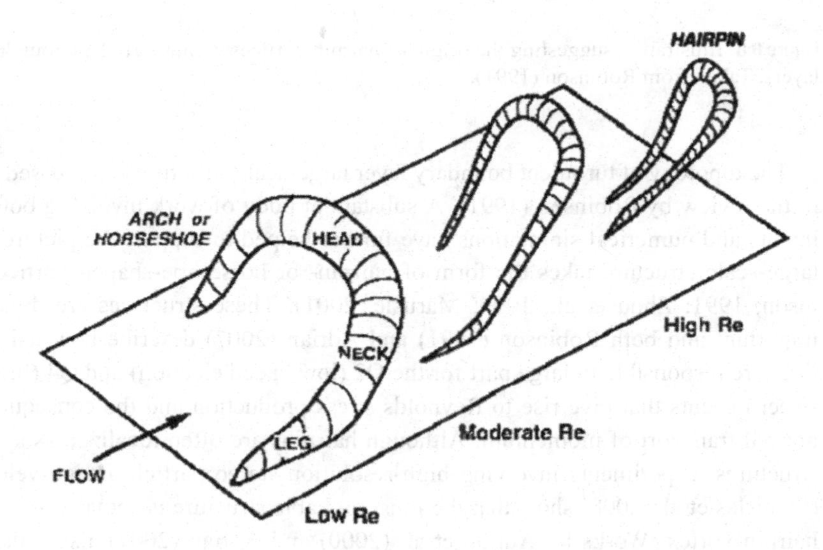

Figure 8.7 Illustration by Head and Bandyopadhyay (1981) of arch- and hairpin-shaped vortical structures embedded within turbulent boundary layers.

estimated that these large-scale structures contributed as much as 80 percent of the Reynolds stress in the outer part of the boundary layer where $y > 0.2\delta$.

Antonia (1972) determined that the maximum $\overline{uv}$ within the turbulent "bulges" represented approximately 45 percent of the wall shear value. This provided support to the Bradshaw (1967) theory that the strength of these large-scale eddies is closely related to the wall shear stress, τ_w.

Head and Bandyopadhyay (1981) presented a picture, shown in Figure 8.7, of the large-scale eddies as being an amalgamation of elongated hairpin vortices. As illustrated in Figure 8.8, their origin could come through an interaction between the spanwise, ω_z, vorticity due to the mean shear, and the streamwise vorticity, ω_x, associated with the wall streak structure.

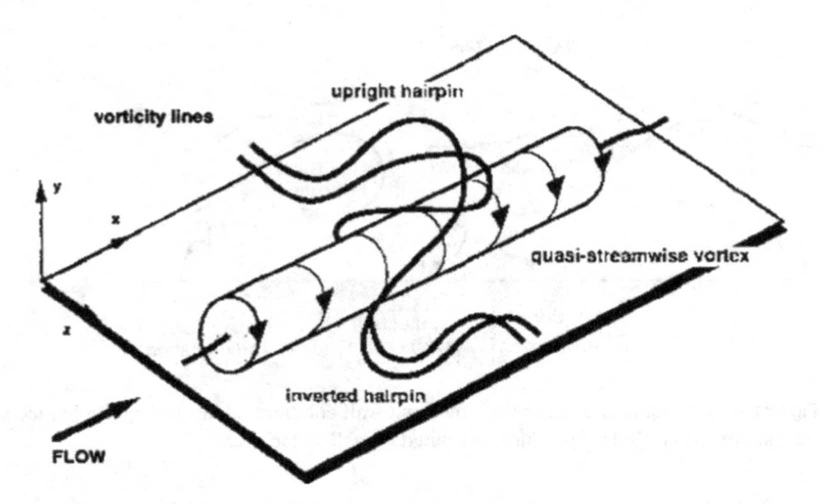

Figure 8.8 Illustration suggesting the origin of hairpin vortices within turbulent boundary layers. Taken from Robinson (1991).

The topology of turbulent boundary layer large-scale structure is discussed in detail in the review by Robinson (1991). A substantial body of work involving both experiments and numerical simulations have further helped to support the picture that the large-scale structure takes the form of hairpin- or horseshoe-shaped vortices (Robinson, 1991; Zhou et al., 1999; Marusic, 2001). These structures are dynamically important and both Robinson (1991) and Adrian (2007) describe how hairpin vortices are responsible in large part for the Q2 (low-speed ejection) and Q4 (high-speed sweep) events that give rise to Reynolds stress production and the consequent wall-normal transport of momentum. Although hairpins are often idealized as symmetric structures, experiments involving high-resolution stereo particle image velocimetry (Stanislas et al., 2008) show that the most probable structure is actually a one-legged hairpin vortex. Works by Adrian et al. (2000) and Adrian (2007) also indicate that hairpin vortices actually form in packets. This is consistent with the earlier observations of Bogard and Tiederman (1986) who showed that a turbulent boundary-layer "burst event" actually consists of a series of sequential Q2 ejection events. In the hairpin-packet model of Adrian (2007), the boundary layer at a given streamwise location consists of newly formed, slowly advecting hairpin packets in the near-wall region, with older, more rapidly moving hairpin packets from the upstream flow being located farther from the wall. Direct numerical simulation studies (Zhou et al., 1999) have demonstrated near-wall generation of nascent hairpin vortices, both upstream and downstream of the parent hairpin, in a process termed autogeneration. This process appears quite robust with new hairpin vortices emerging in spite of artificially elevated levels of fine-scale background turbulence imposed in the simulations. The autogeneration of hairpin vortices and the formation of hairpin packets seem to be *a key aspect of the wall turbulence generation process that might lend itself to flow control.*

The considerable wall-normal extent of the hairpin structures is significant because it suggests the possibility of a coupling of the dynamic processes in the near-wall and outer regions of the boundary layer. Clarifying this near-wall–outer-layer interaction has been the focus of several previous investigations. For example, Kobashi and Ichijo (1986) found that low-frequency pressure fluctuations measured in the near-wall region are correlated with large-scale motions in the outer portion of the boundary layer. Based on these observations, the authors suggested that the turbulent burst sequence is phase-related to the outer large-scale motions. In another study, Aubry et al. (1988) developed a low-order dynamical-system model of the near-wall region of a turbulent boundary layer that was shown to duplicate the experimentally observed dynamical behavior of streamwise vortex formation. The model also included the occurrence of intermittent ejection and burst-like events. That work also found that pressure signals from the outer part of the boundary layer triggered the burst process and determined its frequency. Myose and Blackwelder (1994) also highlighted the importance of spanwise vorticity in the outer part of the boundary layer on the near-wall turbulence production process. Their results show that wall-eddy breakdown could be triggered by the streamwise acceleration associated with the outer region of the turbulent boundary layer. This study seems consistent with observations in canonical turbulent boundary layers, of thin shear layers, separating low-speed and high-speed regions (so-called uniform momentum regions), which have been studied in the last few years (Adrian et al., 2000; de Silva et al., 2016).

Theoretical models for a regenerative bursting process based on the instability of the sublayer were proposed by Hanratty (1956), Einstein and Li (1957), and Black (1966). According to their analysis, the bursting frequency should scale with inner wall variables, ν/u_τ^2. This of course contradicts Rao et al. (1971), which, as previously mentioned, indicated that the burst frequency scaled with outer variables, suggesting an influence of the outer large-scale motions in the boundary layer. Doligalski et al. (1980) proposed a model for the lift-up of low-speed fluid from the wall that was based on the interaction with an eddy structure that originated in the outer part of the boundary layer.

Blackwelder and Woo (1974) investigate the burst process by applying a pressure perturbation that was designed to simulate a passing large eddy. This yielded no change in the burst frequency. Numerical experiments on turbulent channel flow by Jimenez and Pinelli (1999) showed quite convincingly that a near-wall turbulence production cycle exists, which is essentially *independent* of the outer flow. This was demonstrated by artificially manipulating the outer flow in numerical simulations and observing the resulting effect on near-wall turbulence production. The manipulation of the outer flow was found to have little effect on the near-wall turbulence generation process, thereby suggesting its autonomous character. Similarly, it was shown that the turbulence generation mechanism was not directly linked to the presence of the wall; the wall served only to provide the necessary mean shear.

More recently, Schoppa and Hussain (2002) demonstrated that near-wall streamwise coherent structures may originate from a sinuous instability of lifted, vortex-free, and low-speed streaks. It was shown that lifted streaks undergo a sinuous normal mode of instability which exhibits twofold amplification. More significantly, however, the authors described a *streak transient growth* (STG) of x-dependent spanwise velocity disturbances leading to an order-of-magnitude amplification culminating in the formation of streamwise vortices via nonlinear processes. The instantaneous characteristics of the resulting near-wall vortices are consistent with those educed from conditional measurements in fully turbulent flows. Critical to the instability leading to the formation of longitudinal near-wall vortices is the wall-normal vorticity flanking lifted low-speed streaks. The process has the character of a shear layer instability. Its interruption potentially offers an approach to affect turbulence production with a link to viscous drag.

Summarizing all of these observations, there are two types of coherent eddy motions that are important to turbulence production and linked to viscous drag in turbulent boundary layers. One is associated with the wall-layer portion of the boundary layer. The other is associated with the outer portion of the boundary layer. Each of these has been observed to contribute a large percentage of the Reynolds stress. The wall layer is associated with the generation of a low-speed "streak" structure consisting of streamwise vortex pairs. The average spanwise (z) spacing between the "streaks" is approximately $100z^+$ (Kline et al., 1967a). The streak structure undergoes a process of gradual lift-up, interaction with faster moving flow, sudden oscillation, and break-up that accounts for 70 percent of the net turbulence production in the boundary layer. The frequency of occurence of this process was found to scale linearly with the wall shear stress, τ_w.

Away from the wall, the outer turbulent bulges maintain a net circulation that plays an important role in the entrainment of potential fluid, and the extraction of turbulence energy from the mean flow. The strength of these large scale eddies has been found to be related to the wall shear stress, although the mechanism for their generation is still an open question.

In a conceptualized view, it is thought that the outer and wall scales of motion should interact. There have been studies that have sought to couple the two regions, with one of the oldest and most notable being the attached eddy model (Marusic and Perry, 1995; de Silva et al., 2016). Of primary importance is the manner in which these events scale. A majority of the evidence suggests that the frequency of wall-layer "burst" events should scale with wall variables, ν/u_τ^2. However, there is some evidence that the frequency of the "burst" events scale with outer variables, U_∞/δ. Since the "burst" frequency is related to the wall shear stress and therefore to viscous drag (Kline et al., 1967a), the suppression of either or both of these scales of motion presents an objective for flow control.

8.2 Flow Control

In general, the studies cited above fall into two broad groups: (1) those that focus on the influence of outer-layer large-scale structures on the near-wall turbulence generation mechanism and (2) those that view the near-wall mechanism as largely

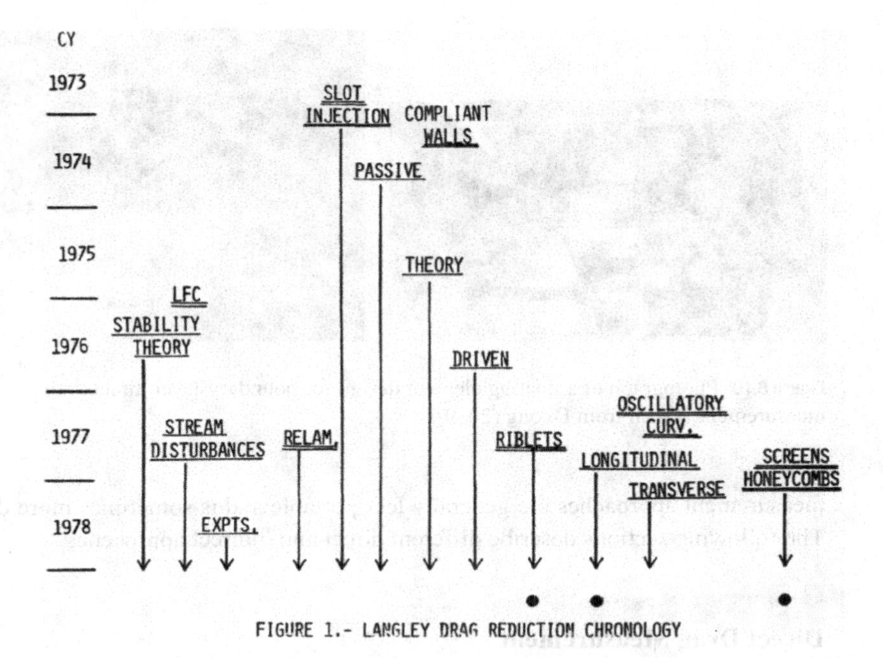

Figure 8.9 Chronology of methods in the 1970s for turbulent boundary-layer drag reduction. Taken from Bushnell (1978).

autonomous. As a result, the flow control approaches for drag reduction have similarly exclusively focused on controlling either the outer large scale eddies or the wall-layer coherent motions. Thus, these two different approaches are presented in separate sections.

Figure 8.9 presents a chronology of boundary-layer flow control for drag reduction in the 1970s (Bushnell, 1978). This was a very prolific period that was motivated by gasoline shortages and the desire to improve aircraft fuel efficiency. Most of these approaches focused on modifying the wall-layer coherent motions. These included slot injection, compliant walls, riblets, and oscillatory wall motions. Screens and honeycombs were intended to alter the outer large-scale motions. These and other techniques are discussed in Sections 8.2.3 and 8.2.4. Before that, experimental methods to quantify the turbulent boundary layer properties are presented.

8.2.1 Boundary-Layer Drag Measurement

A major objective of turbulent boundary control has been toward reducing the viscous drag. Therefore, accurate drag measurement is important. There are a number of ways to measure the viscous drag, both directly and indirectly. The indirect methods often rely on empirical coefficients that are specific to conditions such as pressure gradient and surface roughness. It is therefore important to know when they are appropriate to use. For the most part, direct drag measurement is preferable. However, direct drag

Figure 8.10 Photograph of a floating element design for boundary-layer direct drag measurement. Taken from Duong (2019).

measurement approaches are generally less portable and is sometimes more difficult. The following sections describe different direct and indirect approaches.

Direct Drag Measurement

Turbulent boundary-layer drag measurements generally fall into two approaches, direct and indirect drag measurements. The direct drag measurement usually involves a frictionless floating element that is constrained from moving by a load measuring transducer. An example is shown in Figure 8.10. The floating element is mounted flush with the surface over which the boundary layer develops. It is mounted on a pair of linear air bearings that provide frictionless translation in the mean flow direction. The motion is constrained by a load cell transducer that provides a voltage output proportional to the axial force acting on the platen area. The force should correspond to the product of the shear stress and platen surface area, $\tau_w A_s$. This assumes that the streamwise pressure gradient over the extent of the floating element is negligible and that the boundary layer mean flow is uniform across the span of the floating element, that is, two-dimensional (2-D) mean flow. These assumptions need to be proven in an application by validating the direct drag measurement against other empirical results for similar boundary-layer conditions or against a well-proven indirect drag measurement technique.

Another direct drag measurement technique is oil film interferometry (OFI) that was initially developed by Monson et al. (1993). Other early references to the technique include Zilliac (1996, 1999) and Driver (1998). The principle of OFI is that a drop of oil applied on the wall in a boundary layer will flow due to the wall shear stress. As it flows, the oil thickness reduces. The rate of thinning of the oil can then be related to the local shear stress.

The relation between the oil film thickness and the wall shear stress is

$$\frac{\partial h_0}{\partial t} + \frac{\partial}{\partial x}\left(\frac{\tau_{w,x} h_0^2}{2\mu}\right) + \frac{\partial}{\partial z}\left(\frac{tau_{w,x} h_0^2}{2\mu}\right) = 0, \tag{8.2}$$

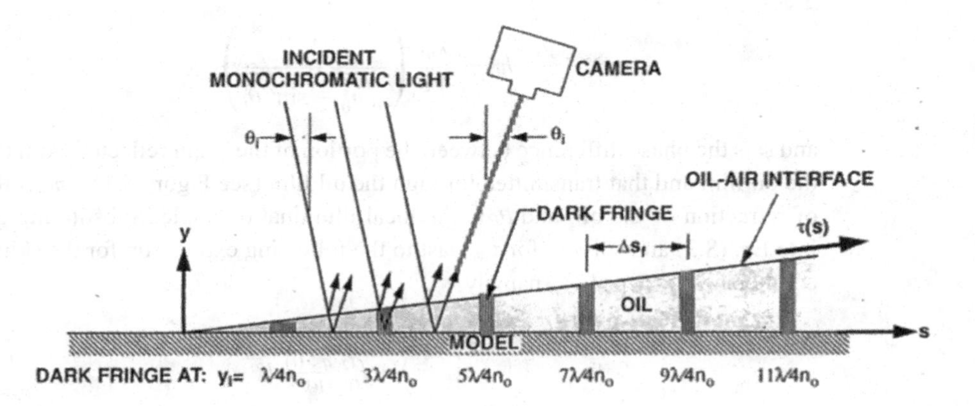

Figure 8.11 Schematic illustrating OFI technique for measuring oil film thickness related to boundary layer wall shear stress. Taken from Monson et al. (1993).

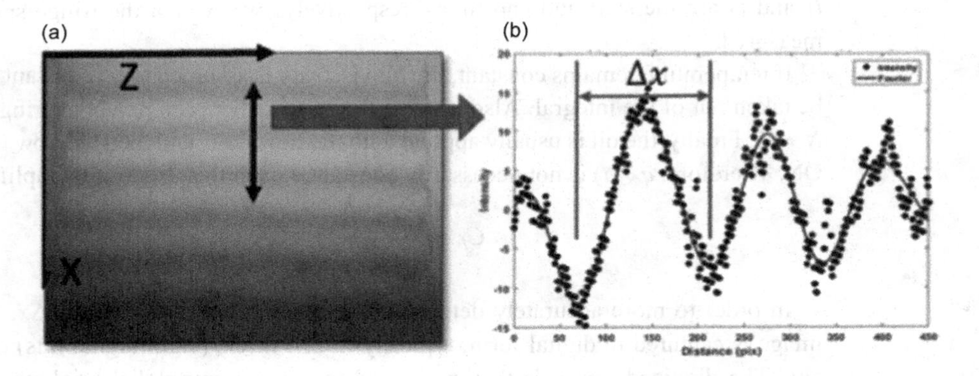

Figure 8.12 Example of 2-D OFI fring pattern (a) and plot of the x variation of the light intensity (b) used in measuring fringe spacing.

where h_0 is the thickness of the oil film, $\tau_{w,x}$ and $\tau_{w,xz}$ are the wall shear stress in the streamwise, x, and spanwise, z, directions, respectively, and t is time. Assuming no variation in the spanwise direction, Eq. (8.2) reduces to

$$\frac{\partial h_0}{\partial t} + \frac{\partial}{\partial x}\left(\frac{\tau_{w,x} h_0^2}{2\mu}\right) = 0. \tag{8.3}$$

In the OFI method, the oil film is illuminated by a monochromatic light source. If the wall surface is reflective, this results in light and dark fringes that are characteristic of interferometry. This is illustrated in Figure 8.11. An example of an OFI fringe pattern is shown in Figure 8.12.

The thickness of the oil film can be found by measuring the spacing between adjacent light and dark bands where

$$h_0 = \frac{\lambda\phi}{4\pi}\left(\frac{1}{\sqrt{n_0^2 - \sin^2\theta_i}}\right) \tag{8.4}$$

and ϕ is the phase difference between the portion of the beam reflected from the top of the oil film and that transmitted through the oil film (see Figure 8.11), n_0 is the index of refraction of the oil, and θ_i is the local illumination angle. Substituting Eq. (8.4) into Eq. (8.3) and solving for τ_w least to the following expression for the skin friction coefficient, $C_f = \tau_w/q_\infty$, namely

$$C_f = \frac{2n_0 \cos\theta_r \Delta x}{N\lambda \int_{t_1}^{t_2} \frac{q_\infty(t)}{\mu(t)} dt}, \tag{8.5}$$

where θ_r is the refracted light angle through the oil, λ is the wavelength of the light source, N is the number of the fringes used in the equation, Δx is the total width of N fringes, μ is the viscosity of the oil, q_∞ is the free-stream dynamic pressure, and t_1 and t_2 are the start and end times, respectively, over which the fringe spacing is measured.

If temperature remains constant, the oil viscosity is assumed to be constant and can be taken out of the integral. Also if Δx is the spacing between neighbor fringes, then $N = 1$. Finally, the oil is usually applied with the flow OFF, and then the flow is turned ON. Therefore, $q_\infty(t)$ is not necessarily constant. Given this, Eq. (8.5) simplifies to

$$C_f = \frac{2\mu n_0 \cos\theta_r \Delta x}{\lambda \int_{t_1}^{t_2} q_\infty(t) dt}. \tag{8.6}$$

In order to more accurately determine the spacing between fringes, Δx, the OFI image is captured in digital form, typically in 256 levels (eight digital bits) of intensity. The digitized image is then processed to remove any global brightness (mean removal) and any linear trends in order to emphasize maxima and minima in the light intensity representing the light and dark bands. An example of the result of this type of processing is shown in Figure 8.12(b). The distance between two neighboring minima is clearly evident and labeled Δx.

Indirect Drag Measurement

Indirect drag measurement techniques most often are based on a measurement of the mean velocity profile. A number of empirical relations exist for determining the wall shear stress from the mean profile. One of the most commonly used approaches involves a best fit of the mean profile to the "log law" relation

$$u^+ = A \ln y^+ + B, \tag{8.7}$$

where A and B are the fitted coefficients. A number of values for these coefficients are reported in the literature. It is now understood that they depend on the streamwise pressure gradient (Nagib and Chauhan, 2008) and possibly Reynolds number. In zero pressure-gradient turbulent boundary layers of moderate Reynolds number, accepted values are $A = 2.439$ and $B = 5.0$. Note that historically $A = 1/\kappa$ where κ is the von

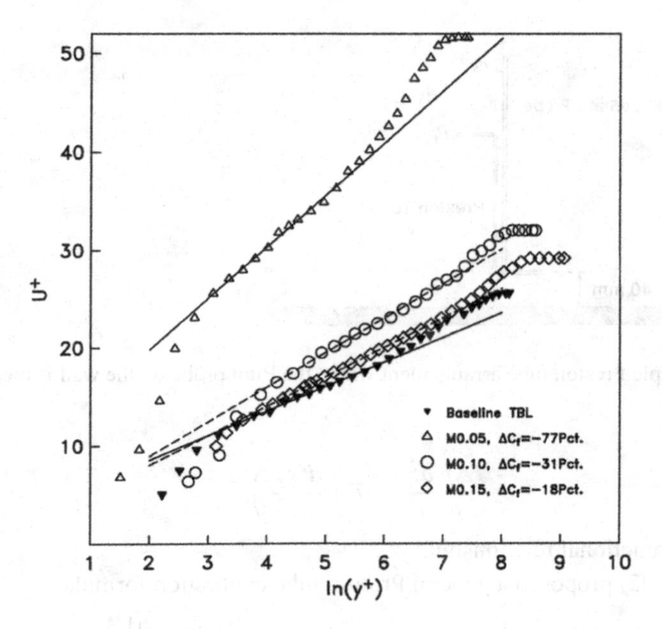

Figure 8.13 Sample Clauser plots used to determine τ_w for different degrees of turbulent boundary-layer drag reduction. Taken from Duong (2019).

Karman coefficient. Therefore, in a zero pressure-gradient turbulent boundary layer, the standard is $\kappa = 0.41$.

A common method for determining the wall shear stress, τ_w, based on the "log law," is attributed to Clauser (1956). For this, the mean velocity profile, $U(y)$, is normalized by wall variables, ν and $u_\tau = \sqrt{\tau_w/\rho}$, giving $u^+ = U/u_\tau$ and $y^+ = (y - y_0)u_\tau/\nu$. Here, y_0 is included as an uncertainty in the wall location, otherwise being $y\ @\ U = 0$. These data are then plotted as u^+ versus $\ln y^+$. Following Eq. (8.7), the slope of a straight line through the log portion, $30 \leq y^+ \leq 500$, is $1/\kappa$. The value of u^+ at, for example, $y^+ = 1$ based on the straight line fit is used to determine the coefficient B. An example of Clauser plots for turbulent boundary layers with different degrees of viscous drag reduction from Duong (2019) is shown in Figure 8.13. The process can also be formulated as a linear least-square curve fit of u^+ versus y^+ with unknowns $1/\kappa$ and B.

Another method for measuring wall shear stress that is based on a universal "law of the wall" is the Preston tube (Preston, 1954), which is widely used due to its simple operation and construction. It involves a total pressure Pitot probe that rests on the wall, and a Pitot-static probe located at the same axial location. A Preston tube arrangement is shown in Figure 8.14. The inside diameter of the Pitot tube located on the wall must be small enough to reside completely inside the log layer, namely $y^+ \leq 100$. Experiments indicate (Patel, 1965) that the ratio of the inside to outside tube diameter has no effect on the measurement, as long as $d_{\text{inside}}/d_{\text{outside}} > 0.2$.

Preston (1954) suggested a nondimensional relationship between the total pressure measured at the wall and the static pressure, Δp, as

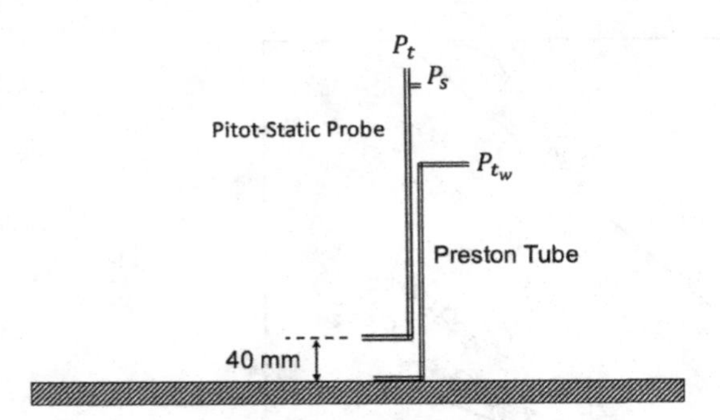

Figure 8.14 Sample Preston tube arrangement where the Pitot probe on the wall is measuring total pressure.

$$\frac{\Delta p}{\rho}\frac{d^2}{\nu^2} = \mathcal{F}\left(\frac{d^2\tau_w}{\rho\nu^2}\right), \tag{8.8}$$

where $\mathcal{F}$ is a functional relationship.

Bechert (1995) proposed a general Preston tube calibration formula

$$\tau^+ = \left[28.44(\Delta p^+)^2 + 6.61 \times 10^6(\Delta p^+)3.5\right]^{1/4}, \tag{8.9}$$

where

$$\tau^+ = \frac{\tau_w d^2}{\rho\nu^2} \tag{8.10}$$

and

$$\Delta p^+ = \frac{\Delta p d^2}{\rho\nu^2}. \tag{8.11}$$

The formula was found to be in excellent agreement with experiments by Head and Ram (1971).

Another less common method for determining τ_w is based on Coles's "Law of the Wake" (Coles, 1958; Coles and Hirst, 1969). This applies to zero pressure-gradient turbulent boundary layers. The relations for the "Law of the Wake" are

$$\frac{U_\infty}{u_\tau} = \frac{1}{\kappa}\ln\left(\frac{\delta u_\tau}{\nu}\right) + B + \frac{2\Pi}{\kappa} \tag{8.12}$$

and

$$\frac{\frac{\delta_1 U_\infty}{\nu} - 65}{\frac{\delta u_\tau}{\nu}} = \frac{1+\Pi}{\kappa}, \tag{8.13}$$

where δ is the boundary-layer thickness, δ_1 is the boundary-layer displacement thickness, and Π is the wake parameter.

Equations (8.12) and (8.13) can be combined to remove the effect of the boundary-layer thickness. This gives

$$\frac{U_\infty}{u_\tau} = \frac{1}{\kappa}\ln\left(\frac{\frac{\delta_1 U_\infty}{\nu} - 65\kappa}{1+\Pi}\right) + B + \frac{2\Pi}{\kappa}. \tag{8.14}$$

For zero pressure-gradient boundary layers, κ and B are the same as for the log law, namely 0.41 and 5.0, respectively. The wake parameter Π is given by Coles (1958) and found to be approximately 0.55 for zero pressure-gradient boundary layers.

The empirical relation of Ludwig and Tillmann (1949)

$$C_f = 0.246\text{Re}_{\delta_2}^{-0.268} 10^{-0.678 H_{12}} \tag{8.15}$$

has also been used to infer the friction coefficient, C_f, from the integral profile properties, which include the momentum thickness, δ_2, and the shape factor, $H_{12} = \delta_1/\delta_2$. This relation was derived by correlating the experimental results from boundary-layer measurements of different investigators.

The issue with the Equations 8.12–8.15 is that they utilize coefficients (κ, B, Π) that are not expected to be constants in drag-reduced boundary layers. Therefore, other than a direct drag measurement, the better approach involves measuring the boundary-layer momentum. The momentum equation for a 2-D, incompressible turbulent boundary layer in integral form is

$$\frac{d}{dx}\left(U_\infty^2 \delta_2\right) + \delta_1 U_\infty \frac{dU_\infty}{dx} = \frac{\tau_w}{\rho}. \tag{8.16}$$

The contribution of turbulent fluctuations to the balance is trivial and neglected in this formulation. The terms δ_1 and δ_2 are again the displacement and momentum thickness, respectively, given by

$$\delta_1 = \int \left(1 - \frac{U(y)}{U_\infty}\right) dy \tag{8.17}$$

and

$$\delta_2 = \int \frac{U(y)}{U_\infty}\left(1 - \frac{U(y)}{U_\infty}\right) dy. \tag{8.18}$$

To apply a momentum balance, mean velocity profiles, $U(y)/U_\infty$, are measured at different streamwise locations. These profiles are integrated to determine δ_1 and δ_2. These are used in solving the momentum balance given by Eq. (8.16) for the shear stress, τ_w. In the 2-D mean flow assumption, the spanwise uniformity needs to be checked. In addition, an uncertainty analysis needs to be performed to determine the accuracy of the shear stress measurement.

8.2.2 Burst Measurement Techniques

As previously discussed, in the context of turbulent boundary layers, "bursting" refers to the process of "gradual lift-up, sudden oscillation, bursting and ejection" first observed as part of the wall-layer "streak structure" by Kline et al. (1967a). The bursting sequence produces a characteristic velocity signature where the "lift-up" brings low momentum fluid away from the surface producing a sharp, localized velocity gradient. Burst detection methods are generally based on (1) the detection of a sharp peak in the local variance in the streamwise, u, component of the velocity time series, or (2) the detection of a sharp increase in the wall-normal, v, velocity component *away* from the wall. The first approach can be performed with a single hot-wire sensor. That approach uses the variable-interval time-averaged (VITA) technique. The second approach requires dual slanted hot-wire sensors. That approach is referred to as the uv-quadrant splitting technique. These two approaches are described in the following sections.

VITA Technique

One "burst" detection method pioneered by Blackwelder and Kaplan (1976) is the VITA technique. This utilizes the velocity time series from a single hot-wire and applies a discriminator, D, such that

$$D(t) = \begin{cases} 1 \text{ if } \widehat{\text{var}} > \text{k}u^2_{\text{rms}} \\ 0 \text{ Otherwise} \end{cases}, \tag{8.19}$$

where

$$\widehat{\text{var}}(x_i, t, T) = \widehat{u^2}(x_i, t, T) - [\widehat{u}(x_i, t, T)]^2 \tag{8.20}$$

is the localized variance of streamwise velocity fluctuations over the interval, T, at a spatial location, x_i. Blackwelder and Kaplan (1976) recommended that $k = 1.2$ and that $Tu_\tau^2/\nu = 10$. They applied the detection criterion at $y^+ = 15$, which is where the turbulence intensity is a maximum in the canonical turbulent boundary layer (Klebanoff, 1954).

A sample ensemble average of the velocity time series based on the VITA detection given in Eq. (8.19) is shown in Figure 8.15(a). Following Blackwelder and Kaplan (1976), the ensemble-averaged velocity, $< u >$, is normalized by $(ku'^2)^{1/2}$, where u' is the long-time RMS of $u(t)$. This corresponds to a baseline turbulent boundary layer, and one in which τ_w was reduced by 76 percent through flow control (Duong, 2019).

The characteristic VITA velocity signature is thought to embody the initial "lift-up" of low momentum fluid from the wall, noted on the plot, that is followed by a faster moving "sweep" of fluid that approaches from outside the wall region (Corino and Brodkey, 1969).

Figure 8.15(b) shows statistics of the "burst" events detected using the VITA technique. These include a probability density of the time period between "burst" events, the average period, the average frequency of the events, and the number of events. The merit of these statistics is that they can be linked to the wall shear stress as demonstrated by Kline et al. (1967a), which is shown in Figure 8.3. In the present example with a 76 percent reduction in τ_w, the frequency of the "burst" events decreased by nearly the same amount (72 percent).

uv-Quadrant Splitting Technique

Another method used to further quantify the effect between uncontrolled and controlled cases is the "quadrant splitting" technique outlined by Lu and Willmarth (1973). It is a conditional sampling technique that can quantify the contributions of different events to the total $\overline{uv}$-Reynolds stress. An example of the result of this technique is shown in Figure 8.16.

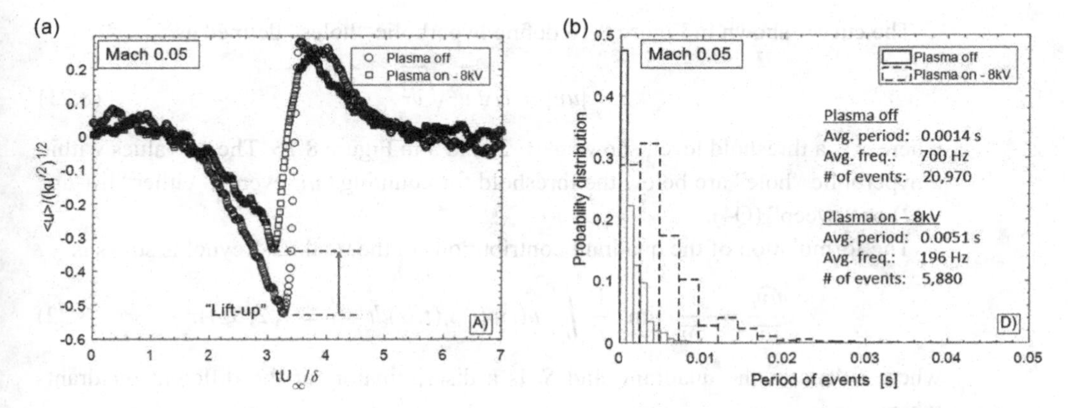

Figure 8.15 Example of VITA signature (a) and corresponding "burst" detection statistics (b) for a baseline boundary layer (circle symbols) and in a boundary layer with 77 percent viscous drag reduction (square symbols). Taken from Duong (2019).

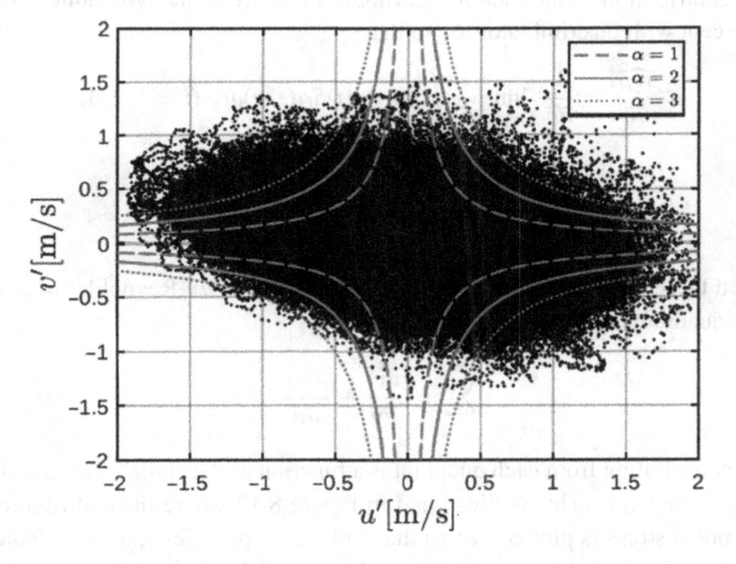

Figure 8.16 Example of uv-quadrant splitting in a turbulent boundary layer at $y^+ = 100$. Taken from Duong (2019).

The two events that are most commonly quantified through the quadrant splitting method are ejection and sweep events, also referred to as Q2 and Q4 events, respectively. With reference to Figure 8.16, Q2 events occur in the so-called second quadrant where $u(t) < 0$ and $v(t) > 0$. The Q4 events occur in the fourth quadrant where $u(t) > 0$ and $v(t) < 0$. Lu and Willmarth (1973) showed that the Q2 and Q4 events were the largest contributors to the $\overline{uv}$-Reynolds stress production in turbulent boundary layers.

The curves shown in Figure 8.16 define hyperbolic "holes" defined as

$$|uv| = \alpha\sqrt{\overline{u^2}}\sqrt{\overline{v^2}}, \tag{8.21}$$

where α is a threshold level, shown as 1, 2, and 3 in Figure 8.16. The uv values within a hyperbolic "hole" are below the threshold for counting "uv-events," either "lift-up" (Q2) or "sweep" (Q4).

The formulation of the quadrant contributions to the total $\overline{uv}$-Reynolds stress is

$$\frac{\widehat{uv_i}}{\overline{uv}} = \frac{1}{\overline{uv}} \lim_{T\to\infty} \frac{1}{T}\int_0^T u(t)v(t)S_i(t,\alpha)dt \quad (i = 1, 2, 3, 4), \tag{8.22}$$

where i denotes the quadrant, and S_i is a discriminator for the different quadrants where

$$S_i(t,\alpha) = \begin{cases} 1 & \text{if } u(t) \text{ and } v(t) \text{ are in quadrant } i \text{ and } |u(t)v(t)| > \alpha\sqrt{\overline{u^2}}\sqrt{\overline{v^2}} \\ 0 & \text{otherwise} \end{cases}. \tag{8.23}$$

The contribution to the total $\overline{uv}$-Reynolds stress from the hyperbolic "hole" is the same except with discriminator S_H so that

$$\frac{\widehat{uv_H}}{\overline{uv}} = \frac{1}{\overline{uv}} \lim_{T\to\infty} \frac{1}{T}\int_0^T u(t)v(t)S_H(t,\alpha)dt \quad (i = 1, 2, 3, 4) \tag{8.24}$$

and

$$S_H(t,\alpha) = \begin{cases} 1 & \text{if } |u(t)v(t)| \le \alpha\sqrt{\overline{u^2}}\sqrt{\overline{v^2}} \\ 0 & \text{otherwise} \end{cases}. \tag{8.25}$$

Given these definitions, the total contributions to the $\overline{uv}$-Reynolds stress including all four quadrants and the hyperbolic "hole" is

$$\sum_{i=1}^{4} = \frac{\widehat{uv_i}}{\overline{uv}} + \frac{\widehat{uv_H}}{\overline{uv}} = 1. \tag{8.26}$$

The contribution from each quadrant is a function of the "hole" size, which is based on the parameter, α. This is illustrated in Figure 8.17 where the contributions to the $\overline{uv}$-Reynolds stress is plotted versus the "hole" size parameter, α. This indicates that the contributions of the Q1 and Q3 to the overall $\overline{uv}$-Reynolds stress is relatively small, and saturates for $\alpha > 2$. The largest contributions come from Q2 (ejection) and Q4 (sweep) events. These, along with the "hole," are sensitive to the hole size parameter, α.

Lu and Willmarth (1973) recommended a large value of $\alpha = 4.5$. As indicated in Figure 8.17, this selection of α, the majority of contributions to the $\overline{uv}$-Reynolds stress, is due to Q2 (ejection) events and that within the "hole."

To relate back to "burst" frequency statistics that comes from the VITA approach, the uv quadrant approach can be used to determine the average time between events as well as the *duration* of events. The duration of an event is defined as the time from when $uv(t)$ exits the hyperbolic "hole" set by α to when it crosses back into the "hole."

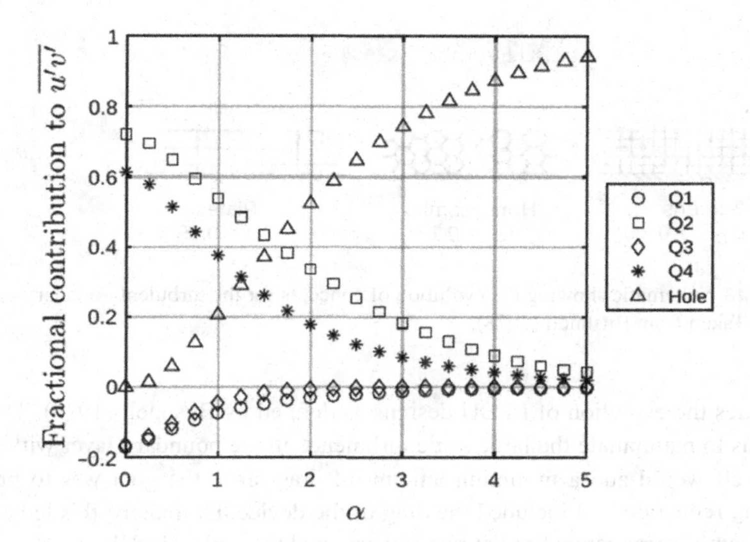

Figure 8.17 Effect of hyperbolic "hole" size parameter, α, on the quadrant contributions to $\overline{uv}$-Reynolds stress. Taken from Duong (2019).

The average time of these events in the ith quadrant is defined as $\overline{\Delta T_i}$. The average time between events is designated $\overline{T_i}$. Therefore, the ratio of these provides a *duty cycle*, $\overline{\Delta T_i}/\overline{T_i}$. The special utility of the event duty cycle is that it is independent of the "hole" size parameter, α. Thus, it is a meaningful method to compare the effect of drag reduction on the turbulent boundary layers. In addition, it provides a comparison to the VITA detection technique.

8.2.3 Outer-Scale Control

Turbulent boundary layer outer-scale control seeks to suppress the large outer scales of motion that are on the order of the boundary-layer thickness. Ultimately, their suppression is intended to impede turbulence production mechanisms in the wall layer through a process referred to as "top down." The following sections describe some of the methods of turbulent boundary-layer outer-scale control.

Large Eddy Break-up Devices

The general approach to drag reduction has been to seek to intervene with the various coherent turbulent motions as discussed in the previous section. Some of these approaches have been passive, and others have been active. An example of the former that was aimed at altering the outer-layer structure are the so-called large eddy break-up (LEBU) devices. These consisted of 2-D thin flat plates aligned with the primary flow direction that were suspended at different heights above the wall in the turbulent boundary layer. The motivation for this came from thick panels with three-dimensional honeycomb-shaped openings that are used in controlling free-stream turbulence in wind tunnels (Loehrke and Nagib, 1972, 1976; Tan-atichat et al., 1982). Figure 8.18

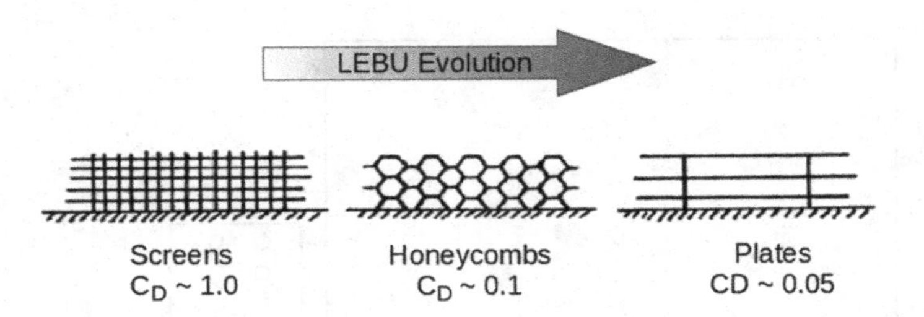

Figure 8.18 Schematic showing the evolution of concepts for the turbulent boundary-layer LEBU. Taken from Bushnell (1978).

illustrates the evolution of LEBU designs as defined by Bushnell (1978). The objective was to manipulate the large-scale turbulence in the boundary layer with a device that itself would add a minimum amount of drag since the goal was to produce a net drag reduction that included the drag of the device. Ultimately, this led to a "2-D honeycomb" arrangement of flat plates developed by Corke (1981b).

In the boundary-layer application, the plates were used singly, vertically stacked with different spacings, or placed in tandem in the streamwise direction. Table 8.1 lists the design parameters for the three configurations that led to the flow visualization images shown in Figure 8.19. In these cases, configurations of four, two, and one LEBU plates were used. All the dimensions are normalized by the boundary-layer thickness at the upstream edge of the LEBU. For all the configurations, the normalized plate thickness was $t/\delta = 0.002$, and the normalized streamwise length was $l/\delta = 0.8$. The normalized heights of the LEBU plates in the boundary layer varied from $0.1 \leq h/\delta \leq 0.7$. Aside from the thickness of the plates, all of the dimensions were of the order of the boundary-layer thickness. All of the plates are located in the constant stress layer of the boundary layer. The theory behind their design was to inhibit wall-normal velocity fluctuations as well as to add small-scale turbulence.

The result these had on the turbulent boundary layer is embodied in the flow visualization images shown in Figure 8.19. These were obtained using a smoke wire (Corke et al., 1977) that was located just downstream of the LEBU. The smoke wire was oriented in the wall-normal direction so as to produce a 2-D sheet of smoke streaklines through the boundary layer when viewed from the side. Figure 8.19(a) shows the large-scale outer envelope of the turbulent boundary layer with potential flow excursions reaching close to the wall. Figure 8.19(b) corresponding to the four-plate M-1 case documents a complete suppression of the large-scale outer structure of the turbulent boundary layer. The single plate (M-2) case documented by Figure 8.19(c) is not as effective although the large-scale outer structure is still significantly reduced. The flow visualization photograph, Figure 8.19(d), for the two-plate case (M-3) shows it to be nearly as effective in suppressing the large-scale outer structure as the four-plate configuration. Minimizing the number of plates was an important metric toward achieving *net* drag reduction since the four-plate configuration is likely to produce more device drag than the two-plate configuration. A design that is not shown, that utilized two single plates at a height in the boundary layer of $h/\delta = 0.8$ and spaced

Table 8.1 LEBU design for images in Figure 8.19.

Designation	t/δ	l/δ	h/δ
M-0	–	–	–
M-1	0.002	0.8	0.7,0.4,0.2,0.1
M-2	0.002	0.8	0.6
M-3	0.002	0.8	0.6,0.2

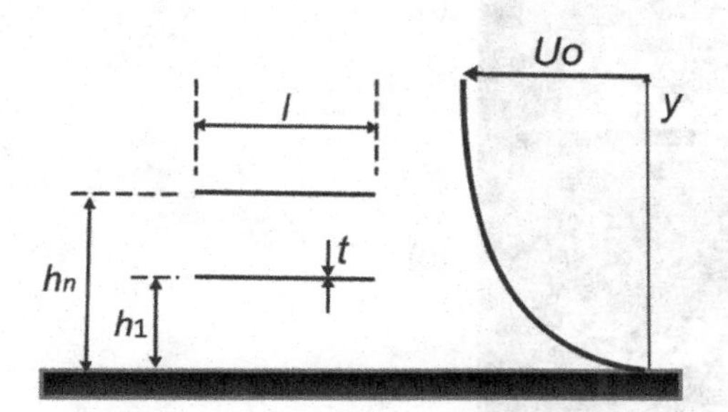

8δ apart in the streamwise direction produced the same degree of large scale suppression but with a lower device drag. That configuration produced a net drag reduction of approximately 20 percent (Corke, 1981b; Corke et al., 1982). The boundary layers with this two-plate LEBU configuration exhibited a 17 percent increase in the sublayer thickness, a 19 percent lower turbulence intensity (u'/U_∞) in the constant stress layer below $y^+ = 100$, and an approximately 17 percent decrease in the "bursting" frequency (Corke, 1981b; Corke et al., 1982). Experiments by others (Walsh and Anders, 1989; Tardu and Binder, 1991) produced in the range of 10 percent local drag reduction using different LEBU configurations.

Anders (1986) utilized NACA 0009 airfoil cross-sectional plates in a tandem LEBU configuration in wind tunnel tests. The airfoil cross section was superior to the flat plate design by having a lower base drag as well as being able to better resist bending. This LEBU design was reported to reduce the skin friction by as much as 30 percent, with a recovery of more than 100δ. With the lower device drag produced by the airfoil section shape, the net drag reduction (accounting for LEBU drag) was up to 7 percent. Anders (1986) also reported that the LEBU performed better at higher Reynolds numbers more representative of flight conditions.

In addition to viscous drag reduction, Keith and Barclay (1993) investigated the effect of LEBUs on the unsteady pressure at the wall in a turbulent boundary layer. The LEBU consisted of a single linear-tapered airfoil shape that was suspended in the boundary layer at a fixed height of 0.6δ. The cross section had a rounded leading edge and a sharp trailing edge. The LEBU was oriented at a positive angle of attack so that the lower surface was parallel to the wall. The chord length was 1.3δ. The configuration was found to produce approximately a 5 percent reduction in τ_w. Concurrent with this was a reduction in the amplitude of lower frequency wall-pressure fluctuations that were associated with the outer-layer and log-law scales of motion. At the same time, the amplitude of higher frequency wall-pressure fluctuations were observed.

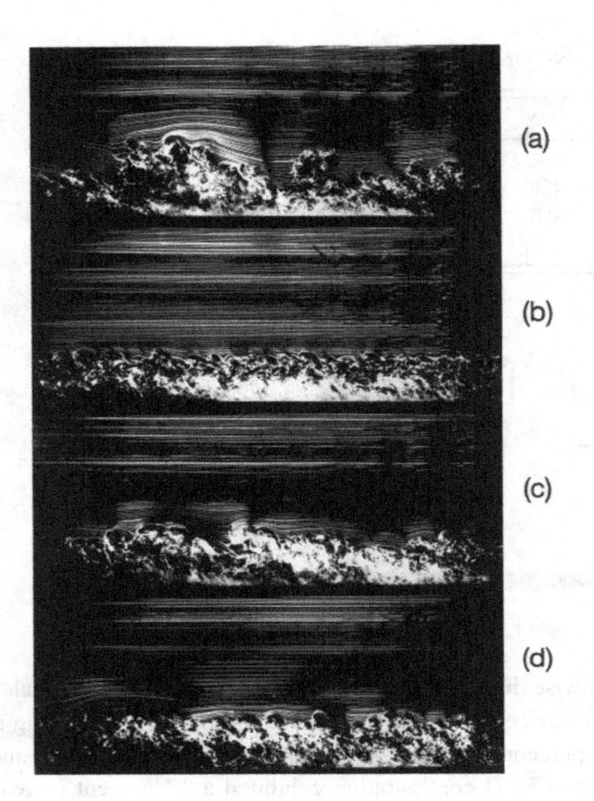

Figure 8.19 Smoke-wire flow visualization of turbulent boundary-layer turbulence manipulators (a), with LEBU designs M-1 (b), M-2 (c), and M-3 (d) listed in Table 8.1. Taken from Corke (1981b).

These experimental results sparked an interest in the underlying mechanisms of LEBUs. Dowling (1985) performed an analytic study that focused on the unsteady shedding of vorticity from an isolated LEBU plate. His analysis indicated that this could lead to a significant reduction in the fluctuating velocity in the vicinity of the wall. He surmised that the optimum position of the plate would be where it was between the wall and the vorticity in the large-scale outer structure of the boundary layer, which is consistent with the optimization performed by Corke (1981b). Savill and Mumford (1988) investigated LEBU devices of various geometries introduced into turbulent boundary layers with $\mathrm{Re}_\theta = 1000 - 3500$. Drag was measured directly by means of a force balance downstream of the device. They indicated that while the suppression of large-scale motions was one mechanism for friction drag reduction, the interaction of vortices inside the wake of the LEBU device with near-wall structure provided the primary mechanism. Support for this came from the fact that maximum local drag reduction always occurred where the wake vortices reached the sublayer. They concluded that the optimum wall-normal height of the device should vary inversely with Reynolds number.

More recently, Chin et al. (2015, 2017) performed large eddy simulation simulations at $\mathrm{Re}_\theta = 4,300$ for a single LEBU plate equivalent to the M-2 configuration of Corke (1981b) and Corke et al. (1982). The LEBU was found to produce skin friction

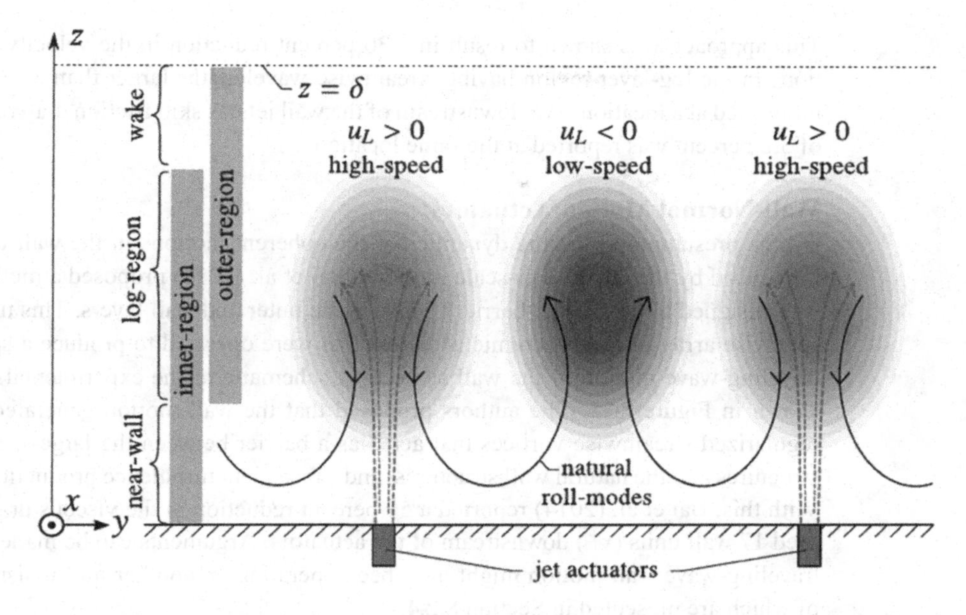

Figure 8.20 Schematic of spanwise array of wall-normal jets designed to operate in an ON/OFF bimodal fashion to the overlying logarithmic region. Taken from Abbassi et al. (2017).

reduction beyond 180δ downstream of the device and a peak local skin friction reduction of 12 percent. However, when accounting for the drag of the LEBU device, no net drag reduction was achieved. Their simulations also indicated a reduction in the large scales, similar to the experiments. In particular, conditional measurements of high and low momentum bulges downstream of the device documented the break-up of the large outer scale motions, compared to that of the baseline turbulent boundary layer. Spanwise energy spectra showed a reduction in energy for long spanwise wavelength motions $\lambda_{z^+} > 1,000$ at all wall-normal locations. Kim et al. (2017) performed a Direct–Navier Stokes (DNS) of a turbulent boundary layer with a single LEBU device also located at the wall-normal location of 0.8δ. They documented a 17 percent reduction in the skin friction coefficient, and a reduction in the energy contained in large-scale spanwise motions at all wall-normal locations that were consistent with results by Chin et al. (2017). Kim et al. (2017) attributed the skin friction reduction to a change in spatial coherence of the outer high- and low-speed structures.

Wall-Normal Blowing

An alternate approach to influence the large-scale outer structure in boundary layers for the purpose of drag reduction has been to use wall-normal blowing. For example, Abbassi et al. (2017) reported the use of a spanwise array of nine wall-normal jets operated in an ON/OFF bimodal fashion with sufficient authority that the jets penetrate the overlying logarithmic region. A schematic of their arrangement is shown in Figure 8.20. The experiments were performed in a zero pressure-gradient turbulent boundary layer with $Re_\tau = 14{,}400$. The jets were triggered by a real-time active controller that sensed the "footprint" of large-scale structures with sensors located upstream. The wall jets were injected into the locations of the large-scale motions.

This approach was shown to result in a 30 percent reduction in the velocity fluctuations in the log-layer region having streamwise wavelengths larger than $\lambda_x = 5\delta$, as monitored at a location 1.7δ downstream of the wall jets. A skin friction drag reduction of 3.2 percent was reported at the same location.

Wall-Normal Motion Actuators

On the presumption that the dynamics of the coherent motions in the wall layer are influenced by the large outer-scale motions, Bai et al. (2014) proposed a method that was designed to provide a "barrier" between the outer and wall layers. This utilized a spanwise array of piezo-ceramic actuators that were operated to produce a spanwise traveling-wave motion of the wall surface. A schematic of the experimental setup is shown in Figure 8.21. The authors proposed that the wall motion generated highly regularized streamwise vortices that acted as a barrier between the large-scale outer structures and the natural wall structures, and subsequent turbulence production cycle. With this, Bai et al. (2014) reported a 50 percent reduction in the viscous drag measured 17 wall units (x^+) downstream of the actuators. Arguments can be made that the traveling-wave wall motion might have been operating on another mechanism, many of which are presented in Section 8.2.4.

8.2.4 Inner-Scale Control

Methods for wall-layer control have generally focused on interacting with the wall-streak structure first observed by Kline et al. (1967a). As indicated in the NASA drag reduction chronology that is shown in Figure 8.9, one of the earliest approaches was "riblets." Other approaches include wall-normal suction and blowing, oscillating wall motion, uniform and oscillating body forces, and lateral sublayer blowing.

Riblets

Riblets consist of small V-grooves applied to the wall surface. The basic design is shown in Figure 8.22. Experiments have indicated an optimum height (h) and lateral spacing (s) of the grooves for drag reduction that corresponds to $h^+ = 8 - 12$ and $s^+ = 15 - 20$. Such scaling is consistent with those of the low-speed wall streaks observed by Kline et al. (1967a), and suggests an underlying stabilizing mechanism. Relative to the average streak spacing of $100z^+$, the optimum groove height is approximately one-tenth of the streak spacing, and the optimum groove spacing is approximately one-fifth of the streak spacing. This optimum design has resulted in a maximum net drag reduction of approximately 8 percent in experiments.

Experimental results by Bechert et al. (1997), shown in Figure 8.23(a), illustrate the degree of riblet drag reduction as a function of s^+. This is expressed as the change in wall shear stress, $\Delta\tau$, normalized by the nominal smooth plate value, τ_0. The results reveal two key regimes.

The "roughness regime" results from having a too large spacing between the riblet grooves. This causes the riblets to act as surface roughness, which ultimately results in a drag increase compared to a smooth wall. To avoid this regime, $s^+ < 27$.

The other is the "viscous regime" that occurs when the spacing between the riblet grooves is too small. The viscous regime ends near the optimum spacing value of

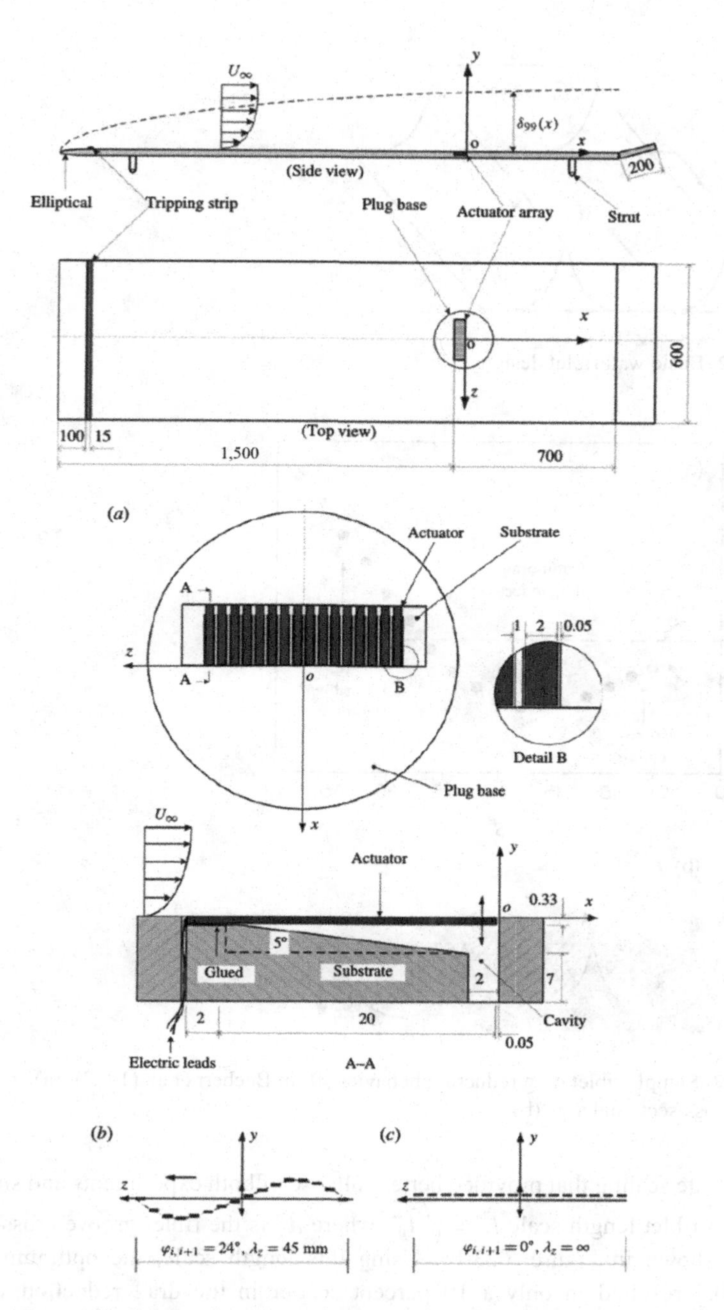

Figure 8.21 Schematic of spanwise array of piezo-ceramic actuators located on the wall in a flat plate over which a turbulent boundary layer develops. Taken from Bai et al. (2014).

$s^+ \simeq 17$. The slope, m, denoted in Figure 8.23(a), namely $m = -(\partial(\frac{\Delta\tau}{\tau_0})/\partial s^+)_{s^+=0}$, was meant to provide a design parameter where the degree of drag reduction would be related to the product $(m \times s^+)$. However, as reported by Garcia-Mayoral and Jimenez (2011), this did not reduce the large (40 percent) scatter of results in experiments.

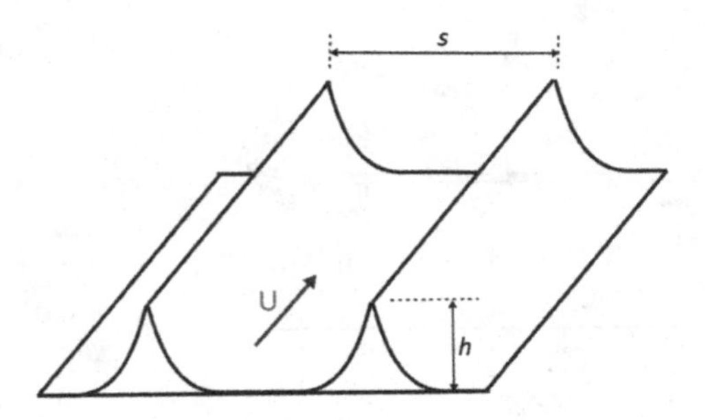

Figure 8.22 Basic wall riblet design.

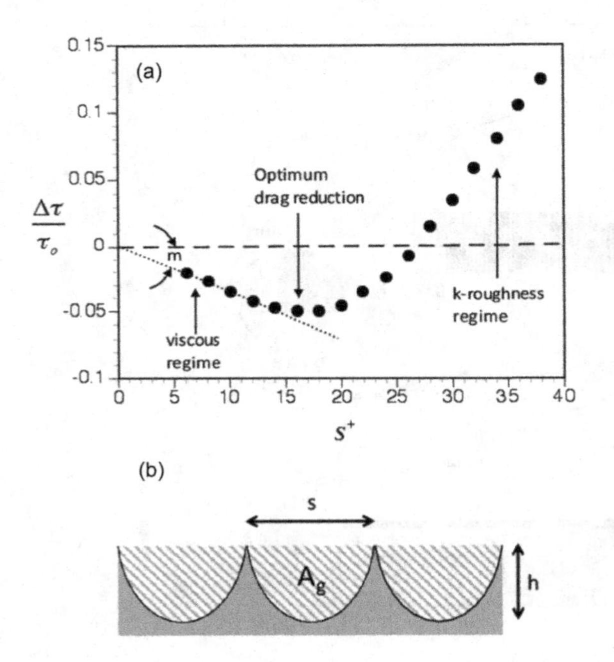

Figure 8.23 Sample riblet drag reduction behavior (from Bechert et al. (1997)) (a) and riblet groove cross-sectional area (b).

Alternate scaling that provided better collapse of both experiments and simulations defined a riblet length scale $l^+ = \sqrt{A_g^+}$, where A_g is the riblet groove cross-sectional area as shown in Figure 8.23(b). Using this length scale, the optimum value of $l_{\text{opt}}^+ = 10.7$ resulted in only a 10 percent scatter in the drag reduction values for a wide range of riblet geometries. For conventional riblets, Garcia-Mayoral and Jimenez (2011) showed that maximum drag reduction ($\text{DR}_{\max}$) was accurately given by $\text{DR}_{\max} = 0.83 m_l l_{\text{opt}}^+$, where $m_l = -(\partial(\frac{\Delta\tau}{\tau_0})/\partial s^+)_{l_{\text{opt}=0}^+}$.

There are a number of theories for the mechanism underlying the drag reduction produced by riblets. All of these focus on some form of interaction with the coherent streamwise vortices thought to be the source of the low-speed wall streaks observed by Kline et al. (1967a). One theory (Bacher and Smith, 1985) is that vorticity generated in

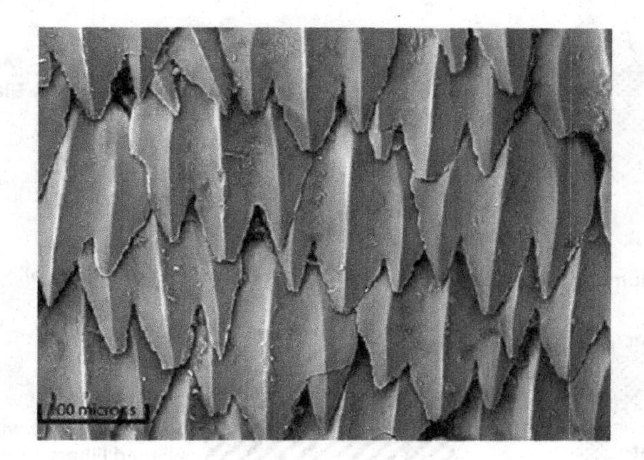

Figure 8.24 Photograph of shark skin surface illustrating riblet-type structure.

the valleys of the riblets interacts with and weakens the coherent streamwise vortices. Another (Smith et al., 1989) postulates that the riblets interfere with the spanwise (sinuous) motion of the low-speed streaks. This mechanism appears consistent with the wall turbulence mechanism of Schoppa and Hussain (2002). Finally, it is thought that the riblets reduce the streamwise extent of the streamwise vortices and increase their spanwise spacing (Karniadakis et al., 1967), and that this has a benefit on reducing the viscous drag.

Finally, riblet-type surfaces occur in nature, particularly on sharks, and are surmised to be adaptation aimed at reducing the viscous drag. An example of a shark skin surface is shown in Figure 8.24. The amount of drag reduction of laboratory-replicated shark skin has shown a large degree of scatter, with a maximal drag reduction of approximately 24 percent (Zhang et al., 2011). This amount of drag reduction is considerably higher than the 8 percent for traditional optimized riblet surfaces.

Compliant Surfaces

Somewhat like riblets, having biological origins from shark skin, the concept of compliant surfaces for drag reduction originated from observations of dolphins. Specifically, Kramer (1957, 1960, 1961b, 1962, 1965) documented drag reduction of up to 50 percent using a compliant wall coating that was modeled after dolphin skin. A series of experiments by others (Looney and Blick, 1966; Blick and Walters, 1968) reported drag reduction of up to 50 percent with an accompanying 25 percent reduction in the Reynolds stress. A comprehensive review of this early work was given by Riley et al. (1988). Examples of compliant surface designs from Banerjee and Jayakumar (2005) are shown in Figure 8.25.

The methodology to selecting an optimum compliant surface is largely attributed to Semenov (1991). Choi et al. (1997) performed an experiment in water using the same test article as Kulik et al. (1991). The test article provided a direct drag measurement. The compliant surfaces were a single-layer 7-mm thick coating of silicone rubber. Their density was 2.14×10^3 kg/m^3, their modulus of elasticity of the coatings ranged

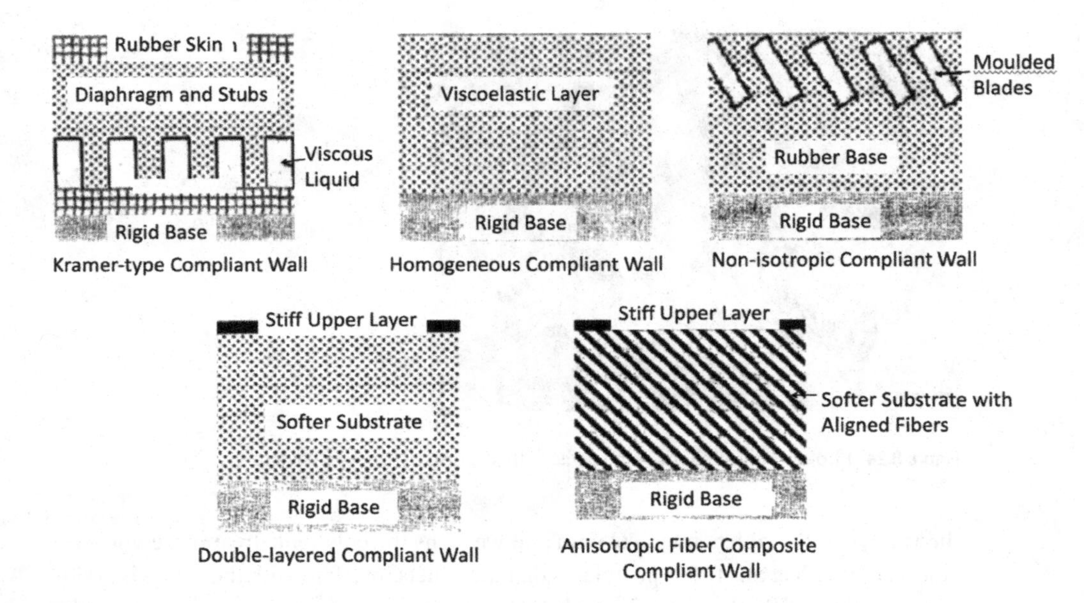

Figure 8.25 Schematic drawing of different designs of compliant walls for turbulent boundary-layer viscous drag reduction. Taken from Banerjee and Jayakumar (2005)

from 1.70×10^6 to 2.81×10^6 Pa, and their values of loss tangent ranged from 0.084 to 0.26 over a range up to 10 kHz. The free-stream speed was 2 m/s. The experiment documented a 7 percent drag reduction, and a 5 percent reduction in the turbulence intensity across the boundary layer.

The experiment by Choi et al. (1997) also indicated that the right combination of material properties is required with at least two requisites. The first of these is that the *dynamic* surface roughness be small enough to be below the value considered to be hydrodynamically smooth. More specifically, the amount of surface deformation of the compliant coating must be much less than the viscous sublayer thickness. The second requirement is that the natural frequency of the compliant surface must be chosen to provide the correct response to the fluctuating pressure forces at the wall. In terms of a nondimensional time, $t_0^+ = f_0^{-1} = t_0 u_\tau^2/\nu$, then $50 \leq f_0^{-1} \leq 150$. Choi et al. (1997) noted that the range of t_0^+ values corresponded to the period between "sweep" events associated with peak unsteady wall-pressure fluctuations. The combination of both static and dynamic properties of a compliant coating is therefore important in order to achieve drag reduction. The various examples of compliant surface designs shown in Figure 8.25 are a means for fine-tuning their characteristics.

Wall Suction and Blowing

There have been several attempts to utilize surface blowing and suction in order to reduce skin friction drag in wall-bounded turbulent flows. Generally, these have sought to intervene in the burst/sweep event cycle that has been shown to be key in Reynolds stress production. As an example, Gad-El-Hak and Blackwelder (1989) used a single

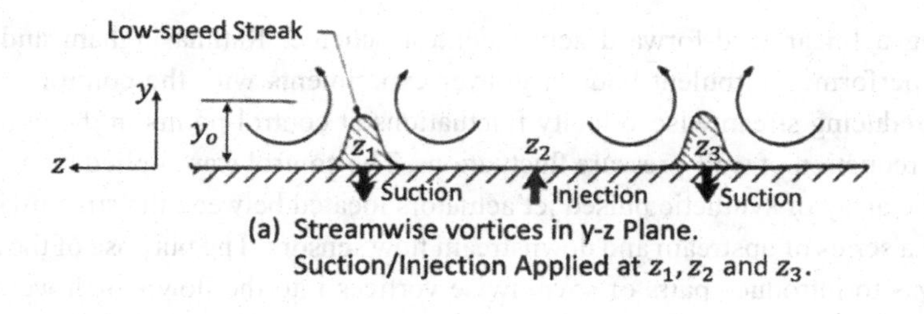

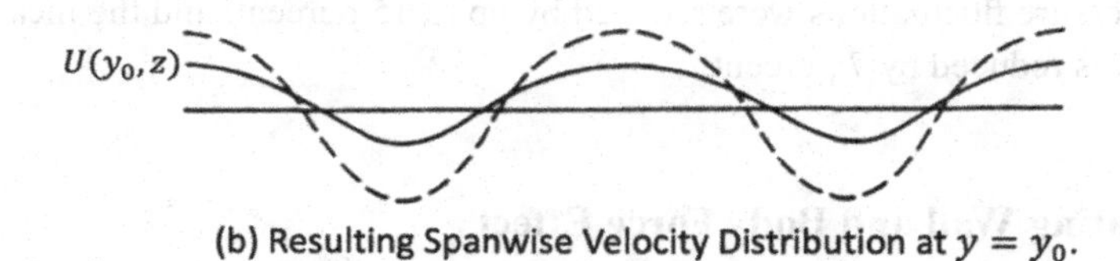

$U(y_o, z)$
(b) Resulting Spanwise Velocity Distribution at $y = y_o$.

Figure 8.26 Proposed suction and blowing to minimize mean flow distortion produced by counter-rotating streamwise vortices associated with wall streak structure. (Source: NASA Contractor Report 4221, 1989)

streamwise-oriented suction slot to reduce bursts in a zero pressure-gradient turbulent boundary layer. The concept is illustrated in Figure 8.26.

The approach was tested by creating artificial wall streaks in a laminar boundary layer using impulsive suction applied to two small holes with a spanwise separation of $z^+ = 100$ (Gad-El-Hak and Hussain, 1986). VITA burst detection was performed downstream. This indicated that a surprisingly small amount of slot suction was needed to eliminate the burst events. Drag measurements were not performed.

In a similar vein, Choi et al. (1994) performed DNS turbulent channel flow simulations that focused on drag reduction through surface boundary conditions intended to suppress dynamically significant coherent motions in the near-wall region. In particular, the focus of the control schemes was the suppression of ejection and sweep events. To that end, wall blowing or suction was applied to oppose the wall-normal velocity detected in a so-called detection plane located at a fixed wall-normal distance above the channel wall. In this scheme, when fluid in the detection plane was moving toward the wall (a sweep event) wall blowing was used to counter it. Similarly, when fluid was detected moving away from the wall (an ejection event), it was countered by an equal wall suction. The associated skin friction drag reduction was characterized in terms of the reduction in mean streamwise pressure gradient required to drive the flow. Since near-wall streamwise vortices also produce strong spanwise velocity, a spanwise out-of-phase boundary condition was also applied in the numerical experiments. The results of the simulation showed drag reductions between 20 percent and 30 percent based on either wall-normal or spanwise wall velocity control. The authors rightly noted that it is *not practical* to place sensors in the flow for most actual drag reduction applications. To this end, they proposed wall sensors that use $\partial/\partial z(\partial w/\partial y)_{\text{wall}}$ to be the best detection criterion for ejection events. When applied, it led to 6 percent drag reduction.

Using a linear feed-forward active control scheme, Rathnasingham and Breuer (2003) performed turbulent boundary-layer experiments with the control objectives of: (1) reducing streamwise velocity fluctuations at control points in the buffer layer and (2) reduction of wall-pressure fluctuations. The control was applied by means of a spanwise array of synthetic pulsed jet actuators located between the streamwise locations of a series of upstream and downstream flow sensors. The purpose of the actuator array was to introduce pairs of streamwise vortices into the flow which would serve to counteract the naturally occurring near-wall structures. Employing this approach, wall-pressure fluctuations were reduced by up to 15 percent, and the mean wall shear stress was reduced by 7 percent.

Oscillating Wall and Body Force Effect

An active flow control approach that involved oscillating the wall in the spanwise direction started in the late 1990s. Figure 8.27 shows a general concept along with a variation using oscillating disks. The approach focused on the effect the oscillating wall had on coherent motions in the wall-layer region that could affect the viscous drag. Experiments and numerical simulations with oscillating walls (Trujillo et al., 1997; Dhanak and Si, 1999; Baron and Quadrio, 1996; Choi and Clayton, 2001; Choi et al., 2002) observed as much as a 40 percent drag reduction, with a corresponding 70 percent reduction in u'/U_∞. As impressive as this seems, the power needed to oscillate the wall *far exceeded* the power savings due to the drag reduction. In laboratory experiments with imposed sinusoidal spanwise velocity forcing amplitude, A, and frequency, ω, the maximum displacement, $D_{\text{max}} = AT/\pi$, is fixed by the wall oscillation mechanism. This is not the case in DNS simulations where D_{max} can vary with A and ω. For a fixed oscillation amplitude normalized by inner wall variables, A^+, DNS simulations have suggested an optimum oscillation period of $T^+_{\text{OPT}} = 100$. Furthermore, Quadrio and Sibilla (2011) showed that the space-averaged spanwise oscillating flow is consistent with the laminar Stokes layer solution. In cases showing drag reduction, the thickness of the Stokes layer was only a few wall units. This points to an interaction between the Stokes layer and the near-wall low-speed streaks as the basis for the reduced drag. It further suggests that the effect should be *confined to the turbulent boundary-layer sublayer* for maximum drag reduction.

Computational flow simulations (Du and Karniadakis, 2000, 2002) indicated that rather than oscillating the wall, the same effect could be achieved through an oscillating body force resembling a traveling wave. In simulations, this resulted in an $\mathcal{O}$(30 percent) drag reduction that was consistent with oscillating wall experiments. The simulations further indicated a strong effect of the oscillation frequency and amplitude on the resulting drag reduction. The optimum amplitude was linked to the choice of the frequency.

The notion that a traveling body force would produce the same effect as the oscillating wall prompted two types of flow control experiments. One was to utilize a Lorentz force produced by an array of magnets and electrodes in an electroconductive solution in water (Pang and Choi, 2004). The other approach utilized the body

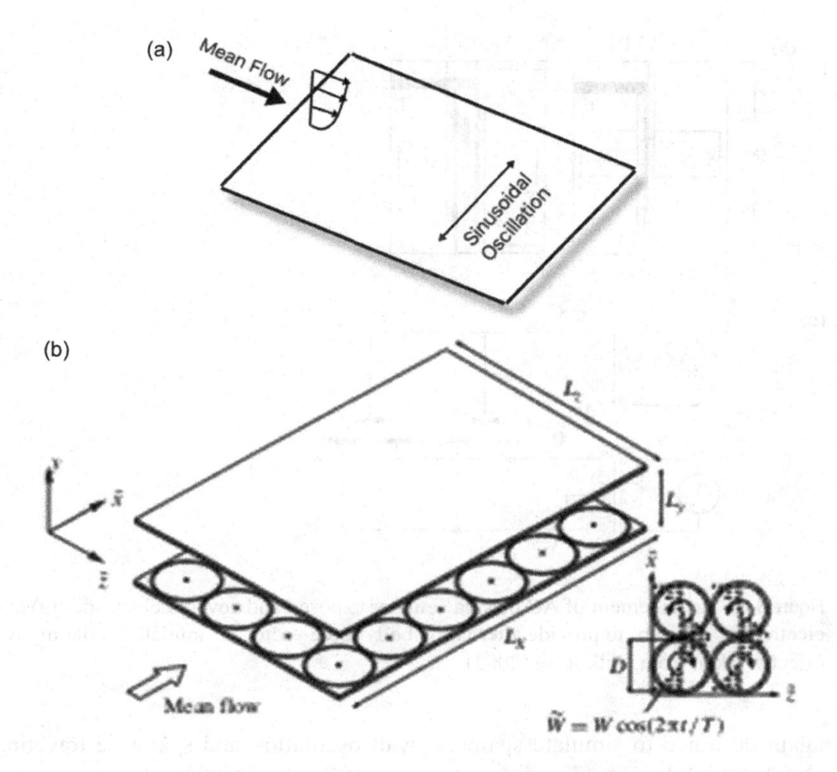

Figure 8.27 Schematic drawing of spanwise oscillating wall concept for wall-layer structure control (a) and a variation on this involving oscillating disks (Wise and Ricco, 2014) (b).

force produced by an array of AC dielectric barrier discharge (DBD) plasma actuators (Wilkinson, 2003). A discussion of plasma actuators is given in Chapter 2 as well as in Section 8.2.4. In the latter case, the plasma actuators consisted of electrodes aligned in the flow direction that were designed to produce an alternating spanwise directed body force equivalent to the oscillating body force applied in simulations (Du and Karniadakis, 2000, 2002). The arrangement of the plasma actuator exposed and covered electrodes, and an electronic circuit used by Wilkinson (2003) to provide alternating body force vector is shown in Figure 8.28. The experiment was performed in air. Unfortunately, the experiment failed to lower the viscous drag. The reasons were attributed to the inability to confine the body force effect to the wall-layer region, and a lack of response of the flow field at the reduced frequencies ($f^+ = f\nu/u_\tau^2 = 0.01$) expected to produce drag reduction in air. To the second point, because the objective was to apply the technology to aircraft, assuming the physics and the nondimensional frequency requirement holds at flight conditions, the required physical frequency would be in excess of 1 MHz. Therefore, the approach of an oscillating body force resembling a traveling wave does not appear to scale well to aircraft flight conditions.

Another attempt at boundary-layer drag reduction using a traveling body force produced by AC-DBD plasma actuators was performed by Choi et al. (2011). This was

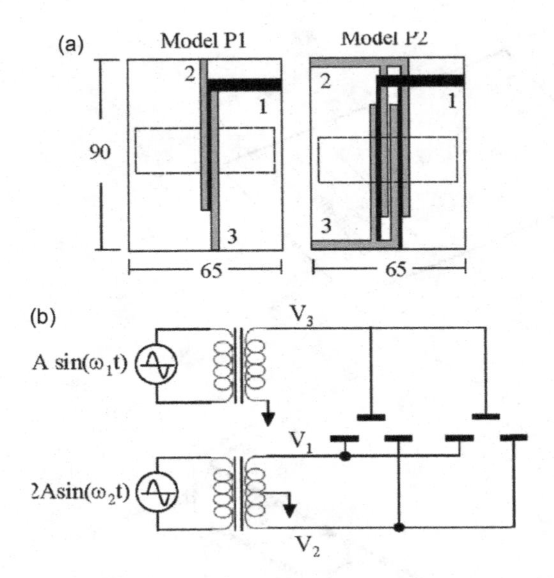

Figure 8.28 Arrangement of AC plasma actuator exposed and covered electrodes (a) and an electronic circuit (b) to provide alternating body force vector to simulate oscillating wall effects. Taken from Wilkinson (2003).

again designed to simulate spanwise wall oscillation and spanwise traveling waves and followed the same general approach as Wilkinson (2003). However, due to a lack of actuator authority, the experiment was limited to a very low free-stream speed of 1.75 m/s. No supporting drag measurements were reported.

Wall-Layer Spanwise Blowing

Around the same period of time when oscillating walls were a focus, Schoppa and Hussain (1998) demonstrated drag reduction in a DNS simulation of a turbulent channel flow where they imposed a streamwise-independent spanwise velocity component along the channel wall through either a pair of counter-rotating streamwise vortices or opposed wall jets. Both proved capable of sustained drag reduction, with a 50 percent reduction with the colliding wall jets and a 20 percent reduction with the counter-rotating vortices.

Later, Schoppa and Hussain (2002) proposed a new mechanism for the production of near-wall streamwise vortices in the self-sustaining process of wall turbulence generation. They designated this as a Streak Transient Growth (STG) in which the wall-normal vorticity component, ω_y, is a critical parameter. The effect of the wall-normal vorticity component, ω_y, on the STG rate is shown in Figure 8.29. The drag reduction they observed in the channel flow simulations is attributed to reducing ω_y through wall-layer spanwise blowing that reduced the mean flow distortion produced by the counter-rotating streamwise vortices associated with the wall streak structure. This subsequently reduced streak lift-up and more generally the "burst" cycle that Kline et al. (1967a) linked to the wall shear stress.

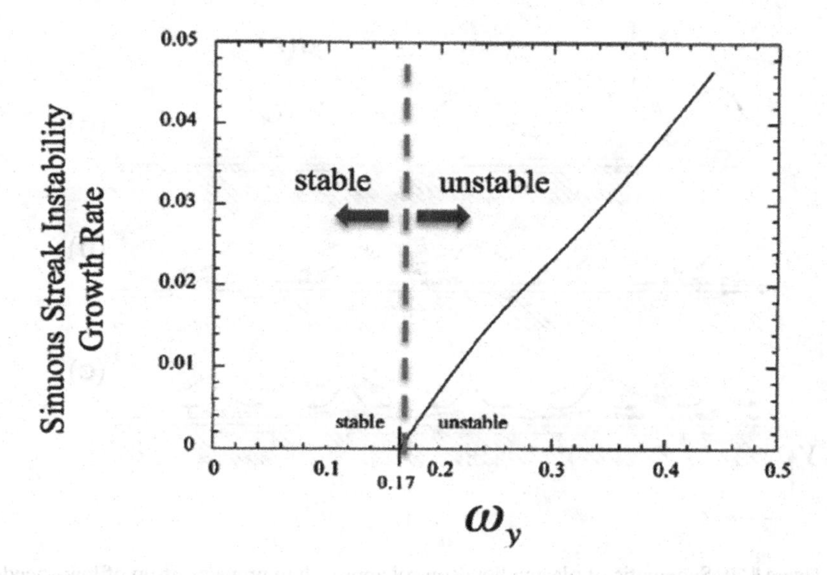

Figure 8.29 Effect of wall-normal vorticity component, ω_y, on streak transient growth rate. Taken from Schoppa and Hussain (2002).

More recently, the effectiveness of the flow control strategy employed by Schoppa and Hussain (1998) was examined as a function of Reynolds number in DNS by Canton et al. (2016). They achieved channel flow drag reduction for a friction Reynolds number $Re_\tau = 104$ corresponding to the earlier work of Schoppa and Hussain and also at $Re_\tau = 180$. However, no drag reduction was observed in simulations performed at $Re_\tau = 550$. Based on this result, they suggested that this flow control approach may be limited to low Reynolds numbers. However, in a subsequent work by Yao et al. (2017), the lack of drag reduction observed by Canton et al. (2016) at the high Re_τ was completely remedied by insuring that the spanwise blowing stays confined to the viscous sublayer which thins with increasing Reynolds number.

Building upon the channel flow drag reduction DNS of Schoppa and Hussain (1998), turbulent boundary-layer drag reduction experiments reported by Corke et al. (2017) and Thomas et al. (2016) demonstrated unprecedented levels of drag reduction of up to 79 percent. These experiments utilized a plasma actuator array similar to that of Wilkinson (2003), but rather than producing an alternating velocity component, they designed it to introduce a steady spanwise velocity component. This steady component was confined to affect only the near wall region of the boundary layer. The purpose of this was to reduce the mean flow distortion resulting from the coherent streamwise vortices near the wall that is attributed with the wall streak structure, and consequently large ω_y. The basic concept is illustrated in Figure 8.30. Figure 8.30(a) shows an idealized view of the spanwise variation of the local streamwise velocity, $u(z)$, that results from the pairs of counter-rotating streamwise vortices in the near-wall region of a turbulent boundary layer. This variation is the leading term in ω_y, which is the critical parameter in the STG mechanism of Schoppa and Hussain (2002) (Fig-

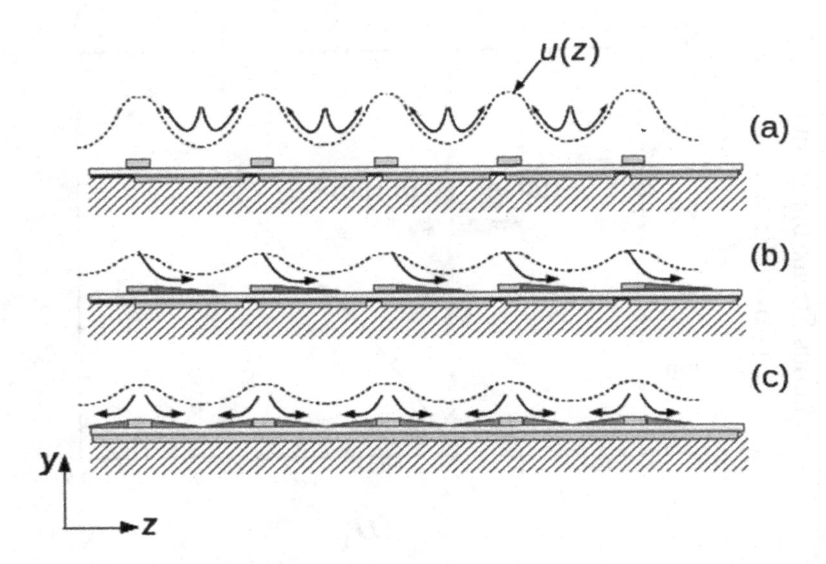

Figure 8.30 Schematic of plasma flow control approach to prevent lift-up of low-speed streaks and thereby reduce ω_y to control streak transient growth where (a) represents the baseline condition with large spanwise mean velocity distortion, (b) and (c) represents the effect of the spanwise blowing and opposed blowing respectively, of the plasma actuator array.

ure 8.29). Figures 8.30(b) and (c) show two concepts that were investigated by Corke et al. (2017) and Thomas et al. (2016). Both are based on flush, surface-mounted plasma actuators. In Concept (b), the plasma actuator was designed to produce a unidirectional spanwise near-wall velocity component (indicated by the curved arrows). In Concept (c), the plasma actuators were designed to produce a series of spanwise opposed wall jets. In either case, following the experience with riblets, one would expect there to be an *optimum spanwise spacing* of the plasma actuators that is related to the scale of the near-wall low-speed streaks.

It can be noted that the ultimate outcome of the spanwise blowing is similar to that presented in Figure 8.26, except that in that approach, wall-normal blowing or suction was utilized, and focused on individual wall streaks. As will be apparent in the following, the spanwise blowing approach of Corke et al. (2017) and Thomas et al. (2016) is capable of affecting a large number of wall streaks simultaneously and does not require feedback control.

The plasma actuators used with the spanwise blowing approach of Corke et al. (2017) and Thomas et al. (2016) were operated using a pulsed-DC approach as opposed to the AC approach used by Wilkinson (2003).

Schematic representations of the AC-DBD plasma actuator used by Wilkinson (2003), and the pulsed-DC plasma actuator used by Corke et al. (2017) and Thomas et al. (2016) are shown in Figure 8.31. Both approaches use a pair of electrodes (staggered as shown) that are separated by a dielectric barrier material. One of the electrodes is exposed to the air. In the AC-DBD, the electrodes are powered by an AC source. In the staggered electrode configuration, the ionized air forms on the surface

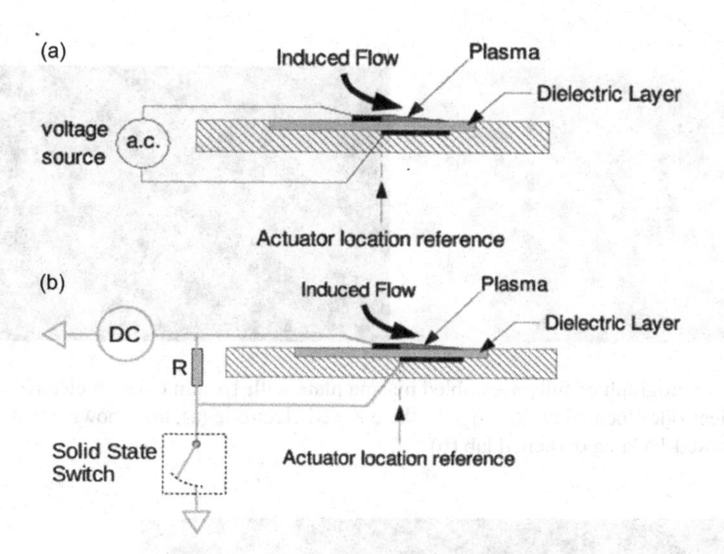

Figure 8.31 Schematic illustration of standard AC-DBD plasma actuator (a) and new pulsed-DC DBD plasma actuator (b).

of the dielectric directly over the covered electrode. The ionization air in the presence of the electric field results in a body force that acts on the neutral air in the direction of the electric field gradient. For the staggered electrode arrangement, the body force is directed away from the exposed electrode toward the covered electrode. This is indicated by the arrow in the plasma actuator schematic labeled "induced flow." In the surface-mounted configuration shown, this produces an effect that is similar to a tangential wall jet.

The pulsed-DC actuator is a hybrid design that utilizes the best feature of AC, which is more efficient in ionizing the air, and that of DC, which is more efficient in generating a body force. It utilizes a DC voltage source that is supplied to both electrodes and remains constant in time for the exposed electrode. The DC source for the covered electrode is periodically grounded by means of a solid-state switch for very short instants on the order of $10^{-5} - 10^{-7}$ s. This is the AC component that ionizes the air. After the switch opens, the DC potential generates a body force on the neutral air. The combination of the DC voltage, pulse frequency, and pulse width (duty cycle) provides better control compared to the AC plasma actuator whose primary control is the AC voltage.

The fully assembled array of plasma actuators designed to produce a uniform velocity component in the spanwise direction (Concept (b)) is shown in Figure 8.32(a). Figure 8.32(b) shows the plasma actuator array while operating with the pulsed-DC power source in a darkened lab. The ionized air (plasma) appears purple due to the nitrogen in the air.

In order to measure the effect of the plasma actuator spanwise blowing on the viscous drag in the boundary layers, the arrays were mounted to the drag force measurement setup that was shown in Figure 8.10. A photograph of the plasma actuator

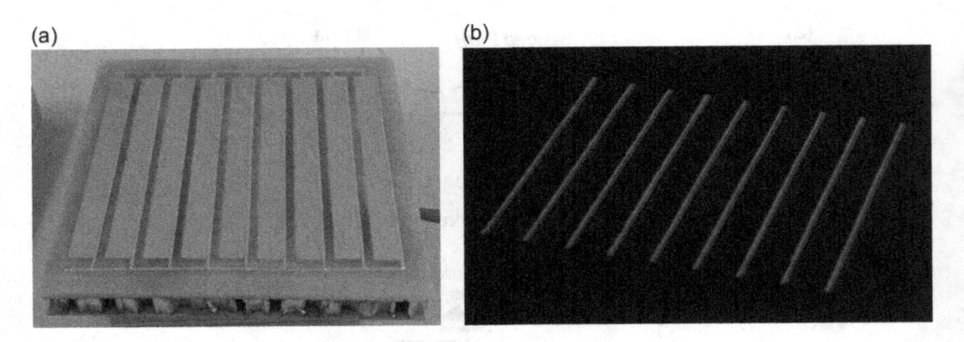

Figure 8.32 Photograph of fully assembled plasma plate with 16-mm covered electrodes and exposed electrodes located on the edge of the covered electrode (a), and shown when operating with the pulsed-DC in a darkened lab (b).

Figure 8.33 Photograph of plasma plate with 16-mm covered electrodes and exposed electrodes located on the edge of the covered electrode in the wind tunnel.

array in the wind tunnel is shown in Figure 8.33. This is viewed looking in the downstream direction toward the exit of the test section. The power to the plasma actuator array was supplied through two 30-gauge coated wire leads, with one each connected to the covered and exposed electrode power bus. These thin wires are visible on the right side of the photograph. The thin wires were chosen to produce a negligible amount of drag on the plasma plate.

The experiments using the plasma array shown in Figure 8.33 were performed by Duong (2019). They were performed in a flat plate, zero pressure-gradient turbulent boundary layer. The boundary-layer thickness was varied by varying the free-stream velocity, with and without a turbulent transition trip. In each of these cases, the DC voltage to the pulsed-DC plasma actuator array was varied. The results are shown in Figure 8.34.

The measured drag depended on the applied voltage, which determined the magnitude of the induced velocity produced by the pulsed-DC actuator. In addition, as presented in Figure 8.35, the amount of drag reduction strongly depends on the spanwise actuator spacing which dictated the number of low-speed streaks that were being simultaneously controlled. Figure 8.35 presents the drag reduction for the three largest DC voltages in Figure 8.34 in terms of the effective spacing between the actuators that came about by changing Re_θ. This shows a clear pattern where the degree of

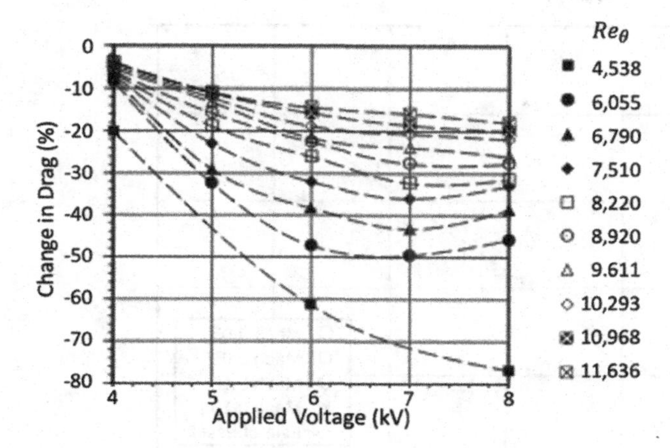

Figure 8.34 Percent change in drag as a function of pulsed-DC actuator voltage for different turbulent boundary-layer conditions. Taken from Duong (2019).

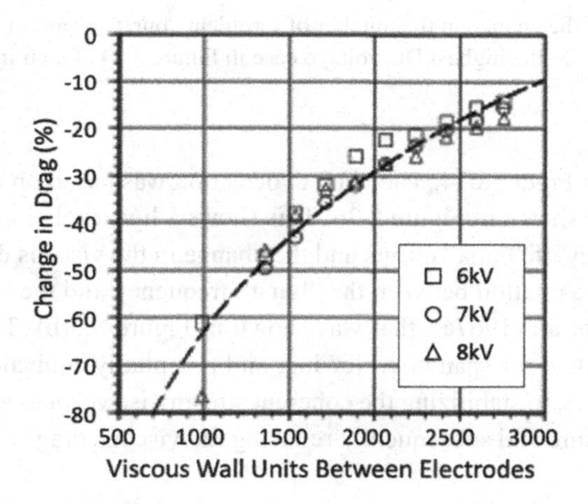

Figure 8.35 Percent reduction in drag as a function of wall viscous unit actuator spacing for results shown in Figure 8.34. Taken from Duong (2019).

drag reduction scales with the electrode spacing in viscous wall units. The dashed curve represents a logarithmic relation between the amount of drag reduction and the viscous-unit spacing of the plasma actuators in the array. Based on an average wall-streak structure spacing of $100z^+$ (Kline et al., 1967a), dividing the values on the abscissa by 100 gives the corresponding number of wall streak structures. For example, 60 percent drag reduction corresponds to an actuator spacing of approximately 10 wall streak structures.

Based on Kline et al. (1967a), lowering the viscous drag of the turbulent boundary layers is expected to have an effect on the turbulence production associated with turbulent "burst" events. This was documented by Duong et al. (2019) and Duong (2019) through hot-wire measurements downstream of the plasma array for the conditions at

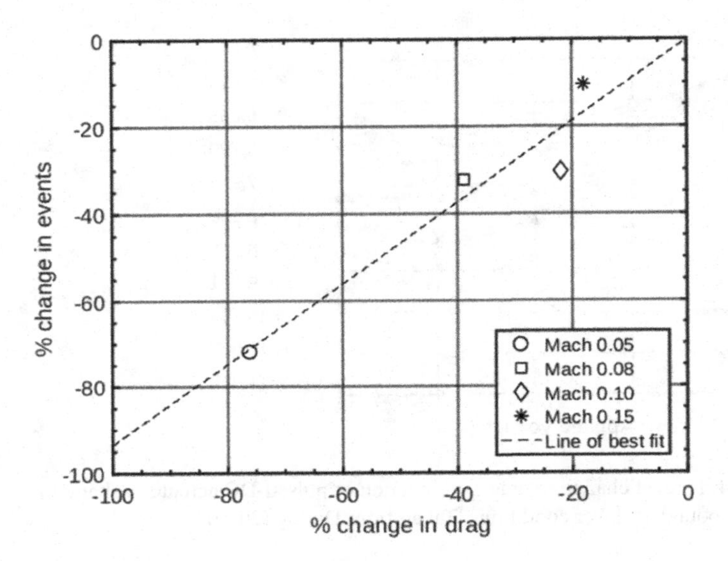

Figure 8.36 Relation between the change in the number of turbulent "burst" events and the change in the drag reduction for the highest DC voltage case in Figure 8.34. Taken from Duong (2019).

the largest DC voltages in Figure 8.34. The "burst" detection was based on the VITA technique. The results are shown in Figure 8.36. This shows a linear relation between the change in the frequency of "burst" events and the change in the viscous drag. This result is reminiscent of the relation between the "burst" frequency and the wall shear velocity, u_τ, from Kline et al. (1967a), that was shown in Figure 8.3(b). This result appears to tie the mechanism for spanwise blowing, and potentially equivalent oscillating walls and body forces, to stabilizing the coherent streamwise vortices associated with the wall streak structure and subsequently reducing the viscous drag.

8.3 Summary: Turbulent Boundary-Layer Drag Reduction

The methods presented are all aimed at controlling identified coherent turbulent motions in the boundary layer, either in the outer part of the boundary layer or near the wall where turbulent production is a maximum. The outer scale control has generally sought to suppress the large-scale intermittent excursions of high-speed potential fluid into the boundary layer. One of the most effective at achieving this has been the parallel plate "manipulators." When optimally designed, and accounting for the drag of the device, net drag reduction of up to 20 percent has been achieved.

The parallel plate manipulator is an example of a passive flow control approach in which no external energy is supplied to the flow control device. Such approaches are intrusive and, as a result, produce a parasitic drag that needs to be exceeded by the overall drag reduction if they are to be practically applied.

Drag reduction that focuses on the near-wall structures has shown more drag reduction benefit than those aimed at outer structures. Most specifically, the focus for these has been on controlling counter-rotating streamwise vortices that are associated with the wall "streak" structure (Kline et al., 1967a). Passive techniques like riblets and compliant coatings have produced from 5 to 8 percent drag reduction. Active approaches that aim at producing a spanwise velocity component through wall motion, body force, or induced flow have yielded from 40 to 80 percent drag reduction. Because these active approaches require external energy input, that has to be factored into deriving the total drag benefit. This eliminates most of the active approaches from practice. To date, only the pulsed-DC plasma actuator approach to produce a spanwise velocity component in the wall layer (Corke et al., 2017; Thomas et al., 2016; Duong, 2019) has yielded a net drag reduction. In most recent works (Duong, 2019), the power supplied to the plasma actuator was as little as one-eighth that of the power gained by the drag reduction ($\Delta D \cdot U_\infty$) at Mach 0.5. The net power gained by this approach increases with increasing Mach number, opening many possible commercial and defense applications.

Problems

8.1 The optimum spacing for a wall riblet to produce the maximum viscous drag reduction is $z^+ \simeq 20$, whereas the average spacing between the wall "streak structure" is $z^+ \simeq 100$.

1. Assuming that riblets somehow intervene with an instability associated with the wall streak structure, given the difference in the spanwise wavelengths, how might this occur? Does Figure 8.23 offer an explanation?
2. If as Schoppa and Hussain (2002) suggest, the level of viscous drag is due to an instability of the fluid flanking the wall streak structure, how might the riblets intervene with that mechanism?
3. In speculating on the previous question, how might riblets modify the mean flow.

8.2 Outer-scale control for drag reduction seeks to reduce the energy in large coherent vortical motions on the scale of the boundary-layer thickness. One such approach discussed is the LEBUs. Consider the design of an LEBU consisting of a single wake-generating shape.

1. If the large-scale motions contribute a majority of the Reynolds stress in the upper 80 percent of the boundary layer, what might be the optimum height within the boundary layer to locate the single LEBU element?
2. If as indicated by Townsend (1956) and Grant (1958), the large-scale outer coherent motions have a scale on the order of the boundary-layer thickness, what would be the appropriate streamwise length of the single LEBU element?
3. If the estimated "lifetime" of large-scale structures is from 5 to 10 boundary-layer thicknesses, what might be the streamwise spacing of the LEBU elements to prevent regeneration of the large-scale structures?

4. There are two theories on the mechanism of the LEBUs. One is that they restrict the vertical (v) motions. The other is that they introduce a wake vortex street (see Chapter 3 that interacts with and overrides the large-scale motions in the boundary layer. If the latter is the case, how might the wake shedding frequency depend on the streamwise length of the LEBU element?

8.3 In the 1970 chronology of methods for turbulent boundary-layer drag reduction shown in Figure 8.9, one of the considered methods was longitudinal wall oscillation.

1. DNS simulations have suggested that the optimum oscillation period, T, occurs with $Tu_\tau^2/\nu = 100$. Based on this, make an estimate of the optimum oscillation frequency, for a boundary layer with $U_\infty = 17$ m/s in air. A good estimate for the shear velocity is $u_\tau = 0.036U_\infty$.
2. What would be the optimum wall oscillating frequency for use on a commercial aircraft with a cruise velocity of 290 m/s? Do you think this is feasible?

8.4 Dynamic wall-normal suction and blowing with spanwise spacing matching that of the wall streak structure has shown some ability to reduce the occurrence of ejection-sweep events and lower the viscous drag. As illustrated in Figure 8.26(b), the suggested effect is to reduce the amplitude of spanwise mean flow distortion. Given this:

1. Discuss how the wall suction and blowing affect the spanwise variation in the wall-normal vorticity, ω_y.
2. Considering that wall oscillation results in a Stokes layer at the wall, what effect would this have on the spanwise mean flow distortion?
3. Compare the effects of these two approaches to the approach shown in Figure 8.30. Could all three be producing the same effect?
4. All three approaches lead to significant amounts of viscous drag reduction. Could this be attributable to interceding with the "streak transient growth" mechanism proposed by Schoppa and Hussain (2002)?
5. From an instability point of view, might this approach involve modifying the mean flow?

8.5 Optimum designs for drag reducing compliant surfaces need to be able to respond to fluctuating pressures in the boundary layer in a range of dimensionless periods that is similar to that of an oscillating wall, namely $50 \leq t^+ \leq 150$.

1. What might be the connection between the two techniques if, as suggested, t^+ in this instance is the period between turbulence producing ejection-sweep events.
2. Is there also a connection to the effect of dynamic wall-normal suction and blowing? If so, how?

9 Shock–Boundary-Layer Interaction

Turcat (2003) describes an incident with flight test 122 of the Concord 001 in which while cruising at Mach 2, the third engine variable inlet ramp was blown out due to "violent pressure fluctuations for about seven seconds." The event is referred to as "inlet buzz," which is a low-frequency, high-amplitude pressure oscillation that is linked to a shock–boundary-layer interaction (SBLI) that can affect engine inlets.

In addition to supersonic engine intakes, shock waves are present in transonic gas turbine blade tip gaps; transonic turbine blade passages; scramjet isolator ducts; transonic and supersonic flight vehicle surfaces; and surfaces of rockets, missiles, and reentry vehicles. The SBLI that can occur in these instances is of particular interest because it can result in large temporal and spatial pressure variations and greatly affect boundary development including causing flow separation that feeds into the flow unsteadiness and subsequently has a large impact on aerodynamic performance. This chapter provides background on sources of SBLI, its characteristics, and various methods of control.

9.1 Background

Most of the research over the last 70 years has focused on three types of SBLIs as illustrated in Figure 9.1. These are (a) an oblique shock wave caused by an elevated body that impinges on a flat-plate boundary layer, (b) a boundary layer on a flat plate that leads up to a compression ramp, and (c) a normal shock wave interacting with a flat-plate boundary layer with a symmetric wall protrusion. In each illustration, the bold lines are compressive shock waves. The dashed curve represents a flow separation bubble. The shorter shock wave segments near the wall represent the unsteadiness in the lower portion of the shock wave, which is a central issue of an SBLI. In Figure 9.1(a), the compression shock that forms upstream of the separation bubble is referred to as a "reflected" shock (Settles and Dodson, 1991). The position of the reflected shock is highly unsteady.

A schematic that illustrates the features of an incident oblique shock wave interacting with a wall boundary layer is shown in Figure 9.2. The incident shock could originate either from an external surface above the wall or from a shock wave that reflected from an upper wall as part of a duct. In this illustration, the incident shock is

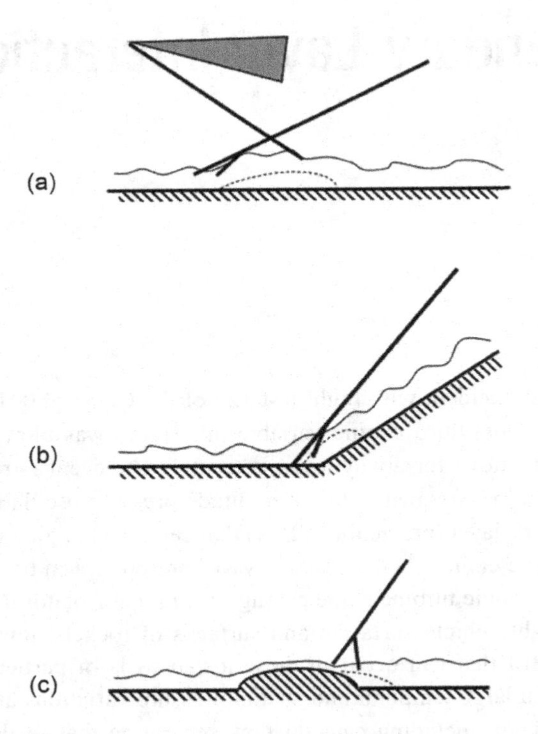

Figure 9.1 Schematic illustration of the types of shock–boundary-layer interactions that have drawn particular interest where (a) is a shock impinging on a turbulent boundary layer, (b) is a shock formed at the junction of a compression ramp, and (c) is a shock that occurs over a wall hump.

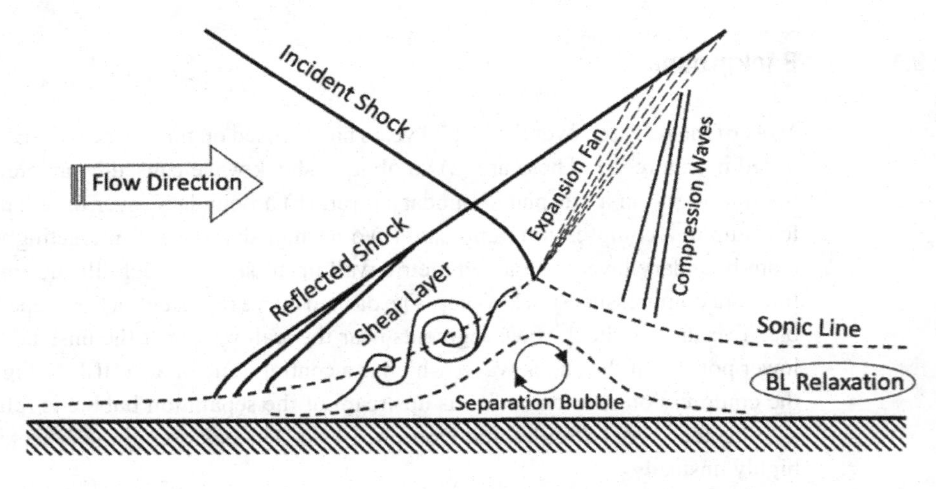

Figure 9.2 Schematic illustration of features of an incident oblique shock wave interacting with a wall boundary layer.

sufficiently strong so that the pressure gradient across the shock causes the boundary layer to separate, forming a recirculating separation bubble (See Chapter 4, "Separated Flows"). A separated shear layer forms over the separation bubble (Pirozzoli and Grasso, 2006; Dupont et al., 2007; Piponniau et al., 2009). As discussed in Chapter 4, free-shear layers are inviscidly unstable to disturbances. The result of the instability is the formation of rolled-up vortices that are depicted in the illustration.

The separation bubble causes the mean flow to deflect away from wall. This results in an adverse pressure gradient that communicates upstream through the subsonic portion of the boundary layer to a point where a "reflected" shock is formed. The location of the reflected shock is well upstream of where it would have been located if the flow were inviscid (Pirozzoli and Grasso, 2006).

An expansion fan forms over the top of the separation bubble. This is quickly followed by compression waves that form near where the separated flow reattaches. Further downstream, the boundary layer can gradually return to a fully developed equilibrium condition. Depending on the size of the separation bubble, the downstream distance needed to reach equilibrium can take an order of 10–20 boundary-layer thicknesses.

Most of the early work on SBLI was experimental (Dolling, 2001). The formation of the separation bubble was observed early on and was instrumental in the development of the Chapman et al. (1958) "free-interaction" concept. The essence of the concept is that *any phenomenon near separation which is independent of the object shape would depend only on the flow that is internal and external to the boundary layer.* Chapman et al. (1958) went on to develop an expression for the pressure rise in the SBLI region given as

$$\frac{\bar{P}_w - \bar{P}_{w0}}{q_0} = \frac{2}{\sqrt{M_0^2 - 1}}\frac{d\delta^*}{dx}, \tag{9.1}$$

where the subscript w refers to the wall region and δ^* is the boundary-layer displacement thickness. They further went on to show that

$$\left(\bar{P}_w - \bar{P}_{w0}\right)/L_i \sim \tau_{w0}/\delta^*, \tag{9.2}$$

where 0 designates conditions at the start of the pressure rise at which $x = x_0$ and L_i is the streamwise extent of the free interaction that is illustrated in Figure 9.3.

Katzer (1989) confirmed through numerical simulations the local scaling laws of the free interaction in the vicinity of the separation point. It distinguishes two mechanisms: a global mechanism that determines the separation-bubble length, L_i, and a local mechanism that controls the free-interaction region, in the vicinity of the separation point. The former was found to depend linearly on the shock strength, defined as the ratio of the downstream to upstream free-stream pressures, p_3/p_1, in Figure 9.2. Katzer (1989) determined that the effect of the Mach number, M, and that of Reynolds number, Re, (based on the distance from the plate leading edge) on L_i scaled as M^{-3} and $\mathrm{Re}^{1/2}$, respectively.

Further confirmation of the free-interaction concept came from Pagella et al. (2004) who numerically investigated cases of a two-dimensional (2-D) compression corner and a 2-D impinging shock at Mach 4.8 in which the separation bubble lengths were matched. They found that the base flow properties were identical, in accordance with the free-interaction theory. They noted that the physics of such flows are not determined by the type of SBLI but rather by the flow field properties at the onset of the interaction. However, Dolling (2001) cautioned that although the free-interaction theory appears successful at predicting the correct pressure scaling, the physics implicit in the theory is not what actually occurs.

Although the early research described the SBLI process as being relatively steady (Dolling, 2001), this is now known to be incorrect. Kistler (1964) reported on the unsteady aspect of shock-induced turbulent separation upstream of a forward-facing step that revealed a characteristic low-frequency unsteadiness. The cause of this unsteadiness is still an open question and remains as a central issue in supersonic and hypersonic SBLIs. From a practical sense, the low-frequency unsteadiness of the reflected shock is a main source of structural fatigue that in turn becomes a major constraint in the choice of materials (Dolling, 2001). This unsteadiness is presumed to be the source of the "inlet buzz" reported prior to the previously mentioned Concord inlet structural failure.

Dussauge et al. (2006) observed that the intermittent motion of the initial shock reflected the initial rise of the mean pressure, similar to free-interaction theory that captured the formation of the separated flow. Thus, it was hoped that the unsteady motion had rather general properties (Dussauge et al., 2006). Based on data from Dussauge et al. (2006), Dupont et al. (2006) found that a Strouhal number for the shock unsteadiness based on the size of the interaction zone, L_i, and the external velocity, $\bar{u}_1$, was relatively constant for a wide range of geometries. Specifically, for the full set of geometries, the Strouhal number fell in a range from 0.02 to 0.5. The interaction length, L_i, was defined as the distance between the mean reflected-shock-foot position and the nominal inviscid impingement location. This length is shown in the schematic in Figure 9.3.

The existence of the low-frequency motions has also been observed in a number of different experiments with different geometries. For impinging shock cases, these include Dupont et al. (2006, 2007), Souverein et al. (2008, 2009), Polivanov et al. (2009), and Humble et al. (2009). For the case of a compression ramp, they include Gramann (1989), McClure (1992) and Ganapathisubramani et al. (2007, 2009).

In specific cases of an oblique shock wave impinging on a turbulent boundary layer at Mach 2.3, Dupont et al. (2006) found the frequency of the unsteady motion had a Strouhal number of approximately 0.03. They also observed that the amplitude of the shock oscillations increased linearly with the shock intensity, p_2/p_1, where p_2 was the free-stream pressure behind the impinging shock, and before the reflected shock. For compression ramps, Gonsalez and Dolling (1993) and Clemens and Narayanaswamy (2014) found similarly scaled Strouhal numbers that ranged from 0.01 to 0.03.

Dupont et al. (2006) attempted to link the unsteadiness to a "breathing motion" of the separation bubble that forms in the interaction zone, and which they found

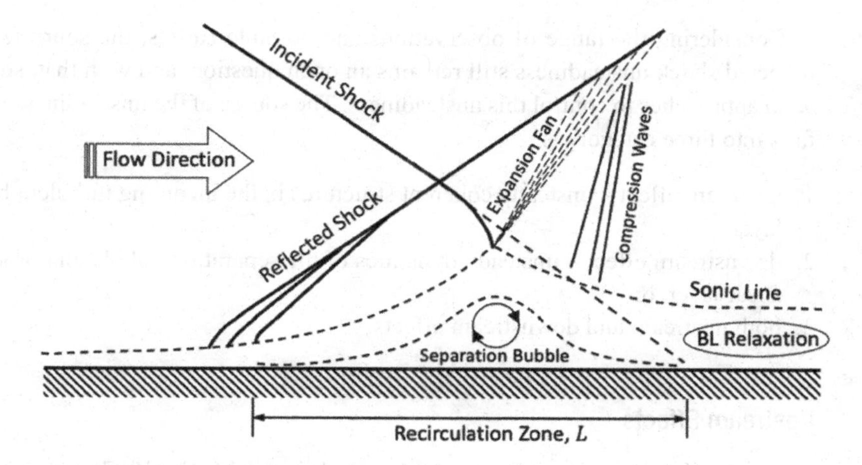

Figure 9.3 Schematic illustration defining the length scale, L, used in formulating a Strouhal number for the reflected shock unsteadiness.

contributed up to 30 percent of the total energy in pressure fluctuations. Piponniau et al. (2009), Grilli et al. (2012), Wu and Martin (2008), and Pirozzoli et al. (2010), each indicated that unsteadiness in the interaction region was related to pulsations of the separation region. For example, with a 28° compression ramp at Mach 5, Gramann (1989) found that the unsteady motion of the separation bubble caused it to extend from two to four times the approaching boundary layer thickness, δ_0.

The causes of intrinsic separation bubble unsteadiness and reattachment point unsteadiness are believed to be either linked to the upstream boundary layer, as proposed by Pirozzoli et al. (2010), or the result of inherent dynamics between the separation bubble and the shock wave, as proposed by Piponniau et al. (2009) and Grilli et al. (2012). Clemens and Narayanaswamy (2014) however, suggested that the degree to which upstream or downstream causes affected the unsteady motion depended on the strength of the impinging shock which determined the strength and scale of the separation bubble.

The observed unsteadiness is manifest in streamwise motion of the reflected shock wave. The question is whether the motion of the reflected shock controls the separation bubble or whether the separation bubble breathing controls the motion of the reflected shock. It is known that a shock wave can move as a result of upstream and downstream conditions. As a result, the response of a shock wave to a disturbance perturbation may provide some insight. For example, Li (2007) considered the linear stability of a steady attached oblique shock wave. The analysis introduced a "sonic point criterion" that determines the stability of an oblique shock to linear disturbances. Specifically, as long as the Mach number downstream of an oblique shock is greater than 1, the shock is predicted to be linearly stable to small disturbances. As possible validation of the analysis, Dussauge et al. (2006) observed that the motion that was evident at the "foot" of the nearly normal (unstable) reflected shock would diminish toward the outer flow where the shock was oblique, stronger, and stable according to the "sonic point criteria." This is depicted in Figure 9.2.

Considering the range of observations and possible causes, the source(s) for the reflected shock unsteadiness still remains an open question, and with that, so do *optimum* approaches to control this unsteadiness. The source of the unsteadiness generally falls into three categories:

1. upstream effect – unsteady coherent structures in the incoming turbulent boundary layer,
2. downstream effect – unsteady dynamics of the separation bubble and/or separated shear layer, or
3. both upstream and downstream effects.

9.1.1 Upstream Effects

With regard to upstream effects, Andreopoulos and Muck (1987) suggest that the frequency of the shock-motion scales on the bursting frequency of the incoming boundary layer. In a later experiment, Thomas et al. (1994) found no discernible statistical relationship between turbulent boundary layer wall layer "burst events" and coherent shock-front motion. Erengil and Dolling (1993) did observe some correlation between small-scale motions of the shock with the passage of turbulence in the boundary layer.

The observations of the SBLI unsteadiness suggest that this occurs at lower frequencies. Turbulent boundary layers have a broad range of scales generally divided into a low frequency range, inertial subrange, and viscous subrange. Within these ranges are coherent and incoherent scales of motions, with an example of the former being the "burst events" that are associated with turbulence production and correlated with the wall shear stress (see Chapter 8). Most of the energy-containing turbulent scales and commensurate convective frequencies are much higher than the band of frequencies of the SBLI unsteady frequencies. However, if the linear stability of the reflected oblique shock is considered, there will be a band of frequencies of disturbances that will be amplified. As long as there is energy in the amplified band, according to linear stability theory, the disturbances will grow exponentially.

In an earlier embodiment of a frequency-filtered response to disturbances later motivated by linear theory, Plotkin (1975) developed a linear model for the motion of the reflected to random upstream disturbances. In this, the shock motion was modeled by a first-order ordinary differential equation subject to stochastic forcing. Plotkin (1975) used experimental data to compute model constants such as motion timescales. The resulting model replicated the wall-pressure spectra well. Poggie and Smits (2001, 2005) found excellent agreement with Plotkin (1975) in cases where the shock undergoes significant low-frequency oscillations.

There appears to be ample evidence and a mechanism by which low-frequency turbulent fluctuations in the incoming boundary layer can result in the unsteadiness in the shock position. Other than turbulence, however, there is speculation that the low-frequency unsteadiness could also be driven by low-frequency thickening and thinning of the upstream boundary layer. This was first speculated by Unalmis and

Dolling (1994) based on experiments on a Mach 5 compression corner. Utilizing the same compression corner, Beresh et al. (2002) found a significant correlation between upstream velocity fluctuations and shock motion that occurred at frequencies that were an order of magnitude lower than the characteristic frequency of the large-scale structures in the incoming turbulent boundary layer. However, they did not believe the shock unsteadiness was the result of thickening and thinning of the boundary layer. In contrast to Beresh et al. (2002), in experiments on a Mach 2 compression-corner, Hou et al. (2003) found a correlation between the shock motion and a thickening/thinning of the upstream boundary layer.

What is the source of the thickening and thinning of the incoming turbulent boundary layer? In the shock-reflection case of Dupont et al. (2006), based on the Strouhal number of the reflected shock oscillations, a length scale would need to be on the order of 115 local boundary-layer thickness, namely frequencies (s^{-1}) of $\omega \sim \bar{u}/115\delta$. It is unlikely that such structures exist in turbulent boundary layers. It is more likely that this is a characteristic of the test facility (wind tunnel), for example, caused by separation bubbles in the inlet contraction or Görtler vortices in the nozzle expansion (Beresh et al., 2002). Therefore, although they illustrate an upstream sensitivity of the shock motion, it is not particularly relevant to steady mean flows.

9.1.2 Downstream Effects

As illustrated in Figure 9.2, an impinging shock of sufficient strength causes a distortion of the mean boundary layer that leads to a separation bubble and an elevated shear layer. The shear layer in particular is inviscidly unstable to disturbances leading to the rapid formation of a coherent vortex sheet. The vortex sheet in particular is a source of acoustic pressure disturbances that can feed upstream through the subsonic portion of the boundary layer. This can induce an oscillatory motion of the flow separation point and a subsequent branching and flapping motion of the reflected shock (Dussauge et al., 2006; Pirozzoli and Grasso, 2006).

Although still a downstream effect, Piponniau et al. (2009) proposed a somewhat different mechanism based on a mass-entrainment timescale that is associated with the separation bubble and shear layer. In particular, as discussed in Chapter 4, the development of the shear layer vortices results in strong entrainment and mixing across the shear layer. Piponniau et al. (2009) suggest that the main parameter controlling the low-frequency shock motion is the spreading rate of the compressible mixing layer. Their model for the process produced timescales that were the same order of magnitude as the shock-motion timescale. The mechanism draws a link to the shock strength, p_2/p_1, which ultimately drives the formation of the separation bubble and mixing layer amplitudes.

Another mechanism for a downstream unsteady disturbance source includes the instability of the separation bubble. Boin et al. (2006) and Boin and Robinet (2004) have indicated that under some assumptions, the laminar separation bubble that forms from an SBLI can result in unsteady self-sustained low-frequency dynamics that can be capable of driving the unsteady shock motion. They indicated that with increasing shock angle, the SBLI goes through a phase where the flow becomes three-dimensional (3-D) and unsteady. Robinet (2007) performed a linearized global stability analysis that indicated that the 3-D character results from a 3-D stationary global instability.

Theofilis et al. (2000) examined the spatial structure of the most amplified 3-D (stationary) BiGlobal eigenmode of the Briley bubble (Briley, 1971). They observed that the dominant disturbance velocity component was along the streamwise direction and peaked in the neighborhood of the primary reattachment, where it gave rise to a sequence of secondary reattachment–separation–reattachment. They point out that the 3-D nature of this instability results in a highly convoluted imprint of the primary reattachment line, while the primary separation line remains unaffected by the amplification of this instability. The scenario of topological changes in the reattachment zone was consistent with the "multiple structurally unstable saddle-to-saddle connections" first postulated by Dallmann (1988). Although originally developed for incompressible flows, Boin and Robinet (2004) believe that this extends to supersonic flows and SBLI and that it can explain the first stage of the establishment of the unsteady low-frequency motion of the reflected shock. In BiGlobal analysis of a laminar shock-reflections condition, Robinet (2007) has documented that a global mode (in the BiGlobal sense) exists for sufficiently strong shock strengths. The most unstable mode is reported to be 3-D, with a wavelength that scales with the separation bubble length.

9.2 SBLI Sensitivity to Initial Conditions

In the development of flow control approaches to mitigate any adverse effects of SBLIs, it is important to understand how such interactions are sensitive to upstream and downstream initial conditions. Such initial conditions can include:

1. wall cooling
2. local changes in wall contour
3. localized wall suction and/or injection
4. modification of the approaching boundary layer

9.2.1 Effect of Wall Temperature

At high Mach numbers relevant to SBLIs, the boundary-layer development strongly depends on the thermal conditions at the wall. More specifically, it depends on the ratio of the wall temperature, T_w, to the adiabatic recovery temperature, T_r. In a laminar boundary layer, T_w/T_a will affect the growth of boundary-layer instabilities leading to turbulence onset.

As discussed in Chapter 6 for 2-D laminar boundary layers at supersonic and hypersonic Mach numbers, wall cooling will suppress the growth of the primary viscous Tollmein–Schlichting (T-S) instability but enhance the growth of the higher inviscid (second, third, etc.) instability. For Mach numbers below Mach 5, the T-S instability is the dominant mechanism leading to turbulence onset. Above Mach 5, the inviscid second-mode instability is dominant. In that case, wall heating is required to delay the turbulence onset.

With regard to the separation bubble that forms downstream of the primary shock, wall cooling can reduce the effect of adverse pressure gradients and delay or reduce boundary-layer flow separation (Kepler and O'Brien, 1962). In a *turbulent* boundary layer, $T_w/T_r < 1$ will lower the viscosity compared to an adiabatic wall condition. This has the effect (Walz, 1969) of increasing the local Reynolds number, reducing the boundary-layer shape factor, $H = \delta/\delta_1$, and, as a result, increasing the wall shear stress, τ_w. By the same arguments, heating the wall has the opposite effect and can therefore hasten a flow separation.

The predicted effect of wall cooling to reduce the size of a shock-induced separation bubble was documented by Frishett (1971) and Spaid and Frishett (1972) for a compression ramp at an approximate Mach number of 2.5. The results are reproduced in Figure 9.4. Figure 9.4(a) documents the effect of the temperature ratio, T_w/T_r, on the separation length, L_s, normalized by the incoming boundary-layer thickness, δ_0, for a range of compression ramp angles. The different temperature ratios affected the displacement thickness Reynolds number for these cases. Therefore, Figure 9.4(b) documents the effect of the change in Reynolds number on the separation length for a fixed temperature ratio. Although there is a noticeable effect of Reynolds number, the overwhelming effect on the separation length comes from the temperature ratio.

The observed trend in Figure 9.4 agrees with the Elfstrom (1971) flow model in which the boundary layer is considered as a rotational inviscid parallel stream (Elfstrom, 1972). According to this assumption, the sole effect of the temperature is to modify the Mach number profile throughout the boundary layer in such a way that for a cooled wall, the Mach number near the wall is raised. Thus, according to Elfstrom's model, the cooled boundary layer develops a greater resistance to separation.

9.2.2 Effect of Wall Contour

The shape of the wall can obviously influence a SBLI. For instance, in a supersonic flow, an appropriate change of the local slope at the impact of an incident shock can theoretically cancel out the reflected shock by producing an expansion of opposite magnitude. This is the approach behind the use of method of characteristics to design a shock-free supersonic nozzle expansion.

It is well known (Gadd, 1961) that transonic boundary-layer separation is delayed on a convex wall. Bohning and Zierep (1976) explain that the post-shock expansion immediately following a transonic shock is strengthened by the convex wall curvature. Near the wall, the expansion influences the boundary-layer development in a manner that delays flow separation. The important parameter is the ratio of the convex radius of curvature to the boundary-layer thickness, namely r/δ_0. This is documented in Figure 9.5 which was reproduced from model results of Bohning and Zierep (1976) for shock-induced boundary-layer separation at transonic Mach numbers.

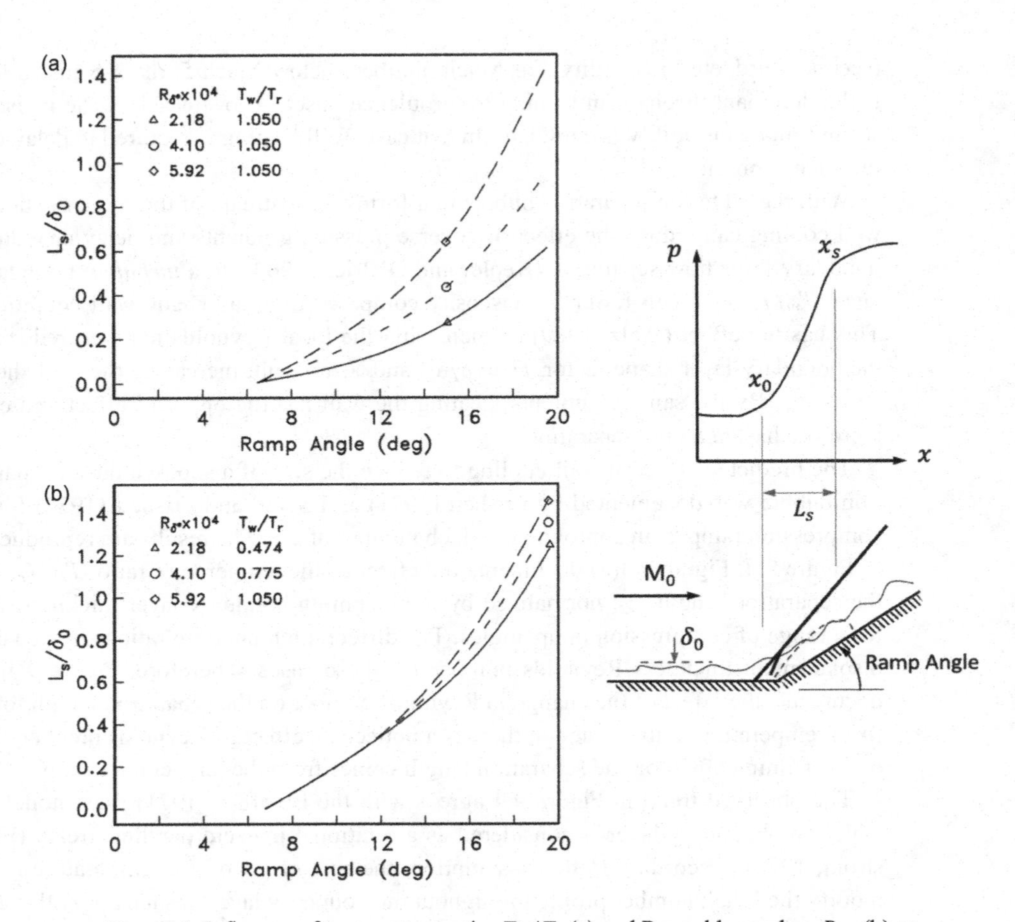

Figure 9.4 Influence of temperature ratio, T_w/T_r (a) and Reynolds number, $R_{\delta*}$ (b) on boundary-layer separation length for increasing angles of a compression ramp with a nominal inflow Mach number of 2.5. Reproduced from Frishett (1971).

This indicates that a wall convex radius of curvature within a couple of orders of magnitude of the boundary-layer thickness can have a significant effect on retarding flow separation.

9.2.3 Effect of Wall Suction and Blowing

The effect of wall suction and blowing on boundary-layer development has been the subject of many experimental and theoretical investigations. The pioneering work includes that of Dorrance and Dore (1954), Jeromin (1968), and Squire (1970). Its initial interest was in the cooling of component surfaces for high stagnation enthalpy turbulent boundary-layer flows where adiabatic wall temperatures can exceed the temperature limitation of most materials. Cool mass injection through a porous wall thereby provided a method of reducing the convective heat transfer from the hot gases to the surface. Such mass injection, or mass removal (suction), has an effect on the mean characteristics of the turbulent boundary layer that can be beneficial to the SBLI problem. Therefore, it is useful to discuss these effects.

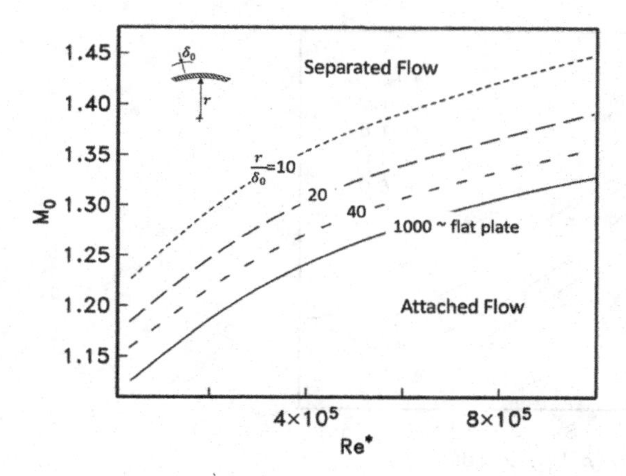

Figure 9.5 Effect of wall curvature on incipient boundary-layer separation in a transonic flow. Reproduced from Bohning and Zierep (1976).

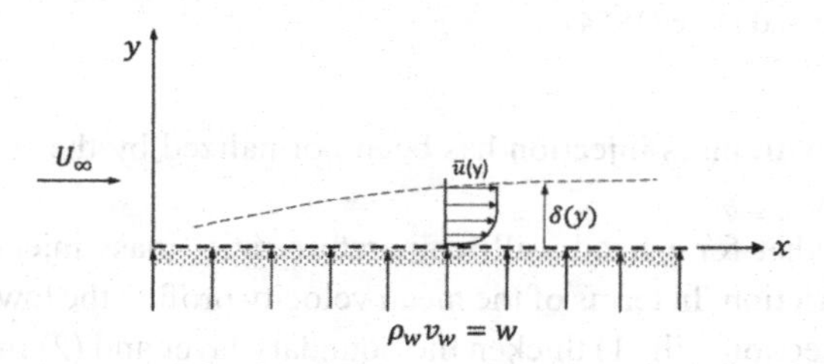

Figure 9.6 Illustration of boundary-layer flow over a plane surface with mass injection normal to the wall surface.

Dorrance and Dore (1954) has presented the theoretical background for compressible turbulent boundary layers with mass transfer at the wall. As illustrated in Figure 9.6, the mass flux per unit area at the wall is $w = \rho_w v_w$, where ρ is the density of the gas at the wall, v is the velocity component in the y (wall-normal) direction, and the subscript, w, corresponds to the wall surface ($y = 0$). Note that w is positive for mass injection and negative for mass suction. For convenience, a mass transfer rate coefficient is defined as $F = \rho_w v_w / \rho_e u_e$, where u is the velocity component in the x direction and the subscript e refers to the edge of the boundary layer.

Among the equations of motion for a thin compressible turbulent boundary layer (see, e.g., Driest (1951)), with mass injection, the continuity equation takes the following form:

$$\overline{\rho v} = w - \int_0^y \frac{\partial(\overline{\rho u})}{\partial x} dy. \tag{9.3}$$

Dorrance and Dore (1954) determined the effect of mass injection and suction on the mean compressible boundary layer wall-normal profiles, wall shear stress, and heat transfer coefficient. Based on their analysis, Figure 9.7 shows the effect of the normalized wall injection, $w/(\rho_\infty U_\infty)$, on the skin friction coefficient with mass injection for a range of wall temperatures, T_w/T_∞, for $M_\infty = 5.0$ and $\text{Re} = \rho_\infty U_\infty x/\mu_\infty = 10^7$.

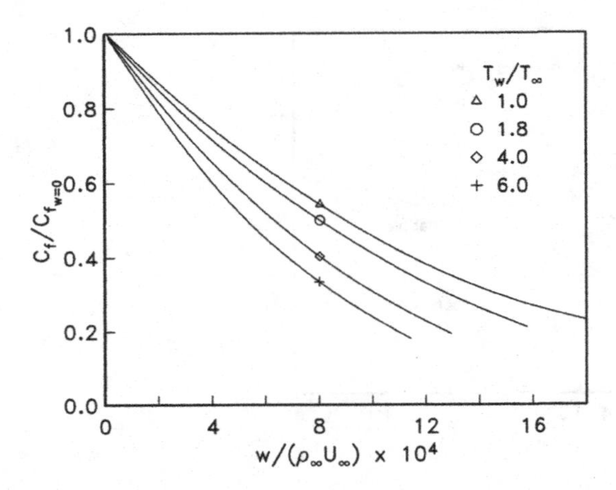

Figure 9.7 The effect of mass injection on the turbulent boundary-layer local skin friction coefficient for different wall temperatures at $M_\infty = 5.0$ and $\mathrm{Re} = \rho_\infty U_\infty x/\mu_\infty = 10^7$. Reproduced from Dorrance and Dore (1954).

The friction coefficient with mass injection has been normalized by the skin friction without mass injection, $C_{f_{w\,=\,0}}$.

Figure 9.7 illustrates that for a fixed wall-temperature ratio, mass injection at the wall decreases the skin friction. In terms of the mean velocity profile, the lower friction coefficient with mass injection will (1) thicken the boundary layer and (2) increase the shape factor, $H = \delta_1/\delta_2$. Wall suction has the opposite effect.

Strategic wall suction could have some benefit to alleviating the adverse effects of a SBLI. If applied downstream of an impinging shock, the increase in the wall shear stress could offset the adverse pressure gradient produced by the primary shock, and thereby limit the downstream thickening of the boundary layer and formation of the separation bubble. As an example, in Mach 2.8 and 3.78 experiments, Seebaugh and Childs (1970) observed that wall suction with a mass flow of 2–5 percent of that of the approaching boundary layer, and applied in the interacted region of an oblique impinging shock, was sufficient to prevent flow separation. Similar results were found in Mach 1.99 experiments by Mathews (1969).

Wall suction applied upstream of the impinging shock could have a detrimental effect by thinning the boundary layer and thereby moving the sonic line closer to the wall, making the shock stronger near the wall. However, as previously discussed, based on the "sonic point criterion," the stronger oblique shock would increase its stability to unsteady disturbances that originate from upstream or downstream sources.

9.2.4 Effect of Wall Tangential Blowing

In the previous section, mass injection was applied in the direction normal to the wall surface. Thus, the *streamwise* velocity component, u, of the injected flow was zero at the surface. As such, if the wall-normal velocity, v, of the injected flow was too large,

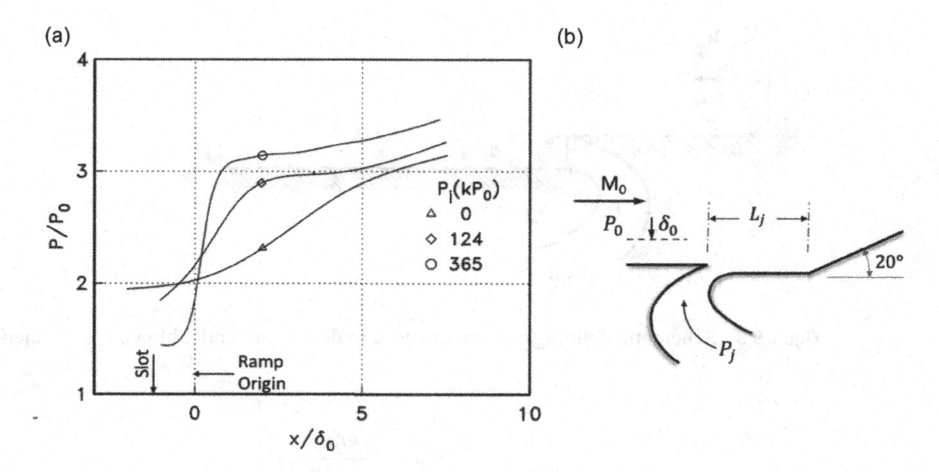

Figure 9.8 Effect of tangential blowing on the flow separation region downstream of a 20° ramp in a Mach 2.5 flow (a) and schematic showing definitions of governing parameters (b). Reproduced from Viswanath et al. (1983).

it could cause the boundary-layer flow to separate as a result of the mass injection. Therefore, an alternative approach involves tangential wall blowing in which the mass injection is directed in the downstream mean flow direction through an angled opening in the wall surface. An example arrangement of a tangential jet is illustrated in Figure 9.8(b). Figure 9.8(a) illustrates that this approach can be quite effective in reducing the flow separation region that forms downstream of a compressive shock wave. This corresponds to surface pressure measurements in the flow separation region induced by a 20° compression ramp in a Mach 2.5 flow presented by Viswanath et al. (1983). The tangential jet produced by the injection would counter the flow direction of the recirculating separation bubble near the surface. The data compare the effect of different injection pressures against that without injection ($P_i = 0$). An indication of the flow reattachment is where the pressure distribution, dP/dx, reaches an asymptote. Without tangential injection, the surface pressure distribution indicates that reattachment occurs near $x/\delta_0 = 5$. The distance to reattachment is significantly reduced by the tangential jet injection to $x/\delta_0 \simeq 2$ for the lowest pressure and to $x/\delta_0 \simeq 1$ for the highest pressure. Thus, such strategically placed tangential jet injection can be highly effective toward eliminating a shock-induced flow separation.

It should be noted that Viswanath et al. (1983), under the same experimental conditions, also investigated the effect of placing the tangential slot upstream of the corner-induced shock. The intention was to add momentum to the lower part of the boundary approaching the ramp. That approach was found to be far less effective in reducing the flow separation that formed downstream of the shock (Viswanath et al., 1983).

Since blowing involves injection of additional mass and momentum into the boundary layer, the parameters affecting its performance include the jet velocity, its density, and the slot width in 2-D flows. These parameters are illustrated in Figure 9.9. The most widely used parameter is the blowing momentum coefficient, C_μ, defined as

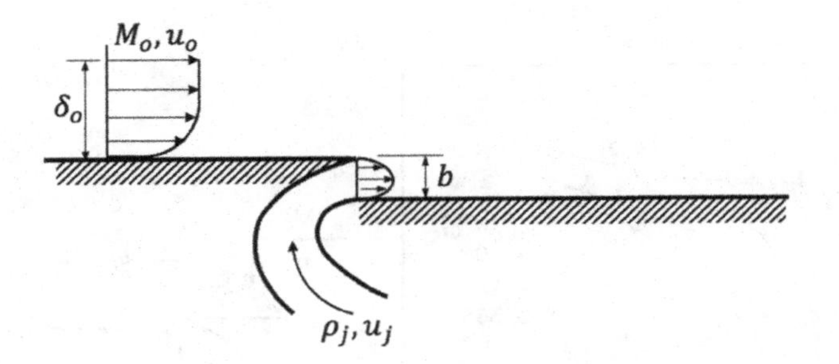

Figure 9.9 Schematic defining parameters used to define tangential blowing coefficients, C_μ and C'_μ.

$$C_\mu = \frac{\dot{m}_j u_j}{\rho_o u_o^2 \theta_o}, \tag{9.4}$$

where $\dot{m}_j$ is the jet mass flow rate, u_j is the jet velocity, ρ_o is the free-stream density, u_o is the free-stream velocity, and θ_o is the momentum thickness of the boundary layer just upstream of the tangential jet. An alternate "excess blowing momentum coefficient," C'_μ, that is sometimes used is

$$C'_\mu = \frac{m_j(u_j - u_o)}{\rho_o u_o^2 \theta_o}. \tag{9.5}$$

Wong and Hall (1975) found a significant reduction in the separation region in a Mach 2.0 inlet with C'_μ values as low as 0.9, and an appreciable improvement in the total pressure recovery and low flow distortion for a $C'_\mu = 1.5$. Based on a series of experiments involving impinging shock waves at Mach numbers of 2.2 and 2.6, Lakshmikantha et al. (1969) presented a correlation for the minimum blowing pressure, $P_{j_{\min}}$, required to suppress separation given as

$$P_{j_{\min}} = 1.25 \Delta p (l_i/\delta_o)^{0.70}, \tag{9.6}$$

where Δp is the observed pressure difference across the shock, l_i is the distance between the tangential jet injection site to the shock intersection point on the wall, and δ_o is the boundary-layer thickness just upstream of the shock intersection point. Lakshmikantha et al. (1969) investigated a range of slot widths with $0.1 \le b/\delta_o \le 0.4$ and subsequently found no dependence on that parameter. These results and correlation suggest that the total mass or momentum injected may not be the most important factors at least, for improving the surface pressures in the separated flow. It is likely that the excess velocity injected is the dominant factor.

9.3 SBLI Flow Control Approaches

As in other applications covered in different chapters in the book, SBLI flow control falls into two categories: passive and active. Both of these exploit the sensitivities to initial conditions that were discussed in the previous section.

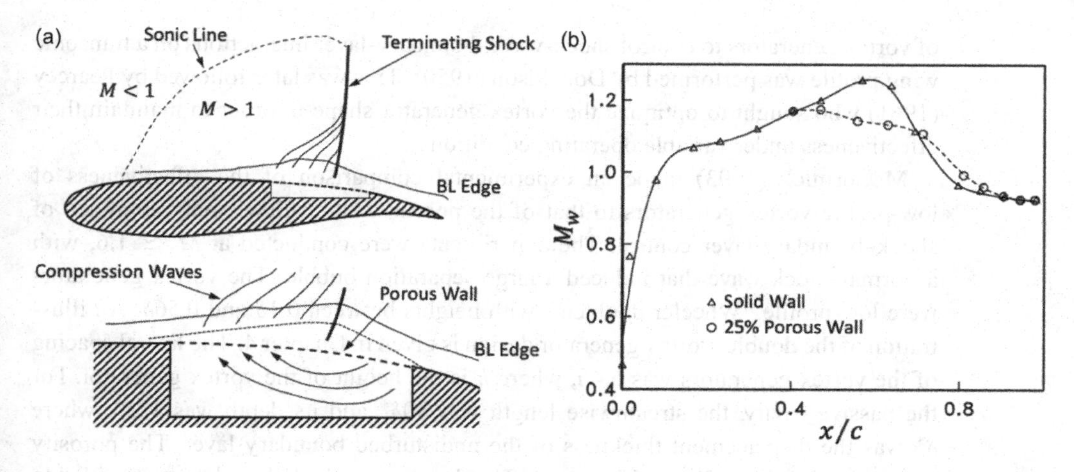

Figure 9.10 Illustration of passive SBLI control based on a porous wall (a) and documentation of the method's ability to weaken shock when applied to the upper surface of a supercritical airfoil (b). Reproduced from Barn et al. (1983).

9.3.1 Passive SBLI Control

As in previous discussions, passive flow control does not require any externally supplied energy. One early example that was first demonstrated by Barn et al. (1983) located a cavity covered by a porous screen at the location of a shock. Figure 9.10(a) illustrates this approach. The concept is that the pressure difference across the shock will drive a flow through the cavity that will exit upstream of the shock wave and energize the incoming boundary layer. This is, therefore, an example of a passive technique that combines both suction and mass injection. The aim is to spread out the compression due to the shock that forms on the upper surface in order to minimize the entropy rise across the shock and thereby reduce wave drag.

Evidence of its effectiveness is presented in Figure 9.10(b), which is data reproduced from Barn et al. (1983). This documents the wall Mach number distribution on the upper surface of an airfoil on which the porous-wall-covered cavity was employed. The free-stream Mach number was 0.806, but as a result of the airfoil thickness, the Mach number on the suction side of the airfoil exceeded Mach 1. The results documented that with the addition of the cavity covered by a porous surface having a 2.5 percent open area, the peak Mack number was lowered, similar to if the airfoil thickness were reduced. Similar results have been reported by Bur et al. (1998) and Raghunathan and McIlwain (1990). The approach has proven to be as effective as active control in improving the performance of supercritical airfoils at off-design conditions (Krogmann et al., 1984, 1985).

Another passive SBLI control approach involves the use of wall-mounted vortex generators that are a common method of boundary-layer separation control that is discussed in Chapter 4. The motivation for their use is to introduce streamwise vortices that enhance mixing throughout the boundary layer and thereby reduce the potential for flow separation caused by an impinging shock. An early investigation on the use of vortex generators to control shock-wave–boundary-layer interactions on a transonic wing profile was performed by Donaldson (1950). This was later followed by Pearcey (1961) who sought to optimize the vortex generator shape in order to maintain their effectiveness under variable operating conditions.

McCormick (1993) made an experimental comparison of the effectiveness of low-profile vortex generators to that of the porous wall-shallow cavity approach of shock–boundary-layer control. The experiments were conducted at $M \simeq 1.6$, with a normal shock wave that induced a large separation bubble. The vortex generators were low-profile "Wheeler doublets" with heights between 0.1δ and 0.56δ. An illustration of the doublet vortex generator design is given in Chapter 4. The lateral spacing of the vortex generators was $6.4h$, where h is the height of the vortex generator. For the passive cavity, the streamwise length was $80\delta^*$ and its depth was $12\delta^*$, where δ^* was the displacement thickness of the undisturbed boundary layer. The porosity (open area) of the wall was 10 percent. The diameter of the holes, d, corresponded to $\delta^*/d = 1.3$. The vortex generators were typically placed $55h$ upstream of the shock location.

Figure 9.11 presents a comparison of the wall static pressure distributions in a SBLI produced by the vortex generators and the porous-wall-covered cavity. The shock was located at $x - x_0 = 0$. The pressure distribution for the baseline boundary layer is shown by the solid curve. The dashed curve corresponds to an ideal inviscid pressure distribution. The boundary-layer separation is determined to reattach where the wall-pressure distribution downstream of the shock is parallel to the inviscid distribution. For the case with the vortex generator boundary layer, reattachment occurred at approximately $10\delta_0$ downstream of the shock. The reattachment location for the porous-wall-covered cavity was considerably further downstream at approximately $57\delta_0$. This was further downstream of the baseline boundary layer that appears to reattach at about $38\delta_0$ from the shock.

The behavior for the porous-wall-covered cavity is consistent with the idea that this approach spreads the pressure rise over a larger axial length. The reduced value in the static pressure relative to the baseline is particularly due to the decrease in the static pressure rise through the more oblique (and weakened) shock such as was evident for the porous wall case in Figure 9.10.

McCormick (1993) observed that the vortex generators resulted in a significant thinning of the boundary layer, with a lower shape factor downstream of the shock wave. In contrast, the porous-wall-covered cavity resulted in a thicker boundary layer with larger displacement and momentum thickness downstream of the shock. This increased thickness is believed to be due to the injection of fluid through the cavity that exits upstream of the shock (see illustration in Figure 9.10(a)). In this regard, the vortex generators are superior to the porous wall.

The advantage of the porous-wall-covered cavity is the reduction of the total pressure loss across the shock (McCormick, 1993). This is documented in Figure 9.12, which shows the normalized mass-averaged total pressure for the baseline boundary layer, vortex generators, and porous-wall-covered cavity. Of the three, the porous wall has the highest mass-averaged total pressure, which signifies that it had the lowest

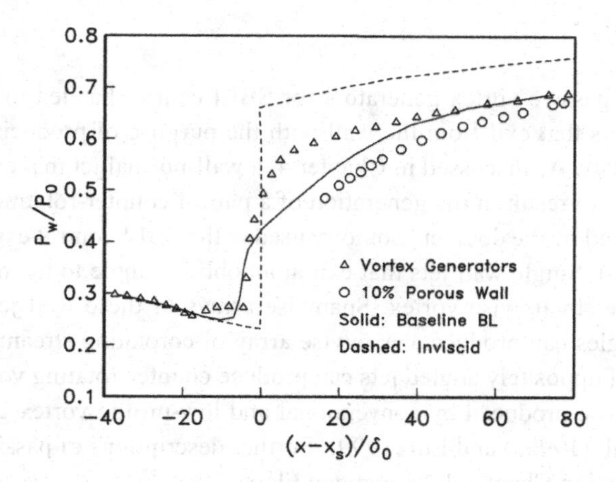

Figure 9.11 Comparison of the wall static pressure distributions in a shock–boundary-layer interaction produced by vortex generators and porous-wall-covered cavity. Reproduced from McCormick (1993).

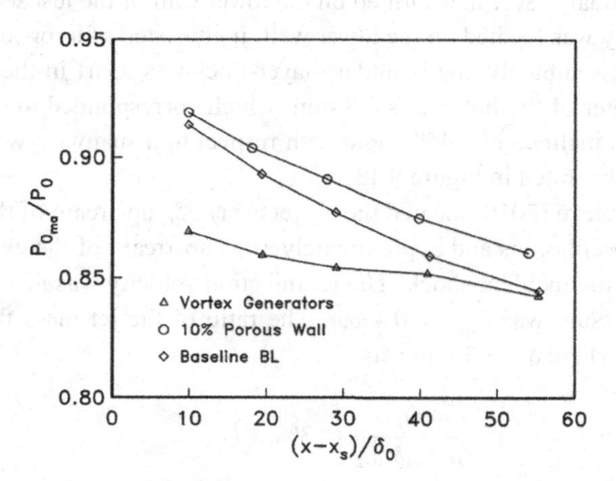

Figure 9.12 Comparison of the mass-averaged total pressure distributions at the downstream location of a shock–boundary-layer interaction produced by vortex generators and porous-wall-covered cavity. Reproduced from McCormick (1993).

shock loss. This is the primary advantage of the porous wall over the vortex generators, which had a higher pressure loss than even the baseline boundary layer.

Based on these results, if a reduction in wave drag, such as on a supercritical wing, is the objective, the porous-wall-covered cavity is the better choice. However, in applications where shock-induced flow separation is to be minimized, such as in a supersonic diffuser, vortex generators would be preferable. In that application, the increased pressure recovery should more than offset the increased total pressure loss across the shock.

9.3.2 Active SBLI Control

The effectiveness of passive vortex generators for SBLI control has led to an active approach using air jets that exit from the wall with the purpose of producing coherent streamwise vorticity. As discussed in Chapter 4, a wall-normal jet that exits into a boundary-layer flow can result in the generation of a pair of counter-rotating vortices similar to that produced by the doublet geometry used in the SBLI control experiments of McCormick (1993). Single wall jets that exit at an oblique angle to the mean flow can result in a single streamwise vortex. Spanwise arrays of these wall jets having the same oblique angles can produce a spanwise array of corotating streamwise vortices. Finally, pairs of oppositely angled jets can produce counter-rotating vortex pairs that are similar to those produced by conventional and low-profile vortex generators described by Lin et al. (1990a) and Lin (1999). Further descriptions of passive vortex generators is contained in Chapter 4, "Separated Flows."

Souverein and Debieve (2010) examined the effect of a spanwise array of continuous air jet vortex generators on the control of a SBLI produced by a suspended 9.5° ramp in a Mach 2.3 flow. The oblique shock wave off of the leading edge of the ramp impinged on the boundary layer that formed on the lower wall of the test section. The vortex generator array was located on the lower wall. It consisted of a row of 10 holes that were spaced approximately one boundary-layer-thickness apart in the spanwise direction. The diameter of the holes was 0.8 mm, which corresponded to $0.1\delta_o$. The axis of the holes was inclined at a 45° angle with respect to a spanwise wall-normal plane. The setup is illustrated in Figure 9.13.

Souverein and Debieve (2010) located the air jet array $5\delta_o$ upstream of the zone of the reflected shock oscillations and approximately $16\delta_0$ upstream of the extrapolated wall impact point of the incident shock. The jet injection velocity was $u_j = 314$ m/s. The jet injection pressure was $P_{j0} = 0.4$ bar. The ratio of the jet mass flow to the boundary mass flow where $\delta^* = 3$ mm was

$$\frac{\dot{m}_j}{\rho_o U_o \delta^* L_z} \approx 3\%, \tag{9.7}$$

where L_z is the spanwise extent of the jet array.

The array of wall jets produced a spanwise array of counter-rotating streamwise vortex pairs of unequal circulation strength. Souverein and Debieve (2010) found that the principal effect of the vortex generating jets was a modification of the mean velocity profile and the integral parameters of the inflow boundary layer that reduced the size of the separation bubble but did not eliminate it. They observed no significant effect on the shock excursion amplitude and position. Also, as a direct consequence of the reduction in the size of the separation bubble size, the shock unsteadiness frequency increased by about 50 percent.

Another experimental study of shock–boundary-layer control using jet vortex generators was conducted by Ramaswamy and Schreyer (2021). The basic flow consisted of a Mach 2.52 flow over a 24° compression ramp. A parametric study in which they varied the jet spacing revealed that an adequate interaction between the vortices is

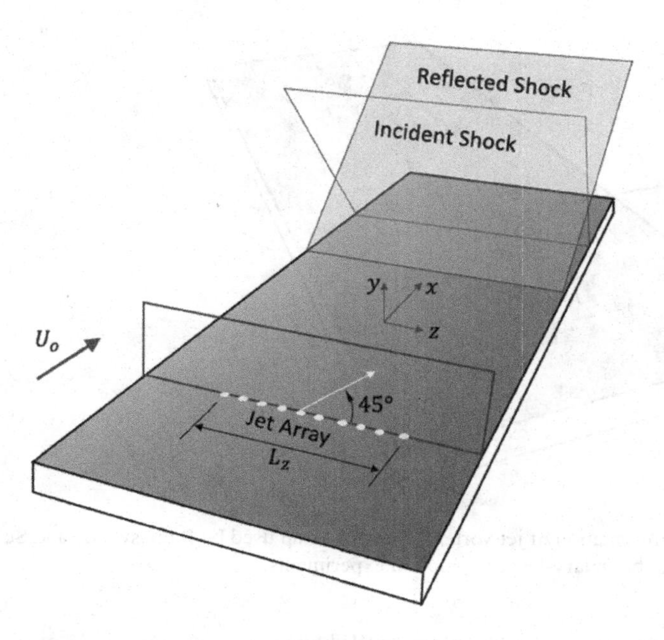

Figure 9.13 Representation of jet vortex generator setup used by Souverein and Debieve (2010) in shock–boundary-layer interaction experiments.

essential for effective control. They found that the most favorable control effect was achieved with a jet spacing of $0.76\delta_o$. The vortex generating wall jets had a diameter of $0.1\delta_o$. The axis of the holes was inclined at a 45° angle with respect to an axial wall-normal plane. This is illustrated in Figure 9.14. These jet parameters were identical to those previously used by Wallis and Stuart (1958), Szwaba (2011) and Souverein and Debieve (2010) and therefore provided an important comparison of results. The injection pressure, P_{oj}, was 0.99 bar, and the mass flow through each jet, $\dot{m}_j$, was 0.0041 kg/s.

An investigation of the effect of the spacing of the jets indicated that with too small of a spacing of $0.38\delta_o$, the shocks that formed upstream of the jet orifices merged into a single bow shock wave. This produced an upstream movement of the SBLI-induced separation line. With an increase in jet spacing to $0.76\delta_o$, the jet-induced shock had a wrinkled pattern that ultimately produced the best control effect. In that case, the separation line moved downstream compared to the uncontrolled case. Further increasing the jet spacing to $2.40\delta_o$ resulted in distinct bow shocks at each of the jet orifices, which indicated a significant reduction in the interaction between the jets. This resulted in a slight upstream movement of the separation line, compared to the optimally spaced jets. Overall, the wall-jet vortex generators produced a considerable decrease in the SBLI-induced flow separation; however, there was no indication that flow separation could be prevented entirely by this control method.

Valdivia et al. (2009) sought to improve the effect of vortex generating wall jets by combining them with the Wheeler doublets used by McCormick (1993). The specific

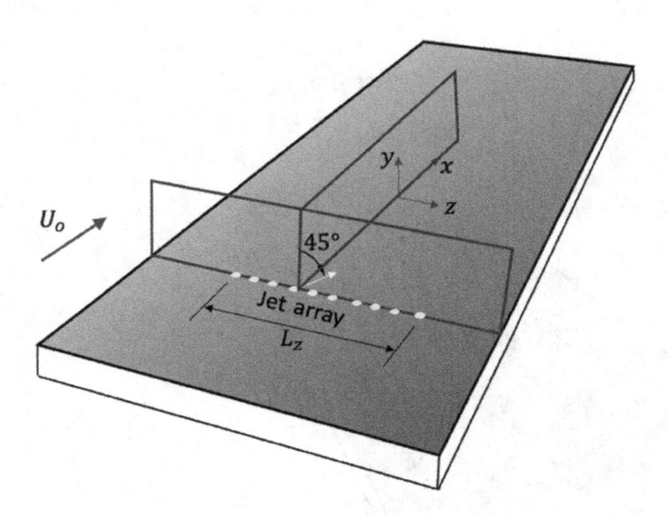

Figure 9.14 Representation of jet vortex generator setup used by Ramaswamy and Schreyer (2021) in shock–boundary-layer interaction experiments.

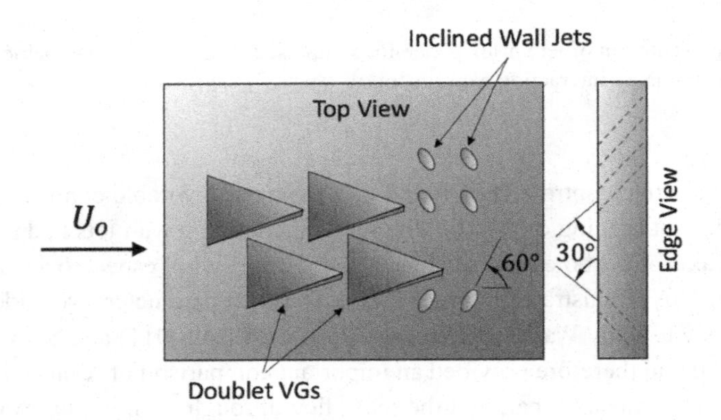

Figure 9.15 Representation of combined jet and passive doublet vortex generator setup used by Valdivia et al. (2009) in shock–boundary-layer interaction experiments designed to control ScramJet "unstart."

motivation was to control "unstart" in a ScramJet inlet isolator at Mach 5. The axis of the vortex generator air jets were skewed at a 60° angle to the mean flow direction, and pitched at a 30° angle in the spanwise direction. This is illustrated in Figure 9.15. A total of six air jets were used. Passive doublet vortex generators were located upstream of the wall jets.

Experiments were performed with the Wheeler doublets alone, the wall-jet vortex generators alone, and the two in combination. The wall-jet pressure, P_{j_o}, and mass flux, $\dot{m}_j$, were not provided. The two types of vortex generators were each found to have a beneficial effect on preventing "unstart." However, the combination of the two was found to be most effective. The provided reason for the better performance was that the doublets placed upstream of the wall jets mitigated the blockage produced by

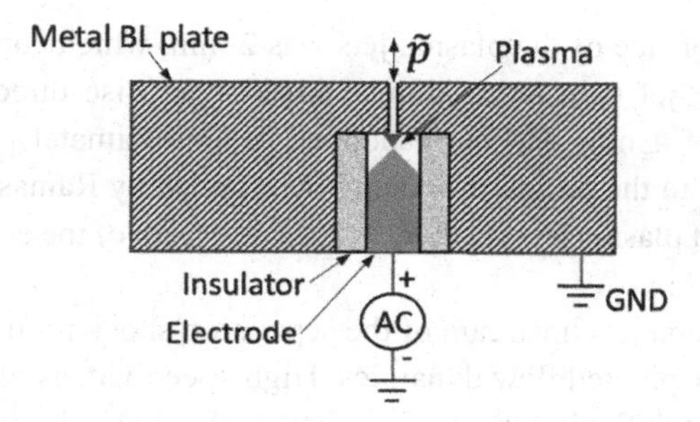

Figure 9.16 Schematic drawing of disturbance actuator used in supersonic boundary-layer experiments of Kosinov et al. (1990).

the bow shock that otherwise formed upstream of the jet orifices. Recall that shock wave formation directly upstream of wall-jet vortex generator orifices was similarly observed by Ramaswamy and Schreyer (2021), and that too if closely spaced could produce a bow shock and subsequent flow blockage.

The previous examples had involved steady blowing. Selig and Smits (1991) found that pulsed injection of air through a slot located inside the separation bubble generated by a compression ramp interaction had the effect of moving the shock upstream but leaving the characteristics of the interaction relatively unchanged. An alternative to unsteady blowing is the use of plasma actuators located in cavities below the wall surface. The concept was discussed in Chapter 6 for the control of 2-D boundary layers at compressible Mach numbers. A schematic of the approach is shown in Figure 9.16. As illustrated, it consists of an electrode that was located in a cavity below the surface of a metal boundary-layer plate. The electrode is powered by an alternating current (AC) voltage source. The reference for the voltage source was earth ground which can be a metal surface over which the boundary layer develops. A sufficiently large AC voltage will cause a plasma discharge to form between the electrode tip and the inside surface of the cavity. This will produce an unsteady pressure in the cavity that communicates to the flow-side surface of the plate through a small hole. The advantage of this approach is that it can produce high-frequency pressure pulses, which is not possible with remotely located high-pressure air sources.

Narayanaswamy et al. (2012) investigated the use of pulsed plasma-jet actuators to control the unsteady motion of the separation shock in a SBLI formed by a 24° compression ramp in a Mach 3 flow. The setup consisted of an array of three plasma wall jets. Two wall-jet orientations were examined: one in which the jet injection was normal to the wall surface, and another in which the jet injection was pitched at a 45° angle and skewed by a 90° angle that is identical to the orientation used by Souverein and Debieve (2010) and shown in Figure 9.13. The typical peak jet exit velocity of the actuators in their case was about 300 m/s, and the pulsing frequencies were a few kilohertz. They specifically used a 2 kHz pulsing frequency which gave a Strouhal number of approximately 0.04 which, as discussed earlier, is a characteristic of the inherent unsteadiness in a SBLI. The instantaneous peak mass flux, $\dot{m}_j$, for an individual plasma jet was reported to be approximately 3.2×10^{-5} kg/m^2/s.

The diameter of the exit orifice of the plasma jets was 2 mm, which corresponded to approximately $0.4\delta_o$. The jet exits were aligned in the spanwise direction, with a center-to-center spacing of 4 mm that corresponded to approximately 0.9δ. This spacing of the jets was close to the optimum spacing determined by Ramaswamy and Schreyer (2021). The array of plasma jets was located $6\delta_o$ upstream of the compression ramp.

The placement of the plasma jets upstream of the separation shock produced a significant modification to the separated flow dynamics. High-speed videos of the shock wave motion visualized through Schlieren imaging showed that the pulsed plasma-jet forcing caused the separation shock to "lock in" to the forcing frequency. In this, the separation shock position oscillated, moving upstream by 0.9δ and then returning to its unforced position during a plasma-jet pulse cycle. With the unsteady forcing, the time-averaged separation shock position was 4 percent upstream of that without the forcing.

When located upstream of the separation shock, periodic plasma-jet forcing at 2 kHz was found to reduce the magnitude of pressure fluctuations in the frequency band associated with SBLI unsteadiness, $\text{St}_L = 0.02 - 0.05$, by about 30 percent. This forcing did not affect the pressure fluctuations in the separation bubble. Plasma jets of comparable strength located within the separation bubble also had no effect on the pressure fluctuations in that region of the flow. However, the effect of the periodic forcing introduced upstream of the separation shock did appear to be amplified by the shear layer that formed above the separation bubble. This points back to the role of the mixing layer as a source of unsteadiness suggested by Piponniau et al. (2009).

A similar study on the use of pulsed plasma jets was performed by Greene et al. (2015). This involved a 20° compression ramp at Mach 3. The plasma actuator design followed the work of Narayanaswamy et al. (2012). Three jet injection angles were investigated: 45° pitch and 0° skew, 20° pitch and 0° skew, and 22° pitch and 45° skew, where the pitch angle is measured from the wall, and the skew angle is toward the spanwise direction. The distance between the jet centerline and the compression ramp corner was varied from $1.7\delta_o$ to $8\delta_o$.

The jet pulsing frequency ranged from 2 to 4 kHz, corresponding to a Strouhal number based on separation length, St_L, from 0.012 to 0.023. The pulse width was varied between 20 and 45 μs, giving a duty cycle of between 5 and 18 percent. The dissipated energy per jet pulse ranged from 22 to 49 mJ for pulse widths of $20 - 50\mu$s.

Greene et al. (2015) found that the most important variable in the effectiveness of the pulsed plasma jets was the distance between the jets and the compression ramp corner, with $1.5\delta_o$ being optimum. The effect of the jets was found to be abruptly diminished at a distance of $3\delta_o$. It was also determined that the optimum jet configuration was a low pitch angle and zero skew angle. The pulsed jets were also most effective with a pulsing frequency corresponding to a Strouhal number of 0.018, which placed it in the lower range of $0.01 \leq \text{St}_L \leq 0.03$ observed by Gonsalez and Dolling (1993) and Clemens and Narayanaswamy (2014) for compression ramps. Based on the optimum conditions, Greene et al. (2015) observed that the distance between the separation shock and the compression ramp corner was reduced by 40 percent.

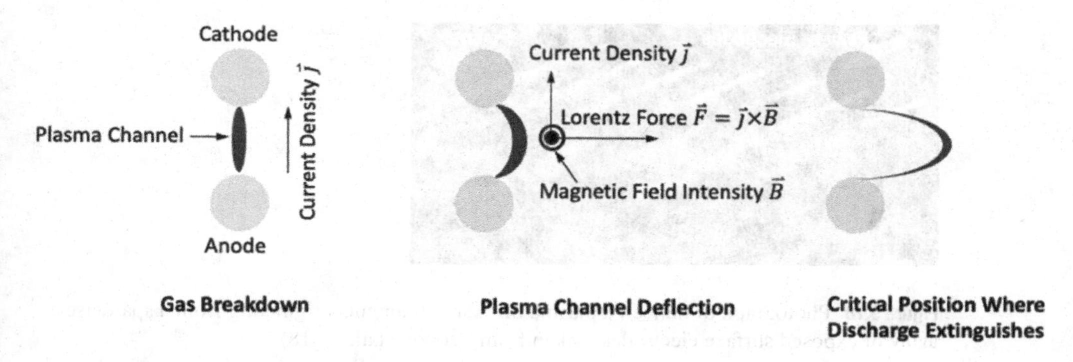

Figure 9.17 Schematic drawing of air breakdown between exposed surface electrodes and Lorentz force under the condition of a magnetic field.

The mechanism behind the pulsed plasma jets is primarily the heating and subsequent expansion of the air in the cavity that generates a pressure pulse at the exit orifice. In an approach to more directly utilize the electromagnetohydrodynamics (EMHD) properties of the plasma, experiments have been performed in which the electrodes are exposed to the primary flow and operated to generate plasma filaments that extend downstream. This is illustrated in Figure 9.17, which shows a sequence beginning with the breakdown of the air between two surface electrodes to form a plasma filament and its evolution as it is stretched by the mean flow field. The addition of a magnetic field, $\vec{B}$, will produce the Lorentz force, $\vec{F} = \vec{j} \times \vec{B}$, that can act on the flow field. The plasma discharge will also generate heat, which is a scaler effect on the flow field.

Using a direct-current (DC) electrical discharge, Leonov et al. (2010, 2005a, 2005b) demonstrated the ability to initiate shock waves or modify the shock wave properties such as oblique wave angle and intensity. An example image of the plasma filaments formed by an array of DC-driven surface electrodes in a Mach 2 boundary layer from Leonov et al. (2018) is shown in Figure 9.18. For this image, the total pressure was $P_0 = 3$ bar and the total plasma power release was $W_{p_i} \approx 20.2$ kW. The plasma discharge did not involve a magnetic field. In this case, the primary effect is the creation of heated zones with a much higher sonic velocity that can significantly change the shock wave structure (Houpt et al., 2017).

An important feature of this electrical discharge is that the gas heating occurs not only at the place of electric current location but also downstream due to both recombination and vibrational–translational relaxation. Under optimal conditions, more than 90 percent of the power deposition can be conserved in a vibrational reservoir and dissociation of molecular gas (Kochetov et al., 2006). Leonov and Yarantsev (2008) suggest that such nonequilibrium and nonuniform plasma discharges can provide a more flexible and higher performance per unit power for SBLI control.

In one of a series of experiments, Leonov et al. (2006b) investigated the use of an array of plasma filaments to control shock waves. A representation of the experimental setup is shown in Figure 9.19, which was intended to represent a simple 2-D inlet to a supersonic air-breathing engine. This was made up of two planar wedges with angles

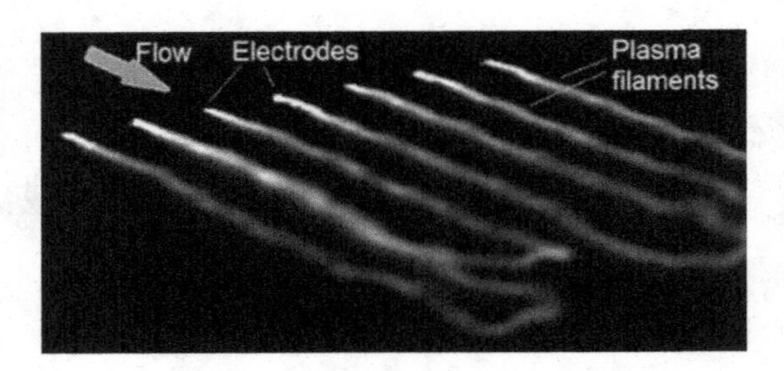

Figure 9.18 Photograph of deflected plasma discharge filaments originating from a spanwise array of exposed surface electrodes. Taken from Leonov et al. (2018).

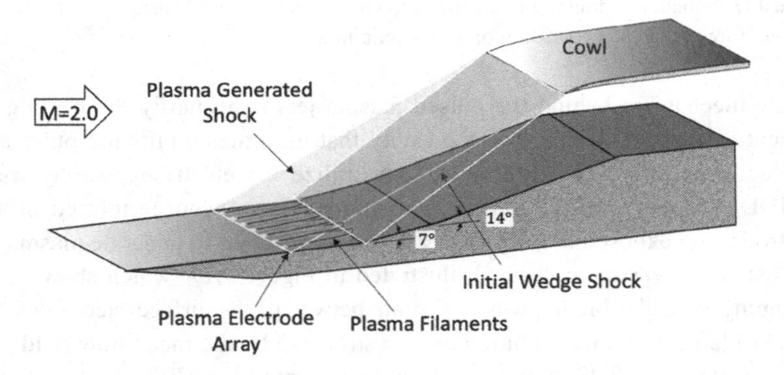

Figure 9.19 Experimental setup of Leonov et al. (2006b) to evaluate shock–boundary-layer interaction control using plasma filaments.

of 7° and 14°. Although not shown, in order to avoid channel blockage, the opposite wall contained backward wedges with wedge angles of 3° and 14°. The inlet Mach number was 2.0. At the plasma array location, the static pressure was 180 Torr and the boundary-layer thickness was $\delta_o = 1.3$ mm.

An array of seven copper surface electrodes for the plasma generation were located upstream of the first compression wedge. The electrodes were arranged by sequence as cathode-anode-· · · -cathode. The array was located 18 mm, or approximately $14\delta_0$, upstream of the first wedge. For the specific electrode and static pressure, the initiation of the plasma filament (air voltage breakdown) occurred at a voltage of approximately 1.7 kV. The total current was approximately 4.5 A, giving a total power ($I \times V$) of 7.7 kW. This formed filaments similar to those shown in Figure 9.18. The measured maximal translational gas temperature in the discharge zone was reported to be $T_g \approx$ 3000 °K, so the Vibrational-Translational (V-T) relaxation length was expected to be a few centimeters (Leonov et al., 2006). As illustrated in Figure 9.17, the V-T relaxation length determines the streamwise extent of the plasma filament before the discharges extinguish.

An example of the effect of the plasma filaments on the SBLI from Leonov et al. (2006), is shown in Figure 9.20. This shows a series of Schlieren images for different plasma filament conditions. Figure 9.20(a) shows the baseline flow, without the

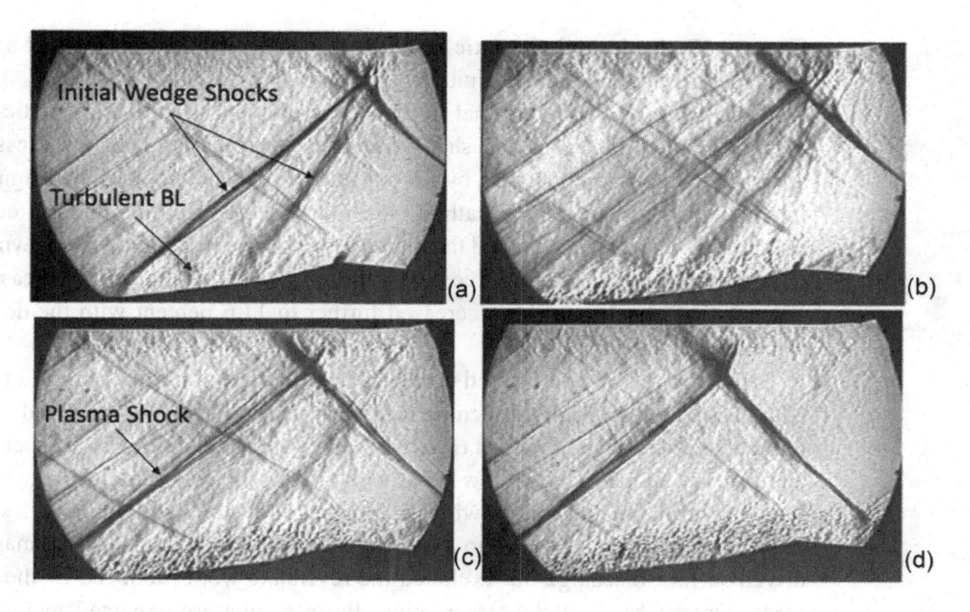

Figure 9.20 Schlieren images for the experimental setup shown in Figure 9.19 (a) is for the baseline flow. Images (b)–(d) are with the plasma filiments at respective power levels of 3.5, 8, and 17 kW. Taken from Leonov et al. (2006).

plasma filaments activated. The image clearly reveals the shock waves that form at the two wedge junctions as well as the turbulent boundary layer on the lower surface. Figures 9.20(b)–(d) show images in which the plasma filaments were operating at increasing power levels of 3.5 (Figure 9.20(b)), 8 (Figure 9.20(c)), and 17 kW (Figure 9.20(d)). At the lowest power level, Figure 9.20(b) indicates that the wedge-generated shocks had been substantially weakened. This continues at the two higher power levels in Figures 9.20(c) and (d). However, the higher plasma power levels caused a shock wave to form at the electrode location, which is not unlike that produced by wall air and plasma jets used in shock–boundary-layer control.

Wang et al. (2009) combined the plasma filament generation with a magnetic field in order to seek to enhance the control of SBLIs. As illustrated in Figure 9.17, the addition of a magnetic field, $\vec{B}$, introduces a Lorentz force that can act on the neutral air. The plasma filament actuator consisted of three cathode–anode pairs of surface electrodes that were spaced at different streamwise distances from a 20° compression ramp at Mach 2.2. The electrodes were made of graphite. They were held in a plate of boron-nitride ceramic which provided electrical insulation between the electrodes. The spanwise spacing between the cathode and anode of each pair was 5 mm.

A static magnetic field was generated by a rubidium–iron–boron magnet that consisted of four pieces. Two of the pieces constructed the north magnetic pole, and the other two pieces constructed the south magnetic pole. The magnets were located below the wall, under the surface electrodes. Wang et al. (2009) estimated the magnetic field strength in the zone of interaction to be approximately 0.4 Tesla. As shown in

Figure 9.17, the desired magnetic field vector was toward the wall. The air breakdown voltage where the plasma was initiated was 3 kV.

Wang et al. (2009) found that in their experiment without the magnetic field, the plasma filaments reduced the shock intensity, corresponding to the pressure drop across the ramp-induced shock, by 1.5 percent. With the addition of the magnetic field, interchanging the anode and cathode locations of the electrode pairs to control the direction of the current changed the direction of the Lorentz force from downstream to upstream. For the upstream force configuration, the shock intensity decreased by 8.8 percent. The shock intensity decreased further to 11.6 percent with the downstream force configuration.

Bisek et al. (2013) performed numerical simulations of plasma control of turbulent SBLI with a magnetically driven surface plasma discharge. This involved a fully turbulent boundary-layer flow that traveled over a 24° wedge at Mach 2.25 that produced an unsteady shock-induced flow separation. Three control scenarios were examined: steady control, pulsing control with a 50 percent duty cycle, and a case with significant Joule heating. The first scenario allowed for momentum transfer by the magnetically driven surface discharge but excluded the reversible work produced by the magnetic force. On the basis of this assumption, those results are expected to represent an upper bound for the expected performance of a magnetically driven surface-discharge controller. The second case was consistent with the perfect controller. In the second scenario, the Strouhal number of the pulsing frequency, based on the length of the separation region and the free-stream velocity, was $St = 0.28$. This, however, falls an order of magnitude above the range of Strouhal numbers associated with the unsteadiness in SBLIs with compression ramps (Gonsalez and Dolling, 1993; Clemens and Narayanaswamy, 2014). The third scenario included Joule heating effects that was considered to be a more realistic situation for the plasma-based control.

In the simulation, the magnetically driven surface plasma-discharge actuator was located on the flat portion of the plate just upstream of the time-mean-averaged separation point that formed downstream of the wedge-generated shock. An example of the results for the different control scenarios is shown in Figure 9.21. This shows the boundary-layer thickness near the compression corner. The location and extent of the surface plasma-discharge actuator is shown by the blue rectangle. The compression wedge is drawn to scale in the plot.

The flow separation that occurs downstream of the shock results in a rapid thickening of the boundary layer. The start of the boundary-layer thickening corresponds to the upstream edge of the flow separation bubble (Bisek et al., 2013). In the simulation, the plasma-discharge actuator extended into the baseline separation bubble. The pulsed and realistic scenarios produced the identical effect of reducing the size of the separation bubble by approximately 50 percent. The first scenario that excluded the reversible work produced by the magnetic force to provide an upper limit to the control reduced the size of the separation bubble by approximately 75 percent.

Analysis of the results (Bisek et al., 2013) suggested that the reduction in the size of the separated region was caused by local streamwise acceleration of the flow near the baseline time-mean separation that resulted from the magnetically driven

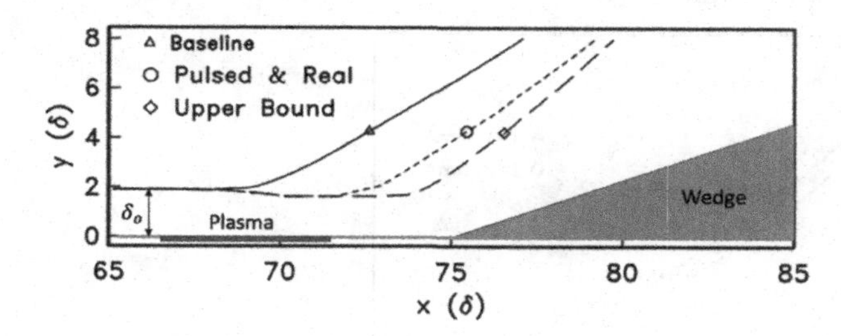

Figure 9.21 Time-averaged boundary-layer thickness near the compression corner for the baseline flow and three control scenarios. Reproduced from Bisek et al. (2013).

surface plasma-discharge actuator. This is evident by the thinning of the boundary layer observed over the plasma discharge region in Figure 9.21.

In a more recent simulation of magnetically driven surface plasma-discharge control, Jiang et al. (2020) considered the placement of the magnetohydrodynamics (MHD) interaction in an oblique shock–turbulent-boundary-layer interaction in a Mach 5 flow field. Four different scenarios were investigated:

1. the MHD interaction located upstream of the uncontrolled separation location;
2. the MHD interaction that crossed the upstream edge of the uncontrolled separation region;
3. the MHD interaction located inside the separation bubble and starting from the uncontrolled separation point;
4. the MHD interaction that encompasses the reattachment point.

In their analysis of the effect on the separation bubble, Jiang et al. (2020) introduced a new MHD parameter

$$S_{\Delta P} = -\frac{\sigma \mathrm{EBL}_{\mathrm{bubble}}}{\Delta P}, \tag{9.8}$$

where σ is the conductivity of the plasma arc column, E is the electric field strength, B is the magnetic field strength, L_{bubble} is the streamwise length of the separation bubble, and ΔP is the pressure increase that occurs after an interaction.

For the electric field, only the z-direction was considered so that

$$E_z = -\frac{V}{l_z} \text{ and } E_x = E_y = 0 \tag{9.9}$$

and where

$$\sigma = I \Big/ \left[\frac{V}{l_z}\frac{\pi l_y^2}{4}\right] \tag{9.10}$$

in which I is the current density, and the plasma arc is considered to have a circular cross section with diameter, l_y, and length, l_z.

The simulations indicated that the most effective control came when the MHD interaction zone was inside the separation bubble. For that case, Jiang et al. (2020) investigated the effectiveness of reducing the length of the separation bubble for a

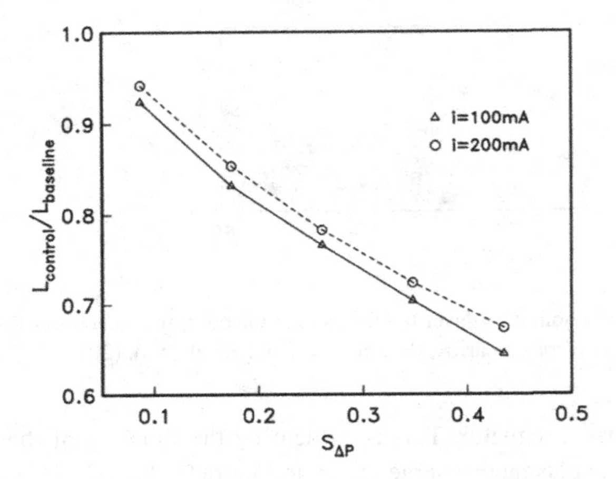

Figure 9.22 Effect of MHD parameter on reducing the length of the separation bubble in an oblique shock–turbulent-boundary-layer interaction at Mach 5 using a magnetically driven surface plasma-discharge located inside the separation bubble. Reproduced from Jiang et al. (2020).

range of $S_{\Delta P}$ values from 0.087 to 0.435. This included the effect of two different plasma currents of 100 and 200 mA. The results are shown in Figure 9.22. These indicted a nearly linear dependence in the reduction in the length of the separation bubble on $S_{\Delta P}$. In addition, there was an adverse effect on using higher current levels, which was presumed to be due to producing higher Joule heating that offset the momentum increase.

9.4 Summary

Even after 70 years of research on SBLI, it still remains an important topic, particularly in applications like ScramJet propulsion. The conditions where such interactions can occur include an oblique shock wave caused by an elevated body that impinges on a flat-plate boundary layer, a boundary layer on a flat plate that leads up to a compression ramp, and a normal shock wave interacting with a flat-plate boundary layer with a symmetric wall protrusion. If an incident shock is sufficiently strong, the pressure gradient across the shock can cause the boundary layer to separate, forming a recirculating separation bubble and a separated shear layer, both of which are unstable to disturbances and can be the source of shock wave unsteadiness that is a hallmark of SBLIs.

The formation of the separation bubble was instrumental in the development of the "free-interaction" concept, the essence of which is that *any phenomenon near separation that is independent of the object shape would depend only on the flow that is internal and external to the boundary layer*. The local scaling laws of the free interaction have distinguished two mechanisms: a global mechanism that determines the separation-bubble length, L, and a local mechanism that controls the free-interaction region, in the vicinity of the separation point. The former was found to depend linearly on the shock strength, defined as the ratio of the downstream to upstream free-stream pressures.

SBLIs have a characteristic low-frequency unsteadiness, with a Strouhal number in the range from 0.02–0.5, based on the size of the interaction zone and external velocity (see Figure 9.3). The lower end of the range appears to occur more frequently with compression ramps. The cause of this unsteadiness is still an open question and remains as a central issue in supersonic and hypersonic SBLIs. From a practical sense, the low-frequency unsteadiness of the reflected shock is a main source of structural fatigue that in turn becomes a major constraint in the choice of materials.

Attempts have been made to link the unsteadiness to a "breathing motion" of the separation bubble that forms in the interaction zone, which may be driven by disturbances in the upstream boundary layer, or the result of inherent dynamics between the separation bubble and the shock wave.

Thoughts on the source of the SBLI unsteadiness has generally fallen into three categories:

1. upstream effect – unsteady coherent structures in the incoming turbulent boundary layer;
2. downstream effect – unsteady dynamics of the separation bubble;
3. or both upstream and downstream effects.

There still remains no consensus, but with *both* upstream and downstream effects being more than likely.

What is known is that the SBLI is sensitive to a variety of initial conditions that include

1. wall cooling;
2. local changes in wall contour;
3. localized wall suction and/or injection;
4. modification of the approaching boundary layer.

This insight is important in the design of flow control strategies.

Both passive and active flow control approaches have been shown to be effective in reducing the adverse effects of SBLI. The most effective passive control approaches include a porous-wall-covered cavity and streamwise vortex generators. The former is to be placed at the location where the shock impinges the wall. The pressure gradient across the shock then drives flow upstream through the cavity and subsequently reduces the shock strength and flow separation. The vortex generators are placed upstream of the impinging shock. Their purpose is to enhance mixing throughout the boundary layer and thereby reduce the potential for flow separation caused by an impinging shock.

Active approaches include air jets that exit from the wall at oblique angles in order to produce streamwise vorticity. The advantage to these over passive streamwise vortex generators is that they can be optimized for different external flow conditions. As with passive vortex generators, the air jet vortex generators are arranged in a spanwise array. As with the passive vortex generators, there is an optimal spacing used for SBLI control.

There have been numerous investigations on the use of pulsed plasma jets for SBLI control. These consist of an electrode located in a cavity below the surface of a metal, earth-grounded boundary-layer plate. A sufficiently large AC voltage supplied to the electrode causes a plasma discharge to form inside the cavity that produces pressure pulses that communicate through a cavity orifice into the flow field. The advantage of this approach is that it can produce high-frequency pressure pulses that is not possible with air-driven jets. Oblique orifice angles are also used in arrays to generate streamwise vorticity into the boundary layer upstream of the impinging shock. Their optimum spanwise spacing is similar to that of wall air jets.

An active control approach that more directly utilizes the electromagnetic-hydrodynamic properties of the plasma places the plasma electrodes flush with the boundary-layer wall surface, exposing them to the primary flow. A large enough DC voltage potential across the electrodes generates plasma filament loops that extend downstream. These introduce Joule heating that can modify the shock wave properties such as oblique wave angle and intensity. In addition, if coupled with a magnetic field, it can produce a Lorentz force that can act on the flow field. Both experiments and simulations have demonstrated both approaches to have significant effects and great potential for SBLI control.

Problems

9.1 As shown in Figure 9.2, a SBLI involves a number of flow modules including an upstream turbulent boundary layer, separated shear layer, and separation bubble.

1. Based on information in previous chapters, list four approaches (two passive and two active) that can be utilized to modify each of these flow modules.
2. Of those listed, which of these might be most effective at reducing the low-frequency unsteadiness associated with SBLI? Consider both upstream and downstream effects.
3. Of those listed, which of the *passive* approaches would be most effective at reducing the low-frequency unsteadiness associated with SBLI?

9.2 Equation (9.2) presents an expression for the pressure rise across an SBLI that leads to a flow recirculation zone over a streamwise distance, L_i, shown in Figure 9.3, which is a function of the incoming boundary-layer displacement thickness, δ^*, and shear stress at the wall, τ_{w0}, where 0 denotes the location of the start in the pressure rise across the shock. Based on Eq. (9.2), in order to minimize the pressure drop that drives the flow separation, it is necessary to minimize τ_{w0}, maximize δ^*, or both.

1. Describe two flow control approaches (one passive and one active) that can minimize the wall shear stress, τ_w, in a turbulent boundary layer.
2. Describe two flow control approaches (one passive and one active) that can maximize the displacement thickness, δ^*, in a turbulent boundary layer.
3. Can the two objectives be achieved simultaneously by one flow control approach?

4. Indicate how that would be implemented in this SBLI application.

9.3 Describe how wall heating or cooling can affect the following:

1. The instability and turbulence onset of the approaching laminar boundary layer and dependence on Mach number.
2. The effect of an adverse pressure gradient to lead to a flow separation.
3. Describe the beneficial effect these would have on the boundary layer δ^* and τ_w to minimize the pressure drop that drives the flow separation.
4. Describe how this might be implemented to mitigate the adverse effects of SBLIs.

9.4 The results in Figure 9.7 indicate that wall-normal fluid injection can reduce τ_w.

1. Explain how this might be used to potentially reduce the low-frequency unsteadiness associated with SBLIs.
2. In answering the previous question, is it based on an upstream or downstream source of unsteadiness?
3. Would there be any further benefit to heating or cooling the injected fluid? If so, how?
4. With respect to mass injection/suction and wall heating/cooling, list the critical parameters and physical mechanisms?

9.5 In Chapter 4, it was shown that a separated flow could be reattached when excited by a periodic disturbance with a dimensionless frequency, $f^+ = fL/U_e = 1$, where L was the streamwise length of the flow separation bubble and U_e was the mean velocity at the edge of boundary layer upstream of the flow separation.

1. Given this background, why do you think that the researchers who applied unsteady excitation have focused on reduced frequencies (Strouhal numbers) on the order of 0.01–0.05?
2. Based on a Mach number of 3, estimate the frequency for excitation at a $f^+ = 1$. Does this seem feasible?
3. What passive approaches are available to cause a separated flow to reattach that might be applied at these higher Mach numbers?

10 Flow Control by Design

The previous chapters have presented examples of different flow fields in which the governing fluid instabilities could be manipulated to achieve a specific end, for example, to prevent turbulent transition, maintain attached flow, increase fluid mixing, or lower viscous drag. In most practical applications, such examples of flow control involve an existing geometry that remains fixed. This chapter considers modifying a geometry to make it more *receptive* to flow control that works through a specific fluid instability. The chapter presents a number of approaches. These range from a simple modification of a geometry to rigorous approaches that utilize an adjoint formulation of the Navier–Stokes equations that seek geometric changes that maximize flow control authority. The development of such flow-control-configured geometries represents the essence of "Flow Control by Design."

10.1 Enhanced Aerodynamic Lift Control

One application of active flow control is to replace mechanical flight control surfaces such as trailing-edge flaps and ailerons, with flow control devices that do not involve moving surfaces. As an example, He and Corke (2009) had demonstrated the concept of lift control on a NACA 0015 section shape airfoil using plasma actuators. The details of plasma actuators as flow control devices are covered in Chapter 2, "Sensors and Actuators." The plasma actuators were located on the upper (suction side) surface of the airfoil trailing edge. The trailing-edge plasma actuator was found to produce a maximum change in the lift coefficient of $\Delta C_l = 0.051$. Using a plain flap model, this was found to be equivalent to a 1.5° deflection of a plain trailing-edge flap. If this were used to replace mechanical ailerons, He and Corke (2009) determined that it would produce a roll moment that was equal to that produced by a 9° aileron flap deflection on a four-place general aviation aircraft.

Swept wings of low-aspect ratio are commonly used on high-speed aircraft because of their favorable wave drag characteristics. Several researchers have employed flow-control methods to improve the aerodynamics of swept wings. Moeller and Rediniotis (2000) demonstrated control of the pitching moment of a 60° swept-delta-wing model at high angles of attack using a series of surface-mounted pneumatic vortex actuators. Amitay et al. (2004) performed experiments on a 1301 unmanned air vehicle (UAV)

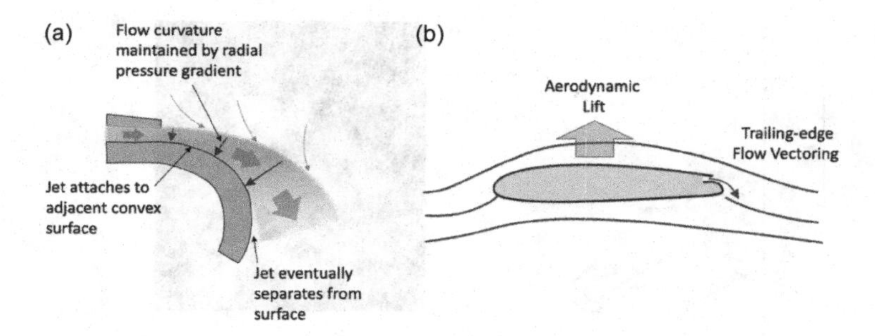

Figure 10.1 Fundamental concept of the Coanda effect (a) and an example use of the Coanda effect for circulation control on an airfoil (b).

design that utilized synthetic jet actuators for flight control. They showed that the flow control could suppress leading-edge flow separation that can occur at higher angles of attack.

Warsop and Crowther (2018) demonstrated active flow control for flight control based on manipulating Coanda flows. As was presented in Chapter 5, Reynolds et al. (2003) had utilized a Coanda effect at the exit of a jet nozzle to radially expand the flow. The Coanda effect is the tendency of a fluid jet to stay attached to a convex surface. This is illustrated in Figure 10.1(a). The first official documents that explicitly mention the Coanda effect were two 1936 patents, Coanda (1936a) and Coanda (1936b). Englar (1975) was the first to apply the concept to an airfoil for circulation control and thereby aerodynamic lift control. A schematic of that concept is illustrated in Figure 10.1(b). A general review of this application is provided by Warsop and Crowther (2018).

In another example of circulation control, Patel et al. (2006) and Lopers et al. (AIAA Paper 2007-636) performed experiments to determine the aerodynamic control effect of a dielectric-barrier-discharge plasma actuator located on the trailing edge of a 47° swept wing of a 1303 (UAV) configuration. Both Patel et al. (2006) and Lopers et al. (AIAA Paper 2007-636) oriented the plasma actuator so that it was parallel to the trailing edge of the airfoil and arranged so that it would induce a flow toward the trailing edge. Figure 10.2 shows an illustration of the half-span planform view of 1303 UAV wind tunnel model that shows the locations of plasma actuators along the trailing edge, as well as an image of airfoil planform with plasma actuators operating (purple lines) in a darkened lab.

As an example of flow control by design, Patel et al. (2006) modified the trailing edge of the airfoil section by adding a 20° ramp that would cause the boundary layer to separate at that location. This is illustrated in Figure 10.3(a). A plasma actuator was located just upstream of the start of the ramp. It was operated to produce an unsteady disturbance (pulsing) at a frequency that would drive the separated flow to reattach. As discussed in Chapter 4, "Separated Flows," an effective approach to attach separated flows is to introduce a periodic disturbance at a frequency that corresponds to a dimensionless frequency, $f^+ = fL/U_e$, where f is the physical frequency, L is the

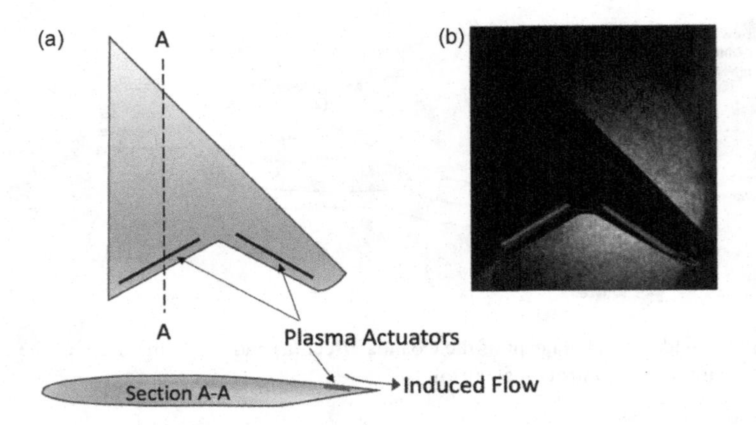

Figure 10.2 Illustration of the half-section planform view of 1303 UAV wing along with airfoil section A-A shape showing locations of trailing-edge plasma actuators (a) and image of airfoil planform with plasma actuators operating (purple lines) in a darkened lab (b). Taken from Patel et al. (2006).

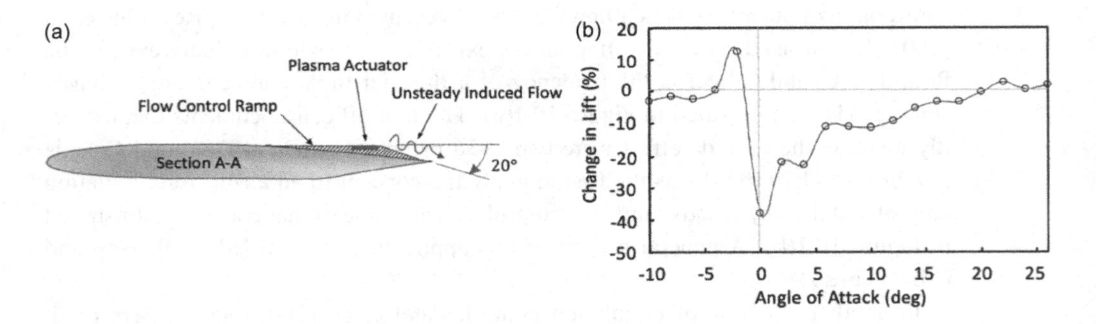

Figure 10.3 Illustration showing trailing-edge separation ramp added to the suction side of the airfoil section (a) and change in the lift coefficient with plasma actuator operating to attach flow over the separation ramp as a function of the airfoil angle of attack (b). Taken from Patel et al. (2006).

streamwise extent of the separated flow region, and U_e is the velocity at the edge of the attached approaching boundary layer.

The purpose of the airfoil trailing-edge separation ramp was to enhance the response of the boundary layer to the plasma actuator control, with the ultimate objective to provide circulation control and thereby lift control, of the wing without a moving surface. The result is shown Figure 10.4(b), which documents the change in aerodynamic lift with the plasma actuator operating as a function of the wing angle of attack. By locating the separation ramp on the suction side of the wing, the effect of controlling the separated flow over the ramp is to reduce the aerodynamic lift. The largest effect occurs with the wing at a zero angle of attack, where the ratio of the lift coefficients with plasma ON and OFF, $C_{L_{ON}}/C_{L_{OFF}}$, is approximately 40 percent. The effect diminishes with increasing angle of attack, which affects the degree to which

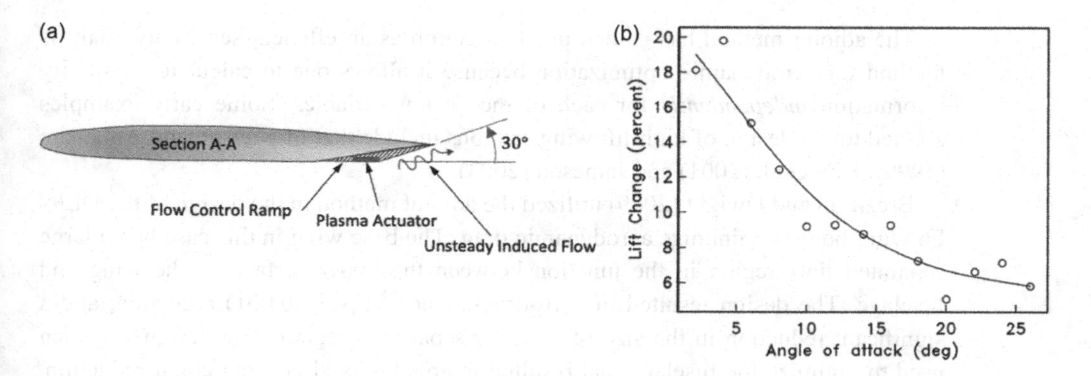

Figure 10.4 Illustration showing trailing-edge separation ramp added to the pressure side of the airfoil section (a) and change in the lift coefficient with plasma actuator operating to attach flow over the separation ramp as a function of the airfoil angle of attack (b). Taken from Lopers et al. (AIAA Paper 2007-636).

the flow separates over the trailing-edge ramp. However, the control effect remains up to the wing stall angle of attack at 20°.

Lopers et al. (AIAA Paper 2007-636) applied the same approach as Patel et al. (2006) but placed the trailing-edge separation ramp on the pressure side of the wing. This is illustrated in Figure 10.4(a). Lopers et al. (AIAA Paper 2007-636) had investigated three ramp angles of 20°, 30°, and 40°. The results shown in Figure 10.4(b) are for the 30° ramp. In that case, because the ramp was located on the pressure side of the wing, the effect was to increase the aerodynamic lift. As with Patel et al. (2006), the effect is largest at a zero angle of attack, with a change in the lift coefficient of nearly 20 percent. The lift change again diminishes with increasing angle of attack, although it still produced a 6 percent change in the lift coefficient up to the 20° stall angle of attack. Based on these two examples, it is conceivable to add similar separation ramps on *both* the pressure and suction sides of the wing to essentially double the degree of lift control.

10.2 Design Optimization

The previous examples in which an existing geometry was modified to be more *receptive* to active flow control motivates a more objective, automated approach. In practical design optimization, based on computational fluid dynamics (CFD) approaches that include viscous effects, the computational costs and run-time are high, especially for high Reynolds numbers. Compounding the cost, an iterative shape optimization could require hundreds of CFD runs, even when only a few shape parameters are involved. Gradient-based methods, can significantly enhance the process. Standard gradient-based or deterministic techniques perform a local search around a current design state, seeking a direction of the steepest descent of the objective functional indicated by a sensitivity derivative or gradient. A formal approach that utilizes a gradient-based nonlinear optimization technique is known as an *adjoint method*.

The adjoint method has gained much attention as an efficient sensitivity analysis method for aerodynamic optimization because it allows one to calculate sensitivity information *independently* for each of the design variables. Some early examples applied to the design of high-lift wing sections include that of Nielson and Anderson (1999), Kim et al. (2001) and Jameson (2003).

Brezillon and Dwight (2009) utilized the adjoint method in the design of the DLR-F6 wing-body to minimize aerodynamic drag. The base wing in this case had a large separated flow region in the junction between the upper surface of the wing and fuselage. The design resulted in a 10-drag-count ($\Delta C_D = 0.001$) reduction, and a significant reduction in the size of the flow separation region. The design was then used to optimize the fuselage that resulted in an additional 20-drag-count reduction. The resulting wing-body geometry that emerged by the adjoint method took the shape of a fairing that is now common to modern aircraft.

Kungurtsev and Juniper (2019) applied an adjoint-based shape optimization to the design of micro-channels in an inkjet print head. The objective was to damp acoustic oscillations that reverberate within micro-channels in the print head when ink drops are ejected. This algorithm converged to a design that had the same viscous dissipation for the steady flow but with a 50 percent larger decay rate of the acoustic oscillations. Kungurtsev and Juniper (2019) perceived that the inkjet manufacturers, using physical insight and a trial and error approach, probably would not have considered the design that emerged from the adjoint-based shape optimization.

10.2.1 Adjoint Approach to Flow Control Design

While active flow control is an established method for controlling flow separation on vehicles and airfoils, the design of the actuation is often done by trial and error. In an answer to this, Carnarius et al. (2011) developed a discrete and a continuous adjoint flow solver for the optimal control of unsteady turbulent flows using steady and unsteady wall suction and blowing. The simulations involved the solution of the incompressible Reynolds-averaged Navier–Stokes equations. They applied this to control the unsteady laminar flow around a circular cylinder at $\mathrm{Re}_D = 100$, with the objective of reducing the drag. For reference, see Chapter 3, "Bluff Body Wakes."

The flow control was produced by pulsed blowing or suction through 15 equally space slots, 18° apart, and covering a circumferential region of the cylinder corresponding to range of angular positions of $54° \leq \theta \leq 306°$, where $\theta = 0$ is the location of the upstream stagnation line. The pulsed velocity was defined as

$$V_n(\theta) = V_a(\theta)\sin[2\pi f(\phi - \phi_0)] - V_a(\theta), \tag{10.1}$$

where V_a is the velocity amplitude, f is the pulsing frequency, and ϕ_0 is a phase shift. Note that the pulsing velocity contained both a mean and fluctuating part. The distribution of the pulsing amplitude, $V_a(\theta)$, was the parameter to be optimized. The frequency was kept constant and corresponded to a Strouhal number of 1 or a frequency of $f = V_\infty/D$. The phase shift was also fixed at $\phi_0 = 0$. The resulting optimum arrangement of pulsed blowing and suction to minimize drag on the circular cylinder

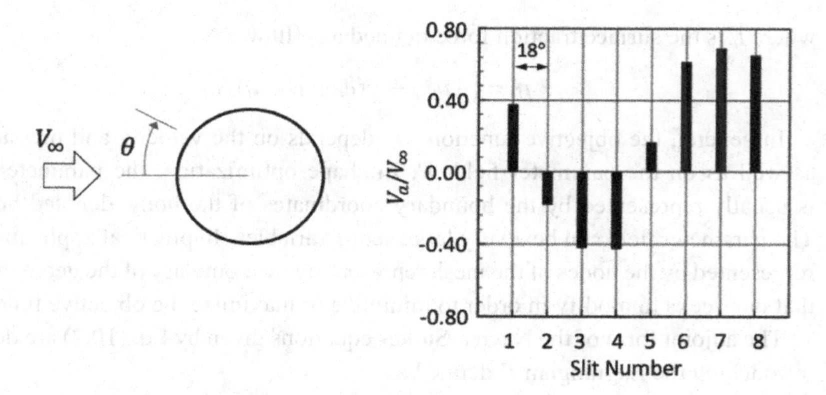

Figure 10.5 Optimum arrangement of pulsed blowing and suction to minimize drag on a circular cylinder at $\mathrm{Re}_D = 100$ based on adjoint flow solver. Slit 1 is at $\theta = 54°$. Compiled from Carnarius et al. (2011).

at $\mathrm{Re}_D = 100$ is shown in Figure 10.5. Here, slit 1 is a $\theta = 54°$, and the slit numbers correspond to increasing angular positions in increments of 18°. This optimum arrangement resulted in a 30 percent drag reduction, where the drag coefficient, C_d, was reduced from 1.336 to 0.899.

Adjoint Method Formulation

The following describes the formulation for the adjoint form of the Navier–Stokes equations that is used in a gradient and gradient-free optimization of an existing geometry to enhance its flow-control sensitivity. This was originally developed by Williams et al. (2015). This formulation is based on the incompressible form of the turbulent Navier–Stokes equations, given in Eq. (10.2). It consists of the incompressible continuity and momentum equations, supplemented by the appropriate turbulence model equations that provide an effective viscosity in the momentum equation. In the momentum equation, f_i is a body force term that can include the effect of a flow control actuator. As usual, these equations need to be accompanied by appropriate boundary conditions and transport properties.

$$\mathcal{A} = \begin{pmatrix} \mathcal{A}_1 \\ \mathcal{A}_2 \end{pmatrix} = \begin{pmatrix} \partial_i u_i \\ u_j \partial_j u_i + \partial_i p - \partial_j \left[(\nu + \nu_t)(\partial_j u_i + \partial_i u_j) \right] + f_i \end{pmatrix}. \tag{10.2}$$

In seeking some optimum, it is necessary to define an objective function which is denoted as $\mathcal{J}$. The objective function is usually chosen to represent a measure of the performance of system. Following the example of Patel et al. (2006) and Lopers et al. (AIAA Paper 2007-636), the object function could concern the aerodynamic lift. In this case, $\mathcal{J}$ could represent the total force, or some component of the force acting on the geometric shape. This can be expressed by the following integral relation:

$$\mathcal{J} = \int_\Gamma t_i n_i d\Gamma, \tag{10.3}$$

where t_i is the surface traction force defined as follows:

$$t_i = \left(-p\delta_{ij} + \nu(\partial_j u_i + \partial_i u_j)\right) n_j. \tag{10.4}$$

In general, the objective function, $\mathcal{J}$, depends on the velocity and pressure fields as well as on the parameter field, $\mathcal{P}$. In shape optimization, the parameter field $\mathcal{P}$ is usually represented by the boundary coordinates of the body, denoted here as x_i. The parameter field can be a very large set of variables. In practical applications, $\mathcal{P}$ is represented by the nodes of the mesh representing the boundary of the geometric shape that one seeks to modify in order to minimize or maximize the objective function.

The adjoint form of the Navier–Stokes equations given by Eq. (10.2) are derived by introducing the Lagrangian $\mathcal{L}$ defined as

$$\mathcal{L} = \mathcal{J} + \langle \mathbf{Q}^*, \mathcal{A} \rangle, \tag{10.5}$$

where $< \cdot, \cdot >$ represents an inner product between two functions, and $\mathbf{Q}^*$ is a vector of adjoint variables. It should be noted that adjoint variables play the role of Lagrange multipliers in the expression for the Lagrangian in Eq. (10.5).

In the case of the incompressible Navier–Stokes equations given in Eq. (10.2), the Lagrangian takes the following form:

$$\mathcal{L} = \mathcal{J} + \int_\Omega \begin{pmatrix} p^* & u_i^* \end{pmatrix} \begin{pmatrix} \mathcal{A}_1 \\ \mathcal{A}_2 \end{pmatrix} d\Omega. \tag{10.6}$$

In order to define the system of adjoint equations, it is necessary to compute the total variation of the Lagrangian, $\mathcal{L}$, by following the rules of the variational calculus, namely

$$\delta\mathcal{L}(u_i, p, x_i) = \delta_{u_i}\mathcal{L} + \delta_p\mathcal{L} + \delta_{x_i}\mathcal{L}. \tag{10.7}$$

The total variation of the Lagrangian as defined in Eq. (10.7) is equivalent to defining the Euler–Lagrange equations, where the Lagrangian is defined in Eq. (10.6). Thus, following the rules of variational calculus, the conservation law defined by the Lagrangian, $\mathcal{L}$, will produce a set of adjoint equations. The adjoint equations are complementary equations with respect to the Lagrangian $\mathcal{L}$. As a result, different forms of adjoint equations are possible depending on the definition of the Lagrangian function. Formally, the requirement that a particular conservation law is being enforced will yield adjoint equations that are consistent with that requirement.

Following Williams et al. (2015), energy conservation is enforced through the selection of the Lagrangian

$$\mathcal{L} = \mathcal{J} + \int_\Omega \begin{pmatrix} p^* & u_i^* \end{pmatrix} \begin{pmatrix} \partial_i u_i \\ \Lambda u_i + \partial_i p + f_i \end{pmatrix} d\Omega, \tag{10.8}$$

where Λ_i is the following operator:

$$\Lambda = \partial_j u_j - \partial_j \left[(\nu + \nu_t)(\partial_j + \partial_i) \right]. \tag{10.9}$$

Formally, the following quasi-linear equation satisfies the definition of the Lagrangian in Eq. (10.8), namely

$$(\mathcal{N}(u_j))\, u_i + \partial_i p + f_i = 0, \tag{10.10}$$

where $(\mathcal{N}(u_j))\, u_i$ is given by

$$(\mathcal{N}(u_j))\, u_i \doteq \Lambda u_i. \tag{10.11}$$

The operator, $\mathcal{N}$, is considered a linear operator by which it is possible to formally define the adjoint operator by enforcing the Lagrange duality principle. This is achieved by taking the total variation of Eq. (10.6) in accordance with Eq. (10.7) and performing integration by parts. The adjoint equations are identified by enforcing the resulting integrals to be equal to zero over the domain, thus resulting in an adjoint system of equations given in Eq. (10.12) and (10.13).

$$\mathcal{A}_1^* = -\partial_i u_i^* \tag{10.12}$$

$$\mathcal{A}_2^* = -u_j \partial_j u_i^* + \partial_j \left[(\nu + \nu_t)(\partial_j u_i^* + \partial_i u_j^*) \right] - \partial_i p^*. \tag{10.13}$$

The main characteristic of the adjoint system given in Eq. (10.12) is that it preserves the time symmetry for the unsteady system (Marchuk, 1995). This is equivalent to the kinetic energy conservation. Furthermore, it should be noted that the forcing term, f_i, that might apply to a flow control actuator does not appear in the adjoint system of equations. This assumes that the equivalent body force produced by a flow control actuator does not depend on coordinates of the test article geometry (airfoil for example) or on the flow field, namely pressure and velocity. Therefore, the total variation of the body force term with respect to independent variables is identically zero.

Through additional simplification in the derivative of the adjoint equations, the turbulent viscosity field does not depend on independent flow variables. This condition is not satisfied in the general case as the turbulent viscosity will depend on pressure, velocity, and coordinates of the airfoil. This assumption has an effect on accuracy of computed derivatives, but it is not expected to be significant for a streamlined body such as the airfoil used in the previous examples of Patel et al. (2006) and Lopers et al. (AIAA Paper 2007-636). Using this assumption greatly simplifies the formulation as the turbulence model does not have to be considered part of the adjoint system.

Boundary conditions for the adjoint system are derived from the surface integral representing the bilinear concomitant related to the adjoint system, namely

$$\int_\Gamma (u_i^* n_i + \partial_p \mathcal{J}) \delta p d\Gamma - \int_\Gamma (s_i + \partial_{u_i} \mathcal{J}) \delta u_i d\Gamma + \int_\Gamma u_i^* (\nu + \nu_t)(\partial_j \delta u_i + \partial_i \delta u_j) n_j = 0, \tag{10.14}$$

where s_i vector is defined as follows:

$$s_i = \left[(u_j u_i^* - p^* \delta_{ij} + (\nu + \nu_t)(u_{i,j}^* + u_{j,i}^*) \right] n_j. \tag{10.15}$$

The adjoint boundary conditions are derived from Eq. (10.14) by enforcing this expression to be equal to zero term by term, for all boundaries. Corresponding boundary conditions for various boundaries at the inlet, outlet, and wall are given by Eq. (10.16)–(10.18), respectively.

$$u_i^* n_i = 0 \tag{10.16}$$

$$u_j u_i^* - p\delta_{ij} + (\nu + \nu_t)(\partial_j u_i^* + \partial_j u_j^*) + \partial_{u_i}\mathcal{J} = 0 \tag{10.17}$$

$$u_i^* n_i + \partial_p \mathcal{J} = 0. \tag{10.18}$$

In addition to the adjoint boundary conditions, the definition or the shape derivative is needed to define the gradient of the objective function, $\mathcal{J}$. The shape derivative is obtained by examining integrals in Eq. (10.14) associated with the surface of the test article and finding the remaining term that is not equal to zero on that boundary. From this consideration, the term that remains nonzero on the boundary is associated with the variation of the coordinates, which is represented in Eq. (10.19).

$$\delta_{x_i}\mathcal{L} = \partial_i \mathcal{L}\delta x_i = -\int_\Gamma s_i \delta u_i d\Gamma = -\int_\Gamma s_j \partial_j u_i \delta x_i d\Gamma. \tag{10.19}$$

The final expression for the shape derivative then takes the form given in Eq. (10.20).

$$\partial_i \mathcal{J} = -\int_\Gamma s_j \partial_j u_i d\Gamma, \tag{10.20}$$

where s_j is given by Eq. (10.21).

$$s_j = \left[(u_j u_i^* - p^* \delta_{ij} + \nu(u_{i,j}^* + u_{j,i}^*) \right] n_j. \tag{10.21}$$

Numerical Method

A standard pressure-based algorithm (Ferziger and Peric, 1996) can be used for the solution of both the flow and adjoint equations. The solution algorithm is based on correcting both primal and adjoint velocity fields to satisfy the corresponding continuity equations. Both the primal and adjoint momentum equations are linearized in the same way to produce a system of linear algebraic equations that take the general form given in Eq. (10.22).

$$A_P^{u_i} u_{i,P}^{n+1} + \sum_l A_l^{u_i} u_{i,l}^{n+1} = f_i^{n=1} - \left(\delta_i p^{n+1}\right)_P. \tag{10.22}$$

Equation (10.22) corresponds to the linearization at point P and index l that is used to denote neighbors of that point in the discretization stencil, where $n + 1$ denotes the current iteration. Linearization coefficients are denoted by the symbol A. They represent matrix entries in the implicit solution method. The notation, δ_i, represents the discretized gradient in the pressure term.

Equation (10.22) can be solved for the unknown velocity field u_i to obtain the expression in Eq. (10.23).

$$u_{i,P}^{n+1} = \frac{1}{A_P^{u_i}}\left(f_i^{n+1} - \sum_l A_l^{u_i} u_{i,l}^{n+1}\right) - \frac{1}{A_P^{u_i}}\left(\delta_i p^{n+1}\right)_P. \tag{10.23}$$

Equation (10.23) makes use of the previously computed pressure p^n in order to decouple the pressure and momentum equations. As a result, an additional step is required to correct the velocity field. This is achieved through the pressure equation given in Eq. (10.24) that is obtained by substituting Eq. (10.23) into the continuity equation.

$$\delta_i \left[\frac{1}{A_P^{u_i}} \left(\delta_i p^{n+1} \right)_P \right] = \delta_i \left(\bar{u}_i^{\dagger} \right)_P . \tag{10.24}$$

The quantity, $\bar{u}_i^{\dagger}$, is an intermediate velocity field that corresponds to the velocity field in which the pressure gradient has been removed, or specifically

$$\bar{u}_i^{\dagger} = f_i^{n+1} - \sum_l A_l^{u_i} u_{i,l}^{n+1} . \tag{10.25}$$

The correction to the velocity field so that the continuity equation is satisfied is given by Eq. (10.26), which is performed at each time step.

$$u_{i,\mathrm{corr}}^{n+1} = \bar{u}_i^{\dagger} - \left[\frac{1}{A_P^{u_i}} \left(\delta_i p^{n+1} \right)_P \right] . \tag{10.26}$$

This predictor–corrector procedure is equally applicable to the primal and the adjoint problems as they are structurally the same. Therefore, both primal and the adjoint problems can be solved using very similar numerical codes, with the difference being in the boundary conditions. The boundary conditions for the primal problem are defined by the physics of the problem, while the adjoint boundary conditions are defined by the objective function and its derivatives.

Implementation of the flow (primal) and adjoint solver has been done in the OpenFOAM library by Jasak et al. (2007). For simplicity, the turbulence in the adjoint problem can be treated as a frozen field. However, adjoint turbulence models are available (Lackner and van Kuik, 2012).

Optimization Procedure

Having defined the shape derivative in Eq. (10.21), the following steps are used as an optimization procedure.

1. Solve the flow (primal) problem to obtain velocity and pressure fields and their gradients.
2. Use the solution of primal problem to solve the dual (adjoint) problem for the adjoint pressure and velocity fields and their gradients.
3. Compute the shape sensitivity according to Eq. (10.20).
4. Deform the geometry by taking the step, Δx_i, in the direction of the local shape derivative according to $x_i^{n+1} = x_i^n + \alpha \partial_i \mathcal{J}$.
5. Perform smoothing of the deformed geometry using the geometrical parameterization through B-splines.
6. Evaluate the value of the objective function, $\mathcal{J}$, and check the convergence conditions.
7. Perform the next optimization step if the convergence criterion is not satisfied.

The success of the optimization procedure highly depends on a number of practical considerations. Smoothing of an updated design allows larger geometric changes between iterations, which reduces the required number of iterations needed to achieve an acceptable solution. The shape derivatives are generally solved at all face center mesh points located on the test article. The updated design is obtained by reconstructing these points after applying perturbations prescribed by the shape derivatives. If no shape smoothing is used, the step change between iterations needs to be less than half of the smallest distance between face centers on the test article in order to ensure a valid reconstruction. If larger step sizes are used, it is possible that the face center grid points overlap after a time step that results in an invalid surface reconstruction.

10.2.2 One-Parameter Design Optimization

The following is an example of a one-parameter design optimization that comes from a wind energy application. It specifically seeks a horizontal wind turbine rotor blade design that maximizes the amount of energy extracted from the wind under varying conditions.

As background (Corke and Nelson, 2020), the power extracted from the wind air stream for a horizontal wind turbine based on actuator disk momentum theory is

$$P = 2\rho A_d V_\infty^3 a[1-a]^2, \tag{10.27}$$

where ρ is the density of the air, A_d is the area of the rotor disk ($A_d = \pi R^2$, where R is the rotor radius), V_∞ is the wind velocity far upstream of the rotor disk, and a is the "axial induction factor" defined as

$$a = \frac{V_\infty - V_d}{V_\infty}, \tag{10.28}$$

and where V_d is the velocity at the face of the rotor disk.

Defining a power coefficient as the ratio of the power extracted from the wind, P, to the available power in the wind air stream gives the following equation:

$$C_p = \frac{P}{\frac{1}{2}\rho A_d V_\infty^3}. \tag{10.29}$$

Substituting for P from Eq. (10.27) gives the power coefficient as an exclusive function of the axial induction factor, namely

$$C_p = 4a[1-a]^2. \tag{10.30}$$

The condition on the axial induction that maximizes the power coefficient is $a = 1/3$. Substituting $a = 1/3$ into Eq. (10.30) gives the maximum theoretical power coefficient of $C_{P_{\max}} = 0.593$, which is referred to as the Betz limit after Albert Betz (1920) who was the first to derive this limit.

Standard horizontal wind turbines use a rotor blade that is a fixed geometry from root to tip. Control comes from changing the pitch angle in order to maintain a constant "design" power level over a range of wind speeds. As a result of the fixed rotor blade geometry, the axial induction factor varies along the rotor span, and most typically is

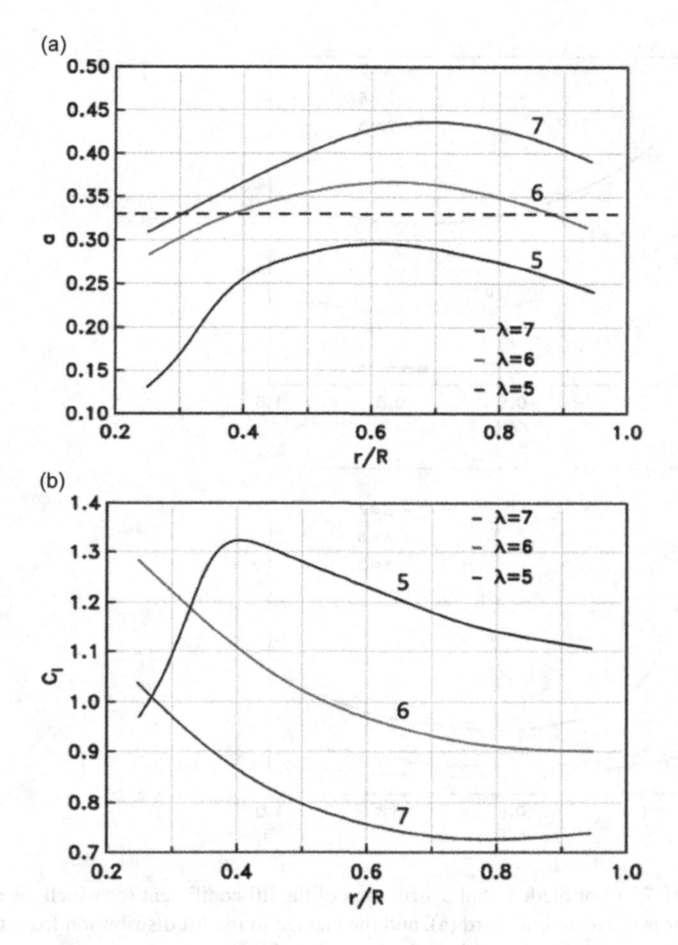

Figure 10.6 JIMP 25 rotor blade radial distribution of the axial induction factor (a) and lift coefficient (b) for different tip speed ratios. Taken from Cooney et al. (2014).

near the optimum one-third-value at only *one* radial location on the rotor blade. As result, the performance of such wind turbines never approach the Betz limit.

As an example from Cooney (2015) Figure 10.6(a) shows the radial distribution of the axial induction factor for a JIMP 25-kW wind turbine for three tip speed ratios, λ. The horizontal dashed line marks the ideal axial induction factor of one-third that will produce the maximum (Betz) power coefficient. This demonstrates that for this wind turbine blade design, in the best case, the optimum axial induction factor occurs only at two radial locations on the rotor blade. Furthermore, depending on the tip speed ratio, no part of the blade may have the optimum inflow induction factor.

Figure 10.6(b) shows the radial distribution of the lift coefficient that corresponds to the axial induction factors shown in Figure 10.6(a). Figure 10.7(a) shows the radial distribution of the lift coefficient that is required to produce the ideal axial induction

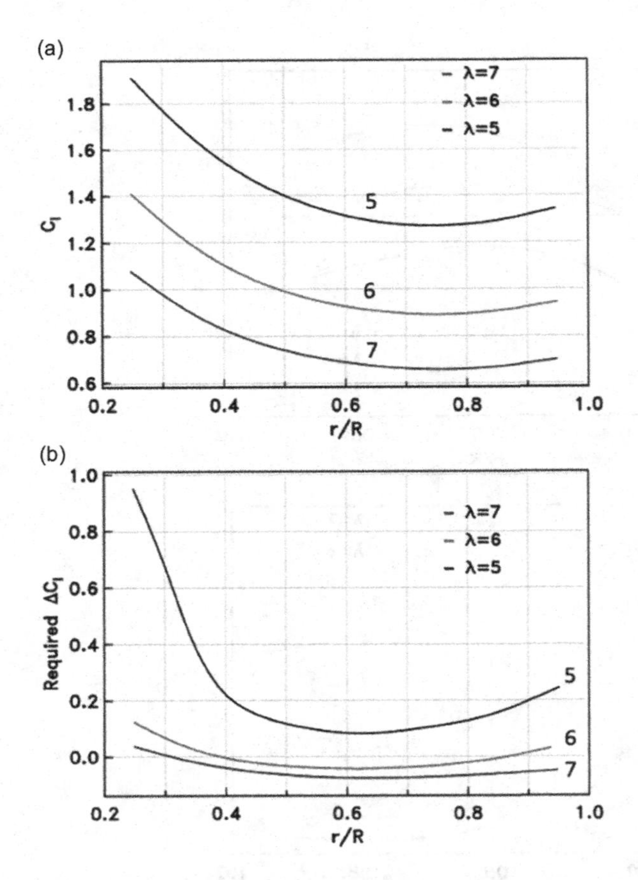

Figure 10.7 JIMP 25 rotor blade radial distribution of the lift coefficient for which the axial induction factor is the ideal one-third (a), and the change in the lift distribution from that in Figure 10.6(b) that is required to achieve the Betz limit (b) for the three tip speed ratios. Taken from Cooney et al. (2014).

factor everywhere along the rotor blade. The necessary difference between ideal radial lift distribution and the actual lift distribution is shown in Figure 10.7(b).

The change in the rotor radial lift distribution needed to produce the ideal inflow induction factor along the JIMP 25 wind turbine became the object function of a one-parameter passive flow control optimization by Williams (2014). The experimental validation was performed by Cooney (2015) and reported on by Cooney et al. (2014).

The optimization procedure followed the method presented in Section 10.2.1. Without otherwise modifying the rotor blade, the objective was to determine the shapes of trailing-edge pieces that could be added to the JIMP 25 rotor to produce the change in the radial lift distribution shown in Figure 10.7(b) that was needed to produce the ideal distribution of the axial induction factor along the rotor.

The sectional geometry of the JIMP 25 rotor is shown in Figure 10.8(a). Seven of these sectional shapes were used in the optimization procedure. These seven are

Table 10.1 Target lift change for the seven selected spanwise rotor locations.

Location	r/R	Target lift change (%)
1	0.42	35
2	0.50	33
3	0.58	31
4	0.65	28
5	0.77	26
6	0.81	25
7	0.88	17

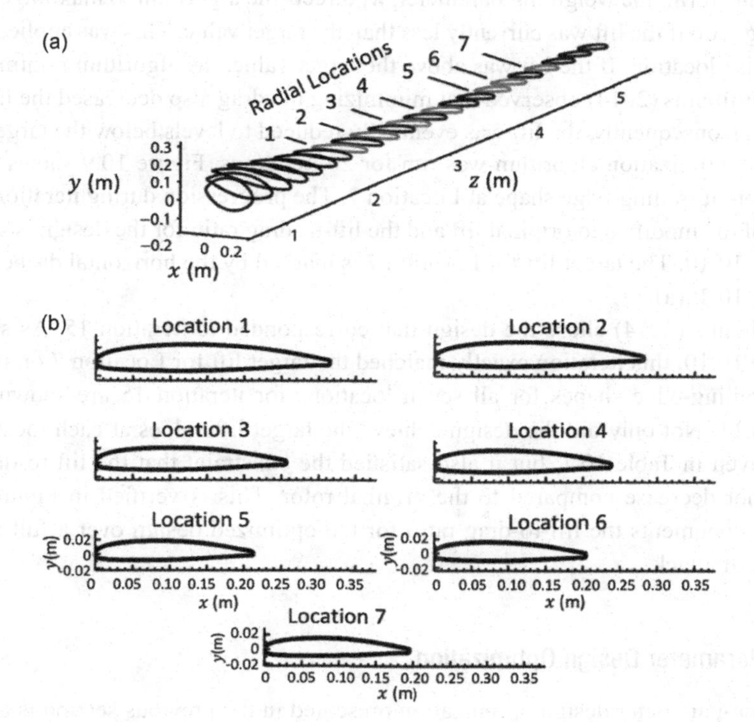

Figure 10.8 Sectional geometry of the JIMP 25 wind turbine rotor (a) and section shapes at seven radial locations that were used in the design optimization procedure (b). Taken from Williams (2014).

shown in Figure 10.8(b). They are labeled as Locations 1–7 and fall in the range $0.42 \leq r/R \leq 0.88$, where r is the local radial position. It is in this range of r/R where a majority of the power is generated by the rotor.

The design objective was to increase the lift coefficients at the locations on the JIMP 25 rotor span that would produce the ideal one-third inflow induction factor. Table 10.1 lists the target lift change for each of the seven spanwise locations. Williams (2014) placed a constraint on the design that in the process of increasing the sectional lift, the design would not increase the aerodynamic drag. This was imposed through the relation

$$\max f = (1 - w)L - wD, \tag{10.31}$$

where f is the objective function, L and D are the respective aerodynamic lift and drag, and w is a weighting parameter defined by

$$w = \begin{cases} 0.5 & \text{if } L < L_{\text{target}} \\ 1.0 & \text{otherwise} \end{cases}. \tag{10.32}$$

In this form, the weighting parameter, w, directs the algorithm to maximize the lift-to-drag ratio if the lift was currently less than the target value. This was applied at each spanwise location. If the lift was above the target value, the algorithm minimized the drag. Williams (2014) observed that minimizing the drag also decreased the lift of the design. Consequently, the lift was eventually reduced to levels below the target value.

The optimization algorithm was run for 20 iterations. Figure 10.9 shows the progression of trailing-edge shape at Location 7. The progression during iterations of the ratio of the modified to original lift and the lift-to-drag ratio for the design is shown in Figure 10.10. The target lift for Location 7 is marked by the horizontal dashed line in Figure 10.10(a).

Williams (2014) chose the design that corresponded to iteration 15. As shown in Figure 10.10, this iteration exactly matched the target lift for Location 7 on the rotor. The trailing-edge shapes for all seven locations for iteration 15 are shown in Figure 10.11. Not only did the design achieve the target lift values at each location that was given in Table 10.1, but it also satisfied the constraint that the lift-to-drag ratio does not decrease compared to the original rotor. This is verified in Figure 10.12, which documents the lift-to-drag ratio for the optimized design over a full range of angles of attack.

10.2.3 Two-Parameter Design Optimization

The one-parameter design optimization presented in the previous section is an example of a passive flow control approach. This section presents an example of a two-parameter design optimization that, in addition to changes in the geometry, includes the location of a flow control actuator.

The example again involves the rotor of a JIMP 25 wind turbine. The object in this case is to develop a design that will provide *dynamic* lift control. The analysis was performed by Williams (2014). Following the previous one-parameter optimization, the example design is focused on the rotor section shape at Location 6, $r/R = 0.81$ in Table 10.1. The original section shape is shown in Figure 10.13. The flow control actuator was a plasma actuator. This had the advantage to have its effect represented by a body force that is based on the physical model developed by Mertz and Corke (2011b). The body force corresponds to f_i in the fluid momentum equation that was given in Eq. (10.2).

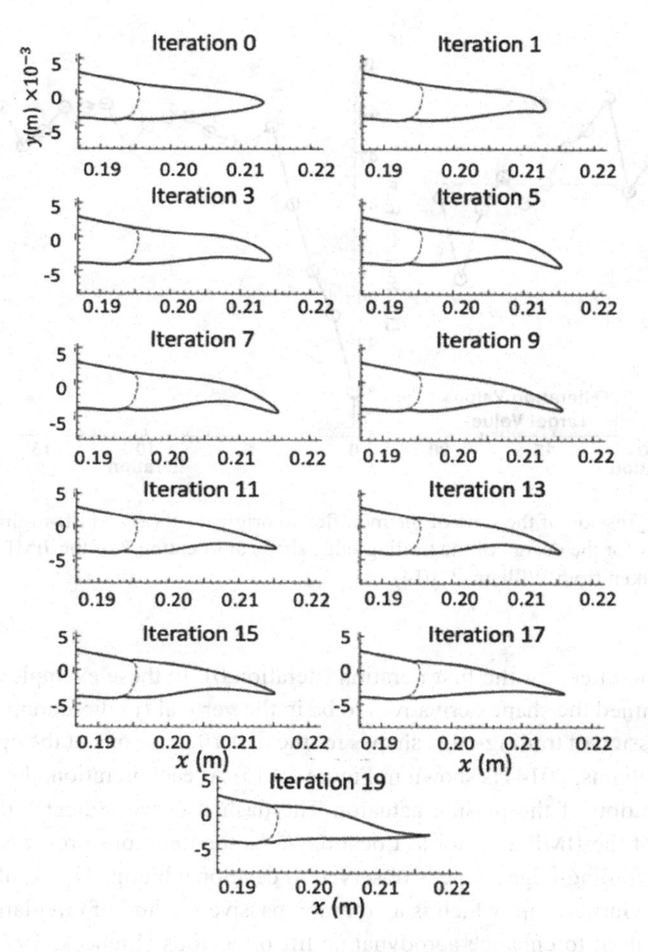

Figure 10.9 Progression of trailing-edge shapes during optimization iterations at Location 7 of the JIMP 25 wind turbine rotor. Taken from Williams (2014).

The objectives for the design were to (1) increase the passive lift, (2) decrease lift when the plasma actuator was operating, and (3) that either effect does not increase the aerodynamic drag. These objectives were represented in the following object function:

$$\max f = w_1 L_{\text{Passive}} - w_2 L_{\text{Plasma}} - w_3 D_{\text{Passive}}, \tag{10.33}$$

where f is the objective function, L_{Passive} and D_{Passive} are the respective aerodynamic lift and drag without the plasma actuator operating, and L_{Plasma} is the lift generated with the plasma actuator operating. The parameter, $w_{1,2,3}$, is a weighting put on the three terms in the object function. Williams (2014) chose $w_1 = 0.2$, $w_2 = 0.4$, and $w_3 = 0.4$.

The shape derivatives for the passive lift, plasma actuator generated lift, and passive drag were calculated at each iteration. Figure 10.14 shows the shape derivative vectors for the three respective weighted object function terms, as well as the composite

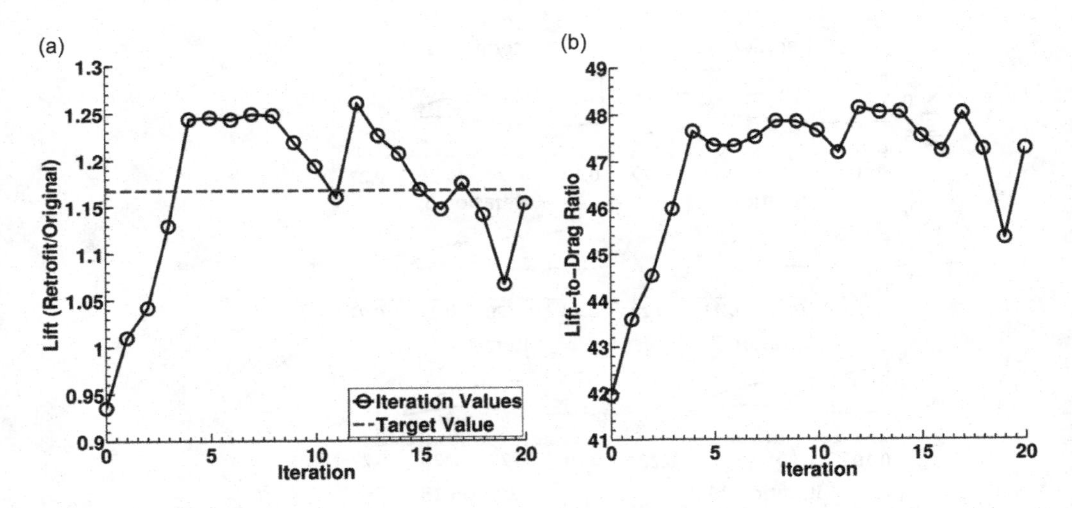

Figure 10.10 Progression of the ratio of the modified to original lift (a) and lift-to-drag ratio (b) during iterations for the design of the trailing-edge shape at Location 7 of the JIMP 25 wind turbine rotor. Taken from Williams (2014).

(average) of the three, for the first iteration (iteration 0). In these examples, Williams (2014) constrained the shape derivatives to be in the vertical (y) direction.

The progression of trailing-edge shapes for the first 10 iterations of the optimization algorithm (Williams, 2014) is shown in Figure 10.15. At each iteration, the blue curve marks the location of the plasma actuator. The dashed curve indicates the original trailing edge of the JIMP 25 rotor at Location 7. As the iterations progress, the lower surface of the trailing-edge shape is observed to develop a bump. The resulting bump is similar to a Gurney flap, which is a common passive method of circulation control that has been used to enhance aerodynamic lift on airfoils (Liebeck, 1978). For the greatest effect, a Gurney flap is placed within the last 15 percent of airfoil's chord, and to minimize drag, its height should be less than the local boundary-layer thickness. The location and height of the developing bump appears to coincide with the Gurney flap norms.

A passive Gurney flap causes the local boundary layer to separate over the flap. This forms a recirculating separation bubble (see Chapter 4) that is located just downstream of the flap. The separation bubble produces a low static pressure region if near the trailing edge and deflects the trailing-edge streamline similar to increasing the airfoil camber. The result is increased lift.

The model for the plasma actuator used in the simulation oriented the generated body force in the flow direction, toward the trailing edge. As discussed in Chapter 4, both steady and unsteady body forces are effective in controlling separated flows. The shape of the bump and the location of the plasma actuator that emerges from the design results from the object function that seeks to maximize the lift control.

Using the Gurney flap analogy, lift control can come from controlling the size and strength of the separation bubble that forms downstream of the bump. This is

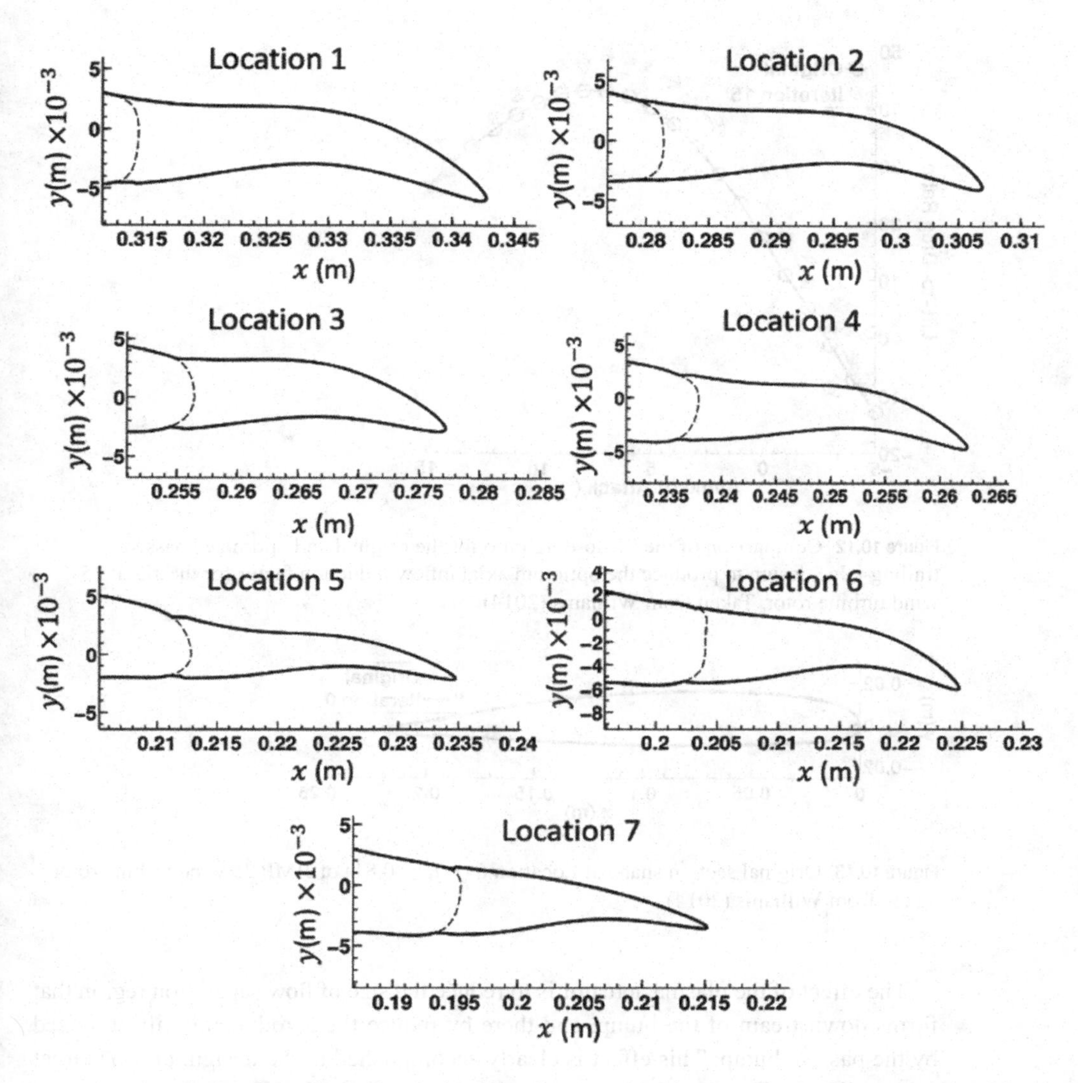

Figure 10.11 Trailing-edge designs for all seven spanwise locations based on the target lift change needed to produce the optimum axial inflow induction factor for the JIMP 25 wind turbine rotor. Taken from Williams (2014).

documented in Figure 10.16, which shows the passive lift and the plasma actuator generated lift compared to the original passive lift for the section shape at Location 6, $r/R = 0.81$, of the JIMP 25 rotor. The horizontal dashed line corresponds to the passive lift target in which the aerodynamic lift on that section of the rotor increased by 25 percent. As indicated in Table 10.1, this value for the passive lift increase at rotor spanwise Location 6 corresponds to the amount that would optimize the inflow induction factor. As shown in Figure 10.16, this target value was reached at iteration 3.

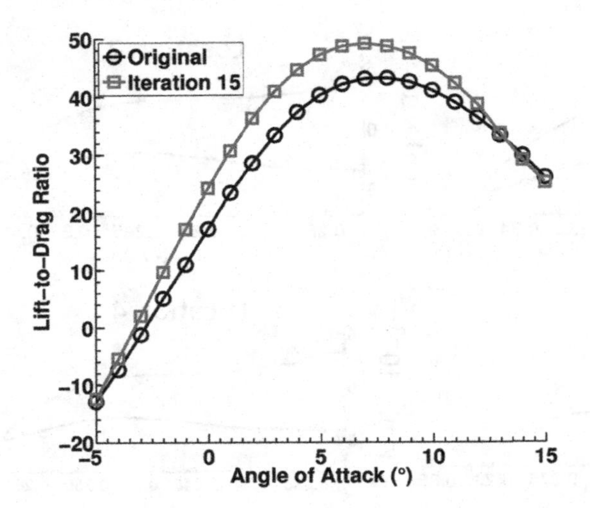

Figure 10.12 Comparison of the lift-to-drag ratio for the original and optimized passive trailing-edge design to produce the optimum axial inflow induction factor for the JIMP 25 wind turbine rotor. Taken from Williams (2014).

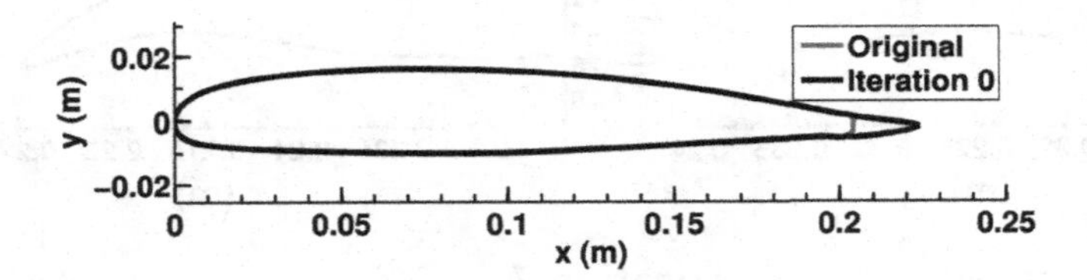

Figure 10.13 Original section shape at Location 6 ($r/R = 0.81$) of JIMP 25 wind turbine rotor. Taken from Williams (2014).

The effect of the plasma actuator is to reduce the size of flow separation region that forms downstream of the bump, and there by reduce the aerodynamic lift produced by the passive bump. This effect is clearly accomplished in the design. For the target passive lift condition reached at iteration 3, the change in the lift between the plasma actuator OFF and ON was $\Delta L = 0.5$. However, for a trailing-edge shape corresponding to iteration 7, $\Delta L = 0.68$ or there is approximately a 50 percent change in the lift.

The object function for the two-parameter optimization that was given in Eq. (10.33) equally weighted the impact of the design on the aerodynamic drag and the plasma lift control. Owing to their importance, these were each weighted twice as high as that of the passive lift generation. To verify that this was realized in the design, Figure 10.17 shows the lift-to-drag ratio as a function of the rotor section angle of attack for the original Location 6, $r/R = 0.81$ section shape, and that of the optimization iteration 3 geometry. This documented that the objective was met, where the lift-to-drag ratio of the passive optimized geometry exceeded that of the original geometry up to the stall angle of attack ($\sim 7°$). The designed loss of lift with the plasma actuator

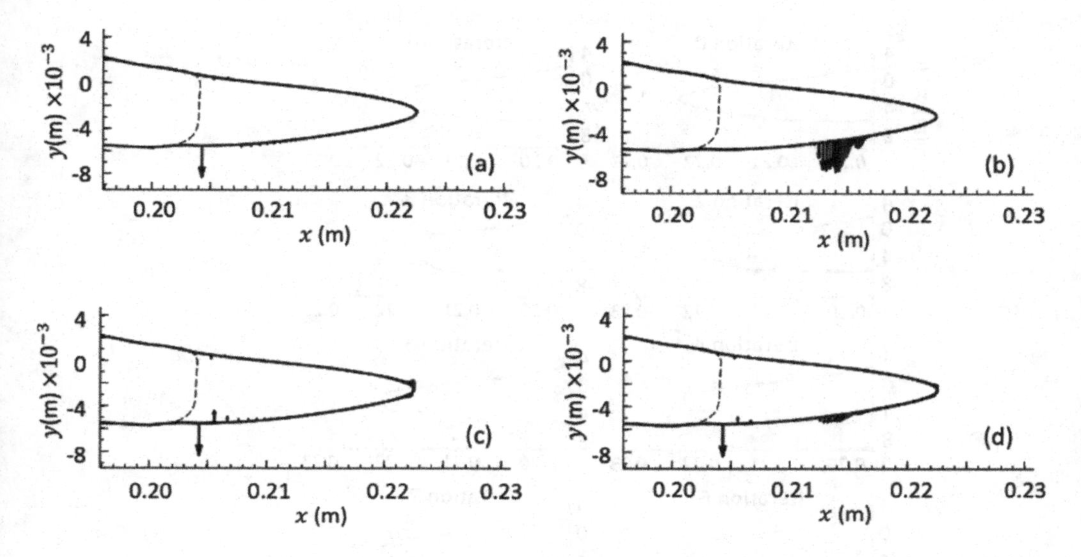

Figure 10.14 Shape derivative vectors for iteration 0 for passive lift (a), plasma actuator lift (b), passive drag (c), and overall composite of all three (d) for Location 6 ($r/R = 0.81$) of the JIMP 25 wind turbine rotor. Taken from Williams (2014).

operating produced the expected decrease in the lift-to-drag ratio that is observed in the figure. Thus, the final design achieved the desired passive lift enhancement as well as the plasma actuator control authority without a concomitant increase in drag.

The two examples illustrate an objective approach to "design for flow control" that can be easily implemented and intended to break the norm where flow control is applied to an existing geometry and/or is based on ad hoc approaches. The methodology is based on an adjoint formulation of the Navier–Stokes equations that seeks optimum geometric designs that enhance the sensitivity of the flow field to controlled actuator input. Although the one example involved a plasma actuator, the methodology is equally suitable for other types of flow actuators.

10.3 Summary

This chapter presented an approach to active flow control in which geometric changes are made to an existing shape in order to introduce a local flow instability that can be more easily manipulated compared to the original primal flow. Ad hoc approaches included adding curvature that, with surface blowing, can exploit a Coanda effect, or adding separation ramps to exploit methods of flow separation control that were presented in Chapter 4.

Such ad hoc approaches may require considerable trial and error to achieve a desired outcome. This motivated the introduction of a more objective approach based on an adjoint formulation of the governing flow field equations. The adjoint formulation is designed to reveal sensitivity of a flow field to geometric changes. The impact

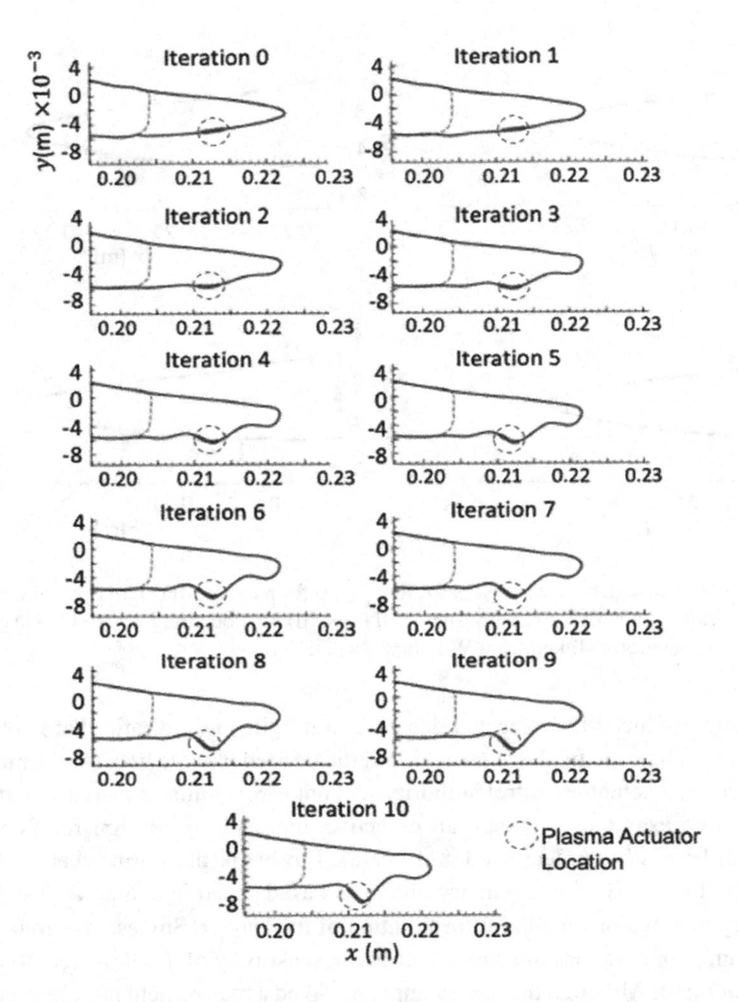

Figure 10.15 Progression of trailing-edge shapes during optimization iterations at Location 6 ($r/R = 0.81$) of the JIMP 25 wind turbine rotor. Circles indicate the plasma actuator location. Taken from Williams (2014).

of these changes is measured against an objective (function) that defines the level of "performance" one is seeking in the design. In aerodynamic applications, this typically involves controlling lift and drag.

A formulation of the adjoint form of the Navier–Stokes equations used in an optimization of an existing geometry to enhance its flow-control sensitivity was presented in Section 10.2.1. This was demonstrated in two examples involving the design of a horizontal wind turbine rotor. One example was a one-parameter shape optimization of the rotor to maximize the coefficient of power. The other example was a two-parameter optimization of the rotor section shape and flow actuator placement to maximize the dynamic lift *control* of the rotor that represents an ultimate application of "flow control by design."

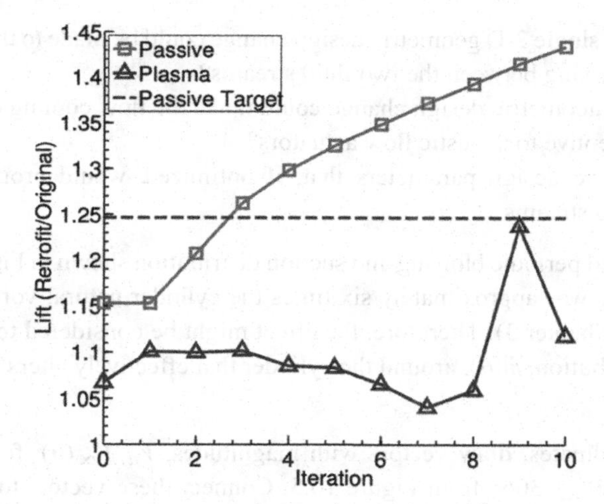

Figure 10.16 Progression of the passively generated lift and the plasma actuator generated lift compared to the original passive lift during iterations of the design of the trailing-edge shape at Location 6 ($r/R = 0.81$) of the JIMP 25 wind turbine rotor. Taken from Williams (2014).

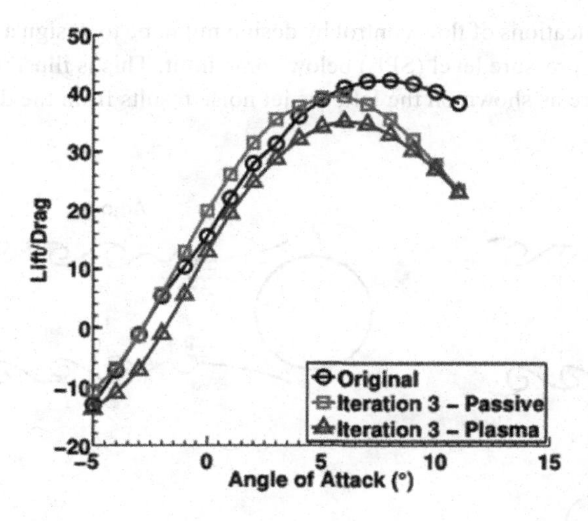

Figure 10.17 Comparison of lift-to-drag ratio versus angle of attack for the original JIMP 25 rotor location 7 ($r/R = 0.81$) to that of the optimization iteration 3 geometry. Taken from Williams (2014).

Problems

10.1 Consider the shear layer instability that forms downstream of a splitter plate that separates two fluid streams of different velocities like those shown in Chapter 5.

1. What single two-dimensional (2-D) geometric design change could be made to the splitter plate to enhance mixing between the two fluid streams?

2. Conversely, what single 2-D geometric design change could be made to the splitter plate to reduce mixing between the two fluid streams?
3. What single 2-D geometric design change could make the flow coming off of the splitter more receptive to acoustic flow actuators?
4. List two geometric design parameters that, if optimized, would promote 3-D mixing of the two streams.

10.2 In the optimized periodic blowing and suction distribution shown in Figure 10.5 the pulsing frequency was approximately six times the cylinder natural vortex shedding frequency (see Chapter 3). Therefore, the effect might be considered to produce a *net* mass flow distribution, $\dot{m}(\theta)$, around the cylinder that effectively alters its shape. Assuming this:

1. In cylinder coordinates, draw vectors with magnitudes, $V_a/V_\infty(\theta)$, for the full range of $54° \leq \theta \leq 306°$ from Figure 10.5. Connect these vectors to form an *effective* shape of the cylinder as a result of the mean mass flow distribution.
2. Starting with a circular cylinder, what do you imagine the shape would become with a single parameter design optimization to minimize drag?
3. Does the effective shape in Part 1 resemble what you imagine in Part 2?

10.3 One of the applications of flow control by design might be to design a jet nozzle that produces a sound pressure level (SPL) below some limit. This is illustrated in the following figure, where as shown on the left, the jet noise results from the developing shear layer instability.

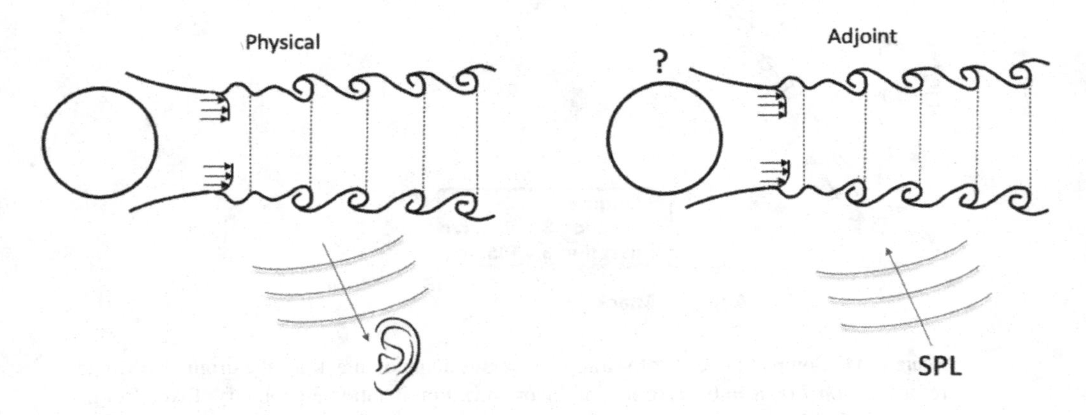

The jet noise control will involve controlling the shear layer instability, where there are a number of methods involving passive shapes of the jet nozzle exit that are discussed in Chapter 5. The process to obtain an optimized design would involve an adjoint formulation of the governing flow equations. The SPL is the metric of merit in the design.

1. List all of the possible passive jet nozzle shapes that can be used to modify the shear layer development (growth and frequency).

2. For each of the candidate nozzle shapes, list the possible design parameters that might be varied in a multiparameter optimization.
3. For each of the candidate nozzle shapes, list possible negative impact on the jet performance. Examples may include loss of thrust, vectored thrust, and cost of manufacturing.
4. Choose one starting candidate shape that has at least two design parameters.
5. Write an objective function with weighting that lists all the objectives of the design. See Eq. (10.33) as an example. Justify your weighting.
6. Starting with your basic candidate shape, what do you imagine the final optimized shape will be?

10.4 Repeat Problem 3 for cases with *active* jet shear layer control.

References

Abbassi, M. R., W. J. Baars, N. Hutchins, and I. Marusic. Skin-friction drag reduction in a high-Reynolds number turbulent boundary layer via real-time control of large-scale structures. *Int. J. Heat Fluid Flow*, 67:30–41, 2017.

Adamson, Jr., T. C., and A. F. Messiter. Analysis of two-dimensional interactions between shock waves and boundary layers. *Annu. Rev. Fluid Mech.*, 12:103–138, 1980.

Adrian, R. Scattering particle characteristics and their effect on pulsed laser measurements of fluid flow: Speckle velocimetry vs particle image velocimetry. *Appl. Opt.*, 23:1690–1691, 1984.

Adrian, R. Hairpin vortex generation in wall turbulence. *Phys. Fluids*, 19:041301, 2007.

Adrian, R., and C.-S. Yao. Development of pulsed laser velocimetry (PLV) for measurement of turbulent flow. *Proc. Symp. Turbul.*, eds. X. Reed, G. Patterson, and J. Zakin, Rolla: University of Missouri-Rolla, pp. 170–186, 1984.

Adrian, R. J. Particle-imaging techniques for experimental fluid mechanics. *Annu. Rev. Fluid Mech.*, 23:261–304, 1991.

Adrian, R. J., C. D. Meinhart, and C. D. Tomkins. Vortex organization in the outer region of the turbulent boundary layer. *J. Fluid Mech.*, 422:1–53, 2000.

Ahmed, A., and B. Bays-Muchmore. Transverse flow over a wavy cylinder. *Phys. Fluids A: Fluid Dyn.*, 4:1959–1967, 1992.

Ahmed, A., M. Kahn, and B. Bays-Muchmore. Experimental investigation of a three-dimensional bluff-body wake. *AIAA J.*, 31:559–563, 1993.

Ahmed, S. R. Influence of base slant on the wake structure and drag of road vehicles. *Trans. ASME, J. Fluids Eng.*, 105:429–434, 1984a.

Ahmed, S. R. Some salient features of the time-averaged ground vehicle wake. *SAE Technical Paper Series*, 1984b *Paper No. 840300.*

Aider, J.-L. Drag and lift reduction of a 3D bluff-body using active vortex generators. *Exp. Fluids*, 48(5), 771–789, 2010.

Amitay, M., A. E. Washburn, S. G. Anders, and D. Parekh. Active flow control on the stingray uninhabited air vehicle: Transient behavior. *AIAA J.*, 42(11): 2205–2215, 2004.

Anders, J. B. Large Eddy breakup devices as low Reynolds number airfoils. *NASA SAE Technical Paper 861769*, 1986.

Andreopoulos, J., and K. C. Muck. Some new aspects of the shock-wave boundary layer interaction in compression ramp corner. *J. Fluid Mech.*, 180: 405–428 1987.

Antonia, R. Conditional sampled measurements near the outer edge of a turbulent boundary layer. *J. Fluid Mech.*, 56(1):1–18, 1972.

Arnal, D. Description and prediction of transition in two-dimensional incompressible flow. *AGARD Report 709, Special course on stability and transition of laminar flow*, 1984.

Arnal, D. Boundary layer transition: Prediction, application to drag reduction. *AGARD Report 786, Special course on skin friction drag reduction*, 1992.

Arnal, D., and J. Juillen. Three-dimensional transition studies at ONERA/CERT. *AIAA Paper 87-1335*, 1987.

Arndt, A., T. C. Corke, E. H. Matlis, and M. Semper. Controlled stationary/traveling cross-flow mode interaction in a Mach 6.0 boundary layer. *J. Fluid Mech.*, 887: 1–25, 2020.

Assi, G., P. Bearman, and N. Kitney. Low drag solutions for suppressing VIV of circular cylinders. *J. Fluids Struct.*, 25:666–675, 2009.

Aubry, N., P. Holmes, J. Lumley, and E. Stone. The dynamics of coherent structures in the wall region of a turbulent boundary layer. *J. Fluid Mech.*, 192:115–173, 1988.

Bacher, E. V., and C. R. Smith. A combined visualization anemometry study of the turbulent drag reduction mechanisms of triangular micro-groove surface modifications. *AIAA Paper 85-0548*, 1985.

Bai, H., Y. Zhou, W. Zhang, et al. Active control of a turbulent boundary layer based on local surface perturbation. *J. Fluid Mech.*, 750:316–354, 2014.

Balakumar, P. Stability of supersonic boundary layers on a cone at an angle of attack. *39th Fluid Dyn. Conf. Exhib., AIAA Paper 2009-3555*, 2009.

Banerjee, N., and P. Jayakumar. Compliant materials for drag reduction of high-speed submerged bodies. *Defense Sci. J.*, 55(1):37–42, 2005.

Barn, L., J. M. Ross, and H. T. Nagamatstu. Passive shock-wave boundary-layer interaction control for transonic airfoil drag reduction. *AIAA Paper 83-137*, 1983.

Baron, A., and M. Quadrio. Turbulent drag reduction by spanwise wall oscillations. *Appl. Sci. Res.*, 55:311–326, 1996.

Bearman, P. The effect of base bleed on the flow behind a two-dimensional model with a blunt trailing edge. *Aeron. Quart.*, XVIII: 207–224, 1967.

Bearman, P. Review: Bluff body flows applicable to vehicle aerodynamics. *J. Fluids Eng.*, 102:265–274, 1980.

Bearman, P., and J. Harvey. Control of circular cylinder flow by the use of dimples. *AIAA J.*, 10:1753–1756, 1993.

Bearman, P., and J. Owen. Reduction of bluff body drag and suppression of vortex shedding by the introduction of wavy separation lines. *J. Fluids Struct.*, 12:123–130, 1998.

Beaudoin, J.-F. Drag and lift reduction of a 3D bluff body using flaps. *Exp. Fluids*, 44:491–501, 2008.

Bechert, D. Sound absorption caused by vorticity shedding demonstrated with a jet flow. *J. Sound Vib.*, 70:389–405, 1980.

Bechert, D. Calibration of preston tube. *AIAA J.*, 34(1):205–206, 1995.

Bechert, D. W., M. Bruse, W. Hage, et al. Experiments on drag-reducing surfaces and their optimization with adjustable geometry. *J. Fluid Mech.*, 338:59–87, 1997.

BenGadri, R. A. Rabehi, F. Massines, and P. Segur. Numerical modelling of atmospheric pressure low-frequency glow discharge between insulated electrodes. *Proc. XIIth ESCAMPIG*, pp. 228–229, Netherlands August 23-26, 1994.

Benney, D., and C. Lin. On the secondary motion induced by oscillations in a shear flow. *Phys. Fluids*, 3:656–657, 1960.

Benney, D. J. Finite amplitude effects in an unstable laminar boundary layer. *Phys. Fluids*, 7:319–326, 1964.

Beresh, S. J., N. T. Clemens, and D. S. Dolling. Relationship between upstream turbulent boundary-layer velocity fluctuations and separation shock unsteadiness. *AIAA J.*, 40(12): 2412–2422, 2002.

Berger, E. Suppression of vortex shedding and turbulence behind oscillating cylinders. *Phys. Fluids Suppl.*, 10:191–193, 1965.

Bernal, L. The coherent structure of turbulent mixing layers: I. Similarity of the primary vortex structure: II. Secondary streamwise vortex structure. Ph.D. Thesis, California Institute of Technology, Pasadena, CA, 1981.

Betz, A. Das maximum der theoretisch moglichen ausnutzung des windes durch windmotoren. *Zeitschrift fur das gesamte Turbinenwesen*, 26:307–309, 1920.

Binder, R. *Advanced fluid dynamics and fluid machinery*. New York: Prentice-Hall, 1951.

Bippes, H. Experiments on transition in three-dimensional accelerated boundary layer flow. *Roy. Aeronaut. Soc. Conf. Boundary Layer Transition Control*, Cambridge, UK, 1991.

Bisek, N. J., D. P. Rizzetta, and J. Poggie. Plasma control of a turbulent shock boundary-layer interaction. *AIAA J.*, 51(8):1789–1804, 2013.

Bishop, R., and A. Hansan. The lift and drag forces on a circular cylinder oscillating in a flowing fluid. *Proc. Roy. Soc. A*, 277:51–75, 1963.

Black, T. The structure of wall turbulence. *Proc. Heat Transfer Fluid Mech. Inst.*, 366, 1966.

Blackwelder, R., and H. Eckelmann. Streamwise vortices associated with the bursting phenomenon. *J. Fluid Mech.*, 94(3):577–594, 1979.

Blackwelder, R., and R. Kaplan. On the wall structure of the turbulent boundary layer. *J. Fluid Mech.*, 76(1):89–112, 1976.

Blackwelder, R., and S. Kovasznay. Time scales and correlations in a turbulent boundary layer. *Phys. Fluids*, 15:15445, 1972.

Blackwelder, R., and H. Woo. Pressure perturbation of a turbulent boundary layer. *Phys. Fluids*, 17:515–519, 1974.

Blick, E. F., and R. R. Walters. Turbulent boundary layer characteristics of compliant surfaces. *J. Aircraft*, 5:11–16, 1968.

Bloor, M. The transition to turbulence in the wake of a circular cylinder. *J. Fluid Mech.*, 19:290–304, 1964.

Bloor, M., and J. Gerrard. Measurements on turbulent vortices in a cylinder wake. *Proc. Roy. Soc. Lond. A*, 294:319–342, 1966.

Bogard, D., and W. Tiederman. Burst detection with single point velocity measurements. *J. Fluid Mech.*, 162:389–413, 1986.

Bohning, R., and J. Zierep. The normal shock at a curved wall in the viscous case. *Symp. Transsonicum II*, pp. 237–243. Berlin: Springer-Verlag, 1976.

Bohnker, J., and K. Breuer. Control of separated flow using actuated compliant membrane wings. AIAA J., 57(9):3801–3811, 2019.

Boiko, A., A. Dogval, and V. Kozlov. Nonlinear interactions between perturbations in transition to turbulence in the zone of laminar boundary-layer separation. *J. Appl. Phys.*, 3:46–52, 1989.

Boin, J.-P., and J.-C. Robinet. Three-dimensional unsteady laminar shock- wave/boundary layer interaction. *NATO RTO-AVT-111 101/1010*, 2004.

Boin, J.-P., J.-C. Robinet, C. Corre, and H. Deniau. 3D steady and unsteady bifurcations in shock-wave/laminar boundary layer interaction: A numerical study. *Theor. Comput. Fluid Dyn.*, 20(3): 163–180, 2006.

Boucinha, V., R. Weber, and A. Kourta. Drag reduction of a 3D bluff body using plasma actuators. *Int. J. Aerodyn.*, 1(3/4):262–280, 2011.

Bowles, P. Wind tunnel experiments on the effect of compressibility on the attributes of dynamic stall. Ph.D. Thesis, University of Notre Dame, Notre Dame, IN, 2012.

Bradshaw, P. The turbulent structure of equilibrium boundary layers. *J. Fluid Mech.*, 29(4):625–645, 1967.

Braslow, A. L. A history of suction-type laminar-flow control with emphasis on flight research. *NASA History Division Monographs in Aerospace History*. NASA, Washington, DC, 1999.

Brayton, D., and W. Goether. New velocity measuring technique using dual scatter laser Doppler shift. *Arnold Engineering Development Center Air Force Systems Command*. Arnold Air Force Station, Tennessee, 1970.

Breidenthal, R. Response of plane shear layers and wakes to strong three-dimensional disturbances. *Phys. Fluids*, 23:1929–1934, 1980.

Brezillon, J., and R. P. Dwight. Aerodynamic shape optimization using the discrete adjoint of the navier-stokes equations: Applications toward complex 3D configurations. Paper No. 36., German Aerospace Center (DLR), Institute of Aerodynamics and Flow Technology, 2009.

Briley, W. R. A numerical study of laminar separation bubbles using the Navier-Stokes equations. *J. Fluid Mech.*, 47:713–736, 1971.

Browand, F. An experimental investigation of the instability of an incompressible separated shear layer. *J. Fluid Mech.*, 26:281–307, 1966.

Browand, F., and P. Weidman. Large scales in the developing mixing layer. *J. Fluid Mech.*, 76:127–144, 1976.

Brown, G., and A. Roshko. On density effects and large structure in turbulent mixing layers. *J. Fluid Mech.*, 64:775–816, 1974.

Brunn, A., E. Wassen, D. Sperbert, W. Nitsche, and F. Thiele. Active drag control for a generic car model. *Active Flow Control (NNFM, Vol. 95)*, King, R. ed., Berlin: Springer Nature, pp. 247–259, 2007.

Buchheim, R., K. Deutenbach, and H. Luckoff. Necessity and premises for reducing the aerodynamic drag of future passenger cars. *SAE Paper 810185*, Society of Automotive Engineers, 1976.

Bur, R., B. Corbel, and J. Delery. Study of passive control in a transonic shock wave/boundary-layer interaction. *AIAA J.*, 36(3):394–400, 1998.

Bushnell, D. Recent turbulent drag reduction at langley research center. *NASA TM-78688*, 1978.

Canton, J., R. Orlu, C. Chin, and P. Schlatter. Reynolds number dependence of large-scale friction control in turbulent channel flow. *Phys. Rev. Fluids*, 1:081501, 2016.

Cardell, G. Flow past a circular cylinder with a permeable splitter plate. Ph.D. Thesis, California Institute of Technology, Pasadena, CA, 1993.

Carnarius, A., F. Thiele, E. Ozkaya, A.Nemili, and N. Gauger. Optimal control of unsteady flows using a discrete and a continuous adjoint approach. *25th Syst. Model. Optim. (CSMO)*, pp. 318–327. Berlin, Germany, September 2011.

Casper, K. M., S. J. Beresh, J. F. Henfling, et al. Hypersonic wind-tunnel measurements of boundary-layer transition on a slender cone. *AIAA J.*, 54(4):1250–1263, 2016.

Castro, P. Wake characteristics of two-dimensional perforated plates normal to an air-stream. *J. Fluid Mech.*, 46:599–609, 1971.

Cattafesta, L., S. Garg, M. Choudhari, and F. Li. Active control of flow-induced cavity resonance. *AIAA Fluid Dyn. Conf., Paper AIAA 1997-1804*, 1997.

Cattafesta, L., S. Garg, and D. Shukla. Development of piezoelectric actuators for active flow control. *AIAA J.*, 39:1562–1568, 2001.

Cermak, J. E. Aerodynamics of buildings. *Annu. Rev. Fluid Mech.*, 8:75–106, 1976.

Chapman, D. R., D. M. Kuehn, and H. K. Larson. Investigation of separated flows in supersonic and subsonic streams with emphasis on the effect of transition. *NACA Report 1356*, 1958.

Chin, C., J. Monty, N. Hutchins, et al. Simulation of a large-eddy-breakup-device (LEBU) in a moderate Reynolds number turbulent boundary layer. *Technical Report, Int. Symp. Turbul. Shear Flow Phenomena*, Melbourne, Australia, June 30–July 3, 2015.

Chin, C., R. Orlu, and J. Monty. Simulation of a large-eddy-break-up device (LEBU) in a moderate Reynolds number turbulent boundary layer. *Flow Turbul. Combust*, 98:445–460, 2017.

Choi, H., P. Moin, and J. Kim. Active turbulence control for drag reduction in wall-bounded flows. *J. Fluid Mech.*, 262:75–110, 1994.

Choi, J., X. Chun-Xiao, and H. Sung. Drag reduction by spanwise wall oscillation in wall-bounded turbulent flows. *AIAA J.*, 40(5):842–850, 2002.

Choi., K.-S., and B. R. Clayton. The mechanism of turbulent drag reduction with wall oscillation. *Int. J. Heat Fluid Flow*, 22:1–9, 2001.

Choi., K.-S., and J.-H. Kim. Plasma virtual roughness elements for cross-flow instability control. *Exp. Fluids*, 59:159–174, 2018.

Choi., K.-S., X. Yang, B. Clayton, et al. Turbulent drag reduction using compliant surfaces. *Proc. Roy. Soc. Lond. A*, 453:2229–2240, 1997.

Choi, K.-S., T. Jukes, and R. Whalley. Turbulent boundary layer control with plasma actuators. *Phil. Trans. Roy. Soc.*, 369:1443–1458, 2011.

Chomaz, J., P. Huerre, and L. Redekopp. Models of hydrodynamic resonances in separated shear flows. *Proc. 6th Symp. Turbul. Shear Flows*, Toulouse, France, pp. 3.2.1–3.2.6, 1987.

Chomaz, J., P. Huerre, and L. Redekopp. Bifurcations to local and global modes in spatially-developing flows. *Phys. Rev. Lett.*, 60:25–28, 1988.

Clauser, F. The turbulent boundary layer. *Adv. Appl. Mech.*, 4:1–51, 1956.

Clemens, N., and M. Mungal. Two- and three-dimensional effects in the supersonic mixing layer. *AIAA J.*, 30(4):973–981, 1992.

Clemens, N. T., and V. Narayanaswamy. Low-frequency unsteadiness of shock wave turbulent boundary layer interactions. *Annu. Rev. Fluid Mech.*, 46:469–492, 2014.

Coanda, H. Device for deflecting a stream of elastic fluid projected into an elastic fluid. *US Patent 2,052,869*, 1936a.

Coanda, H. Lifting device Coanda effect. *US Patent 3,261,162*, 1936b.

Cohen, J., and I. Wygnanski. The evolution of instabilities in the axisymmetric jet: Part 1: The linear growth of disturbances near the nozzle. *J. Fluid Mech.*, 178:191–219, 1987.

Coles, D. The law of the wake in the turbulent boundary layer. *J. Fluid Mech.*, 1:191–226, 1958.

Coles, D. E., and E. A. Hirst. Computation of turbulent boundary layers-1968. *Proc. AFOSR-IFP Stanford Conf.*, 1969.

Comte-Bellot, G. Hot-wire anemometry. Handbook of Fluid Dynamics, ed., R. W. Johnson. New York: CRC Press, 1998.

Cooney, J. Increasing power generation in horizontal axis wind turbines using optimized flow control. Ph.D. Thesis, University of Notre Dame, Notre Dame, IN 2015.

Cooney, J., T. Williams, and T. Corke. Improve power coefficient of horizontal axis wind turbines using optimized lift control. *AIAA-2014-1218*, 2014.

Corino, E. R., and R. S. Brodkey. A visual investigation of the wall region in turbulent flow. *J. Fluid Mech.*, 1:1–30, 1969.

Corke, T. A new view on origin, role and manipulation of large scales in turbulent boundary layers. Technical Report, Ph.D. Thesis, Illinois Institute of Technology, Chicago, IL, 1981a.

Corke, T., and S. Kusek. Resonance in axisymmetric jets with controlled helical-mode input. *J. Fluid Mech.*, 249:307–336, 1993.

Corke, T., and R. Mangano. Resonant growth of three-dimensional modes in transitioning Blasius boundary layers. *J. Fluid Mech.*, 209:93–150, 1989.

Corke, T., and F. Thomas. Dynamic stall in pitching airfoils: Aerodynamic damping and compressibility effects. *Annu. Rev. Fluid Mech.*, 47:479–505, 2011.

Corke, T., and F. Thomas. Dynamic stall in pitching airfoils: Aerodynamic damping and compressibility effects. *Ann. Rev. Fluid Mech.*, 47:479–505, 2015.

Corke, T., D. Koga, R. Drubka, and H. Nagib. A new technique for introducing controlled sheets of smoke streaklines in wind tunnels. *CIASF '77; Int. Congress Instrum. Aerospace Simul. Facilities, Record (A79-15651 04-35).* New York, Institute of Electrical and Electronics Engineers, pp. 74–80, 1977.

Corke, T., H. Nagib, and Y. Guezennec. A new view on origin, role and manipulation of large scales in turbulent boundary layers. *NASA CR 165861*, 1982.

Corke, T., F. Shakib, and H. Nagib. Mode selection and resonant phase locking in unstable axisymmetric jets. *J. Fluid Mech.*, 223:253–311, 1991.

Corke, T., J. Krull, and M. Ghassemi. Three-dimensional-mode resonance in far wakes. *J. Fluid Mech.*, 239:99–132, 1992.

Corke, T., M. Glauser, and G. Berkooz. Utilizing low-dimensional dynamical systems models to guide control experiments. *Appl. Mech. Rev.*, 47(6):S132–S138, 1994.

Corke, T., M. Post, and D. Orlov. SDBD plasma enhanced aerodynamics: Concepts, optimization and applications. *Prog. Aerosp. Sci.*, 43:193–217, 2007a.

Corke, T., M. Post, and D. Orlov. Single dielectric barrier discharge plasma enhanced aerodynamics: Physics, modeling and applications. *Exp. Fluids*, 46:1–26, 2009.

Corke, T., C. L. Enloe, and S. Wilkinson. Dielectric barrier discharge plasma actuators for flow control. *Annu. Rev. Fluid Mech.*, 42:505–529, 2010a.

Corke, T., F. Thomas, A. Duong, and R. McGowan. Turbulent boundary layer drag reduction through plasma streak transient growth instability control. *Invited Oral, AIAA SciTech*, 2017.

Corke, T., A. Arndt, E. Matlis, and M. Semper. Control of stationary cross-flow modes in a Mach 6 boundary layer using patterned roughness. *J. Fluid Mech.*, 856:822–849, 2018.

Corke, T. C. A new view on origin, role and manipulation of large scales in turbulent boundary layers. Ph.D. Thesis, Illinois Institute of Technology, Chicago, IL, 1981b.

Corke, T. C. Three-dimensional mode growth in boundary layers with tuned and detuned subharmonic resonance. *Phil. Trans. Roy. Soc. Lond. A*, 352:453–471, 1995.

Corke, T. C. Plasma actuator array development for cross-flow instability control. *Internal Report, UND-1-01*, 2001.

Corke, T. C., and S. Gruber. Resonant growth of three-dimensional modes in Falkner-Skan boundary layers with adverse pressure gradients. *J. Fluid Mech.*, 320:211–233, 1996.

Corke, T. C., and K. F. Knasiak. Stationary traveling cross-flow mode interactions on a rotating disk. *J. Fluid Mech.*, 355:285–315, 1998.

Corke, T. C., and R. C. Nelson. *Wind energy design*, first ed. CRC Press, Taylor & Francis, New York, 2020.

Corke, T. C., D. A. Cavalieri, and E. Matlis. Boundary-layer instability on sharp cone at Mach 3.5 with controlled input. *AIAA J.*, 40:1015–1018, 2002.

Corke, T. C., E. H. Matlis, and H. Othman. Transition to turbulence in rotating-disk boundary layers-convective and absolute instabilities. *J. Eng. Math.*, 57:253–272, 2007b.

Corke, T. C., C. L. Enloe, and S. P. Wilkinson. Dielectric barrier discharge plasma actuators for flow control. *Ann. Rev. Fluid Mech.*, 42:505–529, 2010b.

Corrsin, S., and A. Kistler. The free stream boundaries of turbulent flows. *Technical Report 3133, NACA Technical Note*, 1954.

Coustols, E., and J. Cousteix. Performance of riblets in the supersonic regime. *AIAA J.*, 32(2): 431–433, 1994.

Craik, A. *Wave interactions and fluid flows*. Cambridge: Cambridge University Press, 1985.

Craik, A. D. D. Nonlinear resonant instability in boundary layers. *J. Fluid Mech.*, 50:393–413, 1971.

Crow, S., and F. Champagne. Orderly structure in jet turbulence. *J. Fluid Mech.*, 48:547–591, 1971.

Culley, D. Variable frequency diverter actuation for flow control. *AIAA Flow Control Conf., Paper AIAA 2006-3034*, 2006.

Dagenhart, J., W. Saric, M. Mousseux, and J. Stack. Crossflow-vortex instability and transition on a 45-degree swept wing. *AIAA Paper 89-1892*, 1989.

Dallmann, U. Three-dimensional vortex structures and vorticity topology. Fluid Dyn. Res., 3(1–4):183–189, 1988.

Darekar, R., and S. Sherwin. Flow past a bluff body with a wavy stagnation face. *J. Fluids Struct.*, 15:587–596, 2001a.

Darekar, R., and S. Sherwin. Flow past a square-section cylinder with a wavy stagnation face. *J. Fluid Mech.*, 426:263–295, 2001b.

Davidson, G., and R. O'Neil. Optical radiation from nitrogen and air at high pressure excited by energetic electrons. *J. Chem. Phys.*, 41:3946–3949, 1964.

Davies, C., and P. W. Carpenter. Global behaviour corresponding to the absolute instability of the rotating-disc boundary layer. *J. Fluid Mech.*, 486:287–329, 2003.

de Silva, C., N. Hutchins, and I. Marusic. Uniform momentum zones in turbulent boundary layers. *J. Fluid Mech.*, 786:309–331, 2016.

Dhanak, M., and C. Si. On reduction of turbulent wall friction through spanwise oscillations. *J. Fluid Mech.*, 383:175–195, 1999.

Dimotakis, P., and G. Brown. The mixing layer at high Reynolds number: Large-structure dynamics and entrainment. *J. Fluid Mech.*, 78:535–560, 1976.

Dobre, A., H. Hangan, and B. Vickery. Wake control based on spanwise sinusoidal perturbations. *AIAA J.*, 44:485–492, 2006.

Dogval, A., V. Kozlov, and A. Michalke. Laminar boundary layer separation: Instability and associated phenomena. *Prog. Aerosp. Sci.*, 30:61–94, 1994.

Doligalski, T., C. Smith, and J. Walker. Production mechanism for turbulent boundary layer flows, viscous drag reduction. *Prog. Aeron. Astron.*, 72:47–72, 1980.

Dolling, D. S. Fifty years of shock-wave/boundary-layer interaction research: What next? *AIAA J.*, 39(8):1517–1531, 2001.

Donaldson, C. D. Investigation of a simple device for preventing separation due to shock and boundary-layer interaction. NACA RM L 50 B02a, 1950.

Dorr, P., M. Kloker, and A. Hanifi. Effect of upstream flow deformation using plasma actuators on crossflow transition induced by unsteady vortical free-stream disturbances. *AIAA Fluid Dyn. Conf., Paper 2017-3114*, 2017.

Dorrance, W. H., and F. J. Dore. The effect of mass transfer on the compressible turbulent boundary layer skin friction and heat transfer. *J. Aeron. Sci.*, 21(6):404–410, 1954.

Dowling, A. The effect of large-eddy breakup devices on oncoming vorticity. *J. Fluid Mech.*, 160:447–463, 1985.

Driest, E. R. V. Turbulent boundary layer in compressible fluids. *J. Aeron. Sci.*, 18(3):145–161, 1951.

Driver, D. Application of oil film interferometry skin friction to large wind tunnels. *Advanced Aerodynamic Measurement Technology, AGARD Conf. Proc. CP-601, Paper 25*, 1998.

Drubka, R. Instability in the near field of turbulent jets and their dependence on initial conditions and Reynolds number. Ph.D. Thesis, Illinois Institute of Technology, Chicago, IL, 1981.

Drubka, R., P. Reisenthel, and H. Nagib. The dynamics of low initial disturbance turbulent jets. *Phys. Fluids*, A1:1723–1735, 1989.

Dryden, H. L. Review of published data on the effect of roughness on transition from laminar to turbulent flow. *J. Aeron. Sci.*, 20:477–482, 1953.

Du, Y., and G. Karniadakis. Suppressing wall turbulence via a transverse traveling wave. *Science*, 288:1230–1234, 2000.

Du, Y., and G. Karniadakis. Drag reduction in wall-bounded turbulence via a transverse traveling wave. *J. Fluid Mech.*, 457:1–34, 2002.

Duan, L., and M. Choudhari. Effect of riblets on skin friction in high-speed turbulent boundary layers. *AIAA Paper 2012-1108*, 2012.

Duan, L., X. Wang, and X. Zhong. Stabilization of a Mach 5.92 boundary layer by two-dimensional finite height roughness. *AIAA J.*, 51(1):266–270, 2013.

Duong, A. Active turbulent boundary layer control: An experimental evaluation of viscous drag reduction using pulsed-DC plasma actuators. Ph.D. Thesis, University of Notre Dame, Notre Dame, IN, 2019.

Duong, A., T. Corke, F. Thomas, and K. Disser. Active turbulent boundary layer drag reduction using pulsed-DC plasma actuator. APS Division of Fluid Dynamics, 2019 abstract id. S26.003.

Dupont, P., C. Haddad, and J.-F. Debieve. Space and time organization in a shock-induced separated boundary layer. *J. Fluid Mech.*, 559:255–277, 2006.

Dupont, P., S. Piponniau, A. Sidorenko, and J.-F. Debieve. Investigation of an oblique shock reflection with separation by PIV measurements. *45th AIAA Aerosp. Sci. Meet. Exhib., Paper 2007-0119*, 2007.

Dussauge, J. P., P. Dupont, and J.-F. Debieve. Unsteadiness in shock wave boundary layer interaction with separation. *Aerosp. Sci. Tech.*, 10:85–91, 2006.

Einstein, H. A., and H. Li. The viscous sublayer along a smooth boundary. J. Eng. Mech.-ASCE, 82: 1–7, 1956.

Eisenlohr, H., and H. Eckelmann. Vortex splitting and its consequences in the vortex street wake of cylinders at low Reynolds numbers. *Phys. Fluid*, A1:189–192, 1989.

Elfstrom, G. M., Turbulent separation in hypersonic flow. *I.C. Aero Report 71 16*, 1971.

Elfstrom, G. M., Turbulent hypersonic flow at a wedge compression corner. *J. Fluid Mech.*, 53(1):113–127, 1972.

Elliott, G., M. Samimy, and S. Arnette. The evolution of large scale structure in compressible mixing layers. *9th Symp. Turbulent Shear Flows*, Kyoto, Japan 1993.

Englar, R. J. Experimental investigation of the high velocity Coanda wall jet applied to bluff trailing edge circulation control airfoils. *Technical Report 4708*, David W. Taylor Naval Ship Research and Development Center, Bethesda 1975.

Enloe, L., T. McLaughlin, R. VanDyken, et al. Mechanisms and responses of a single-dielectric barrier plasma actuator: Plasma morpholog. *AIAA J.*, 42(3):589–594, 2004a.

Enloe, L., T. McLaughlin, R. VanDyken, et al. Mechanisms and responses of a single-dielectric barrier plasma actuator: Geometric effects. *AIAA J.*, 42(3):595–604, 2004b.

Erengil, M. E., and D. S. Dolling. Effects of sweepback on unsteady separation in Mach 5 compression ramp interactions. *AIAA J.*, 31:302–311, 1993.

Erturk, E., and T. C. Corke. Boundary layer leading-edge receptivity to sound at incidence angles. *J. Fluid Mech.*, 444:383–407, 2001.

Fage, A., and J. H. Preston. On transition from laminar to turbulent flow in the boundary layer. *Proc. Roy. Soc. Lond., Ser. A*, 178:201–227, 1941.

Falkenstein, Z., and J. Coogan. Microdischarge behaviour in the silent discharge of nitrogen-oxygen and water-air mixtures. *J. Phys. D: App. Phys.*, 30:817–825, 1997.

Faller, A. J. Instability and transition of disturbed flow over a rotating disk. *J. Fluid Mech.*, 230:245–269, 1991.

Faller, A. J., and R. E. Kaylor. A numerical study of the instability of the laminar Ekman boundary layer. *J. Atmos. Sci.*, 23:466–480, 1966.

Fan, X. Laminar flow control models with neural networks. Ph.D. Thesis, The Ohio State University, Columbus, OH, 1995.

Fan, X., T. Herbert, and J. H. Haritonidis. Transition control with neural networks. *AIAA Paper 1995-0674* 1995.

Federov, A. V. Laminar turbulent transition in a hypersonic boundary layer. *European Office of Aerospace Research and Development Special Project SPC-96-4024*, 1997.

Fedorov, A. V., and N. D. Malmuth. Stabilization of hypersonic boundary layers by porous coatings. *AIAA J.*, 39(4):605–610, 2001.

Federov, B., G. Plavnik, I. Prokhorov, and L. Zhukhovitskii. Transitional flow conditions on a rotating disk. *Inzh. -Fiz. Zh.*, 31:1060–1067, 1976.

Feindt, E. G. Untersuchungen uber die abhangigkeit des umschlages laminar-turbulent von der oberflachenrauhigkeit und der druckverteilung. Thesis, Technische Universität Braunschweig, Technische Universitat Braunschweig, 1957.

Ferziger, J. H., and M. Peric. Solution of the Navier-Stokes equations. *Computational methods in fluid dynamics*, eds., J. H. Ferziger and M. Peric, pp. 149–208. Berlin: Springer.

Fiedler, H., B. Dziomba, P. Mensing, and T. Rosgen. Initiation, evolution and global consequences of coherent structures in turbulent shear flows. *The role of coherent structures in modelling turbulence and mixing, Lecture notes in physics*, ed. J. Jimenez, pp. 219–251, Springer-Verlag Berlin Heidelberg New York 1981.

Fisher, J. Flight test report use of vortex generators to improve performance of lcb systems. *Report 134-1*, Cessna Aircraft Co., Wichita, KS, 1954.

Fong, K. D. A numerical study of 2-D surface roughness effects on the growth of wave modes in hypersonic boundary layers. Ph.D. Thesis, Los Angeles: University of California, 2017.

Forliti, D., P. Strykowsky, and K. Debatin. Bias and precision errors of digital particle image velocimetr. *Exp. Fluids*, 28:436–447, 2000.

Freymuth, P. On transition in a separated laminar boundary layer. *J. Fluid Mech.*, 25:683–704, 1966.

Friehe, C., and W. Schwartz. Deviations from the cosine law for yawed cylindrical anemometer sensors. *J. Appl. Mech.*, 35:177–182, 1968.

Frishett, J. C. Incipient separation of a supersonic turbulent boundary-layer including effects of heat transfer. Ph.D. Dissertation, University of California, Los Angeles, 1971.

Funaki, J., G. Mizuno, M. Kondo, and K. Hirata. Oscillation mechanism of a flip-flop jet nozzle based on the flow which flows through a connecting tube. *Trans. JSME (Ser. B)*, 65(631):928–933, 1999.

Gad-El-Hak, M., and A. K. M. F. Hussain. Coherent structure in a turbulent boundary layer, part 1: Generation of artificial bursts. *Phy. Fluids*, 29:2124–2139, 1986.

Gad-El-Hak, M., and R. F. Blackwelder. Selective suction for controlling burst events in the boundary layer. *AIAA J.*, 27:308–314, 1989.

Gadd, G. E. Interaction between normal shock-waves and turbulent boundary-layers. *ARC R&M 3262*, 1961.

Ganapathisubramani, B., N. T. Clemens, and D. S. Dolling. Effects of upstream boundary layer on the unsteadiness of shock-induced separation. *J. Fluid Mech.*, 585:369–394, 2007.

Ganapathisubramani, B., N. T. Clemens, and D. S. Dolling. Effects of upstream boundary layer on the unsteadiness of shock-induced separation. *J. Fluid Mech.*, 585:397–425, 2009.

Garcia-Mayoral, R., and J. Jimenez. Drag reduction by riblets. *Phil. Trans. Roy. Soc. A*, 369:1412–1427, 2011.

Gaster, M. A note on a relation between temporally increasing and spatially increasing disturbances in hydrodynamic stability. *J. Fluid Mech.*, 14:222–224, 1962.

Gaster, M. On the flow along swept leading edges. *Aeron. Quart.*, 18(5):165–184, 1967.

Gaudet, L. Properties of riblets at supersonic speed. *Appl. Sci. Res.*, 46:245–254, 1989.

Gerrard, J. The mechanics of the vortex formation region of vortices behind bluff bodies. *J. Fluid Mech.*, 25:401–413, 1966a.

Gerrard, J. The three-dimensional structure of the wake of a circular cylinder. *J. Fluid Mech.*, 25:143–164, 1966b.

Gilarranz, J., L. Traub, and O. Rediniotis. A new class of synthetic jet actuators-part I: Design, fabrication and bench top characterization. J. Fluids Eng., 127(2):367–376, 2005.

Gilev, V., A. Dovgal, and Y. Kachanov. Development of three-dimensional disturbances in a boundary layer with pressure gradient. *Fluid Dyn. (translated from Russian)*, 23:393–399, 1988.

Glauser, M., X. Zheng, and C. Doering. Turbulence and coherent structures. *The dynamics of organized structures in the axisymmetric jet mixing layer*, eds., M. Lesieur and O. Metais, pp. 253–265. Kluwer Academic, 1991.

Glezer, A. The formation of vortex rings. *Phys. Fluids*, 31:3532–3542, 1988.

Gokoglu, S., M. Kuczmarski, and D. Culley. Numerical studies of a fluidic diverter for flow control. *AIAA Fluid Dyn. Conf., AIAA Paper 2009-4012*, 2009.

Goldstein, M. E. The evolution of Tollmien-Schlichting waves near a leading edge. *J. Fluid Mech.*, 127:59–81, 1983.

Goldstein, M. E. Scattering of acoustic waves into Tollmien-Schlichting waves by small streamwise variation in surface geometry. *J. Fluid Mech.*, 154:509–529, 1985.

Goldstein, M. E., and L. S. Hultgren. Boundary-layer receptivity to long-wave free-stream disturbances. *Annu. Rev. Fluid Mech.*, 21:137–166, 1989.

Goldstein, M. E., and S. S. Lee. Fully coupled resonant-triad interaction in an adverse-pressure-gradient boundary layer. *J. Fluid Mech.*, 245:523–551, 1992.

Gonsalez, J. C., and D. S. Dolling. Correlation of interaction sweepback effects on the dynamics of shock-induced turbulent separation. *31st AIAA Aerosp. Sci. Meet. Exhib., AIAA Paper 1993-0776*, Reno, NV 1993.

Gramann, R. A. Dynamics of separation and reattachment in a Mach 5 unswept compression ramp flow. Ph.D. Dissertation, Deptartment of Aerospace Engineering and Engineering Mechanics, University of Texas, Austin, TX, 1989.

Grant, H. L. The large eddies of turbulent motion. *J. Fluid Mech.*, 4:149–190, 1958.

Greenblatt, D., K. Pascha, C.-S. Yao, and J. Harris. A separation control CFD validation test case part 2. Zero efflux oscillatory blowing. *43rd AIAA Aerosp. Sci. Meet. Exhib., Paper 0485*, 2005.

Greene, B. R., N. T. Clemens, P. Magari, and D. Micka. Control of mean separation in shock boundary layer interaction using pulsed plasma jets. *Shock Waves*, 25:495–505, 2015.

Gregory, J., J. Ruotolo, A. Byerley, and T. McLaughlin. Switching behavior of a plasma-fluidic actuator. *45th AIAA Aerosp. Sci. Meet. Exhib., Paper AIAA 2007-785*, 2007a.

Gregory, J., P. Sullivan, W. Lafayette, G. Raman, and S. Raghu. Characterization of the microfluidic oscillator. *AIAA J.*, 45:568–578, 2007b.

Gregory, J., E. Gnanamanickam, J. Sullivan, and S. Raghu. Variable-frequency fluidic oscillator driven by a piezoelectric bender. *AIAA J.*, 47:2717–2725, 2009.

Gregory, N., and W. S. Walker. Experiments on the flow due to a rotating disc. *Fluid Motion Sub-Committee*, 1960.

Gregory, N., J. T. Stuart, and W. S. Walker. On the stability of three-dimensional boundary layers with application to the flow due to a rotating disk. *Phil. Trans.*, 248:155–199, 1955.

Grilli, M., P. J. Schmid, S. Hickel, and N. A. Adams. Analysis of unsteady behavior in shockwave turbulent boundary layer interaction. *J. Fluid Mech.*, 700:16–28, 2012.

Guo, F., and S. Zhong. A PIV investigation of the characteristics of micro-scale and macro-scale synthetic jets. *AIAA Paper 2006-3183*, 2006

Gutmark, E., and C.-M. Ho. Preferred modes and the spreading rates of jets. *Phys. Fluids*, 26:2932–2938, 1983.

Gutmark, E. K. Schadow, and K. Yu. Mixing enhancement in supersonic free shear flows. *Annu. Rev. Fluid Mech.*, 27:375–417, 1995.

Haddad, O. M., and T. C. Corke. Boundary layer receptivity to free-stream sound on parabolic bodies. *J. Fluid Mech.*, 368:1–26, 1998.

Haddad, O. M., E. Erturk, and T. C. Corke. Acoustic receptivity of the boundary layer over parabolic bodies at angles of attack. *J. Fluid Mech.*, 536:377–400, 2005.

Haley, C. L., K. M. Casper, and X. Zhong. Joint numerical and experimental investigation of roughness effect on hypersonic 2nd mode instability and transition. *Sandia Laboratory SAND2018-13919C*, 2018.

Hall, P., M. R. Malik, and D. I. A. Poll. On the stability of an infinite swept attachment-line boundary layer. *Proc. Roy. Soc. Lond. Ser. A*, 395:29–245, 1984.

Hama, F. R. Boundary-layer transition induced by a vibrating ribbon on a flat plate. *Proc. 1960 Heat Transfer Fluid Mechan. Instit.*, p. 92. Stanford University Press, Redwood City, CA, 1960.

Hamilton, J., J. Kim, and F. Waleffe. Regeneration mechanisms of near wall turbulence structure. *J. Fluid Mech.*, 287:317–348, 1995.

Hammache, M., and M. Gharib. An experimental study of the parallel and oblique shedding from circular cylinders. *J. Fluid Mech.*, 232:567–590, 1991.

Hanratty, T. Turbulent exchange of mass and momentum with a boundary. *J. Am. Inst. Chem. Engrs.*, 2(3):359–362f, 1956.

Hao, J., and C.-Y. Wen. Stabilization of a two-dimensional hypersonic boundary layer using a shallow cavity. *AIAA J.*, 59(2):430–438, 2021.

Hargather, M., M. Lawson, G. Settles, and L. Weinstein. Seedless velocimetry measurements by schlieren image velocimetry. *AIAA J.*, 49(3):611–620, 2011.

He, C. Plasma slats and flaps: An application of plasma actuators for hingeless aerodynamic control. Ph.D. Dissertation, University of Notre Dame, Notre Dame, IN, 2008.

He, C., and T. Corke. Numerical and experimental analysis of plasma flow control over a hump model. *45th Aerosp. Sci. Meet., Paper 2007-0935*, 2007.

He, C., and T. C. Corke. Plasma flaps and slats: An application of weakly ionized plasma actuators. *J. Aircraft*, 46(3):864–873, 2009.

Head, M., and P. Bandyopadhyay. New aspects of turbulent boundary layer structure. *J. Fluid Mech.*, 107:297–338, 1981.

Head, M., and V. Ram. Simplified presentation of preston tube calibration. *Aeron. Quart.*, 222:295–300, 1971.

Herbert, T. Subharmonic three-dimensional disturbances in unstable plane shear flows. *AIAA Paper 83-1759*, 1983.

Herbert, T. Secondary instability of boundary layers. *Annu. Rev. Fluid Mech.*, 20:487–526, 1988.

Herbert, T., and M. V. Morkovin. Dialogue on bridging some gaps in stability and transition research. *Laminar-Turbulent Transition*, 47–72 1980.

Ho, C.-M., and E. Gutmark. Vortex induction and mass entrainment in a small-aspect-ratio elliptic jet. *J. Fluid Mech.*, 179:383–405, 1987.

Ho, C.-M., and L. Huang. Subharmonics and vortex merging in mixing layers. *J. Fluid Mech.*, 1(19):443–473, 1982.

Ho, C.-M., and P. Huerre. Perturbed free shear layers. *Annu. Rev. Fluid Mech.*, 16:365–424, 1984.

Ho, C.-M., and N. Nosseir. Dynamics of an impinging jet. Part 1: The feedback phenomenon. *J. Fluid Mech.*, 105:119–142, 1981.

Ho, J. Experimental investigation of the effect of wall suction on cross-flow absolute instability in a rotating disk boundary layer. Ph.D. Thesis, University of Notre Dame, Notre Dame, IN, 2014.

Ho, J., T. C. Corke, and E. H. Matlis. Effect of wall suction on rotating disk absolute instability. *J. Fluid Mech.*, 791:704–737, 2016.

Holman, R., Y. Utturkar, R. Mittal, B. Smith, and L. Cattafesta. Formation criterion for synthetic jets. *AIAA J.*, 43(10):2110–2116, 2005.

Hou, Y. X., N. T. Clemens, and D. S. Dolling. Wide-field PIV study of shock- induced turbulent boundary layer separation. *41st Aerosp. Sci. Meet. Exhib., Paper 20030441*, 2003.

Houpt, A., B. Hedlund, S. Leonov, T. Ombrello, and C. Carter. Quasi-DC electrical discharge characterization in a supersonic flow. *Exp. Fluids*, 58(4):25–42, 2017.

Hsiao, F.-B., C.-F. Liu, and J.-Y. Shyu. Control of wall-separated flow by internal acoustic excitation. *AIAA J.*, 28:1440–1446, 1990.

Huang, J., T. Corke, and F. Thomas. Unsteady plasma actuators for separation control of low-pressure turbine blades. *AIAA J.*, 44(1):51–57, 2006a.

Huang, J., T. Corke, and F. Thomas. Unsteady plasma actuators for separation control of low-pressure turbine blades. *AIAA J.*, 44(7):1477–1487, 2006b.

Huang, X. Feedback control of vortex shedding from a circular cylinder. *Exp. Fluids*, 20:218–224, 1996.

Hucho, W., and G. Sovran. Aerodynamics of road vehicles. *Annu. Rev. Fluid Mech.*, 25:485–537, 1993.

Huerre, P., and P. Monkewitz. Local and global instabilities in spatially developing flows. *Annu. Rev. Fluid Mech.*, 22:473–537, 1990.

Humble, R. A., G. E. Elsinga, F. Scarano, and B. W. Oudheusden. Three-dimensional instantaneous structure of a shock wave/turbulent boundary layer interaction. *J. Fluid Mech.*, 622:33–62, 2009.

Ingard, U., and S. Labate. Acoustic circulation effects and the non-linear impedance of orifices. *J. Acoust. Soc. Am.*, 22(2):211–218, 1950.

Jackson, C. A finite-element study of the onset of vortex shedding in flow past variously shaped bodies. *J. Fluid Mech.*, 182:23–45, 1987.

James, R., J. Jacobs, and A. Glezer. A round turbulent jet produced by an oscillating diaphragm. *Phys. Fluids*, 8(9):2484–2495, 1996.

Jameson, A. Aerodynamic shape optimization using the adjoint method. *Von Karman Institute for Fluid Dynamics Lecture Series*, 2003.

Jasak, H., A. Jemcov, and Z. Tukovic. Openfoam: A C++ library for complex physics simulations. *Int. Worksh. Coupled Meth. Numer. Dyn.*, 2007.

Jeromin, L. O. F. An experimental investigation of the compressible turbulent boundary-layer with air injection. *ARC R&M No. 3526*, 1968.

Jewell, J., R. Wagnild, I. Leyva, G. Candler, and J. Shepherd. Transition within a hypervelocity boundary layer on a 5-degree half-angle cone in air/Co2 mixtures. *AIAA Paper 2013-0523*, 2013.

Jiang, H., J. Liu, S. Luo, J. Wang, and W. Huang. Hypersonic flow control of shock wave/turbulent boundary layer interactions using magnetohydrodynamic plasma actuators. *App. Phys. Eng.*, 21(9):745–760, 2020.

Jimenez, J., and P. Moin. The minimal flow unit in near wall turbulence. *J. Fluid Mech.*, 225:213–240, 1991.

Jimenez, J., and A. Pinelli. The autonomous cycle of near-wall turbulence. *J. Fluid Mech.*, 389:335–359, 1999.

Johnston, J., and M. Nishi. Vortex generator jets means for flow separation control. *AIAA J.*, 28:989–994, 1990.

Kachanov, Y. S., and V. Y. Levchenko. The resonant interaction of disturbances at laminar-turbulent transition in a boundary layer. *J. Fluid Mech.*, 138:209–247, 1984.

Kachanov, Y. S., V. V. Koslov, and V. Y. Levchenko. Nonlinear development of a wave in a boundary layer. *Fluid Dyn.*, 12:383–390, 1977.

Kaplan, R., and J. Laufer. The intermittantly turbulent region of the boundary layer. *Proc. 12th Int. Congr. Mech.*, p. 236, 1969.

Karman, T. V. Uber den mechanismus des widerstandes, den ein bewegter korper in einer flussigkeit erfahrt. *Nachrichten von der Koniglichen Gesellschaft der Wissenschaften zu Gottingen*, pp. 547–556, 1912.

Karman, T. V. Uber laminare und turbulente reibung. *ZAMM*, 1:233–252, 1921.

Karniadakis, G., M. Israeli, and S. Orzag. Mechanisms on transverse motions in turbulent wall flows. *Annu. Rev. Fluid Mech.*, 35:45–62, 1967.

Katzer, E. On the length scales of laminar shock/boundary-layer interaction. *J. Fluid Mech.*, 206:477–496, 1989.

Keane, R., and R. Adrian. Theory of cross-correlation analysis of PIV images. *Appl. Sci. Res.*, 49:191–215, 1992.

Kegelman, J. T., R. C. Nelson, and T. J. Mueller. The boundary layer on an axisymmetric body with and without spin. *AIAA J.*, 21(11):1485–1491, 1983.

Kegerise, M., R. Cabell, and L. Cattafesta. Real-time feedback control of flow-induced cavity tones, part 1: Fixed-gain control. *J. Sound Vib.*, 307:906–923, 2007.

Keith, W., and J. Barclay. Effects of large eddy breakup device on the fluctuating wall pressure field. *J. Fluids Eng.*, 115:389–397, 1993.

Kell, D. A surface flow visualization technique for use in cryogenic wind tunnels. *Aeron. J.*, 82:484–487, 1978.

Kelley, C., P. Bowles, J. Cooney, et al. Leading-edge separation control using alternating-current and nanosecond-pulse plasma actuators. *AIAA J.*, 52(9):1871–1884, 2014.

Kelley, C., T. Corke, and F. Thomas. Design and scaling of plasma streamwise vortex generators for flow separation control. *AIAA J.*, 54(11):3397–3408, 2016.

Kelly, R. On the resonant interaction of neutral disturbances in two inviscid shear flows. *J. Fluid Mech.*, 31:789–799, 1968.

Kendall, J. An experimental investigation of leading-edge shock-wave: Boundary-layer interaction at Mach 5.8. *J. Aeron. Sci.*, 24:47–56, 1957.

Kepler, C. E., and R. L. O'Brien. Supersonic turbulent boundary-layer growth over cooled walls in adverse pressure gradients. *Report ASD TDR-62-87*, Aeronautical System Division, 1962.

Keser, H., M. Unal, and P. Bearman. Simulation of wake from a circular cylinder with spanwise sinusoidal waviness. *Proc. 2nd Intl. Conf. Vortex Methods*, ed. M. Unal, pp. 131–137. Istanbul, Turkey, 2001.

Kibens, V. Discrete noise spectrum generated by an acoustically excited jet. *AIAA J.*, 18(4):434–441, 1980.

Kibens, V., and A. Glezer. Excitation of supersonic shear layers by piezoelectric actuators. *Bull. Am. Phys. Soc.*, 37(8): 1764, 1992.

Kim, C. S., C. Kim, and O. H. Rho. Sensitivity analysis for the navier-stokes equations with two-equation turbulence models. *AIAA J.*, 39(5):838–845, 2001.

Kim, D., H. Do, and H. Choi. Drag reduction on a three-dimensional model vehicle using a wire-to-plate dbd plasma actuator. *Exp. Fluids*, 61:135–149, 2020.

Kim, H., S. Kline, and W. Reynolds. The production of turbulence near a smooth wall in a turbulent boundary layer. *J. Fluid Mech.*, 50(1):133–160, 1971.

Kim, J. Physics and control of wall turbulence for drag reduction. *Phil. Trans. Roy. Soc. A*, 369:1396–1411, 2011.

Kim, J.-S., J. Hwang, M. Yoon, J. Ahn, and H. J. Sung. Influence of a large-eddy breakup device on the frictional drag in a turbulent boundary layer. *Phys. Fluids*, 29:065103, 2017.

Kimura, T., and T. Tsutahara. Fluid dynamic effects of grooves on circular cylinder surface. *AIAA J.*, 29(12):2062–2068, 1991.

King, L. V. On the convection of heat from small cylinders in a stream of fluid: Determination of the convection constants of small platinum wires with applications to hot-wire anemometry. *Phil. Trans. Roy. Soc. Lond. A*, 214:373–432, 1914.

King, R. A. Three-dimensional boundary-layer transition on a cone at Mach 3.5. *Exp. Fluids*, 856:822–849, 1992.

Kirshner, J. *Fluid amplifiers*. McGraw-Hill, New York, 1966.

Kistler, A. L. Fluctuating wall pressure under a separated supersonic flow. *J. Acoust. Soc. Am.*, 36(3):543–550, 1964.

Kit, E., I. Wygnanski, and D. Friedman. On the periodically excited plane turbulent mixing layer emanating from a jagged partition. *J. Fluid Mech.*, 589:479–507, 2007.

Klaassen, G., and W. Peltier. The onset of turbulence in finite-amplitude Kelvin-Helmholtz billows. *J. Fluid Mech.*, 155:1–35, 1985.

Klebano, P. S. Characteristics of turbulence in a boundary layer with zero pressure gradient. *NACA Technical Note 3178*, 1954.

Klebano, P. S., and K. D. Tidstrom. Evaluation of amplified waves leading to transition in a boundary layer with zero pressure gradient. *NASA TND-195*, 1959.

Klebano, P. S., K. D. Tidstrom, and L. M. Sargent. The three-dimensional nature of boundary-layer instability. *J. Fluid Mech.*, 12:1–34, 1962.

Klevin, C. Vane separation control in a linear cascade with area expansion using AC DBD plasma actuators. Ph.D. Thesis, University of Notre Dame, Notre Dame, IN, 2016.

Kline, S., W. Reynolds, F. Schraub, and P. Runstadler. The structure of turbulent boundary layers. *J. Fluid Mech.*, 30:741–773, 1967a.

Kline, S., W. Reynolds, F. Schraub, and P. Runstadler. The structure of turbulent boundary layers. *J. Fluid Mech.*, 30(4):741–733, 1967b.

Knutson, A., G. S. Sidharth, and G. Candler. Direct numerical simulation of Mach 6 flow over a cone with highly swept fin. *AIAA Paper 2018-0379*, 2018.

Kobashi, Y., and M. Ichijo. Wall pressure and its relation to turbulent structures of the turbulent boundary layer. *Exp. Fluids*, 4:49–55, 1986.

Kobayashi, R. Linear stability theory of boundary layer along a cone rotating in axial flow. *Bull. Jpn. Soc. Mech. Eng.*, 24:934–940, 1981.

Kobayashi, R., and H. Izumi. Boundary layer transition on a rotating cone in still fluid. *J. Fluid Mech.*, 127:353–364, 1983a.

Kobayashi, R., and H. Izumi. Boundary-layer transition on a rotating cone in still fluid. *J. Fluid Mech.*, 27:353–364, 1983b.

Kobayashi, R., and Y. Kohama. *Spiral vortices in boundary-layer transition on a rotating cone*, pp. 573–580. Springer-Verlag, New York, 1985.

Kobayashi, R., Y. Kohama, and C. Takamadate. Spiral vortices in boundary layer transition regime on a rotating disk. *Acta Mech.*, 71:71–82, 1980.

Kobayashi, R., Y. Kohama, T. Arai, and M. Ukaku. The boundary-layer transition on rotating cones in axial flow with free-stream turbulence. *JSME Int. J.*, 30(261):423–429, 1987.

Koch, W. Local instability characteristics and frequency determination of self-excited shear flows. *J. Sound Vibr.*, 99:53–83, 1985.

Kochetov, A., S. Napartovich, and S. Leonov. Plasma ignition of combustion in a supersonic flow of fuel-air mixtures: Simulation problems. *J. High Energy Chem.*, 40:98–104, 2006.

Kohama, Y. Study on boundary layer transition of a rotating disk. *Acta Mech.*, 50:193–199, 1984.

Kohama, Y. Flow structures formed by axisymmetric spinning bodies. *AIAA J.*, 23:1445–1454, 1985.

Kohama, Y. Cross-flow instability in rotating disk boundary-layer. *AIAA Paper 87-1340*, 1987.

Kohama, Y., and R. Kobayashi. Boundary-layer transition and the behavior of spiral vortices on rotating spheres. *J. Fluid Mech.*, 137:153–164, 1983a.

Kohama, Y., and R. Kobayashi. Behavior of spiral vortices on rotating axisymmetric bodies. *Rep. Inst. High Speed Afech.*, 47:27–38, 1983b.

Kohama, Y., W. S. Saric, and J. A. Hoos. A high frequency secondary instability of cross flow vortices that leads to transition. In *Boundary layer transition and control*, Roy. Aeronautical Society, Lond., 1991.

Koratagere, S. Broad-band 3-D mode development in Blasius boundary layers by resonant mode detuning. M. S. Thesis, Illinois Institute of Technology, Chicago, IL, 1990.

Kosinov, A., A. Maslov, and S. Shevelkov. Experiments on the stability of supersonic laminar boundary layers. *J. Fluid Mech.*, 219:621–633, 1990.

Koso, T., S. Kawaguchi, M. Hojo, and H. Hayami. Flow mechanism of a self-induced oscillating jet issued from a flip-flop jet nozzle. *5th JSMEKSME Fluids Eng. Conf.*, Nagoya, Japan, 2002.

Kourta, A., and C. Leclerc. Characterization of synthetic jet actuation with application to ahmed body wake. *Sens. Actuators A*, 192:262–280, 2013.

Kovasznay, L. Hot-wire investigation of the wake behind cylinders at low Reynolds numbers. *Proc. Roy. Soc. A*, 198:174–190, 1949.

Kovasznay, L. Structure of the turbulent boundary layer. *Phys. Fluids*, 10:S25–S30, 1967.

Kovasznay, L., V. Kibbens, and R. Blackwelder. Large scale motions in the intermittent region of a turbulent boundary layer. *J. Fluid Mech.*, 41:283–325, 1970.

Kozlov, A., and F. Thomas. Flow control via two types of dielectric barrier discharge plasma actuation. *AIAA J.*, 49(9):1919–1931, 2011.

Kramer, H. Heat transfer from spheres to flowing media. *Physica*, 12:61–80, 1946.

Kramer, K. Uber die wirkung von stolperdrahten auf den grenzschichtumschlag. *Z. Flugwiss*, 9:20–27, 1961a.

Kramer, M. Boundary layer stabilization by distributed damping. *J. Aeron. Sci.*, 24(6):459–460, 1957.

Kramer, M. Boundary layer stabilization by distributed damping. *J. Am. Soc. Nav. Eng.*, 72:25–33, 1960.

Kramer, M. The dolphins secret. *J. Am. Soc. Nav. Eng.*, 73:103–107, 1961b.

Kramer, M. Boundary layer stabilization by distributed damping. *J. Am. Soc. Nav. Eng.*, 74:111–130, 1962.

Kramer, M. Hydrodynamics of the dolphin. *Advances in hydroscience*, vol. 2, ed. V. T. Chow, pp. 111–130. Academic, New York, 1965.

Krehl, P., and S. Engemann. August toepler the first who visualized shock waves. *Shock Waves*, 5:1–18, 1995.

Kreith, F., D. Ellis, and J. Giesing. An experimental investigation of the flow engendered by a rotating cone. *Appl. Sci. Res. A*, 11:430–440, 1962.

Krentel, D., R. Muminovic, A. Brunn, W. Nitsche, and R. King. Application of active flow control on generic 3-D car models. *Active flow control II (NNFM, vol. 108)*, pp. 223–239., Springer Nature, Berlin, 2010.

Krishnani, P., and D. Zhou. CFD analysis of drag reduction for a generic SUV. *Proc. 2009 ASME Int. Mechan. Eng. Congr. Expo.*, 13:589–598, 2010.

Krogmann, P., E. Stanewsky, and P. Theide. Transonic shock/boundary-layer interaction control. *ICAS Paper 84-2-3-2*, 1984.

Krogmann, P., E. Stanewsky, and P. Theide. Effect of suction on shock boundary-layer interaction and shock-induced separation. *J. Aircraft*, 22:37–42, 1985.

Krothapalli, A., D. Baganoff, and K. Karamcheti. On the mixing of a rectangular jet. *J . Fluid Mech.*, 107:201–220, 1981.

Kuethe, A. Effect of streamwise vortices on wake properties associated with sound generation. *J. Aircraft*, 9(10):715–719, 1972.

Kulik, V. M., I. S. Poguda, and B. N. Semenov. Experimental investigation of one-layer visco-elastic coatings action on turbulent friction and wall pressure pulsations. *Recent Develop. Turbul. Manage.*, 20:263–289, 1991.

Kungurtsev, P. V., and M. P. Juniper. Adjoint-based shape optimization of the microchannels in an inkjet printhead. *J. Fluid Mech.*, 871:113–138, 2019.

Kusek, S., T. Corke, and P. Reisenthel. Control of two and three dimensional modes in the initial region of an axisymmetric jet. *AIAA 2nd Shear Flow Conf. Paper 89-0968*, 1989.

Lackner, M., and G. van Kuik. Topological derivative formulation for shape sensitivity in incompressible turbulent flows. *9th Int. Conf. CFD Minerals Process Ind.*, 2012.

Lakshmikantha, H., K. Y. Narayan, and G. Srinivasan. Effect of fluid injection on shock-wave boundary layer interaction. *Report 69 FM 7*, Department of Aerospace Engineering, Indian Institute of Science Bangalore, 1969.

Lam, K., and Y. Lin. Drag force control of flow over wavy cylinders at low Reynolds number. *J. Mech. Sci. Technol.*, 21:1331–1337, 2007.

Lam, K., and Y. Lin. Large eddy simulation of flow around wavy cylinders at a subcritical Reynolds number. *Int. J. Heat Fluid Flow*, 29:1071–1088, 2008.

Lam, K., and Y. Lin. Effects of wavelength and amplitude of a wavy cylinder in cross-flow at low Reynolds numbers. *J. Fluid Mech.*, 620:195–220, 2009.

Lam, K., F. Wang, J. Li, and R. So. Experimental investigation of the mean and fluctuating forces of wavy (varicose) cylinders in a cross-flow. *J. Fluids Struct.*, 19:321–334, 2004a.

Lam, K., F. Wang, and R. So. Three-dimensional nature of vortices in the near wake of a wavy cylinder. *J. Fluids Struct.*, 19:815–833, 2004b.

Lasheras, J., and H. Choi. Three-dimensional instability of a plane free shear layer: An experimental study of the formation and evolution of streamwise vortices. *J. Fluid Mech.*, 189:53–86, 1988.

Laufer, J., and P. Monkewitz. On turbulent jet flow in a new perspective. *AIAA Paper 80-0962*, 1980.

Laufer, J., and M. Narayanan. Mean period of the turbulent production mechanism in a boundary layer. *Phys. Fluids*, 14:182–183, 1971.

Lebedeva, I. Experimental study of acoustic streaming in the vicinity of orifices. *Soviet Phys. Acoust.*, 26(4):331–333, 1980.

Lee, S., and A. Nguyen. Experimental investigation on wake behind a wavy cylinder having sinusoidal cross-sectional area variation. *Fluid Dyn. Res.*, 39:292–304, 2007.

Lees, L. The stability of the laminar boundary layer in a compressible fluid. *Technical Report 876, NACA*, 1947.

Lees, L., and E. Reshotko. Stability of the compressible laminar boundary layer. *J. Fluid Mech.*, 12:555–590, 1962.

Lehugeur, B., and P. Gillieron. Active control of vortex breakdown phenomenon in the wake of a simplified car geometry. *ASME Paper FEDSM2006-98349*, 2006.

Lekoudis, S. Resonant wave interactions on a swept wing. *AIAA J.*, 18:122–124, 1979.

Leonov, S., and D. Yarantsev. Near-surface electrical discharge in supersonic airflow: Properties and flow control. *J. Propul. Power*, 24(6):1168–1181, 2008.

Leonov, S., V. Soloviev, and D. Yarantsev. High-speed inlet customization by surface electrical discharge. *AIAA Paper 2006-0403*, 2006.

Leonov, S., A. A. Firsov, and D. A. Yarantsev. Active steering of shock waves in compression ramp by nonuniform plasma. *AIAA Paper 2010-0260*, 2010.

Leonov, S. B., V. A. Bityurin, and D. A. Yarantsev. High-speed flow control due to interaction with electrical discharges. *AIAA Paper 2005-3287*, 2005a.

Leonov, S. B., D. A. Yarantsev, and Y. Isaenkov. Properties of filamentary electrical discharge in high-enthalpy flow. *AIAA Paper 2005-0159*, 2005b.

Leonov, S. B., A. A. Firsov, and A. W. Houpt. Suppression of reflected oblique shock wave by multi-filamentary plasma. *XVII Worksh. Magneto-Plasma Aerodyn. IOP Publishing IOP Conf. Ser.*, 2018.

D. Li. Shock reflection and oblique shock waves. *J. Math. Phys.*, 48(12):123102-1–123102-20, 2007.

Li, Y., and M. Gaster. Active control of boundary-layer instabilities. *J. Fluid Mech.*, 550:185–205, 2006.

Li, F., M. Choudhari, C.-L. Chang, and J. White. Analysis of instabilities in non-axisymmetric hypersonic boundary layers over cones. *10th AIAA/ASME Joint Thermophys. Heat Transfer Conf., Paper AIAA 2010-4643*, 2010.

Liebeck, R. H. Design of subsonic airfoils for high lift. *J. Aircraft*, 15(9):547–561, 1978.

Liepmann, H. W., and D. M. Nosenchuck. Active control of laminar-turbulent transition. *J. Fluid Mech.*, 118:201–204, 1982.

Liepmann, H. W., G. C. Brown, and D. M. Nosenchuck. Control of laminar-instability waves using a new technique. *J. Fluid Mech.*, 118:187–200, 1982.

Lin, J. Control of turbulent boundary-layer separation using micro-vortex generators. *30th AIAA Fluid Conf., Paper AIAA 99-3404*, Norfolk, VA, 1999.

Lin, J., F. Howard, and G. Selby. Small submerged vortex generators for turbulent flow separation control. *J. Spacecr. Rockets*, 27(5):503–507, 1990a.

Lin, J. C. Review of research on low-profile vortex generators to control boundary-layer separation. *Prog. Aerosp. Sci.*, 38(4):389–420, 2002.

Lin, N., H. Reed, and W. S. Saric. Boundary layer receptivity to sound: Navier-stokes computation. *Appl. Mech. Rev.*, 43:S175–S180, 1990b.

Lingwood, R. J. Absolute instability of the boundary layer on a rotating disk. *J. Fluid Mech.*, 299:17–33, 1995.

Lingwood, R. J. An experimental study of absolute instability of the rotating-disk boundary-layer flow. *J. Fluid Mech.*, 314:373–405, 1996.

Loehrke, R., and H. Nagib. Experiments on management of free-stream turbulence. *AGARD Report R-598*, 1972.

Loehrke, R., and H. Nagib. Control of freestream turbulence by means of honeycombs: A balance between suppression and generation. *J. Fluids Eng.*, 98:342–353, 1976.

Lombardi, A. Closed-loop dynamic stall control using a plasma actuator. M.S. Thesis, University of Notre Dame, Notre Dame, IN, 2011.

Lombardi, A., P. Bowles, and T. Corke. Closed-loop dynamic stall control using a plasma actuator. *AIAA J.*, 51:1130–1141, 2013.

Looney, W. R., and E. F. Blick. Skin friction coefficients of compliant surfaces in turbulent flow. *J. Spacecraft Rockets*, 3:1562–1564, 1966.

Lopers, J., T. Ng, M. Patel, S. Vasudevan, and T. Corke. Aerodynamic control of 1303 UAV using windward surface plasma actuators on a separation ramp. *AIAA Paper 2007-636*.

Lowell, H. Design and application of hot-wire anemometers for steady-state measurements at transonic and supersonic airspeeds. *Technical Report Technical Note 2117*, NACA, Washington, 1950.

Lu, S. S., and W. W. Willmarth. Measurement of the structure of the Reynolds stress in a turbulent boundary layer. *J. Fluid Mech.*, 60:481–581, 1973.

Ludwig, H., and W. Tillmann. Untersuchungen uber die wandschubspannung in turbulenten reibungsschichten. *Ing. Arch. (also NACA TM 1285)*, 17:288–299, 1949.

Mack, L. Boundary-layer stability theory. *Document No. 900-277, Rev. A*, Jet Propulsion Laboratory, Pasadena, CA, 1969.

Mack, L. Review of linear compressible stability theory. *Stability of time dependent and spatially varying flows*, Springer-Verlag, pp. 164–187, New York, NY, 1985.

Mack, L. M. Boundary-layer linear stability theory. *AGARD Report 709, Special course on stability and transition of laminar flow, part 3, NASA Contract NAS7-918*, 1984.

Mahulikar, S., and H. Herwig. Fluid friction in incompressible laminar convection: Reynolds' analogy revisited for variable fluid properties. *Euro. Phys. J. B: Condens. Matter & Complex Syst.*, 62(1):7–86, 2008.

Malik, M., S. Wilkinson, and S. Orszag. Instability and transition in rotating disk flow. *AIAA J.*, 19:1131–1138, 1981.

Malik, M., F. Li, and C.-L. Chang. Crossflow disturbances in three-dimensional boundary layers: Nonlinear development, wave interaction and secondary instability. *J. Fluid Mech.*, 268:1–36, 1994.

Maltby, R., and R. Keating. The surface oil flow technique for use in low speed wind tunnels. *AGARDograph*, 70:29–38, 1962.

Mankbadi, R. R. Critical layer nonlinearity in the resonance growth of three dimensional waves in boundary layer. *NASA TM-103639*, 1990.

Mankbadi, R. R. Asymptotic analysis of boundary-layer transition. *4th Int. Symp. CFD*, 1991.

Mankbadi, R. R., X. S. Wu, and S. S. Lee. A critical-layer analysis of the resonant triad in boundary-layer transition: Nonlinear interactions. *J. Fluid Mech.*, 256:85–106, 1993.

Marchuk, G. I. *Adjoint equations and analysis of complex systems*. Kluwer Academic, 1995 Dordrecht/Boston/London.

Marusic, I. On the role of large-scale structures in wall turbulence. *Phys. Fluids*, 13:735–743, 2001.

Marusic, I., and A. Perry. A wall-wake model for the turbulence structure of boundary layers. *J. Fluid Mech.*, 298:389–407, 1995.

Maslowe, S. Weakly nonlinear stability of a viscous free shear layer. *J. Fluid Mech.*, 79:689–702, 1977.

Massines, F., A. Rabehi, P. Decomps, et al. Experimental and theoretical study of a glow discharge at atmosperic pressure controlled by dielectric barrier. *J. Appl. Phys.*, Vol. 83, No. 6, 2950–2957.

Mathews, D. C. Shock-wave/boundary-layer interactions in two-dimensional and axially-symmetric flows including the influence of suction. Ph. D. Thesis, University of Washington, Seattle, WA, 1969.

Mathis, C., M. Provansal, and L. Boyer. The Benard-Von Karman instability: An experimental study near the threshold. *J. Phys. Lett.*, 45:483–491, 1984.

Matlis, E. H. Controlled experiments on instabilities and transition to turbulence on a sharp cone at Mach 3.5. Ph.D. Thesis, University of Notre Dame, Notre Dame, IN, 2004.

Matlis, E. H. Wavenumber analysis and resonance of stationary and traveling cross-flow modes on a rotating disk. M.S. Thesis, Illinois Institute of Technology, Chicago, IL, 1997.

Mattingly, G., and C. Chang. Unstable waves on an axisymmetric jet column. *J. Fluid Mech.*, 65:541–560, 1974.

McClure, W. B. An experimental study of the driving mechanism and control of the unsteady shock induced turbulent separation in a Mach 5 compression corner flow. Ph.D. Dissertation, Department of Aerospace Engineering and Engineering Mechanics, University of Texas, Austin, TX, 1992.

McCormick, D. Shock–boundary layer interaction control with low-profile vortex generators and passive cavity. *AIAA Aerosp. Sci. Meet. Exhib., AIAA Paper 92-0064*, Reno, NV, , 1992.

McCormick, D. C. Shock/boundary-layer interaction control with vortex generators and passive cavity. *AIAA J.*, 31(1):91–96, 1993.

McCroskey, W. Unsteady airfoils. *Annu. Rev. Fluid Mech.*, 14:285–311, 1982.

Mednikov, E., and B. Novitskii. Experimental study of intense acoustic streaming. *Soviet Phys.-Acoust.*, 21(2):152–154, 1975.

Mertz, B. and T. C. Corke. Single-dielectric barrier discharge plasma actuator modeling and validation. *J. Fluid Mech.*, 669:557–583, 2011.

Merzkirch, W. Techniques of flow visualization. *NATO AGARDograph No. 302*, 1984.

Michalke, A. Instability of compressible circular free jet with consideration of the influence of the jet boundary layer thickness. *Z. Flugwiss.*, 19(8/9):319–328, 1971.

Michalke, A. On the instability of wall-boundary layers close to separation. *IUTAM Symp.*, eds. V. Kozlov and A. Dovgal, pp. 557–564, Springer, Berlin, 1991.

Middlebrooks, J. Crossflow transition control on a swept fin boundary layer at Mach 6. Ph.D. Thesis, University of Notre Dame, Notre Dame, IN, 2022.

Miksad, R. Experiments on the nonlinear stages of free shear layer transition. *J. Fluid Mech.*, 56:695–719, 1972.

Milling, R. W. Tollmien-Schlichting wave cancellation. *Phys. Fluids*, 24(5):979–981, May 1981.

Moeller, E. B. and O. K. Rediniotis. Hingeless flow control over a delta wing platform. *AIAA Paper 2000-117*, 2000.

Monkewitz, P. On the effect of the phase difference between fundamental and subharmonic instability in a mixing layer. *Internal Report*, University of California, Los Angeles, 1982.

Monkewitz, P. The absolute and convective nature of instability in two-dimensional wakes at low Reynolds numbers. *Phys. Fluids*, 31:999–1006, 1988.

Monkewitz, P., and P. Huerre. The influence of the velocity ratio on the spatial instability of mixing layers. *Phys. Fluids*, 25:1137–1143, 1982.

Monkewitz, P., and L. Nguyen. Absolute instability in the near-wake of two-dimensional bluff bodies. *J. Fluids Struct.*, 1:165–184, 1987.

Monson, D., G. Mateer, and F. Menter. Boundary-layer transition and global skin friction measurements with an oil-fringe imaging technique. *SAE Paper 932550*, 1993.

Morkovin, M., and S. Paranjape. Acoustic excitation of shear layers. *Z. Flugwiss*, 9:328–335, 1971.

Morkovin, M. V. Flow around a circular cylinder: A kaleidoscope of challenging fluid phenomena. *Proc. ASME Symp. Fully Separated Flows*. p. 102, 1964.

Morkovin, M. V. Bypass transition to turbulence and research desiderata. *NASA Lewis Research Center Document ID 19850023129*, 1985.

Morkovin, M. V. Recent insights into instability and transition to turbulence in open-flow systems. *Technical Report, NASA Contractor Report 181693*, NASA, ICASE, Langley VA: NASA Langley Research Center, 1988.

Morkovin, M. V., and W. Bradfield. Probe interference in measurements in supersonic laminar boundary layers. *J. Aeron. Sci.*, 21(11):785–787, 1954.

Morris, N. *An introduction to fluid logic*. McGraw-Hill, London, 1973.

Mueller, T. On the historical development of apparatus and techniques for smoke visualization of subsonic and supersonic flow. *AIAA Paper 80-0430-CP*, 1980.

Mueller, T. Flow visualization by direct injection. In *Fluid Mechanics Measurements*, ed., R. Goldstein, 1st edn., Hemisphere, Washingot D. C., 1983.

Mueller, T. J., R. C. Nelson, J. T. Kegelman, and M. V. Morkovin. Smoke visualization of boundary-layer transition on a spinning axisymmetric body. *AIAA J.*, 19:607–608, 1981.

Muller, B., and H. Bippes. Experimental study of instability modes in a three-dimensional boundary layer. *Technical Report 18, AGARD-CP-438*, 1988.

R. Myose and R. Blackwelder. On the role of the outer region in the turbulent boundary layer bursting process. *J. Fluid Mech.*, 259:345–373, 1994.

Nagib, H., and K. Chauhan. Variations of Von Karman coefficient in canonical flows. *Phys. Fluids*, 20:101518, 2008.

Nagib, H., T. Corke, K. Helland, and J. Way. Computer analysis of flow visualization records obtained by the smoke-wire technique. *Proc. Dyn. Flow Conf. 1978 Dyn. Meas. Unsteady Flows*. Springer, Dordrech, 1978.

Narayanaswamy, V., L. R. Laxminarayan, and N. T. Clemens. Control of unsteadiness of a shock wave/turbulent boundary layer interaction by using a pulsed-plasma-jet actuator. *Phys. Fluids*, 24:076101, 2012.

Nielson, E. J., and W. K. Anderson. Aerodynamic design optimization on unstructured meshes using the Navier-Stokes equations. *AIAA J.*, 37(11):1411–1419, 1999.

Nishioka, M., M. Asai, and S. Yoshida. Control of flow separation by acoustic excitation. *AIAA J.*, 28(11):1909–1915, 1990.

Nishizawa, A., S. Takagi, H. Abe, R. Maeda, and H. Yoshida. Toward smart control of separation around a wing – development of an active separation control system. *Proc. 4th Symp. Smart Control of Turbulence*, pp. 13–21. University of Tokyo, Tokyo 2003.

Obremski, H. T., M. V. Morkovin, and M. T. Landahl. A portfolio of stability characteristics of incompressible boundary layers. *NATO AGARDograph No. 134*, 1969.

Oertel, H. Wakes behind blunt bodies. *Annu. Rev. Fluid Mech.*, 22:539–564, 1990.

Okamoto, T., M. Yagita, and Y. Kamijima. Experimental investigation on the boundary-layer flow over rotating cone cylinder body in a uniform stream. *Bull. Jpn. Soc. Mech. Eng.*, 19:930–937, 1976.

Othman, H. Experimental study of absolute instability over a rotating disk. Ph.D. Thesis, University of Notre Dame, Notre Dame, IN, 2005.

Othman, H., and T. C. Corke. Experimental investigation of absolute instability of a rotating-disk boundary layer. *J. Fluid Mech.*, 565:63–94, 2006.

Otto, C., P. Tewes, J. Little, and R. Woszidlo. Comparison of various fluidic oscillators for separation control on a wall-mounted hump. *AIAA SciTech 2019, Paper AIAA 2019-0884*, 2019.

Owen, J., A. Szewczyk, and P. Bearman. Suppression of Kármán vortex shedding. *Phys. Fluids*, 12:S9, 2000.

Pagella, A., A. Babucke, and U. Rist. Two-dimensional numerical investigations of small-amplitude disturbances in a boundary layer at Ma = 4.8: Compression corner versus impinging shock wave. *Phys. Fluids*, 16(7):2272–2281, 2004.

Pang, J., and K.-S. Choi. Turbulent drag reduction by Lorentz force oscillation. *Phys. Fluids*, 16(5):L35–L38, 2004.

Parekh, D., A. Leonard, and W. Reynolds. Bifurcating jets at high Reynolds numbers. *Technical Report AFOSR-TR-89-0282, Air Force Office of Scientific Research*, 1989.

Park, P. H., L. Dongkon, J. Woo-pyung, et al. Drag reduction in flow over a two-dimensional bluff body with a blunt trailing edge using a new passive device. *J. Fluid Mech.*, 563:389–414, 2006.

Patel, M. P., Z. H. Sowle, T. C. Corke, and C. He. Autonomous sensing and control of wing stall using a smart plasma slat. *AIAA Paper 2006-1207*, 2006.

Patel, V. Calibration of the Preston tube and limitations on its use in pressure gradients. *J. Fluid Mech.*, 23(1):185–208, 1965.

Pearcey, H., and D. Holder. Examples of the effects of shock-induced boundary layer separation in transonic flight. In *Ministry of Technology Aeronautical Research Council Reports and Memoranda* Her Majesty's Stationary Office, London, R&M 3510, 1967 (Replaces A.R.C. 16 446), 1954.

Pearcey, H. H. Shock-induced separation and its prevention by design and boundary layer control. *Boundary layer and flow control*, vol. 2, ed. G. V. Lachmann, Oxford: Pergamon Press, 1961.

Perry, A., and M. Chong. The vortex shedding process behind two-dimensional bluff bodies. *J. Fluid Mech.*, 116:77–90, 1982.

Pfenninger, W. Note about the spanwise turbulent contamination of swept low drag suction wings. *Northrop Norair Intern*, 1963.

Pfenninger, W. Some results from the x-21 program: Part 1. Flow phenomena at the leading edge of swept wings. *AGARDograph No. 97*, 1965.

Philips, E., R. Woszidlo, and I. Wygnanski. Separation control on a dynamic trailing-edge flap. *AIAA flow control conference, Paper AIAA-2010-4246*, 2010.

Pier, B. Finite-amplitude crossflow vortices, secondary instability and transition in the rotating-disk boundary layer. *J. Fluid Mech.*, 487:315–343, 2003.

Piponniau, S., J. P. Dussauge, J. F. Debieve, and P. Dupont. A simple model for low-frequency unsteadiness in shock-induced separation. *J. Fluid Mech.*, 629:87–108, 2009.

Pirozzoli, S., and F. Grasso. Direct numerical simulation of impinging shock wave/turbulent boundary layer interaction at m = 2.25. *Phys. Fluids*, 18:065113, 2006.

Pirozzoli, S., J. Larsson, J. W. Nichols, et al. Analysis of unsteady effects in shock/boundary layer interactions. *Proc. Summer Program 2010*. Center for Turbulence Research, Stanford University, Stanford, pp. 153–164, 2010.

Plotkin, K. J., Shock wave oscillation driven by turbulent boundary layer fluctuations. *AIAA J.*, 13:1036–1040, 1975.

Poggie, J., and A. J. Smits. Shock unsteadiness in a reattaching shear layer. *J. Fluid Mech.*, 429:155–185, 2001.

Poggie, J., and A. J. Smits. Experimental evidence for plotkin model of shock unsteadiness in separated flow. *Phys. Fluids*, 17(1): 018107-1–018107-4, 2005.

Polivanov, P. A., A. A. Sidorenko, and A. A. Maslov. Experimental study of unsteady effects in shock wave/turbulent boundary layer interaction. *47th AIAA Aerosp. Sci. Meet.*, Orlando, FL, 2009.

Poll, D. I. A. Leading edge transition on swept wings. *AGARD CP No. 224, Paper 21*, 1977.

Poll, D. I. A. Some aspects of the flow near a swept attachment line with particular reference to boundary layer transition. *College of Aeronautics Report 7805*, Cranfield Institute of Technology, Bedford, England, 1978.

Poll, D. I. A. Transition in the infinite swept attachment line boundary layer. *Aeron. Quart.*, 30:607–629, 1979.

Poll, D. I. A. Some observations of the transition process on the windward face of a long yawed cylinder. *J. Fluid Mech.*, 150:329–356, 1985.

Post, M. Plasma actuators for separation control on stationary and unstationary airfoils. Ph.D. Thesis, University of Notre Dame, Notre Dame, IN, 2004.

Prandtl, L. Bemerkungen uber die enstehung der turbulenz. *ZAMM*, 1:431–436, 1921.

Preston, J. The determination of turbulent skin friction by means of Pitot tubes. *J. Roy. Aeron. Soc.*, 14:109–121, 1954.

Provansal, M., C. Mathis, and L. Boyer. Benard-Von Karman instability: Transient and forced regimes. *J. Fluid Mech.*, 182:1–22, 1987.

Quadrio, M., and S. Sibilla. Numerical simulation of flow in a pipe oscillating around its axis. *J. Fluid Mech.*, 424:217–241, 2011.

Rabehi, A., R. BenGadri, P. Segur, F. Massines, and P. Decomps. Numerical modelling of high pressure glow discharges controlled by dielectric barrier. Proc. Conf. Electr. Insul. Dielectr. Phenom. pp. 840–845. Arlington, TX, October 23–26, 1994.

Radeztsky, R. H., M. S. Reibert, and W. S. Saric. Effect of isolated micron-sized roughness on transition in swept-wing flows. *AIAA J.*, 37(11):1370–1377, 1999.

Radeztsky, R. H., M. S. Reibert, W. S. Saric, and S. Takagi. Effect of micron-sized roughness on transition in swept-wing flows. *AIAA Paper 93-0076*, 1993.

Raffel, M., M. Willert, and J. Kompenhan. *Particle image velocimetry: A practical guide.* Springer-Verlag, Berlin, 1998.

Raghu, S. Feedback-free fluidic oscillator and method. U.S. Patent No. 6,253,782, 2001.

Raghu, S., and G. Raman. Miniature fluidic devices for flow control. *ASME-JSME Joint Fluids Eng. Conf., ASME FEDSM 99-7256*, 1999.

Raghunathan, S., and S. T. Mcllwain. Further investigations of transonic shock wave/boundary-layer interaction with passive control. *J. Aircraft*, 27(1):60–65, 1990.

Raman, G., and A. Cain. Innovative actuators for active flow and noise control. *Proc. Inst. Mech. Eng. G. J. Aerosp. Eng.*, 216:303–324, 2002.

Raman, G., and E. Rice. Supersonic jet mixing enhancement using impingement tones from obstacles of various geometries. *AIAA J.*, 33(3):454–462, 1995.

Raman, G., E. Rice, and D. Cornelius. Evaluation of flip-flop jet nozzles for use as practical excitation devices. *J. Fluids Eng.*, 116:508–515, 1994.

Ramaswamy, D.P., and A.-M. Schreyer. Control of shock-induced separation of a turbulent boundary layer using air-jet vortex generators. *AIAA J.*, 59(3):927–939, 2021.

Rampunggoon, P. Interaction of a synthetic jet with a flat plate boundary layer. Ph.D. Dissertation, Department of Mechanical Engineering, University of Florida, Gainesville, FL, 2001.

Rao, D., and T. Kariya. Boundary-layer submerged vortex generators for separation control: An exploratory study. *1st National Fluid Dynamics Congress*, Cincinnati, OH, 1988.

Rao, K., R. Narashima, and M. Narayanan. The bursting phenomenon in a turbulent boundary layer. *J. Fluid Mech.*, 48(2):339–352, 1971.

Rasheed, A., H. G. Hornung, A. V. Fedorov, and N. D. Malmuth. Experiments on passive hypervelocity boundary-layer control using an ultrasonically absorptive surface. *AIAA J.*, 40(3):481–489, 2002.

Rathnasingham, R., and K. Breuer. Active control of turbulent boundary layers. *J. Fluid Mech.*, 495:209–233, 2003.

Rayleigh, L. Aeolian tones. *Philos. Mag.*, 29:433–444, 1915.

Reibert, M., W. Saric, R. Carrillo, and K. Chapman. Experiments in nonlinear saturation of stationary crossflow vortices in a swept-wing boundary layer. *AIAA Paper 94-0001*, 1996.

Reibert, M. S. Nonlinear stability, saturation, and transition in crossflow-dominated boundary layers. Ph.D. Thesis, Arizona State University, Tempe, AZ, 1996.

Reisenthel, P. Hybrid instability in an axisymmetric jet with enhanced feedback. Ph.D. Thesis, Illinois Institute of Technology, Chicago, IL, 1988.

Reynolds, W., D. Parekh, P. Juvet, and M. Lee. Bifurcating and blooming jets. *Annu. Rev. Fluid Mech.*, 35:295–315, 2003.

Rice, E., and G. Raman. Enhanced mixing of a rectangular supersonic jet by natural and induced screech. *Technical Report, NASA Technical Memorandum 106245*, Langley, VA: NASA Lewis Research Center, 1993.

Riley, J., M. Gad el Hak, and R. Metcalfe. Compliant coatings. *Annu. Rev. Fluid Mech.*, 20:393–420, 1988.

Rixon, G., and H. Johari. Development of a steady vortex generator jet in a turbulent boundary layer. *J. Fluids Eng.*, 125(6):1006–1015, 2003.

Robinet, J.-C. Bifurcations in shock-wave/laminar-boundary-layer interaction: Global instability approach. *J. Fluid Mech.*, 579:85–112, 2007.

Robinson, S. Coherent motions in the turbulent boundary layer. *Annu. Rev. Fluid Mech.*, 23:601–639, 1991.

Rosenhead, L. *Laminar boundary layers*, pp. 467–475. Clarendon University Press, Cambridge, 1963.

Roshko, A. On the development of turbulent wakes from vortex streets. *NACA Report 1191*, 1954.

Roshko, A. Perspectives on bluff body aerodynamics. *J. Wind Ind. Aerodyn.*, 49:79–100, 1993.

Roumeas, M., P. Gillieron, and A. Kourta. Drag reduction by flow separation control on a car after body. *Int. J. Numer. Methods Fluids*, 60(11):1222–1240, 2009.

Salzberg, F., and S. P. Kezios. Mass transfer from a rotating cone in axisymmetric flow. *J. Heat Transfer*, 87:469–476, 1965.

Samimi, M., J.-H. Kim, J. Kastner, I. Adamovich, and Y. Utkin. Active control of high-speed and high-Reynolds-number jets using plasma actuators. *J. Fluid Mech.*, 578:305–330, 2007.

Samper, T. Separation control in 2-D channel bends. Ph.D. Thesis, University of Notre Dame, Notre Dame, IN, 2019.

Sandborn, V. A. *Resistance temperature transducers* 1st edn., Metrology Press, Fort Collins, CO, 1972.

Sandham, N., and W. Reynolds. Compressible mixing layer: Linear theory and direct simulation. *AIAA J.*, 28:618–624, 1990.

Santhanakrishnan, A., and J. Jacob. Flow control with plasma synthetic jet actuators. *J. Phys. D: Appl. Phys.*, 40:637–651, 2007.

Saric, W. S. Laminar-turbulent transition: Fundamentals. *AGARD Report 786, Special course on skin friction drag reduction*, 1992.

Saric, W. S. Private communication, 2001.

Saric, W. S. Personal communication, 2008.

Saric, W. S., and A. H. Nayfeh. Non-parallel stability of boundary layer flows. *Phys. Fluids*, 18:945–950, 1975.

Saric, W. S., and B. K. Rasmussen. Boundary-layer receptivity: Free-stream sound on an elliptical leading edge. *Bull. Am. Phys. Soc.*, 37:1720, 1992.

Saric, W. S., and H. L. Reed. Supersonic laminar flow control on swept wings using distributed roughness. *AIAA Paper 2002-0147*, 2002.

Saric, W. S., and H. L. Reed. Crossflow instabilities-theory & technology. *AIAA Paper 2003-0771*, 2003.

Saric, W. S., and A. S. W. Thomas. Experiments on the subharmonic route to transition. *Turbulence and chaotic phenomena in fluids*, North-Holland, 1984.

Saric, W. S., and L. G. Yeates. Experiments on the stability of crossflow vortices in swept-wing flows. *AIAA Paper 85-0493*, 1985.

Saric, W. S., W. Wei, and B. K. Rasmussen. Effect of leading edge on sound receptivity. *Laminar-Turbul. Transition IV: Proc. IUTAM Symp. Sendai*, Japan, 1994.

Saric, W. S., R. B. Carrillo, and M. S. Reibert. Nonlinear stability and transition in 3-D boundary layers. *Meccanica*, 33:469–487, 1998a. www.springerlink.com/index/Q5706086218340P2.pdf.

Saric, W. S., R. Carrillo, and M. S. Reibert. Leading-edge roughness as a transition control mechanism. *AIAA Paper AIAA-98-0781*, 1998b.

Saric, W. S., H. L. Reed, and D. W. Banks. Flight testing of laminar flow control in high-speed boundary layers. *RTO AVT Specialists Meeting on "Enhancement of NATO Military Flight Vehicle Performance by Management of Interacting Boundary Layer Transition and Separation," Published in RTO-MP-AVT-111*, 2004.

Sato, H. Further investigation on the transition of two-dimensional separated layers at subsonic speed. *J. Phys. Soc. Japan*, 14:1797–1810, 1959.

Savill, A. M., and J. C. Mumford. Manipulation of turbulent boundary layers by outer-layer devices: Skin friction and flow-visualization results. *J. Fluid Mech.*, 191:389–418, 1988.

Schardin, H. Die schlierenverfahren und ihre anwendungen ergebnisse der exakten naturwissenschaften. 20:303–439, 1942.

Schatzman, D., and F. Thomas. An experimental investigation of an unsteady adverse pressure gradient turbulent boundary layer: Embedded shear layer scaling. *J. Fluid Mech.*, 815:592–642, 2017.

Schlichting, H. Zur entstehung der turbulenz bei der plattenstromung, nachr. ges. wiss. gottingen. *Math.-Phys. Kasse*, 1993:181–208, 1933a.

Schlichting, H. Berechnung der anfachung kleiner storungen bei der plattenstromung. *ZAMM*, 13:171–174, 1933b.

Schlichting, H. Amplitudenverteilung und energiebilanz der kleinen storungen bei der plattengren-zschicht, nachr. ges. wiss. gottingen. *Math.-Phys. Kasse*, 1:47–78, 1935.

Schlichting, H. Uber die theoretische berechnung der kritischen reynoldschen zahl einer reibungaschicht in beschleunigter ind verzogerter stromung. *J. deutsch. Luftfahrtfor.* I: 97–112, 1940.

Schmidt, H., R. Woszidlo, C. Nayeri, and C. Paschereit. Separation control with fluidic oscillators in water. *Exp. Fluids*, 58:106-1–106-17, 2017.

Schubauer, G. B., and H. K. Skramstad. Laminar boundary layer oscillations and transitions on a flat plate. *J. Aeron. Sci.*, 14:69–76, 1947.

Schuele, C.-Y. Control of stationary cross-flow modes in a Mach 3.5 boundary layer using passive and active roughness. Ph.D. Thesis, University of Notre Dame, Notre Dame, IN, 2011.

Schuele, C.-Y., T. C. Corke, and E. H. Matlis. Control of stationary cross-flow modes in a Mach 3.5 boundary layer using patterned passive and active roughness. *J. Fluid Mech.*, 718:5–38, 2013.

Sears, W. R. The boundary layer of yawed cylinders. *J. Aeron. Sci.*, 15:49–52, 1948.

Seebaugh, W. R. and M. E. Childs. Conical shock-wave/turbulent boundary-layer interaction including suction effects. *J. Aircraft*, 7(4):334–340, 1970.

Seele, R., P. Tewes, R. Woszidlo, et al. Discrete sweeping jets as tools for improving the performance of the v-22. *J. Aircraft*, 46(6):2098–2106, 2009.

Seifert, A., A. Darabi, and I. Wygnanski. Delay of airfoil stall by periodic excitation. *J. Aircraft*, 33(4):691–699, 1996.

Seifert, A., S. Eliahu, D. Greenblatt, and I. Wygnanski. Use of piezoelectric actuators for airfoil separation control. *AIAA J.*, 36:1535–1537, 1998.

Selig, M. S., and A. J. Smits. Effect of periodic blowing on attached and separated supersonic turbulent boundary layers. *AIAA J.*, 29(10):1651–1658, 1991.

Semenov, B. On conditions of modelling and choice of viscoelastic coatings for drag reduction. In *Recent developments in turbulence management*, ed. K.-S. Choi, pp. 241–262. Kluwer, Dordrecht, 1991.

Semionov, N. V., and A. D. Kosinov. Method laminar-turbulent transition control of supersonic boundary layer on a swept wing. *Thermophys. Aeromech.*, 14(3):337–341, 2007.

Semionov, N. V., A. D. Kosinov, and V. Y. Levchenko. Experimental study of turbulence beginning and transition control in a supersonic boundary layer on swept wing. *6th IUTAM Symp. Laminar Turbul. Transition*, 2006.

Settles, G. *Schlieren and shadowgraph techniques: Visualizing phenomena in transparent media*, 1st edn. Springer, Berlin, 2001.

Settles, G., and M. Hargather. A review of recent developments in Schlieren and shadowgraph techniques. *Meas. Sci. Technol.*, 28:1–25, 2017.

Settles, G. S., and L. J. Dodson. Hypersonic shock/boundary layer database. *NASA CR 177577*, 1991.

Shadmani, S., S. Nainiyan, R. Ghasemiasl, M. Mirzaei, and S. Pouryoussefi. Experimental study of flow control over an ahmed body using plasma actuator. *Mech. Mech. Eng.*, 22(1):239–251, 2018.

Shapiro, P. The influence of sound upon laminar boundary layer instability. *Acoustics and Vibration Laboratory Report 83458-83560-1*, Cambridge, MA: Massachusetts Institute of Technology, 1977.

Shaw, L., B. Smith, and S. Saddough. Full scale flight demonstration of active flow control of a pod wake. *AIAA Paper 20063185*, 2006.

Schoppa, W., and F. Hussain. A large-scale control strategy for drag reduction in turbulent boundary layers. *Phys. Fluids*, 10:1049–1051, 1998.

Schoppa, W., and F. Hussain. Coherent structure generation on near-wall turbulence. *J. Fluid Mech.*, 453:57–108, 2002.

Shukla, S., R. Govardhan, and J. Arakeri. Flow over a circular cylinder with a hinged splitter plate. *J. Fluids Struct.*, 25:713–720, 2009.

Shukla, S., R. Govardhan, and J. Arakeri. Dynamics of a flexible splitter plate in the wake of a circular cylinder. *J. Fluids Struct.*, 41:127–134, 2013.

Shuster, J., and D. Smith. Experimental study of the formation and scaling of a round synthetic jet. *Phys. Fluids*, 19(4):045109, 2007.

Sinha, S. Flow separation control with microflexural wall vibrations. *J. Aircraft*, 38(3):496–503, 2001.

Smith, B. L. Synthetic jets and their interaction with adjacent jets. Ph.D. Thesis. Georgia Institute of Technology, Atlanta, GA, 1999.

Smith, B. L., and A. Glezer. The formation and evolution of synthetic jets. *Phys. Fluids*, 3:2281–2297, 1998.

Smith, B. L., and G. Swift. Synthetic jet at large Reynolds number and comparison to continuous jets. *AIAA Paper 20013030*, 2001.

Smith, C., J. Walker, A. Haidari, and B. Taylor. Hairpin vortices in turbulent boundary layers: The implications for reducing surface drag. *Proc. IUTAM Symp. Struct. Turbul. Drag Reduction*, Springer-Verlag, 1989.

Smith, N. Exploratory investigation of laminar boundary layer oscillations on a rotating disk. *NACA TN 1227*, 1946.

Souverein, L. J., and J. F. Debieve. Effect of air jet vortex generators on a shock wave boundary layer interaction. *Exp. Fluids*, 49(5):1053–1064, 2010.

Souverein, L. J., B. W. van Oudheusden, F. Scarano, and P. Dupont. Unsteadiness characterization in a shock wave turbulent boundary layer interaction through dual-PIV. *38th Fluid Dyn. Conf. Exhib.*, Seattle, Washington D. C., 2008.

Souverein, L. J., P. Dupont, J. F. Debieve, et al. Effect of interaction strength on the unsteady behavior of shock wave boundary layer interactions. *29th AIAA Fluid Dyn. Conf., AIAA Paper 2009-3715*, San Antonio, TX, 2009.

Spaid, F. W., and J. C. Frishett. Incipient separation of a supersonic, turbulent boundary-layer, including effects of heat transfer. *AIAA J.*, 10(7):91–92, 1972.

Spalding, D. B., A single formula for the law of the wall. *J. Appl. Mech.*, 28:455–457, 1961.

Squire, L. C. Further experimental investigations of compressible turbulent boundary-layers with air injection. *ARC R&M No. 3267*, 1970.

Sreenivasan, K., P. Strykowski, and D. Olinoer. Hopf bifurcation, Landau equation, and vortex shedding behind circular cylinders. *Proc. Forum Unsteady Flow Separation*, ed. K. N. Ghia, American Society of Mechanical Engineers, pp. 1–13. 1987.

Sreenivasan, K., S. Raghu, and D. Kyle. Absolute instability in variable density jets. *Exp. Fluids*, 7:309–317, 1989.

Stainbeck, P. Effect of unit Reynolds number, nose bluntness, angle of attack, and roughness on transition on a 5 degree half-angle cone at Mach 8. *NASA Technical Report TND-4961*, 1969.

Stanbrook, A. The surface oil flow technique for use in high speed wind tunnels. *AGARDograph*, 70:39–49, 1962.

Stanislas, M., L. Perret, and J.-M. Foucaut. Vortical structures in the turbulent boundary layer: A possible route to universal represenation. *J. Fluid Mech.*, 602:327–382, 2008.

Steele, R., P. Tewes, R. Woszidlo, et al. Discrete sweeping jets as tools for improving the performance of the v-22. *J. Aircraft*, 46:(6)2098–2106, 2009.

Stephens, J., T. Corke, and S. Morris. Turbine blade tip leakage flow control: Thick/thin blade effects. *Proc. 45th Aerosp. Sci. Meet., Paper 2007-0646*, 2007.

Stephens, J., T. Corke, and S. Morris. Turbine blade tip leakage flow control: Thick/thin blade effects and separation line control. *Proc. GT2008, ASME Turbo Expo. 2008, Paper 50705*, 2008.

Stetson, K. Nosetip bluntness effects on cone frustum boundary layer transition in hypersonic flow. *AIAA Paper 83-1763*, 1983.

Stetson, K., and G. Rushton. Shock tunnel investigation of boundary-layer transition at m=5.5. *AIAA J.*, 5(5):899–906, 1967.

Stetson, K., E. Thompson, J. Donaldson, and L. Siler. Laminar boundary layer stability experiments on a cone at Mach 8, part 3: Sharp cone at angle of attack. *AIAA Paper 1985-492*, 1985.

Strouhal, V. Uber eine besondere art der tonerregung. *Ann. Phys. Chem. (Liepzig)*, 5:(241) 216–251, 1878.

Strykowski, P. The control of absolutely and convectively unstable shear flows. Ph.D. Thesis, Engineering and Applied Science, Yale University, New Haven, CT, 1986.

Strykowski, P. J., and K. R. Sreenivasan. On the formation and suppression of vortex shedding at low Reynolds numbers. *J. Fluid Mech.*, 218:71–107, 1990.

Stuart, J. T. Nonlinear effects in hydrodynamic stability. *Proc. 10th Int. Congr. Appl. Mech.*, eds. F. Rona and W. T. Koiter, Elsevier Publishing Company, pp. 63–97. 1962.

Suzuki, H., N. Kasagi, and Y. Suzuki. Active control of an axisymmetric jet with distributed electromagnetic flap actuators. *Exp. Fluids*, 36:498–509, 2004.

Szwaba, R. Comparison of the influence of different air-jet vortex generators on the separation region. *Aerosp. Sci. Tech.*, 15(1): 45–52, 2011.

Tam, C. Excitation of instability waves in a two-dimensional shear layer by sound. *J. Fluid Mech.*, 89(2):357–371, 1978.

Tan-atichat, J., H. Nagib, and R. Loehrke. Interaction of free-stream turbulence with screens and grids: A balance between turbulence scales. *J. Fluid Mech.*, 114:501–528, 1982.

Taneda, S. The stability of two-dimensional laminar wakes at low Reynolds numbers. *J. Phys. Soc. Japan*, 18(2):288–296, 1963.

Tani, I., R. Hama, and S. Mituisi. On the permissible roughness in the laminar boundary layer. *Report 199*, Aeronautical Research Institute, Tokyo Imperial University, Tokyo, Japan, 1940.

Tardu, S., and G. Binder. Review: Effect of the olds on near wall coherent structures; discussion and need for future work. *Recent developments in turbulence management*, ed. K.-S. Choi, pp. 147–160. 1991.

Taylor, G. I. Eddy motion in the atmosphere. *Phil. Trans. Roy. Soc. A*, 215:1–26, 1915.

Taylor, H. The elimination of diffuser separation by vortex generators. *Report R-4012-3*, United Aircraft Corporation, 1947.

Tewes, P., L. Taubert, and I. Wygnanski. On the use of sweeping jets to augment the lift of a lambda-wing. *AIAA Flow Control Conf., Paper AIAA-2010-4689*, 2010.

Theofilis, V., S. Hein, and U. Dallman. On the origins of unsteadiness and three-dimensionality in a laminar separation bubble. *Proc. Roy. Soc. Lond. Ser. A, Math. Phys. Sci.*, 358(1777): 3229–3246, 2000.

Thomas, A. S. W. The control of boundary-layer transition using a wave- superposition principle. *J. Fluid Mech.*, 137:233–250, 1983.

Thomas, C., and C. Davies. The effects of mass transfer on the global stability of the rotating-disk boundary layer. *J. Fluid Mech.*, 663:401–433, 2010.

Thomas, F., and A. Koslov. Plasma flow control of cylinders in a tandem configuration. *AIAA Paper 2010-4703*, 2010.

Thomas, F., A. Koslov, and T. Corke. Plasma actuators for landing gear noise reduction. *AIAA Paper 2005-3010*, 2005.

Thomas, F., A. Koslov, and T. Corke. Plasma actuators for cylinder flow control and noise reduction. *AIAA J.*, 46(8):1921–1931, 2008.

Thomas, F., T. Corke, M. Iqbal, A. Kozlov, and D. Schatzman. Optimization of dielectric barrier discharge plasma actuators for active aerodynamic flow control. *AIAA J.*, 47(9):2169–2178, 2009.

Thomas, F., T. Corke, F. Hussain, A. et al. Turbulent boundary layer drag reduction by active control of streak transient growth. *69th Meet. Am. Phys. Soc. Fluid Dyn. Div., Session D7*, 9, 2016.

Thomas, F. O., C. M. Putnam, and H. C. Chu. On the mechanism of unsteady shock wave/turbulent boundary layer interactions. *Exp. Fluids*, 18(1/2):69–81, 1994.

Tien, C. L., and D. T. Campbell. Heat and mass transfer from rotating cones. *J. Fluid Mech.*, 17:105–112, 1963.

Tollmien, W. Uber die entstehung der turbulenz, nachr. ges. wiss. gottingen. *Math.-Phys. Klasse*, 21–44, 1929. http://eudml.org/doc/59276.

W. Tollmien. Ein allgemeines kriterium der instabilitat laminarer geschwindigkeitsverteilungen, nachr. ges. wiss. gottingen. *Math.-Phys. Klasse*, 50:79–114, 1935.

Tombazis, N., and P. Bearman. A study of three-dimensional aspects of vortex shedding from a bluff body with a mild geometric disturbance. *J. Fluid Mech.*, 330:85–112, 1997.

Townsend, A. A. *The structure of turbulent shear flow*. Cambridge: Cambridge University Press, 1956.

Townsend, A. A. Entrainment and structure of turbulent flow. *J. Fluid Mech.*, 41(1):13–46, 1970.

Triantafyllou, G., and A. Dimas. The low froude number wake of floating bluff objects. *Internal Report MITSG*, 1989.

Triantafyllou, G., and G. Karniadakis. Computational reducibility of unsteady viscous flows. *Phys. Fluids A: Fluid Dyn.*, 2:653–656, 1990.

Trujillo, S., D. Bogard, and K. Ball. Turbulent boundary layer drag reduction using an oscillating wall. *AIAA Paper 97-187*, 1997.

Turcat, A. *Concorde: Essais dhier, batailles d'aujourd'hui.*. Le Cherche midi, 2003.

Tyndall, J. *Sound*, P.F. Collier, New York, 1864.

Unalmis, O. H., and D. S. Dolling. Decay of wall pressure field and structure of a Mach 5 adiabatic turbulent boundary layer. *AIAA Paper 1994-2363*, 1994.

Utturkar, Y. Numerical investigation of synthetic jet flow fields. M.S. Thesis, Department of Mechanical Engineering, University of Florida, Gainesville, FL, 2002.

Valdivia, A., K. B. Yuceil, J. L. Wagner, N. T. Clemens, and D. S. Dolling. Active control of supersonic inlet unstart using vortex generator jets. *39th AIAA Fluid Dyn. Conf., Paper 2009-4022*, San Antonio, TX, June 22–25, 2009.

Valtchanov, H., J. Brinkerhoff, and M. Yaras. Numerical study of passive forcing on the secondary instability of a laminar planar free shear layer. *J. Turb.*, 21(5-6):259–285, 2020.

VanDercreek, C., M. Smith, and K. Yu. Focused Schlieren and deflectometry at AEDC hypervelocity wind tunnel no. 9. *AIAA Paper 2010-4209*, 1993.

VanDyke, M. *An album of fluid motion*, 1st edn. Stanford, CA: Parabolic Press, 1982.

Viets, H. Flip-flop jet nozzle. *AIAA J.*, 13:1375–1379, 1975.

Viswanath, P. R., L. Sankaran, R. Narasimha, A. Prabhu, and P. M. Sagdeo. Injection slot location for boundary-layer control in shock-induced separation. *AIAA J.*, 20(8):726–732, 1983.

Viswanathan, K., and P. Morris. Predictions of turbulent mixing in axisymmetric compressible shear layers. *AIAA J.*, 30(6):529–536, 1992.

Waleffe, F., J. Kim, and J. Hamilton. On the origin of streaks in turbulent boundary layers. *Turbul. Shear Flows*, 8:37–49, 1993.

Wallis, R. A., and C. M. Stuart. On the control of shock-induced boundary layer separation with discrete air jets. *Aeronautical Research Council CP 595*, 1958.

Walsh, M., and B. Anders. Riblet/LEBU research at NASA Langley. *Appl. Sci. Res.*, 46:255–262, 1989.

Walz, A. Boundary-layers of flow and temperature. Massachusetts Institute of Technology Press, Cambridge, MA, p. 113, 1969.

Wanderley, J. B. V., and T. C. Corke. Boundary layer receptivity to freesteam sound on elliptic leading edges of flat plates. *J. Fluid Mech.*, 429:1–29, 2001.

Wang, J., Y.-H. Li, B.-Q. Cheng, et al. Effects of plasma aerodynamic actuation on oblique shock wave in a cold supersonic flow. *J. Phys. D: Appl. Phys.*, 42:165503, 2009.

Warsop, C., and W. Crowther. Fluidic flow control effectors for flight control. *AIAA J.*, 56(10):3808–3824, 2018.

Wartemann, V., H. Ludeke, S. Willems, and A. Gulhan. Stability analyses and validation of a porous surface boundary condition by hypersonic experiments on a cone model. *7th Aerothermodyn. Symp.*, Brugge, Belgium, 2011.

Wartemann, V., S. Willems, and A. Gulhan. Mack mode damping by micropores on a cone in hypersonic flow. *Hypersonic Laminar-Turbul. Transition Conf.*, AVT-200-15, San Diego, CA, 2012.

Wartemann, V., A. Wagner, T. Eggers, and K. Hannemann. Passive hypersonic boundary layer transition control using an ultrasonically absorptive coating with random microstructure - part 2: Computational analysis. *Procedia IUTAM 00 (2014) 000000*, 2014. www.elsevier.com/locate/procedia.

Wehrmann H. Influence of vibrations on the flow field behind a cylinder. *Phys. Fluids Suppl.*, 10:187–190, 1967.

Wehrmann, O. Reduction of velocity fluctuations in a Kármán vortex street by a vibrating cylinder. *Phys. Fluids*, 8:760–761, 1965.

Weinstein, L. Large-field high-brightness focusing schlieren system. *AIAA J.*, 31(7):1250–1255, 1993.

Westerweel, J. Fundamentals of digital particle image velocimetry. *Meas. Sci. Technol.*, 8:1379–1392, 1997.

Westerweel, J., A. Draad, J. T. van der Hoeven, and J. van Oord. Measurement of fully-developed turbulent pipe flow with digital particle image velocimetry. *Exp. Fluids*, 20:165–177, 1996.

Westerweel, J., G. Elsinga, and R. Adrian. Particle image velocimetry for complex turbulent flows. *Annu. Rev. Fluid Mech.*, 45:409–436, 2013.

Whitefield, D. L. Analytical description of the complete turbulent boundary layer velocity profile. *AIAA Paper 78-1150*, 1979.

Wicks, M., F. Thomas, T. Corke, and A. Cain. Plasma flow control on a landing gear model. *53rd AIAA Aerosp. Sci. Meet., Paper AIAA 2015-1036*, 2015.

Wilkinson, S. Investigation of an oscillating surface plasma for turbulent drag reduction. *AIAA Paper 2003-1023*, 2003.

Wilkinson, S., and M. Malik. Stability experiments in the flow over a rotating disk. *AIAA J.*, 23:588–595, 1985.

Wille, R. Beitrage zur phanomenologie der freistrahlen. *Z. Flugwiss*, 11:222–233, 1963.

Willert, C. The fully digital evaluation of photographic piv recordings. *Appl. Sci. Res.*, 56:79–102, 1992.

Willert, C. E., and M. Gharib. Digital particle image velocimetry. *Exp. Fluids*, 10:181–193, 1991.

Williams, T. Compliant flow designs for optimum lift control of wind turbine rotors. Ph.D. Thesis, University of Notre Dame, Notre Dame, IN, 2014.

Williams, T., A. Jemcov, and T. Corke. Shape optimization for dielectric barrier discharge plasma compliant flows. *AIAA J.*, 53(10):3125–3128, 2015.

Williamson, C. Oblique and parallel modes of vortex shedding of a circular cylinder. *J. Fluid Mech.*, 206:579–627, 1989.

Willmarth, W., and S. Lu. Structure of the Reynolds stress near the wall. *J. Fluid Mech.*, 55(1):65–92, 1972.

Wiltse, J., and A. Glezer. Manipulation of free shear flows using piezoelectric actuators. *J. Fluid Mech.*, 249:261–285, 1993.

Winant, C., and F. Browand. Vortex pairing, the mechanism of turbulent mixing? Layer growth at moderate Reynolds number. *J. Fluid Mech.*, 63:237–255, 1974.

Wise, D., and P. Ricco. Turbulent drag reduction through oscillating discs. *J. Fluid Mech.*, 746:536–564, 2014.

Wishart, D., A. Krothapalli, and M. Mungal. Supersonic jet control via point disturbances inside the nozzle. *AIAA J.*, 31(7):1340–1341, 1993.

Wlezien, R., and V. Kibens. Passive control of jets with indeterminate origins. *AIAA J.*, 24(8):1263–1270, 1986.

Wlezien, R., and V. Kibens. Influence of nozzle asymmetry on supersonic jets. *AIAA J.*, 26(1):27–33, 1988.

Wolf, S., and J. Laub. 1997 NASA Ames laminar flow supersonic wind tunnel (LFSWT) tests of a 10 degree cone at Mach 1.6. *NASA-TM 110438*, 1997.

Wong, W. F., and G. R. Hall. Suppression of strong shock/boundary-layer interaction in supersonic inlets by boundary-layer blowing. *AIAA Paper 1975-1209*, 1975.

Wood, C. The effect of base bleed on a periodic wake. *J. Roy. Aeron. Soc.*, 68:477–482, 1964.

Woszidlo, R., and I. Wygnanski. Parameters governing separation control with sweeping jet actuators. *AIAA Fluid Dyn. Conf., Paper AIAA-2011-3172*, 2011.

Woszidlo, R., H. Nawroth, S. Raghu, and I. Wygnanski. Parametric study of sweeping jet actuators for separation control. *AIAA Flow Control Conf., Paper AIAA-2010-4247*, 2010.

Wu, M., and M. P. Martin. Analysis of shock motion in shockwave and turbulent boundary layer interaction using direct numerical simulation data. *J. Fluid Mech.*, 594:71–83, 2008.

Yao, J., X. Chen, F. Thomas, and F. Hussain. Large-scale control for drag reduction in turbulent channel flows. *Phys. Rev. Fluids*, 2:062601, 2017.

Yu, K., E. Gutmark, R. Smith, and K. Schadow. Supersonic jet excitation using cavity-actuated forcing. *AIAA Paper 94-0185*, 1994.

Zabib, A. Stability of viscous flow past a circular cylinder. *J. Eng. Maths*, 21:155–165, 1987.

Zelman, M. B., and I. I. Maslennikova. Tollmien-Schlichting-wave resonant mechanism for subharmonic-type transition. *J. Fluid Mech.*, 252:449–478, 1993.

Zhang, D.-y., Y.-h. Luo, X. Li, and H.-w. Chen. Numerical simulation and experimental study of drag-reducing surface of a real shark skin. *J. Hydrodyn., Ser. B*, 23(2):204–211, 2011.

Zhang, W., C. Dai, and S. Lee. PIV measurements of the near-wake behind a sinusoidal cylinder. *Exp. Fluids*, 38:824–832, 2005.

Zhong, S., M. Jabbal, H. Tang, et al. Towards the design of synthetic-jet actuators for full-scale flight conditions, part 1: The fluid mechanics of synthetic-jet actuators. *Flow, Turbul. Combust.*, 78(3–4):283–307, 2007.

Zhou, J., R. Adrian, R. Balachandar, and T. Kendall. Mechanisms for generating coherent packets of hairpin vortices in channel flow. *J. Fluid Mech.*, 387:356–396, 1999.

Zhou, J., H. Tang, and S. Zhong. Vortex roll-up criterion for synthetic jets. *AIAA J.*, 47(5):1252–1262, 2009.

Zilliac, G. Further developments of the fringe-imaging skin friction technique. *NASA Technical Report 110425*, 1996.

Zilliac, G. The fringe-imaging skin friction technique application user's manual. *NASA Technical Report 208794*, 1999.

Index

Printed in the United States
by Baker & Taylor Publisher Services